# New Functional Coatings and Thin Films for Sensor and Green Energy Technologies

# New Functional Coatings and Thin Films for Sensor and Green Energy Technologies

Guest Editors

**Gergana Alexieva**
**Rositsa Gergova**
**Tsvetanka Babeva**

MDPI

Basel • Beijing • Wuhan • Barcelona • Belgrade • Novi Sad • Cluj • Manchester

*Guest Editors*
Gergana Alexieva
Faculty of Physics
Sofia University
Sofia
Bulgaria

Rositsa Gergova
Central Laboratory of
Solar Energy and New
Energy Sources
Bulgarian Academy of Sciences
Sofia
Bulgaria

Tsvetanka Babeva
Institute of Optical Materials
and Technologies
Bulgarian Academy of Sciences
Sofia
Bulgaria

*Editorial Office*
MDPI AG
Grosspeteranlage 5
4052 Basel, Switzerland

This is a reprint of the Special Issue, published open access by the journal *Coatings* (ISSN 2079-6412), freely accessible at: https://www.mdpi.com/journal/coatings/special_issues/Functional_Energy.

For citation purposes, cite each article independently as indicated on the article page online and as indicated below:

| Lastname, A.A.; Lastname, B.B. Article Title. *Journal Name* **Year**, *Volume Number*, Page Range. |
| --- |

**ISBN 978-3-7258-6720-2 (Hbk)**
**ISBN 978-3-7258-6721-9 (PDF)**
**https://doi.org/10.3390/books978-3-7258-6721-9**

# Contents

*coatings*

*Editorial*

# Special Issue: "New Functional Coatings and Thin Films for Sensor and Green Energy Technologies"

**Gergana Alexieva** [1,2]

[1] Faculty of Physics, University of Sofia, 5 James Bourchier Blvd., 1164 Sofia, Bulgaria; gerry@phys.uni-sofia.bg; Tel.: +359-2-8161-646

[2] Institute of Optical Materials and Technologies "Akad. J. Malinowski", Bulgarian Academy of Sciences, Akad. G. Bonchev Str., bl. 109, 1113 Sofia, Bulgaria

## 1. Introduction

The new challenges we face in modern life, such as the increasing need to monitor health, food quality, and environmental control, along with the strategic areas related to clean energy, require the development of a wide range of products and technological applications. These include sensors, photoelectronic devices, low-cost and fast electronics, energy conversion, harvesting, and storage [1–3]. The growth of the high-tech industry, the rising demand for high-performing and sustainable products, and the advances in functional materials, particularly in nanotechnology, have led to significant research progress in the fabrication of high-quality coatings used in a wide range of technological applications [3–5]. The material composition, structure, and dimensionality have decisive impacts on their properties [6,7]. Furthermore, the ability to dope or to introduce side chemical groups can significantly alter material sensing properties, functionality, and high performance [4,6]. The design of nanostructured organic, inorganic, and composite materials, when specially controlled by the aid of various techniques, can also lead to a reduction in fabrication costs for sensing and photovoltaic devices, as well as their practical applications [1,2,6,8].

This Special Issue gathers and highlights recent research achievements in the design and use of functional materials for active coatings, in a wide range of applications, from photoelectronic to chemically sensing devices.

## 2. Statistics of the Special Issue

The Special Issue consists of 21 full papers authored by 137 authors. The geographical distribution of authors is presented in Figure 1. The authors originate from 11 countries in four continents—Europe, Asia, and North and South America. The average number of authors per paper is approximately 7.

 https://doi.org/10.3390/coatings16010127

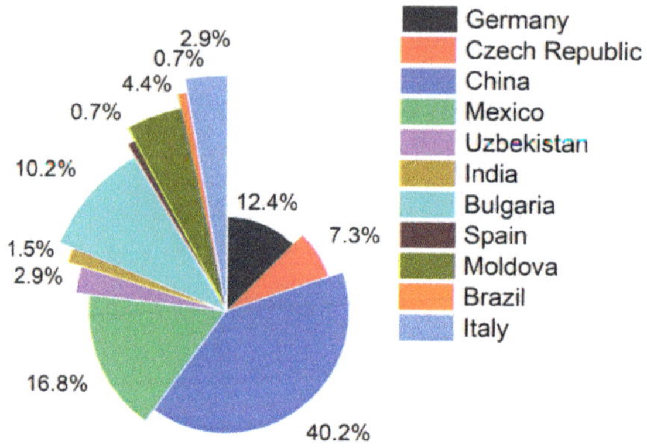


**Figure 1.** The distribution of authors by country.

## 3. Brief Overview of the Contributions to This Special Issue

Kapran et al. [9] utilize the potential of a pulsed hollow-cathode discharge to deposit mixed cobalt nickel oxide films. During annealing at 600 °C, the mixed metal oxide phase is separated into two crystalline cubic $Co_3O_4$ and cubic NiO subphases. The annealed films are opaque at small wavelengths, while at longer wavelengths they become increasingly more transparent. In Wang et al. [10], the electrochemical performance of the high-nickel electrode material $LiNi_{0.8}Co_{0.1}Mn_{0.1}O_2$, used in lithium-ion batteries, is significantly improved through $Y_2O_3$ coating and sintering. The microstructure, morphology, and electrochemical properties of the material are precisely explored, as well as the underlying mechanism of coating modification. In Du et al. [11], the droplet formation behavior of Poly3-hexylthiophene inks across various Weber numbers is systematically investigated. The findings confirm the crucial role of ink's rheological properties in determining the precision of inkjet printing. A framework for prediction and control of droplet formation fidelity is proposed. In the paper of Zhang et al. [12], a pre-crystallization approach is employed during the multi-step fabrication process of $CsPbBr_3$ films for perovskite solar cells. The prepared films possess larger crystal sizes, lower grain boundary density, enhanced crystallinity, and higher phase purity. Compared with an untreated device, the fabricated photovoltaic device achieves 27% higher photovoltaic conversion efficiency and demonstrates superior stability. In another study on perovskite materials, Jiang et al. [13] propose a method for fabrication of high-quality $FAPbI_3$ thin films using a vacuum flash-assisted coating process. The results demonstrate that solid Lewis base additives can enable control and scalability of perovskite crystal growth, which is of significant importance for the development of perovskite film solar devices. The paper of Baltierra-Costeira et al. [14] presents a study on the effects of various factors on the anodizing process of Al6061 Aluminum. The effects of temperature, time, and atmospheric conditions on the thickness, porosity, electrochemical behavior, and purity of the anodized layers are systematically examined. The optimal conditions for reaching a corrosion-resistant anodized aluminum are determined. In the paper of Donaev et al. [15], GaInP nanostructures are obtained by implanting $In^+$ ions into GaP(111). By using a complex of secondary and photoelectron spectroscopy, the effects of $In^+$ ions implantation at different doses, followed by subsequent annealing of the composition, the electronic and crystal structure of the GaP(111) surface are investigated. In order to elucidate the effects of various inorganic salts on the foamability of SDS surfactant solution, the paper of Raykundaliya et al. [16] reports interesting results and a number of contradictory findings about the correlations between system foamability


https://doi.org/10.3390/coatings16010127

and its surface tension. Lizama-Tzec et al. [17] demonstrated the improvement of electrodeposited black nickel by coating with spray-pyrolyzed silica film. The fabrication process is scaled up to 193 cm × 12 cm copper fins, which are incorporated in commercial-size flat-plate solar collectors. The proposed selective coating outperforms both the commercial black paint system and the system without modified silica film. Optimization of the deposition conditions in Silicon-Rich Silicon Nitride films, aimed at improvements of their photoluminescence and optical properties, is reported in the paper of González et al. [18]. The coatings, obtained by LPCVD at various temperatures and precursor gas pressure ratio, show a wide photoluminescence in the visible region, as well as proper refractive index and absorption in the UV, suitable for use in silicon solar cells. A comparative study of the impact of GaAs and Si substrates on the topology, luminescence, and wettability properties of ZnMgO films deposited by spray pyrolysis, is presented by Monaico et al. [19]. The potential uses, like in photocatalytic, plasmonic, and gas sensor applications, are discussed.

The work of Li et al. [20] is devoted to deposition of Tungsten coatings on CuCrZr alloy from binary oxide molten salt. The pulsed current electrodeposition method is applied and the coating microstructure is characterized by density, porosity, hardness, adhesion to the matrix, bonding strength, surface stress, stability and durability. The study is aimed to contribute to the development of controlled nuclear fusion reactors. Polymeric composites, formed between polyurethane (PU) and chemically sensing polymers, are proposed by M. de Carvalho et al. [21] as a new class of sensing materials for SAW sensors. The microstructure of the films, deposited on the complex surface of the piezoelectric sensor element, and the effects of the chemical resistance and adhesion test (CAT) on the coating layers, are presented and analyzed. The role of polyurethane as a composite binder and as an adhesion promoter to the $SiO_2$ interface is highlighted. Subsequently to the work in [20], the problematic combination of steel and tungsten is studied in the article of Dong et al. [22]. The effects of variations in current density on the morphology, thickness, and hardness of the layers are reported. The preference of electrodeposition by a pulsed current over a direct current mode is demonstrated. The composite material of $Nb_2O_5$ embedded carbon nanofibers is investigated by Zhu et al. [23]. A critical composition–structure-property–performance relationship is established in order to enable the development of high-performance sodium-ion batteries with a flexible self-supporting framework. Comprehensive studies on the preparation of composite CuCoMnOx coatings, their thermogravimetric and spectral characterization, as well as the results from plasma-spray deposition over 316L Stainless Steel, are presented in Zeng et al. [24]. An advanced metal–ceramic coating material with high-temperature stability is proposed for solar selective absorption. The paper of Gärtner et al. [25] describes the manufacturing and characterization of silica nanostructures, specifically designed as 355 nm antireflectors. Hollow silica nanostructures with transmittance of 99.8% are made and proposed as UV laser-resistant coatings in advanced optical systems. The development of optical and gravimetric sensors of humidity is presented by Larazova et al. [26]. Poly(vinyl alcohol) derivatives comprising grafted poly(N,N-dimethylacrylamide) chains with varied length and graft density are synthesized, characterized, and deposited as thin films on silicon wafer substrate, as well as on the active area of quartz crystal resonators. The potential of these new nanoscale materials for humidity sensing applications is thoroughly discussed. The third contribution from the University of Science and Technology Beijing, Dong et al. [27], reports the electrodeposition of Tungsten coatings on CuCrZr substrates by the use of another binary oxide molten salt. The phase composition, microstructure, and properties of the resulting fine-grained tungsten coatings are systematically studied in order to optimize the deposition process and coating performance. A comprehensive

 https://doi.org/10.3390/coatings16010127

critical review of deposition techniques for embedded and harsh-environment applications, highly specialized to the functional coatings in fiber Bragg grating sensors, is presented in the paper by Vendittozi et al. [28]. The state-of-the-art in the development of these highly important sensors is suitably presented and credibly discussed by consolidating advanced coating materials, deposition techniques, and environment relationships into unified design rules. The contribution by Dimov et al. [29] compares the characteristics of the newly synthesized MOF MIL-101(Cr) crystals to those of the commercially available material. The crystals obtained in this work are around 20 nm in size, much smaller than the commercial ones (300 nm), and exhibit enhanced stability as Langmuir films. The different approaches to deposition of LB MOF assemblies in sensors design are outlined. The paper addresses the lack of research on depositing presynthesized MOFs by using one of the best methods for nanoarchitectonics—the LB method.

## 4. Conclusions

The World of Coatings is vast and heterogeneous in materials, methods, techniques, and applications. There is a large literature in this field, including many reviews and special issues. In our Special Issue, we have collected the most recently made works, in accordance with the authors' choice. Most of its papers deal with the design of innovative functional materials and new solutions for the optimization of deposition processes. All of them are aimed at making functional coatings with improved performance that bridge the gap between materials design and applications. Although principally new and AI-driven materials are not presented, our Special Issue clearly demonstrates the steady growth of interest in functional coatings for sensor and green energy technologies. I hope that readers will find the published papers interesting and helpful.

**Acknowledgments:** Working on this Special Issue, I had the pleasure of contacting renowned authors from around the world and receiving high-quality submissions. I express my gratitude to all the authors for their contributions, as well as to the Editorial Staff of the Journal for constant help and support.

**Conflicts of Interest:** The author declares no conflict of interest.

## References

1. Elanjeitsenni, V.P.; Vadivu, K.S.; Prasanth, B.M. A review on thin films, conducting polymers as sensor devices. *Mater. Res. Express* **2022**, *9*, 022001. [CrossRef]
2. Abdel-Wahed, M.S.; Sayed, M.H.; Gomaa, M.M. Thin films nanocomposite: Multifunctional materials for energy and water purification. *Discover Nano* **2025**, *20*, 198. [CrossRef]
3. Agresti, F.; Angella, G.; Arshad, H.; Barison, S.; Barreca, D.; Bassani, P.; Battiston, S.; Biffi, C.A.; Buscaglia, M.T.; Canu, G. Sustainable Materials for Energy. *Nanomaterials* **2025**, *15*, 1388. [CrossRef] [PubMed]
4. Xu, K.; Xiao, X.; Wang, L.; Lou, M.; Wang, F.; Li, C.; Ren, H.; Wang, X.; Chang, K. Data-Driven Materials Research and Development for Functional Coatings. *Adv. Sci.* **2024**, *11*, 2405262. [CrossRef] [PubMed]
5. Beyrami, H.; Golshan, M.; Zardehi-Tabriz, A.; Salami-Kalajahi, M. Smart Coatings: Fundamentals, Preparation Approaches, and Applications. *Adv. Mater. Technol.* **2025**, *10*, e00574. [CrossRef]
6. Dallaev, R. Conductive Polymer Thin Films for Energy Storage and Conversion: Supercapacitors, Batteries, and Solar Cells. *Polymers* **2025**, *17*, 2346. [CrossRef] [PubMed]
7. Sakthinathan, S.; Meenakshi, G.A.; Vinothini, S.; Yu, C.-L.; Chen, C.-L.; Chiu, T.-W.; Vittayakorn, N. A Review of Thin-Film Growth, Properties, Applications, and Future Prospects. *Processes* **2025**, *13*, 587. [CrossRef]
8. Hossain, M.I.; Mansour, S. A critical overview of thin films coating technologies for energy applications. *Cogent Eng.* **2023**, *10*, 2179467. [CrossRef]
9. Kapran, A.; Hippler, R.; Wulff, H.; Olejnicek, J.; Volfova, L.; Pisarikova, A.; Nepomniashchaia, N.; Cada, M.; Hubicka, Z. Crystalline Structure and Optical Properties of Cobalt Nickel Oxide Thin Films Deposited with a Pulsed Hollow-Cathode Discharge in an Ar+$O_2$ Gas Mixture. *Coatings* **2024**, *14*, 319. [CrossRef]

 https://doi.org/10.3390/coatings16010127

10. Wang, X.; Wang, F.; Zheng, M.; Rong, M.; Wang, J.; Deng, J.; Liu, P.; Liu, D. Realizing Dual Functions through $Y_2O_3$ Modification to Enhance the Electrochemical Performance of $LiNi_{0.8}Co_{0.1}Mn_{0.1}O_2$ Material. *Coatings* **2024**, *14*, 443. [CrossRef]
11. Du, Z.; Zhang, L.; Du, Y.; Wei, X.; Du, X.; Lin, X.; Liu, J.; Huang, Y.; Xue, Y.; Zhao, N.; et al. Controlling the Polymer Ink's Rheological Properties to Form Single and Stable Droplet. *Coatings* **2024**, *14*, 600. [CrossRef]
12. Zhang, Y.; Jiang, Z.; Li, J.; Meng, G.; Guo, J.; Zhang, W. Preparation of Highly Efficient and Stable All-Inorganic $CsPbBr_3$ Perovskite Solar Cells Using Pre-Crystallization Multi-Step Spin-Coating Method. *Coatings* **2024**, *14*, 918. [CrossRef]
13. Jiang, Z.; Li, J.; Li, K.; Zhang, R.; Qin, Z.; Zhang, Y.; Wu, B.; Ma, Q.; Zhang, Y.; Zhang, W. Controllable Crystallization of Perovskite Films during the Blade-Coating Fabrication Process for Efficient and Stable Solar Cells. *Coatings* **2024**, *14*, 1113. [CrossRef]
14. Baltierra-Costeira, G.; Camporredondo-Saucedo, J.E.; García-Rentería, M.A.; Falcón-Franco, L.A.; Castruita-Ávila, L.G.; García-Lara, A.M. Effect of the Atmosphere on the Properties of Aluminum Anodizing. *Coatings* **2024**, *14*, 1166. [CrossRef]
15. Donaev, S.; Shirinov, G.; Umirzakov, B.; Donayev, B.; Wang, S. Effect of Low-Energy Implantation of $In^+$ Ions on the Composition and Electronic Structure of Single-Crystal GaP(111). *Coatings* **2024**, *14*, 1231. [CrossRef]
16. Raykundaliya, N.; Karakashev, S.I.; Parsana, V.M.; Grozev, N.A.; Mircheva, K.; Ivanova-Stancheva, D. Inorganic Salts as Chemical Foam Suppressors. *Coatings* **2024**, *14*, 1358. [CrossRef]
17. Lizama-Tzec, F.I.; Cetina-Dorantes, M.d.J.; Herrera-Zamora, D.M.; Alvarado-Gil, J.J.; Rodríguez-Gattorno, G.; Estrella-Gutiérrez, M.A.; García-Valladares, O.; Vales-Pinzón, C.; Oskam, G. A Spray-Deposited Modified Silica Film on Selective Coatings for Low-Cost Solar Collectors. *Coatings* **2024**, *14*, 1368. [CrossRef]
18. Uribe González, F.; Monfil Leyva, K.; Moreno Moreno, M.; Morales Sánchez, A.; Muñoz Zurita, A.L.; Luna López, J.A.; Carrillo López, J.; Hernández de la Luz, J.A.; Salazar Valdez, A.S. Optimization of LPCVD Deposition Conditions of Silicon-Rich Silicon Nitride to Obtain Suitable Optical Properties for Photoluminescent Coating. *Coatings* **2024**, *14*, 1383. [CrossRef]
19. Monaico, E.V.; Morari, V.; Buiucli, S.; Zalamai, V.V.; Ursaki, V.V.; Tiginyanu, I.M. Impact of Substrate upon Morphology, Luminescence, and Wettability of ZnMgO Layers Deposited by Spray Pyrolysis. *Coatings* **2024**, *14*, 1395. [CrossRef]
20. Li, Y.; Dong, X.; Liu, Q.; You, Y.; Gao, Z.; Zhang, Y. Preparation and Properties of Thick Tungsten Coating Electrodeposited from $Na_2WO_4$-$WO_3$-KCl-NaF Molten Salt System. *Coatings* **2024**, *14*, 1471. [CrossRef]
21. de Carvalho, M.D.S.; Rapp, M.; Voigt, A.; Dirschka, M.; Geckle, U. Structural Aspects and Adhesion of Polyurethane Composite Coatings for Surface Acoustic Wave Sensors. *Coatings* **2025**, *15*, 139. [CrossRef]
22. Dong, X.; Li, Y.; You, Y.; Gao, Z.; Zhang, Y. Electrodeposition of Metallic Tungsten Coating on 9Cr-ODS Steel Substrate from Binary Oxide Molten Salt. *Coatings* **2025**, *15*, 257. [CrossRef]
23. Zhu, X.; Lin, D.; Liu, S.; Li, J.; Tang, Y.; Pan, Y.; Deng, Q.; Yao, L. Flexibility-Engineered $Nb_2O_5$ Carbon Nanofiber Film Anodes: Concentration-Dependent Optimization for Mechanically Robust and Stable Sodium Storage. *Coatings* **2025**, *15*, 374. [CrossRef]
24. Zeng, X.; Long, Z.; Liu, Z.; Cao, Q.; Cheng, X. Preparation and Characterization of Co-CuCoMnOx Solar Selective Absorption Coatings. *Coatings* **2025**, *15*, 547. [CrossRef]
25. Gärtner, A.; Mureșan, M.G.; Mühlig, C.; Herffurth, T.; Felde, N.; Wagner, H.; Schulz, U.; Bingel, A.; Schröder, S.; Mocek, T.; et al. Manufacturing Process and Characteristics of Silica Nanostructures for Anti-Reflection at 355 nm. *Coatings* **2025**, *15*, 556. [CrossRef]
26. Lazarova, K.; Bozhilova, S.; Docheva, M.; Pavlova, K.; Alexieva, G.; Christova, D.; Babeva, T. Hygrosensitive Response and Characteristics of Copolymer Coatings with Potential for Humidity Monitoring. *Coatings* **2025**, *15*, 954. [CrossRef]
27. Dong, X.; Liu, W.; Li, Y.; Gao, Z.; Zhang, Y. Electrodeposition of Fine-Grained Tungsten Coatings on CuCrZr Alloy Substrates from Relatively Low Temperature KF-KCl-WO3 Molten Salt System. *Coatings* **2025**, *15*, 1219. [CrossRef]
28. Vendittozzi, C.; Di Micco, E.; Caponero, M.A.; D'Amato, R. Functional Coatings for Fiber Bragg Gratings: A Critical Review of Deposition Techniques for Embedded and Harsh-Environment Applications. *Coatings* **2025**, *15*, 1268. [CrossRef]
29. Dimov, A.; Ivanov, G.R.; Keil, L.; Terfort, A.; Liu, J.; Strijkova, V. Langmuir and Langmuir–Blodgett Monolayers from 20 nm Sized Crystals of the Metal–Organic Framework MIL-101(Cr). *Coatings* **2025**, *15*, 1449. [CrossRef]




 https://doi.org/10.3390/coatings16010127

coatings

MDPI

Article

# Crystalline Structure and Optical Properties of Cobalt Nickel Oxide Thin Films Deposited with a Pulsed Hollow-Cathode Discharge in an Ar+$O_2$ Gas Mixture

Anna Kapran [1,2], Rainer Hippler [1,3,*], Harm Wulff [3], Jiri Olejnicek [1], Lenka Volfova [1], Aneta Pisarikova [1,4], Natalia Nepomniashchaia [1,5], Martin Cada [1] and Zdenek Hubicka [1]

1 Institute of Physics, Czech Academy of Sciences, Na Slovance 2, 18221 Prague, Czech Republic; kapran@fzu.cz (A.K.); olejn@fzu.cz (J.O.); volfl@fzu.cz (L.V.); hrubantova@fzu.cz (A.P.); nepomni@fzu.cz (N.N.); cada@fzu.cz (M.C.); hubicka@fzu.cz (Z.H.)

2 Faculty of Mathematics and Physics, Charles University, Ke Karlovu 3, 12116 Prague, Czech Republic

3 Institut für Physik, Universität Greifswald, Felix-Hausdorff-Str. 6, 17489 Greifswald, Germany; wulff@uni-greifswald.de

4 Joint Laboratory of Optics, Palacky University Olomouc, Av. 17. listopadu 50A, 77207 Olomouc, Czech Republic

5 Faculty of Nuclear Sciences and Physical Engineering, Czech Technical University in Prague, Brehova 7, 11519 Prague, Czech Republic

* Correspondence: hippler@physik.uni-greifswald.de

**Abstract:** Cobalt nickel oxide films are deposited on Si(111) or fluorine-doped tin-oxide-coated (FTO) glass substrates employing a pulsed hollow-cathode discharge. The hollow cathode is operated with argon gas flowing through the nozzle and with $O_2$ gas admitted to the vacuum chamber. Three different cathode compositions (Co20Ni80, Co50Ni50, and Co80Ni20) are investigated. Deposited and annealed thin films are characterized by X-ray diffraction, infrared (Raman) spectroscopy, and ellipsometry. As-deposited films consist of a single mixed cobalt nickel oxide phase. Upon annealing at 600 °C, the mixed cobalt nickel oxide phase separates into two cystalline sub-phases which consist of cubic NiO and cubic $Co_3O_4$. Annealed films are investigated by spectroscopic ellipsometry and the optical bandgaps are determined.

**Keywords:** hollow-cathode discharge; mixed cobalt nickel oxide film; refractive index; absorption coefficient; optical bandgap

## 1. Introduction

Mixed metal oxides (MMO) consist of two or more metal atoms in combination with oxygen. Mixed cobalt nickel oxides (CoNiO) have received renewed interest in the past years [1]. Non-stoichiometric CoNiO and stoichiometric $NiCo_2O_4$ are of interest for many applications, e.g., as batteries [2–6], supercapacitors [7,8], battery supercapacitor hybrids [9], sensors [10], solar cells [11], and catalysts [12–16]. Here, we are guided by an interest in the crystallographic and optical properties of deposited CoNiO films with varying Co/Ni ratios. Both cobalt oxide and nickel oxide are p-type semiconductors. The optical bandgaps of cobalt oxide and nickel oxide are rather different. NiO has a direct optical bandgap of about 3.8–4.1 eV [17–20]. Smaller optical bandgaps of about 2 eV have been reported for cobalt oxide [21,22]. The bandgap of 2.0 eV is frequently attributed to single-phase $Co_3O_4$ while mixed CoO/$Co_3O_4$ phases have larger bandgaps [23]. In general, cobalt oxide shows more than one optical bandgap, which may depend on the exact composition and on the morphology [23–26].

In the present paper, we utilize a pulsed hollow-cathode (PHC) discharge with a cobalt nickel cathode to deposit mixed cobalt nickel oxide films. The potential of PHC discharges for deposition of thin films has not been fully explored yet. In particular, rather few investigations regarding the deposition of multi-component films have been reported yet. In this context, PHC discharges using multi-component cathodes are particularly useful. Here, we employ mixed CoNi cathodes with different stoichiometric compositions, e.g., Co20Ni80, Co50Ni50, and Co80Ni20. Films are deposited on Si(111) wafers and on soda lime glass coated with fluorine-doped tin oxide (FTO). The structural, morphological, and optical properties of the deposited films are studied with the help of scanning electron microscopy, X-ray diffraction, Raman spectroscopy, and ellipsometry.

## 2. Experiment

The experimental set-up has been described elsewhere [27,28]. It consists of a cylindrical hollow cathode (HC) inside a vacuum chamber. The HC is made from cobalt nickel alloy with a purity of 99.95% [28–31]. Several cathodes with different compositions have been employed for film deposition, e.g., Co20Ni80, Co50Ni50, and Co80Ni20, where the numbers indicate the atomic percentage of the cathode material. We use argon and oxygen as working gases at gas flow rates of 200 sccm and 100 sccm, repsectively.

The pulsed hollow cathode is operated with a discharge current of 0.25 A at a repetition frequency of 5 kHz and a pulse length of 100 μs.

Films are deposited on Si(111) or FTO substrates at two different gas pressures (2.8 Pa and 50 Pa) in the vacuum chamber. More details may be found elsewhere [27]. Typical deposition times are 90–120 min.

As-deposited and annealed films are characterized by X-ray diffraction, Raman spectroscopy, and ellipsometry. Annealing is carried out in ambient air using temperatures of 410 °C and 600 °C.

Energy dispersive X-ray spectroscopy (EDX) is employed for the elemental composition analysis of the deposited thin films. The TESCAN Ferra 3 scanning electron microscope (SEM) equipped with an EDAX Octane super 60 $mm^2$ is employed. Chemical composition analysis is performed with an energy of 15 kV at a working distance of 9 mm. Thickness measurements are performed with an energy of 5 kV at the same distance.

The crystallographic structure of the samples is investigated with the help of the X-ray diffraction (XRD) technique. The Grazing Incidence (GIXRD) geometry with Cu $K\alpha$ radiation ($\lambda$ = 0.154 nm) is employed [32]. Raman spectroscopy is carried out at room temperature using a Renishaw Raman Microscope RM 1000.

The optical transmission is measured with a UV-VIS spectrophotometer (LAMBDA 1050 UV/Vis/NIR, Perkin-Elmer). Spectroscopic ellipsometry is conducted at room temperature (300 K) in a dry nitrogen atmosphere with a variable angle ellipsometer (J.A. Woollam VUV-VASE) in the photon energy range 0.8–7.5 eV at three angles of incidence (60°, 65°, and 70°). Refraction index $n$, extinction coefficient $k$, and absorption coefficient $\alpha$ are obtained with the help of commercial WVASE software.

The photoelectrochemical (PEC) activity is investigated in an electrochemical cell in a three-electrode configuration. Details may be found elesewhere [27].

## 3. Results and Discussion

### 3.1. Film Composition

Deposited films have been analyzed with the help of scanning electron microscopy (SEM). The film thickness is obtained from the cross sectional view of the sample which is cut at the center. The extracted film thickness is 1.98 μm, 3.68 μm, and 1.56 μm for the films deposited with Co20Ni80, Co50Ni50, and Co80Ni20 cathodes, respectively. The film composition of annealed films is measured at two positions (center and near the edge) by EDX. The two results agree well with each other and only the average results are shown in Table 1. The measured film composition and, in particular, the extracted Co/(Co+Ni) ratios agree reasonably with the nominal composition of the considered cathodes.

**Table 1.** Composition of annealed (600 °C) films on Si(111) substrates for different cathode materials. Gas pressure 50 Pa.

| Cathode | Co (at %) | Ni (at %) | O (at %) | Co/(Co+Ni) Ratio |
|---|---|---|---|---|
| Co20Ni80 | 14.1 | 30.7 | 55.1 | 0.32 |
| Co50Ni50 | 24.7 | 26.9 | 48.4 | 0.48 |
| Co80Ni20 | 54.1 | 11.9 | 35.5 | 0.82 |

### 3.2. Crystal Structure

#### 3.2.1. Co50Ni50 Cathode

Figure 1 shows GIXRD results for as-deposited and annealed (at 410 °C and 600 °C) films on Si(111) substrates deposited with the Co50Ni50 cathode at gas pressures of 2.8 Pa and 50 Pa. These films, hence, should contain an approximately equal amount of Co and Ni. The X-ray diffractograms of the as-deposited films show strongly broadened reflections. For the as-deposited films, it was not possible to properly distinguish between pure nickel oxide, cobalt oxide, and/or cobalt nickel oxide phases. The annealed films (410 °C) contain a single crystalline phase which is assigned to cubic nickel oxide (c-NiO) with space group (SG) 225 (ICSD CIF 9866 [33]). We mention in passing that this does not rule out the existence of a mixed $Ni_{1-x}Co_xO$ phase with a c-NiO structure [34] or of an amorphous cobalt-rich phase. In fact, mixed $Ni_{1-x}Co_xO$ phases with a c-NiO lattice structure and $x$ as large as 0.4 (40 at%) have been reported before [34].

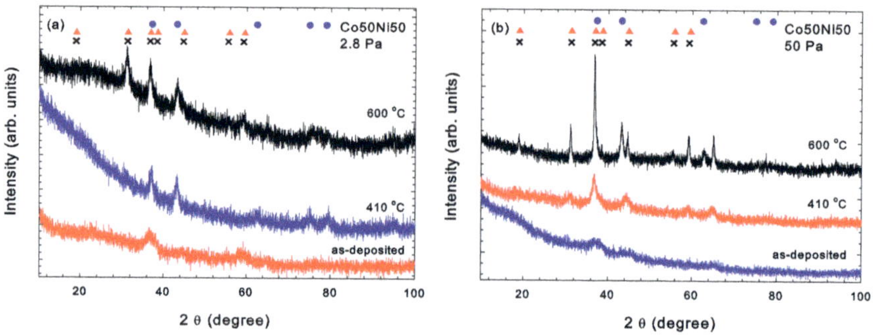


**Figure 1.** GIXRD of as-deposited and annealed (410 °C, 600 °C) CoNiO films deposited with a Co50Ni50 cathode at (**a**) 2.8 Pa and (**b**) 50 Pa. Reflections from c-NiO (•), c-$Co_3O_4$ (▲), and c-$NiCo_2O_4$ (×) are indicated.

Annealing at 600 °C results in a phase separation into c-NiO (SG 225) and spinel-type c-$Co_3O_4$ (SG 227, ICSD CIF 28158 [33]) crystalline phases and leads to an increase in the mean grain size $D$. Here, *grain* reflects the structural homogenous part in the direction of the diffraction vector and its size is given as a length unit (nm). The mean grain size was evaluated using the Rietveld refinement method (TOPAS software, Bruker). A separation of crystalline phases during annealing has been observed before for CuNiO films [27]. The crystalline phase fractions $f$ of 50% each reflect well the initial composition of the employed Co50Ni50 cathode. No evidence of crystalline CoO or mixed $NiCo_2O_4$ phases was found. Table 2 summarizes the results of our analyses.

#### 3.2.2. Co20Ni80 Cathode

Figure 2 shows the GIXRD results for the nickel-rich films deposited with the Co20Ni80 cathode. Diffractograms of the as-deposited film and of the film annealed at 410 °C show reflections from c-NiO but not from cobalt oxide. Hence, there is no indication of a second crystalline phase. The film deposited at 2.8 Pa has an almost perfect NiO(200) preferred orientation whereas the film deposited at 50 Pa does not show this preference. The reference value for the lattice parameter of pure c-NiO is 0.4178 nm [33]. The slightly smaller lattice parameters (Table 2) can be explained by a partial incorporation of $Co^{2+}$ (and, eventually, $Co^{3+}$) ions into the $Ni^{2+}$ lattice sites and the smaller ion radii of Co ions compared to Ni ions [35].

**Table 2.** Identified crystalline phases and lattice parameters of as-deposited and annealed (410 °C, 600 °C) films deposited at 2.8 Pa and 50 Pa. c = cubic, $f$ = crystal phase fraction, $D$ = mean grain (particle) size. † = preferred c-NiO(200) orientation, ‡ = cubic phase modification (space group 216), n/e = not evaluable, S/N = poor signal-to-noise ratio.

| Film | Crystalline Phase | Lattice Parameter (nm) | $f$ (%) | $D$ (nm) |
|---|---|---|---|---|
| | | (a1) Co50Ni50—2.8 Pa | | |
| as-deposited | c-$Co_3O_4$/c-NiO | n/e | n/e | S/N |
| 410 °C | c-NiO | 0.4191 | 100 | 12 |
| 600 °C | c-NiO | 0.418 | 50 | 13 |
| | c-$Co_3O_4$ | 0.810 | 50 | 13 |
| | | (a2) Co50Ni50—50 Pa | | |
| as-deposited | c-$Co_3O_4$/c-NiO | n/e | n/e | S/N |
| 410 °C | c-$NiCo_2O_4$ | 0.8165 | 100 | 9 |
| 600 °C | c-NiO | 0.4176 | 50 | 320 |
| | c-$Co_3O_4$ | 0.8099 | 50 | 200 |
| | | (b1) Co20Ni80—2.8 Pa | | |
| as-deposited | c-NiO † | 0.4185 | 100 | 12 |
| 410 °C | c-NiO † | 0.4170 | 100 | 11 |
| 600 °C | c-NiO † | 0.4172 | 87 | 26 |
| | c-$Co_3O_4$ | 0.8104 | 13 | 15 |
| | | (b2) Co20Ni80—50 Pa | | |
| as-deposited | c-NiO | 0.415 | 100 | 14 |
| 410 °C | c-NiO | 0.415 | 100 | 17 |
| 600 °C | c-NiO | 0.4182 | 83 | 34 |
| | c-$Co_3O_4$ | 0.8078 | 17 | 15 |
| | | (c1) Co80Ni20—2.8 Pa | | |
| as-deposited | c-$Co_3O_4$ ‡ | 0.8139 | 100 | 22 |
| 410 °C | c-$Co_3O_4$ ‡ | 0.8098 | 100 | 56 |
| | | (c2) Co80Ni20—50 Pa | | |
| as-deposited | c-$Co_3O_4$/c-NiO | n/e | n/e | S/N |
| 410 °C | c-$Co_3O_4$‡ | 0.8078 | 100 | 25 |
| 600 °C | c-NiO | 0.4205 | 25 | 200 |
| | c-$Co_3O_4$ | 0.8098 | 75 | 200 |

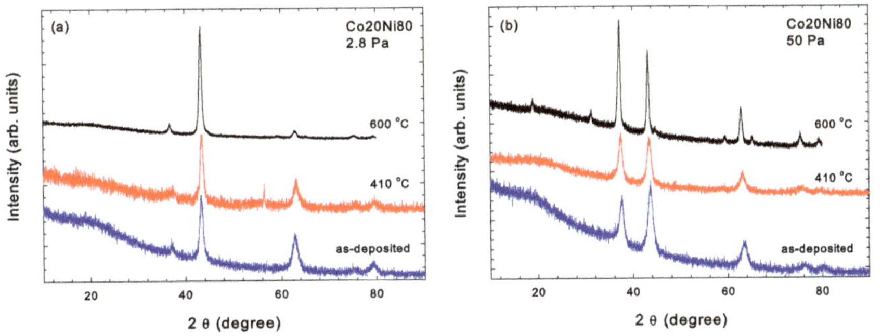


**Figure 2.** GIXRD of as-deposited and annealed (410 °C, 600 °C) CoNiO films deposited with a Co20Ni80 cathode at (**a**) 2.8 Pa and (**b**) 50 Pa.

Typical crystallite (grain) sizes of the as-deposited and annealed (410 °C) samples are in the range of 11–17 nm. Both diffraction patterns from the as-deposited and annealed

(410 °C) sample deposited at 2.8 Pa differ from the 50 Pa sample only in the peak intensities. Annealing at 600 °C leads to a complete separation into cobalt oxide and nickel oxide phases. The results of a quantitative phase analysis with the Rietveld method correspond well with the composition of the employed cathode.

#### 3.2.3. Co80Ni20 Cathode

Figure 3a shows the GIXRD results for the cobalt-rich film deposited at 2.8 Pa with the Co80Ni20 cathode. The diffractograms of the as-deposited film and of the film annealed at 410 °C show the typical reflections from a c-$Co_3O_4$ phase modification with space group 216 (ICSD CIF 9362). There is no indication of a second crystalline phase. The X-ray diffractogram of the as-deposited film deposited at 50 Pa shows a single reflection in the angular range 36.8–38.1°, which can be attributed to c-NiO, c-$Co_3O_4$, and/or c-$NiCo_2O_4$ (Figure 3b). It was not possible to distinguish between these crystalline phases due to the strongly broadened line profile, however. Upon annealing at 410 °C, the crystalline phase changes to c-$Co_3O_4$ (SG 216, ICSD CIF 9362). Mixed $Co_{3-x}Ni_xO_4$ phases with a c-$CO_3O_4$ lattice structure and $x$ as large as 0.8 (80 at %) have been reported before [34]. Further annealing at 600 °C again leads to a phase separation into c-NiO (SG 225) and c-$Co_3O_4$ (SG 227, ICSD CIF 28158) and gives rise to large grain sizes of about 200 nm. The extracted Ni/Co ratio of 25/75 is in reasonable agreement to the measured composition (Table 1) and close to the one expected for a Co80Ni20 cathode.

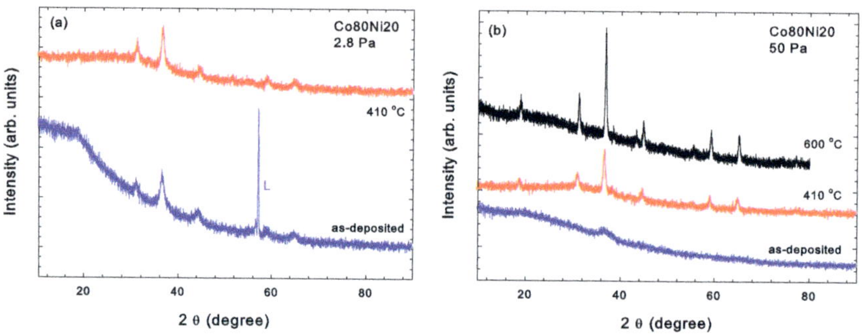


**Figure 3.** GIXRD of as-deposited and annealed (410 °C, 600 °C) CoNiO films deposited with a Co80Ni20 cathode at (**a**) 2.8 Pa and (**b**) 50 Pa. L = Laue peak.

### 3.3. *Raman Spectroscopy*

The GIXRD results are supported by Raman spectroscopy. The as-deposited film from the nickel-rich Co20Ni80 cathode shows a pronounced broad peak at about 545 $cm^{-1}$ and a smaller peak at 1080 $cm^{-1}$ (Figure 4). The peaks are frequently assigned to one-phonon longitudinal optical (LO) mode and two-phonon vibrational modes associated with NiO [15,36,37]. In particular, the one-phonon LO mode at 545 $cm^{-1}$ mode is strongly suppressed in bulk NiO, while it is enhanced in nano-crystalline NiO [38,39]. Similar structures were observed for copper nickel oxide (Cu50Ni50) films [27]. The center wave number of the 545 $cm^{-1}$ peak broadens and shifts to 565 $cm^{-1}$ if the Co50Ni50 cathode is employed. The appearance of two weak shoulders at 480 $cm^{-1}$ and 660 $cm^{-1}$ is indicated. The film deposited with the cobalt-rich Co80Ni20 cathode in addition displays several narrow peaks at 196 $cm^{-1}$, 480 $cm^{-1}$, and 680 $cm^{-1}$ which are attributed to $Co_3O_4$ [27].

The Raman spectra of the annealed films display several narrow peaks at 193 $cm^{-1}$, 480 $cm^{-1}$, 525 $cm^{-1}$, and 680 $cm^{-1}$, which are attributed to $Co_3O_4$ (Figure 4) [40,41]. The Raman spectrum obtained with the nickel-rich Co20Ni80 cathode additionally shows a pronounced shoulder at 560 $cm^{-1}$ and a weak peak at 1080 $cm^{-1}$ indicating the presence of NiO [15,36,37]. The pronounced peak at 680 $cm^{-1}$ is frequently assigned to the $A_{1g}$ mode of $Co_3O_4$ [29]. The weak band at 193 $cm^{-1}$ is assigned to tetrahedral sites of $Co_3O_4$ [15].

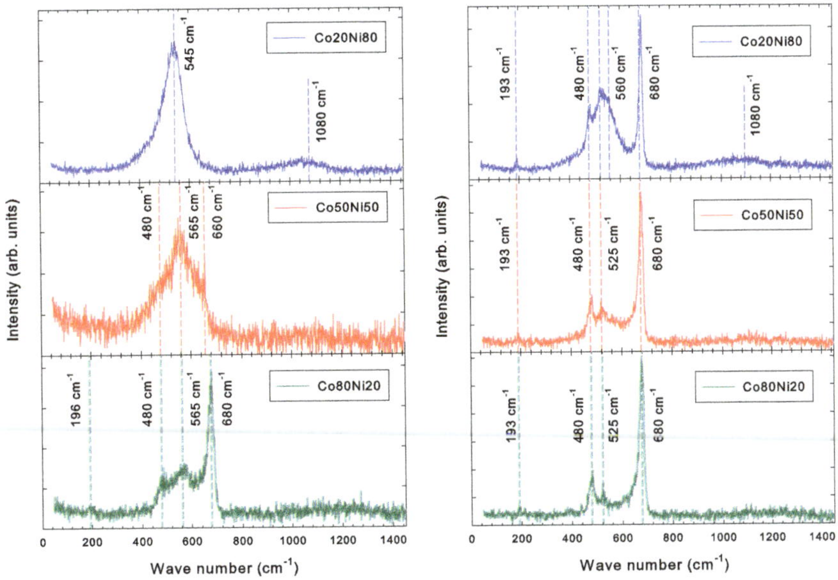


**Figure 4.** Raman spectroscopy of as-deposited (**left**) and annealed (**right**) films deposited on Si(111) substrates. PHC discharge with three different (Co20Ni80, Co50Ni50, and Co80Ni20) cathodes. Gas pressure 50 Pa.

### 3.4. Optical Properties

The optical properties of annealed films are investigated by UV-VIS measurements and by ellipsometry. The transmittance of annealed cobalt nickel oxide films deposited on FTO-coated glass using different cathodes obtained from the UV-VIS measurements is displayed in Figure 5a. No corrections regarding the different film thicknesses have been applied. It is evident that films are opaque at small wavelengths below 400 nm and become increasingly transparent at larger wavelengths. This result is further confirmed by our ellipsometry measurements; see below. The transmittance of non-stoichiometric $Co_xNi_{1-x}O$ films was investigated by Roffi et al. [42]. Accordingly, the measured transmittance decreases with increasing cobalt content $x$, while the absorption edge moves to longer wavelengths (smaller photon energies). These observations are in fair agreement with our results (Figure 5).

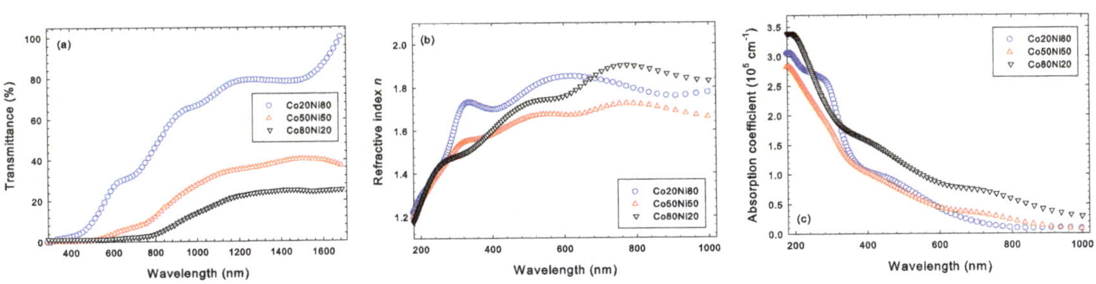


**Figure 5.** (**a**) Transmittance, (**b**) refractive index $n$, and (**c**) absorption coefficient $\alpha$ of annealed films (600 °C) on FTO glass obtained with different cathodes: ○ Co20Ni80, △ Co50Ni50, and ▽ Co80Ni20.

The optical properties of films deposited at 50 Pa on FTO-coated glass and annealed at 600 °C are further obtained from ellipsometry measurements. Figure 5b displays the refractive index $n$, which increases (decreases) with wavelength $\lambda$ (photon energy $E_{ph}$) from about $n \approx 1.2$ at $\lambda = 180$ nm ($E_{ph} = 6.9$ eV) to $n \approx 1.8$ at $\lambda = 1000$ nm ($E_{ph} = 1.24$ eV). Only minor differences are observed for the three samples deposited with different cathode compositions (Co20Ni80, Co50Ni50, and Co80Ni80). The absorption coefficient for these films is displayed in Figure 5c as function of wavelength. Again, the present ellipsometry results show that all films become opaque at small wavelengths and increasingly transparent at larger wavelengths. A similar behavior of the absorbance was observed before [43]. The absorption behavior is rather complicated, however, showing more than one absorption edge corresponding to several optical bandgaps [44–46].

A more thorough analysis of the absorption behavior employing a so-called Tauc plot making use of

$$(\alpha\, E_{ph})^2 = E_{ph} - E_g \tag{1}$$

is shown in Figure 6, where $\alpha$ is the absorption coefficient and $E_{ph}$ is the photon energy [47]. A direct optical bandgap as in the case of CoO, $Co_3O_4$, and NiO is assumed [48–50]. From this plot we derive optical bandgaps of 4.14 eV, 4.20 eV, and 4.14 eV for copper nickel oxide films obtained with the Co20Ni80, Co50Ni50, and Co80Ni20 cathodes, respectively. There is an indication of a second optical bandgap at 3.50 eV for the Co20Ni80 cathode. NiO has a direct optical bandgap of about 3.8 eV [18–20,51–53]. A similar optical bandgap is observed for CoO [53]. Slightly larger bandgap energies of 3.88 eV/4.07 eV [54] and 4.38 eV [8] are reported for $NiCo_2O_4$. However, our XRD data for the annealed films reveal no evidence of crystalline CoO or $NiCo_2O_4$ phases. The bandgap of ≈4.15 eV observed here is, hence, assigned to the nickel oxide composition of the annealed films.

Smaller optical bandgaps are reported for cobalt oxide [21,22]. In general, cobalt oxide shows several optical bandgaps which depend on the composition and on the morphology. An expanded view of the photon energy region below 3.5 eV is displayed in the inset of Figure 6. Direct optical bandgaps of 2.10 eV, 2.23 eV, and 2.19 eV for the Co20Ni80, Co50Ni50, and Co80Ni20 cathodes, respectively, are extracted from this graph. The extracted bandgaps agree well with the reported optical bandgap of about 2.17 eV for $Co_3O_4$ [23–26,55]. The result is further supported by our XRD analysis.

Finally, we mention in passing that films deposited on FTO-coated glass are also tested for photoelectrochemical activity using the same set-up as for CuNiO films [27]. $Na_2SO_4$ and NaOH are employed as electrolytes. No photoelectrochemical activity of the present copper nickel oxide has been observed, however.

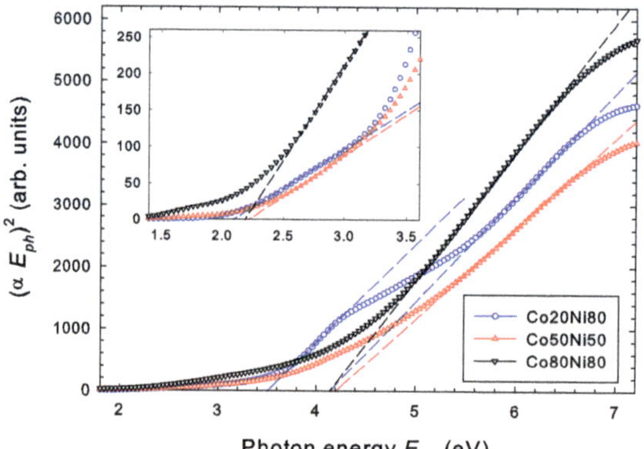


**Figure 6.** Tauc plot assuming a direct bandgap for annealed films (600 °C) deposited on FTO glass with different cathodes: ○ Co20Ni80, △ Co50Ni50, and ▽ Co80Ni20. Inset shows expanded view between 1.5 eV and 3.5 eV.

## 4. Conclusions

Results are reported for mixed cobalt nickel oxide films deposited with the help of a pulsed hollow-cathode (PHC) discharge operated at 5 kHz. Deposited films are composed of a mixed metal oxide (MMO) phase. During annealing at 600 °C, the MMO phase separates into two crystalline cubic $Co_3O_4$ and cubic NiO subphases. The results are confirmed by Raman spectroscopy. The refraction index and absorption coefficient are derived from ellipsometry measurements. The refractive index increases from 1.2 at 200 nm to about 1.8 at 1000 nm. Annealed films are opaque at small wavelengths and become increasingly more transparent at longer wavelengths. Optical bandgaps of about 4.15 eV are linked to the NiO phase of the annealed films, while a second optical bandgap at about 2.17 eV is linked to the $Co_3O_4$ phase. Photoelectrochemical activity was not observed.

**Author Contributions:** Conceptualization, R.H., M.C. and Z.H.; Methodology, R.H. and L.V.; Formal analysis, N.N.; Investigation, A.K., R.H., J.O., A.P. and N.N.; Resources, M.C. and Z.H.; Data curation, A.K., R.H. and H.W.; Writing—original draft, R.H.; Writing—review & editing, H.W., M.C. and Z.H. All authors have read and agreed to the published version of the manuscript.

**Funding:** This research was funded by Technology Agency of the Czech Republic grant number FW06010462.

**Institutional Review Board Statement:** Not applicable.

**Informed Consent Statement:** Not applicable.

**Data Availability Statement:** Data are contained within the article.

**Conflicts of Interest:** The authors declare no conflicts of interest.

## References

1. Poonam; Sharma, K.; Singh, N.; Tripathi, S.K. Characterization of nickel cobalt oxide: A potential material for supercapacitor. *Mater. Res. Express* **2018**, *6*, 025502. [CrossRef]
2. Huang, X.; Wu, J.; Guo, R.; Lin, Y.; Zhang, P. Aligned nickel–cobalt oxide nanosheet arrays for lithium ion battery applications. *Int. J. Hydrogen Energy* **2014**, *39*, 21399–21404. [CrossRef]
3. Masoud, E.M. Improved initial discharge capacity of nanostructured Ni-Co spinel ferrite as anode material in lithium ion batteries. *Solid State Ionics* **2013**, *253*, 247–252. [CrossRef]
4. He, Q.; Gu, S.; Wu, T.; Zhang, S.; Ao, X.; Yang, J.; Wen, Z. Self-supported mesoporous $FeCo_2O_4$ nanosheets as high capacity anode material for sodium-ion battery. *Chem. Eng. J.* **2017**, *330*, 764–773. [CrossRef]
5. Sharma, Y.; Sharma, N.; Subbarao, G.; Chowdari, B. Studies on spinel cobaltites, $FeCo_2O_4$ and $MgCo_2O_4$ as anodes for Li-ion batteries. *Solid State Ionics* **2008**, *179*, 587–597. [CrossRef]
6. Julien, C.; Gendron, F.; Amdouni, A.; Massot, M. Lattice vibrations of materials for lithium rechargeable batteries. VI: Ordered spinels. *Mater. Sci. Eng. B* **2006**, *130*, 41–48. [CrossRef]
7. Chavhan, M.P.; Sethi, S.R.; Ganguly, S. Mixed metal oxides in synergy at nanoscale: Electrospray induced porosity of in situ grown film electrode for use in electrochemical capacitor. *Electrochim. Acta* **2020**, *347*, 136277. [CrossRef]
8. Pradeepa, S.S.; Rajkumar, P.; Diwakar, K.; Sutharthani, K.; Subadevi, R.; Sivakumar, M. A Facile One-Pot Hydrothermal Synthesis of Zn, Mn Co-Doped NiCo2O4 as an Efficient Electrode for Supercapacitor Applications. *ChemistrySelect* **2021**, *6*, 6851–6862. [CrossRef]
9. Hong, W.-L.; Lin, L.-Y. Design of nickel cobalt oxide and nickel cobalt oxide@nickel molybdenum oxide battery-type materials for flexible solid-state battery supercapacitor hybrids. *J. Power Sources* **2019**, *435*, 226797. [CrossRef]
10. Banno, S.; Imanaka, N.; Adachi, G. Selective nitrogen dioxide sensor based on nickel copper oxide mixed with rare earths. *Sens. Actuators B Chem.* **1995**, *25*, 619–622. [CrossRef]
11. Lee, J.H.; Noh, Y.W.; Jin, I.S.; Park, S.H.; Jung, J.W. Efficient perovskite solar cells with negligible hysteresis achieved by sol–gel-driven spinel nickel cobalt oxide thin films as the hole transport layer. *J. Mater. Chem. C* **2019**, *7*, 7288–7298. [CrossRef]
12. Vidales, A.G.; Choi, K.; Omanovic, S. Nickel-cobalt-oxide cathodes for hydrogen production by water electrolysis in acidic and alkaline media. *Int. J. Hydrogen Energy* **2018**, *43*, 12917–12928. [CrossRef]
13. Yentekakis, I.V.; Panagiotopoulou, P.; Artemakis, G. A review of recent efforts to promote dry reforming of methane (DRM) to syngas production via bimetallic catalyst formulations. *Appl. Catal. B Environ.* **2021**, *296*, 120210. [CrossRef]
14. Kongsawatvoragul, K.; Kalasina, S.; Phattharasupakun, N.; Sawangphruk, M. A single energy conversion and storage cell of nickel-doped cobalt oxide under UV and visible light illumination. *Electrochim. Acta* **2019**, *328*, 135120. [CrossRef]
15. Jirátová, K.; Čada, M.; Naiko, I.; Ostapenko, A.; Balabánová, J.; Koštejn, M.; Maixner, J.; Babii, T.; Topka, P.; Soukup, K.; et al. Plasma Jet Sputtering as an Efficient Method for the Deposition of Nickel and Cobalt Mixed Oxides on Stainless-Steel Meshes: Application to VOC Oxidation. *Catalysts* **2022**, *13*, 79. [CrossRef]
16. Jirátová, K.; Soukal, P.; Kapran, A.; Babii, T.; Balabánová, J.; Koštejn, M.; Čada, M.; Maixner, J.; Topka, P.; Hubička, Z.; et al. Nickel-Copper Oxide Catalysts Deposited on Stainless Steel Meshes by Plasma Jet Sputtering: Comparison with Granular Analogues and Synergistic Effect in VOC Oxidation. *Catalysts* **2023**, *13*, 595. [CrossRef]

17. Peng, T.; Xiao, X.; Han, X.; Zhou, X.; Wu, W.; Ren, F.; Jiang, C. Characterization of DC reactive magnetron sputtered NiO films using spectroscopic ellipsometry. *Appl. Surf. Sci.* **2011**, *257*, 5908–5912. [CrossRef]
18. Duraisamy, N.; Kandiah, K.; Rajendran, R.; S, P.; R, R.; Dhanaraj, G. Electrochemical and photocatalytic investigation of nickel oxide for energy storage and wastewater treatment. *Res. Chem. Intermed.* **2018**, *44*, 5653–5667. [CrossRef]
19. Awais, M.; Rahman, M.; MacElroy, J.D.; Coburn, N.; Dini, D.; Vos, J.G.; Dowling, D.P. Deposition and characterization of NiOx coatings by magnetron sputtering for application in dye-sensitized solar cells. *Surf. Coat. Technol.* **2010**, *204*, 2729–2736. [CrossRef]
20. Sato, H.; Minami, T.; Takata, S.; Yamada, T. Transparent conducting p-type NiO thin films prepared by magnetron sputtering. *Thin Solid Films* **1993**, *236*, 27–31. [CrossRef]
21. Hong, T.; Liu, Z.; Zheng, X.; Zhang, J.; Yan, L. Efficient photoelectrochemical water splitting over $Co_3O_4$ and $Co_3O_4$/Ag composite structure. *Appl. Catal. B Environ.* **2017**, *202*, 454–459. [CrossRef]
22. Moridon, S.N.F.; Salehmin, M.N.I.; Arifin, K.; Minggu, L.J.; Kassim, M.B. Synthesis of Cobalt Oxide on FTO by Hydrothermal Method for Photoelectrochemical Water Splitting Application. *Appl. Sci.* **2021**, *11*, 3031. [CrossRef]
23. Jung, S.; Nandi, D.K.; Yeo, S.; Kim, H.; Jang, Y.; Bae, J.-S.; Hong, T.E.; Kim, S.-H. Phase-controlled growth of cobalt oxide thin films by atomic layer deposition. *Surf. Coat. Technol.* **2018**, *337*, 404–410. [CrossRef]
24. Al-Tuwirqi, R.M.; Al-Hazmi, F.; Alnowaiser, F.; Al-Ghamdi, A.A.; Aal, N.A.; El-Tantawy, F. Synthesis and physical properties of mixed Co3O4/CoO nanorods by microwave hydrothermal technique. *Superlattices Microstruct.* **2011**, *50*, 437–448. [CrossRef]
25. Barakat, N.A.M.; Khil, M.S.; Sheikh, F.A.; Kim, H.Y. Synthesis and Optical Properties of Two Cobalt Oxides (CoO and Co3O4) Nanofibers Produced by Electrospinning Process. *J. Phys. Chem. C* **2008**, *112*, 12225–12233. [CrossRef]
26. Prieto, P.; Marco, J.; Serrano, A.; Manso, M.; de la Figuera, J. Highly oriented (111) CoO and $Co_3O_4$ thin films grown by ion beam sputtering. *J. Alloys Compd.* **2019**, *810*, 151912. [CrossRef]
27. Kapran, A.; Hippler, R.; Wulff, H.; Olejnicek, J.; Pisarikova, A.; Cada, M.; Hubicka, Z. Characteristics of a pulsed hollow cathode discharge operated in an Ar+O2 gas mixture and deposition of copper nickel oxide thin films. *Vacuum* **2023**, *215*, 112272. [CrossRef]
28. Hippler, R.; Cada, M.; Hubicka, Z. A positively biased external anode for energy control of plasma ions: Hollow cathode and magnetron sputtering discharge. *Plasma Sources Sci. Technol.* **2021**, *30*, 045003. [CrossRef]
29. Olejníček, J.; Šmíd, J.; Perekrestov, R.; Ksirova, P.; Rathouský, J.; Kohout, M.; Dvořáková, M.; Kment, Š.; Jurek, K.; Čada, M.; et al. Co3O4 thin films prepared by hollow cathode discharge. *Surf. Coat. Technol.* **2019**, *366*, 303–310. [CrossRef]
30. Olejníček, J.; Šmíd, J.; Čada, M.; Kšírová, P.; Kohout, M.; Perekrestov, R.; Tvarog, D.; Kment, Š.; Kmentová, H.; Hubička, Z. High rate deposition of photoactive TiO2 films by hot hollow cathode. *Surf. Coat. Technol.* **2020**, *383*, 125256. [CrossRef]
31. Olejníček, J.; Hrubantová, A.; Volfová, L.; Dvořáková, M.; Kohout, M.; Tvarog, D.; Gedeon, O.; Wulff, H.; Hippler, R.; Hubička, Z. $WO_3$ and $WO_{3-x}$ thin films prepared by DC hollow cathode discharge. *Vacuum* **2021**, *195*, 110679. [CrossRef]
32. Wulff, H.; Steffen, H. Characterization of Thin Films. In *Low Temperature Plasmas*; Hippler, R., Kersten, H., Schmidt, M., Schoenbach, K.-H., Eds.; Wiley-VCH: Berlin, Germany, 2008; pp. 329–362.
33. ICSD, Inorganic Crystal Structure Database, FIZ Karlsruhe—Leibniz-Institut für Informationsinfrastruktur GmbH. Eggenstein-Leopoldshafen, Germany. Available online: https://icsd.products.fiz-karlsruhe.de/ (accessed on 25 February 2024).
34. De Faria, L.; Prestat, M.; Koenig, J.-F.; Chartier, P.; Trasatti, S. Surface properties of Ni+Co mixed oxides: A study by X-rays, XPS, BET and PZC. *Electrochim. Acta* **1998**, *44*, 1481–1489. [CrossRef]
35. Shannon, R.D. Revised effective ionic radii and systematic studies of interatomic distances in halides and chalcogenides. *Acta Cryst.* **1976**, *A32*, 751–766. [CrossRef]
36. Javadian, A.; Fadavieslam, M.R. Impact of copper doping in NiO thin films deposited by spray pyrolysis on their physical properties. *J. Mater. Sci. Mater. Electron.* **2022**, *33*, 23362–23374. [CrossRef]
37. Mahala, C.; Basu, M. Nanosheets of NiCo2O4/NiO as Efficient and Stable Electrocatalyst for Oxygen Evolution Reaction. *ACS Omega* **2017**, *2*, 7559–7567. [CrossRef]
38. Duan, W.J.; Lu, S.H.; Wu, Z.L.; Wang, Y.S. Size Effects on Properties of NiO Nanoparticles Grown in Alkalisalts. *J. Phys. Chem. C* **2012**, *116*, 26043–26051. [CrossRef]
39. Mironova-Ulmane, N.; Kuzmin, A.; Sildos, I.; Puust, L.; Grabis, J. Magnon and Phonon Excitations in Nanosized NiO. *Latv. J. Phys. Tech. Sci.* **2019**, *56*, 61–72. [CrossRef]
40. Hippler, R.; Cada, M.; Ksirova, P.; Olejnicek, J.; Jiricek, P.; Houdkova, J.; Wulff, H.; Kruth, A.; Helm, C.; Hubicka, Z. Deposition of cobalt oxide films by reactive pulsed magnetron sputtering. *Surf. Coat. Technol.* **2021**, *405*, 126590. [CrossRef]
41. Hadjiev, V.G.; Iliev, M.; Vergilov, I.V. The Raman spectra of $Co_3O_4$. *J. Phys. C Solid State Phys.* **1988**, *21*, L199–L201. [CrossRef]
42. Roffi, T.M.; Uchida, K.; Nozaki, S. Structural, electrical, and optical properties of $Co_xNi_{1-x}O$ films grown by metalorganic chemical vapor deposition. *J. Cryst. Growth* **2015**, *414*, 123–129. [CrossRef]
43. Fernández, C.d.J.; Mattei, G.; Sada, C.; Battaglin, C.; Mazzoldi, P. Nanostructural and optical properties of cobalt and nickel-oxide/silica nanocomposites. *Mater. Sci. Eng. C* **2006**, *26*, 987–991. [CrossRef]
44. Cui, B.; Lin, H.; Liu, Y.-Z.; Li, J.-B.; Sun, P.; Zhao, X.-C.; Liu, C.-J. Photophysical and Photocatalytic Properties of Core-Ring Structured NiCo2O4 Nanoplatelets. *J. Phys. Chem. C* **2009**, *113*, 14083–14087. [CrossRef]
45. Bitla, Y.; Chin, Y.-Y.; Lin, J.-C.; Van, C.N.; Liu, R.; Zhu, Y.; Liu, H.-J.; Zhan, Q.; Lin, H.-J.; Chen, C.-T.; et al. Origin of metallic behavior in NiCo2O4 ferrimagnet. *Sci. Rep.* **2015**, *5*, 15201–15201. [CrossRef] [PubMed]

46. Nashaat, A.M.; Abu El-Fadl, A.; Nakamura, H.; Kassem, M.A. Glassy magnetic freezing and exchange bias effect in NiCo2O4/(Co-Ni)O nanoparticles composite. *J. Magn. Magn. Mater.* **2023**, *567*, 170366. [CrossRef]
47. Tauc, J.; Grigorovici, R.; Vancu, A. Optical Properties and Electronic Structure of Amorphous Germanium. *Phys. Status Solidi B* **1966**, *15*, 627–637. [CrossRef]
48. Mane, A.U.; Shivashankar, S. MOCVD of cobalt oxide thin films: Dependence of growth, microstructure, and optical properties on the source of oxidation. *J. Cryst. Growth* **2003**, *254*, 368–377. [CrossRef]
49. Pandey, V.; Adiba, A.; Munjal, S.; Ahmad, T. Optical bandgap tuning of cubic spinel $Co_3O_4$ by annealing temperature. *Materialia* **2022**, *26*, 101554. [CrossRef]
50. Manjunatha, K.N.; Paul, S. Investigation of optical properties of nickel oxide thin films deposited on different substrates. *Appl. Surf. Sci.* **2015**, *352*, 10–15. [CrossRef]
51. Rajesh, M.; Vengatesan, K.; Aly, M.H.; Sitharthan, R.; Dhanabalan, S.S.; Karthikeyan, M. Electrical and optical properties of sol–gel-deposited NiO films and corresponding response to annealing temperature. *Opt. Quantum Electron.* **2023**, *55*, 1167. [CrossRef]
52. Lahiji, F.A.F.; Bairagi, S.; Magnusson, R.; Sortica, M.A.; Primetzhofer, D.; Ekström, E.; Paul, B.; le Febvrier, A.; Eklund, P. Growth and optical properties of NiO thin films deposited by pulsed dc reactive magnetron sputtering. *J. Vac. Sci. Technol. A* **2023**, *41*, 063402. [CrossRef]
53. Powell, R.J.; Spicer, W.E. Optical Properties of NiO and CoO. *Phys. Rev. B* **1970**, *2*, 2182–2193. [CrossRef]
54. Silambarasan, M.; Ramesh, P.S.; Geetha, D. Facile one-step synthesis, structural, optical and electrochemical properties of NiCo2O4 nanostructures. *J. Mater. Sci. Mater. Electron.* **2016**, *28*, 323–336. [CrossRef]
55. Louardi, A.; Rmili, A.; Chtouki, T.; Elidrissi, B.; Erguig, H.; El Bachiri, A.; Ammous, K.; Mejbri, H. Effect of annealing treatment on $Co_3O_4$ thin films properties prepared by spray pyrolysis. *J. Mater. Environ. Sci.* **2017**, *8*, 485.

coatings

MDPI

*Article*

# Realizing Dual Functions through $Y_2O_3$ Modification to Enhance the Electrochemical Performance of $LiNi_{0.8}Co_{0.1}Mn_{0.1}O_2$ Material

**Xintao Wang, Feng Wang *, Meiqi Zheng, Maohua Rong, Jiang Wang *, Jianqiu Deng, Peng Liu and Daosheng Liu**

School of Materials Science and Engineering & Guangxi Key Laboratory of Information Materials, Guilin University of Electronic Technology, Guilin 541004, China; jqdeng@guet.edu.cn (J.D.)
* Correspondence: wf@guet.edu.cn (F.W.); waj124@guet.edu.cn (J.W.)

**Abstract:** In recent years, the remarkable energy density of high-nickel ternary materials has captured considerable attention. Nevertheless, the high-nickel ternary cathode material encounters several challenges, including cationic mixing, microcrack formation, poor cycling capability, and limited thermal stability. Coating, as a viable approach, proves to be effective in enhancing the material properties. In this study, the $LiNi_{0.8}Co_{0.1}Mn_{0.1}O_2$ (NCM811) sample underwent a dry grinding process, followed by $Y_2O_3$ coating and subsequent sintering at varying temperatures. The microstructure, morphology, and electrochemical properties of the materials were meticulously examined, and the underlying mechanism of coating modification was meticulously explored. The outcomes demonstrate the attainment of dual coating and doping effects through $Y_2O_3$ modification. $Y_2O_3$ coating mitigates the direct interaction between the NCM811 surface and the electrolyte, thereby inhibiting undesired side reactions at the interface. Moreover, the Y element infiltrates the crystal structure, imparting stability at elevated sintering temperatures. Remarkably, the $Y_2O_3$-coated cathode materials exhibit significantly enhanced cycling stability, discharge capacity, and rate performance. These findings can provide novel insights that can be harnessed to improve the energy density cathode material of NCM811.

**Keywords:** lithium-ion battery; $LiNi_{0.8}Co_{0.1}Mn_{0.1}O_2$ cathode material; $Y_2O_3$ modification; dual functions; cycle stability

## 1. Introduction

Long range is the primary technical index of new energy vehicles. Under the premise of ensuring safety, it is an inevitable development trend to continuously improve the energy density of batteries [1,2]. The cathode material is one of the decisive factors that determine the energy density of lithium-ion batteries. Layered transition metal oxides for lithium-ion batteries have the advantages of high energy density, good cycle stability, and rapid charge and discharge [3]. The earliest $LiCoO_2$ materials have relatively low capacity, and Co is expensive, highly toxic, and scarce [4]. $LiNiO_2$ is prone to thermal runaway reactions during overcharge and overdischarge. $LiMnO_2$ is prone to short cycle life and capacity decay at high temperatures. The ternary high-nickel cathode materials have the advantages of low cost, good safety performance, and high capacity. To a certain extent, the advantages of various cathode materials are comprehensively used to improve the performance and stability of batteries. The increase in nickel content is helpful to increase the capacity of the battery. However, $Ni^{2+}$ and $Li^+$ easily switch positions, which adversely affects its cycle performance, because the radius of $Li^+$ is very close to that of $Ni^{2+}$. High-nickel materials are also prone to react with $H_2O$ and $CO_2$ in the air to produce LiOH and $Li_2CO_3$, which can cause side reactions with electrolytes and lead to surface instability [5]. In addition, lattice oxygen precipitation, microcrack formation, and other problems lead to the destruction of its structure and capacity attenuation.

According to the characteristics of high-nickel materials, coating and doping methods are generally used to improve their properties. The commonly used coating materials are phosphates, fluorides, metal oxides, etc., such as $Al_2O_3$ [6–8], MgO [9,10], $CeO_2$ [11], $TiO_2$ [12], $AlPO_4$ [6,13], $AlF_3$ [14], $CaF_2$ [15], LiF [16], etc. Fluoride coating has strong corrosion resistance and effectively protects the performance of high-nickel materials in corrosive media [17–19]. However, it is easy to fail in a high-temperature reduction environment; the coating layer is relatively brittle and easy to crack or peel, which affects the stability and protection effect of the coating layer [20]. Phosphate coating has good electrochemical properties, but its structural stability is poor, and it is prone to phase transition or reaction to form precipitation, which affects the stability and protection effect of coating [21–23]. The metal oxide coating can isolate the cathode material from the electrolyte and minimize the harmful side reactions, thus protecting the surface of the material from the erosion of the electrolyte during charge and discharge [21,24]. As a rare earth element, yttrium has high oxygen affinity. At the same time, yttrium oxide ($Y_2O_3$) has excellent chemical inertia, high corrosion resistance, outstanding thermal stability, promoted electron transfer, and so on [25]. Liu et al. reported that coating $LiNi_{0.5}Co_{0.2}Mn_{0.3}O_2$ with yttrium oxide can successfully improve its cycle and rate performance and reduce the internal resistance of the material [26]. Therefore, using yttrium oxide coating in a high-nickel cathode material system will be a beneficial attempt.

Among the various coating methods, wet coating is widely used. The existence of media such as water may lead to lithium leaching and a change in positive electrode surface stoichiometry [27,28]. ALD coating is a method of depositing material layer by layer through gas phase reaction, but the process is complex, slow, and it requires highly precise control [29]. CVD is formed by sending gaseous reactants into the reaction chamber to form a coating on the surface of the substrate [30]. This coating process requires vacuum conditions and high temperatures, and the equipment cost is high [31]. Dry coating does not require a liquid phase medium and the process is relatively simple. It is a method that can achieve high-speed and efficient production and is suitable for large-scale preparation. In order to better meet the needs of practical production, this study adopts $Y_2O_3$ coating via the dry milling method to prepare $LiNi_{0.8}Co_{0.1}Mn_{0.1}O_2$ cathode material and investigates the effect of $Y_2O_3$ modification on the cathode material at different sintering temperatures. $Y_2O_3$ can improve the electrochemical performance of lithium-ion batteries through the synergistic improvement of coating and doping at higher temperatures.

## 2. Materials and Methods

### 2.1. Materials Preparation

The materials and reagents used in this experiment are commercial materials. Nano-scale $Y_2O_3$ powder was purchased from the Shanghai Pacific Technology limited company. $LiNi_{0.8}Co_{0.1}Mn_{0.1}O_2$ powder was purchased from the Dongguan Oriental Sunshine Scientific Research and Development limited company. $LiNi_{0.8}Co_{0.1}Mn_{0.1}O_2$ was taken as the coated object, and $Y_2O_3$ was taken as the coating material. A total of 0.5 g of $LiNi_{0.8}Co_{0.1}Mn_{0.1}O_2$ and 0.01 g (coating amount is 2 wt %) of $Y_2O_3$ were weighed into an agate mortar and fully milled. The target samples were obtained via calcination at different temperatures (400 °C, 700 °C, and 750 °C) for 12 h. The samples were labeled NCM@YO400, NCM@YO700, and NCM@YO750, respectively. The sample and the assembled button battery used for synthesis in this experiment are shown in Figures S1 and S2.

### 2.2. Material Characterization

The material characterization methods used in this experiment are X-ray diffraction (XRD, Rigaku D/max2550 VB, Showima City, Tokyo, Japan), scanning electron microscopy (SEM, Quanta 450 FEG, FEI, Hillsboro, OR, USA), energy-dispersive X-ray detection (EDX), high-resolution transmission electron microscopy (HRTEM, Hillsborough, OR, USA), and X-ray photoelectron spectroscopy (XPS, ESCALAB250xi, Waltham, MA, USA). The specific parameters used can be found in the literature previously reported by our research group [32]. The XRD data was analyzed using JADE software (JADE 9).

### 2.3. Electrochemical Measurement

The electrochemical performance of the Land CT2001A battery testing system (1 C = 200 mA $g^{-1}$) was tested at room temperature in the voltage range of 2.8–4.3 V and 2.8–4.5 V, respectively. The scanning speed of the cyclic voltammetry curve was 0.1 mV $S^{-1}$, and the scanning range was 2.8 V–4.5 V (relative to Li/$Li^{+}$). In the amplitude of 5 mV and the frequency range of 0.01 hz~100 KHz, the electrochemical impedance spectroscopy (EIS) test was performed. EIS curves of 150 cycles were obtained.

## 3. Results

The XRD spectrum and amplification spectrum of pure NCM811 and $Y_2O_3$-coated NCM811 samples belong to the R3m space group as shown in Figure 1a, which is a hexagonal lamellar structure of α-$NaFeO_2$. The pure NCM811 is consistent with PDF standard card #87-1562. At the same time, sharp visible peaks are observed at $2\theta \approx 29.15°$ and $2\theta \approx 33.79°$ of the $Y_2O_3$-modified NCM811 sample, which is consistent with PDF card #791257 through analysis and belongs to the Ia-3 space group. The (108)/(110) and (006)/(102) peaks split obviously, indicating that pure NCM811 and $Y_2O_3$-modified NCM811 maintain a highly ordered layered hexagonal structure [33]. The Rietveld refinement parameters of bare and $Y_2O_3$-modified NCM811 samples are shown in (Figure 1b) and Table 1. The strength ratios of $I_{(003)}/I_{(004)}$ of the samples are 1.5066, 1.5506, 1.5124, and 1.5263, respectively, which are much higher than 1.20, confirming the low cation ordered layer structure mixing. The amplification of the doped (003) peak shifts to smaller angles, and the shifted angles are 0.04°(NCM811@YO400), 0.08°(NCM811@YO700), and 0.10°(NCM811@YO750), respectively. The above results may be due to the fact that the ion radius of $Y^{3+}$ is larger than that of transition metal ions M (M=Ni, Co, Mn), and some Y is doped into NCM811 materials, which increases the host cell parameters and enlarges the plane spacing of the materials. Increasing the plane spacing can effectively improve the diffusion rate of $Li^{+}$ [34].

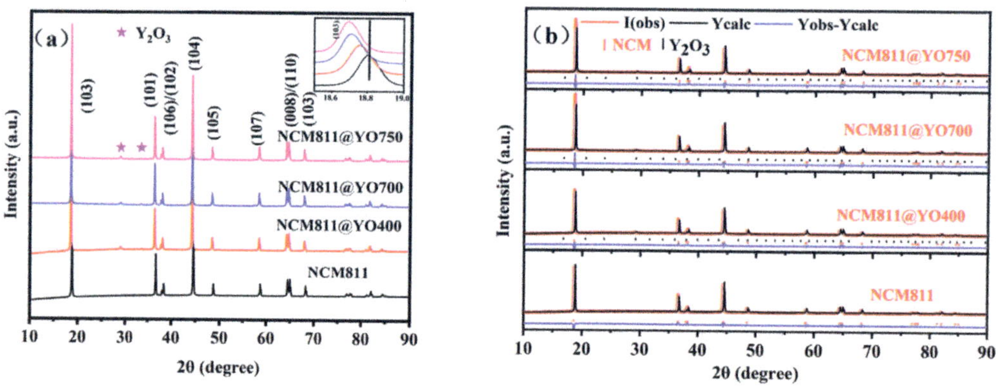


**Figure 1.** (**a**) XRD patterns of pure NCM811 and $Y_2O_3$-modified NCM811 samples. (**b**) Patterns of the Rietveld refinement parameter for NCM811 and $Y_2O_3$-modified NCM811 samples.

**Table 1.** Rietveld refinement parameters of bare and $Y_2O_3$-coated NCM811 samples.

| Sample | a (Å) | C (Å) | c/a | V (Å3) | Rp (%) | I (003)/I (104) |
|---|---|---|---|---|---|---|
| NCM811 | 2.8709 | 14.2044 | 4.9477 | 101.4 | 3.95 | 1.5066 |
| NCM811@YO400 | 2.8715 | 14.2042 | 4.9466 | 101.4 | 4.45 | 1.5506 |
| NCM811@YO700 | 2.8740 | 14.2116 | 4.9449 | 101.7 | 4.01 | 1.5123 |
| NCM811@YO750 | 2.8748 | 14.2136 | 4.9442 | 101.7 | 4.23 | 1.5263 |

The scanning electron microscope image is shown in Figure 2a–d. Through the observation of NCM811, the sample presents a similar spherical appearance, and the diameter of the particles is between 10 and 15 μm. The particles are composed of original particles with smaller particle size [35]. The enlarged SEM image is shown in Figure 2a1–d1. The surface of the original sample is relatively smooth, while the surface of the $Y_2O_3$ sample coated at 400 °C is rough. This may be due to the coating effect of $Y_2O_3$ on the surface of the bare NCM811 sample. The surface of the modified sample is smooth at 700 °C and 750 °C. Compared with NCM811 particles without $Y_2O_3$ modification, the particles of the modified samples are denser. The reduction in voids may be due to the diffusion of $Y_2O_3$ coating into the particle voids at high temperatures, which is more beneficial for reducing the chances of electrolytes entering the gaps and then restraining the harmful reaction. The surface of NCM@YO700 was analyzed using an EDS test to further understand the elemental distribution of the coating. The main element components Ni, Co, and Mn are uniformly distributed on the surface of the sample [36], as shown in Figure 2e–g. In addition, Y can be observed in Figure 2h, which reaffirms the success of the $Y_2O_3$ coating in the NCM811 sample.

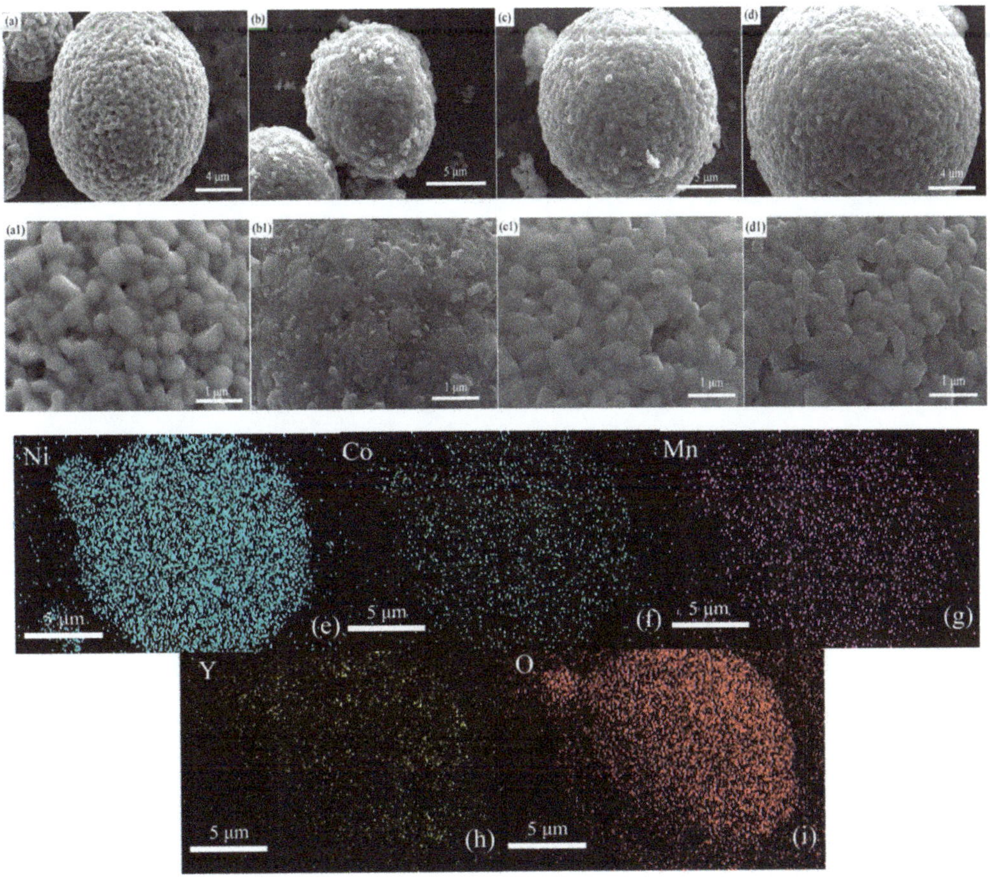


**Figure 2.** SEM images of pure NCM811 and $Y_2O_3$-modified NCM811 materials; (**a**) NCM811; (**b**) NCM@YO400; (**c**) NCM@YO700; (**d**) NCM@YO750; (**a1–d1**) the enlarged images of (**a–i**) EDS mappings of NCM@ YO700.

The transmission electron microscope images in Figure 3 show the crystal structure of NCM811 and NCM@YO700 samples, with continuous and straight lattice stripes, indicating high crystallinity and good layered structure. In Figure 3a, the spacings of the

(003), (006), and (101) planes of the NCM sample are 0.476 nm, 0.237 nm, and 0.245 nm, respectively. Figure 3b shows that $Y_2O_3$ successfully coats the surface of the substrate, and the composition analysis of the coating is carried out. In Figure 3c, the (003) plane spacing of the NCM-@YO700 sample is 0.481 nm. The crystal plane spacing of NCM-@YO700 is slightly wider than bare NCM811 [37]. It is proved that a certain amount of Y enters the lattice of NCM811. Since the ionic radius of the transition metal ion M (M = Ni, Co, Mn) is smaller than $Y^{3+}$, the sample lattice spacing increases, and the cell volume of the sample increases, which is consistent with the results analyzed with the X-ray diffractometer. The magnification analysis of Figure 3I,II shows that the distance between crystal planes is 0.251 nm, 0.209 nm, and 0.160 nm, corresponding to (411), (134), and (622) of $Y_2O_3$, respectively, indicating that some $Y_2O_3$ coating is realized on the surface of the NCM811 samples to some extent.

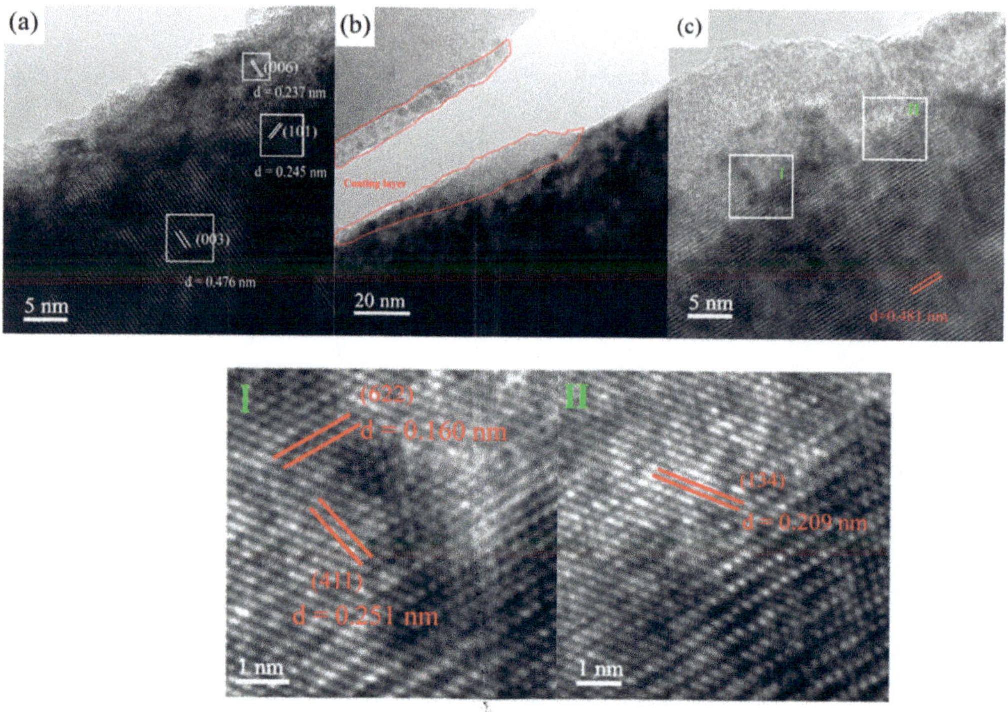


**Figure 3.** HRTEM images of the bare NCM811 and NCM@YO700; (**a**) bare NCM811; (**b**,**c**) NCM@YO700; and (**I**,**II**) the enlarged images of (**c**).

The XPS analysis of the surface elements of the materials reveals important information about their oxidation valence states. In Figure 4a, Ni $2p_{3/2}$ and $2p_{1/2}$ exhibit two peaks at 854.8 eV and 872.2 eV [38,39]. The Ni $2p_{3/2}$ spectrum shows deconvolution peaks at 854.6 eV and 855.7 eV, indicating the presence of $Ni^{2+}$ and $Ni^{3+}$ valence states. According to Figure 4b,c, the Co $2p_{1/2}$ and Mn $2p_{1/2}$ peaks are observed at 794.9 eV and 654.2 eV, respectively. The Co $2p_{3/2}$ and Mn $2p_{3/2}$ peaks are observed at 780.0 eV and 642.6 eV, respectively. This indicates that the valence states of Co and Mn are mainly trivalent and tetravalent [40]. In Figure 4d, the decrease in lithium peak intensity may be attributed to the $Y_2O_3$ modification. The O 1s peak at 529.4 eV is related to the lattice oxygen in the metal skeleton in Figure 4e. The intensity of the lattice oxygen peaks is enhanced compared to the untreated $Y_2O_3$ samples. Similar conclusions have been previously reported in the literature [38,41]. This enhancement indicates an increase in oxygen vacancy peak area, which effectively inhibits the loss of oxygen and improves the structural stability of the material [42]. As shown in Figure 4f, the binding energies for Y $2p_{5/2}$ and Y $2p_{3/2}$ of the modified samples are measured at 156.48 eV and 158.48 eV, respectively, indicating that the valence state of Y is +3.

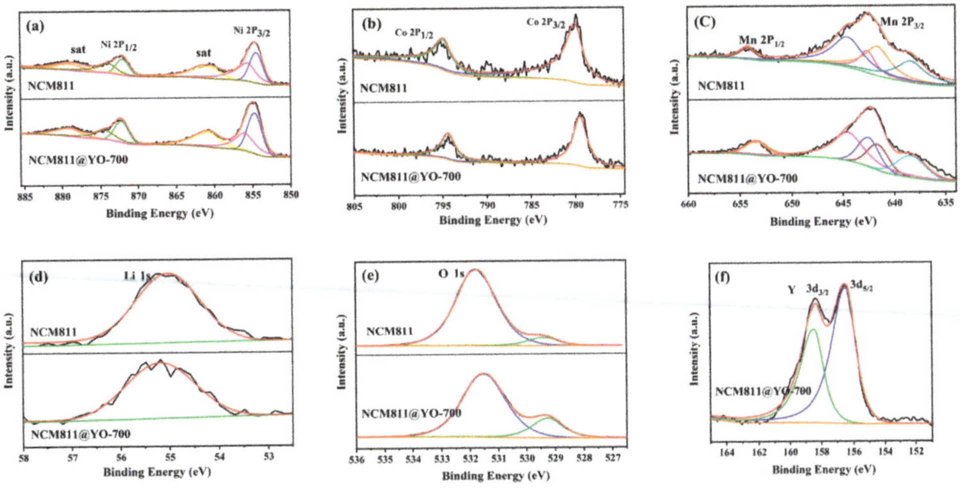

**Figure 4.** (**a–e**) XPS data and the fitting results of Ni, Co, Mn, Li, and O of the NCM811 and NCM@YO700 samples; (f) XPS spectra of Y for NCM@ YO700.

Figure 5a shows the charge and discharge data of NCM811, NCM811@YO400, NCM81-1@YO700, and NCM811@YO750 at a sub-constant current of 0.2 C (1 C = 200 mA $g^{-1}$) and room temperature. The naked NCM811 and the $Y_2O_3$-coated sample have a similar discharge platform. At the current density of 0.2 C, the discharge capacities of NCM811, NCM811@YO400, NCM811@YO700, and NCM811@YO750 are measured to be 189.4 mAh $g^{-1}$, 198.1 mAh $g^{-1}$, 201.9 mAh $g^{-1}$, and 203.1 mAh $g^{-1}$, respectively. $Y_2O_3$ coating improves the specific discharge capacity of NCM811. At higher sintering temperatures of 700 °C and 750 °C, the discharge capacity increases more obviously since some $Y_2O_3$ diffuses into the gap of the primary intergranular pores of NCM811 against electrolyte attack, and other $Y_2O_3$ is doped into the layered structure of NCM811, thereby enlarging the $Li^+$ ion transport channels. The dual effect of coating and doping is realized through $Y_2O_3$ modification, leading to the enhanced discharge capacity and improved electrochemical performance of the material. Figure 5b presents the cycling performance of the samples at 1 C and the corresponding electrochemical parameters are shown in Table 2. The capacity of NCM811@YO700 after 150 cycles is measured to be 183.6 mAh $g^{-1}$, and the retention efficiency is 96.58%. This improvement in cycling stability can be attributed to the role of Y doping in stabilizing the crystal structure of NCM811 and enhancing its stability. Additionally, a portion of Y remains on the surface of NCM811, which helps resist electrolyte erosion and improve electrochemical stability. The rate performance of NCM811 and $Y_2O_3$-coated samples is tested in the voltage range of 2.7 V–4.3 V at room temperature, as depicted in Figure 5c. At 5 C, the discharge capacity of NCM811 and NCM811@YO700 is measured to be 107.54 mAh $g^{-1}$ and 138.88 mAh $g^{-1}$, respectively, indicating a significant improvement in rate performance. This can be attributed to the effective inhibition of harmful reactions between the electrolyte and active substances through the $Y_2O_3$ coating. Figure 5d shows the cycling data of the samples at 2.7 V–4.5 V, which follows a similar trend to Figure 5b. After modification with $Y_2O_3$, NCM811@YO700 exhibits a capacity of 147.20 mAh $g^{-1}$ and a retention efficiency of 76.44% after 150 cycles. The coating of $Y_2O_3$ can also effectively reduce the capacity attenuation. However, for the same sample, the cyclic attenuation at 4.5 V is significantly worse than that at 4.3 V. Under the condition of deep lithium removal at high potential, $Ni^{3+}$ and $O^{2-}$ in high-nickel ternary cathode materials oxidize at the same time to form $Ni^{3+}$ and highly active oxygen [O] (such as $O^{-}$, $O^{2-}$). The electrolyte irreversibly reacts with it to produce gas, which is not conducive to surface charge transfer, resulting in fast capacity decay [41,43].

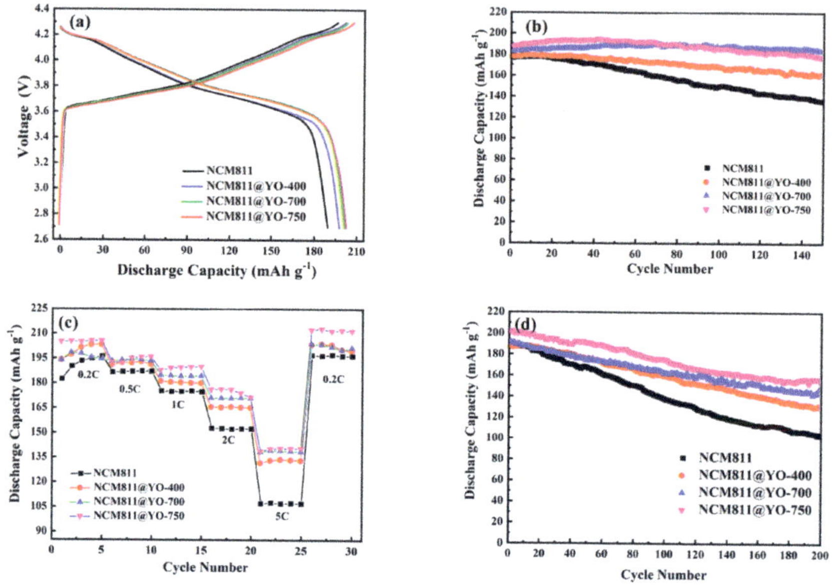


**Figure 5.** (**a**) Charge and discharge curves for NCM811 and NCM@YO400, NCM@YO700, and NCM@YO750 in the range of 2.7–4.3 V at 0.2 C. (**b**) Cycling performance for the NCM811 and NCM811@YO samples at 1 C for 150 cycles. (**c**) Specific capacity of NCM811 and NCM811@YO samples at different multipliers from 0.2 C to 5 C. (**d**) Cycling performance of NCM811 and NCM811@YO at 1 C over the high voltage range of 2.7–4.5 V.

**Table 2.** Electrochemical performance of the pristine and $Y_2O_3$-modified samples in the range of 2.7–4.3 V.

| Systems | 1st Discharge Capacity (mAh $g^{-1}$) | Maximum Discharge Capacity (mAh $g^{-1}$) | 150 st Discharge Capacity (mAh $g^{-1}$) | Capacity Retention (%) |
|---|---|---|---|---|
| Original | 176.3 | 177.9 | 135.7 | 76.28 |
| NCM811@YO400 | 177.7 | 179.3 | 161.5 | 90.07 |
| NCM811@YO700 | 182.9 | 190.1 | 183.6 | 96.58 |
| NCM811@YO750 | 187.3 | 194.7 | 177.8 | 91.32 |

In order to further confirm the results, the SEM analyses of NCM811 and NCM@YO700 after 200 cycles were conducted in Figure 6. Obvious microcracks can be observed in the particles of bare NCM811 after cycling in Figure 6a1. The morphology of the NCM@YO700 particles remains relatively intact in Figure 6b1, and there are no obvious microcracks. It can be demonstrated that coating $Y_2O_3$ on the surface of particles can result in decreased electrolyte corrosion, preserved crystal structure stability, and ultimately improved material properties. This inference is similar to the explanation reported by Du et al. [44].

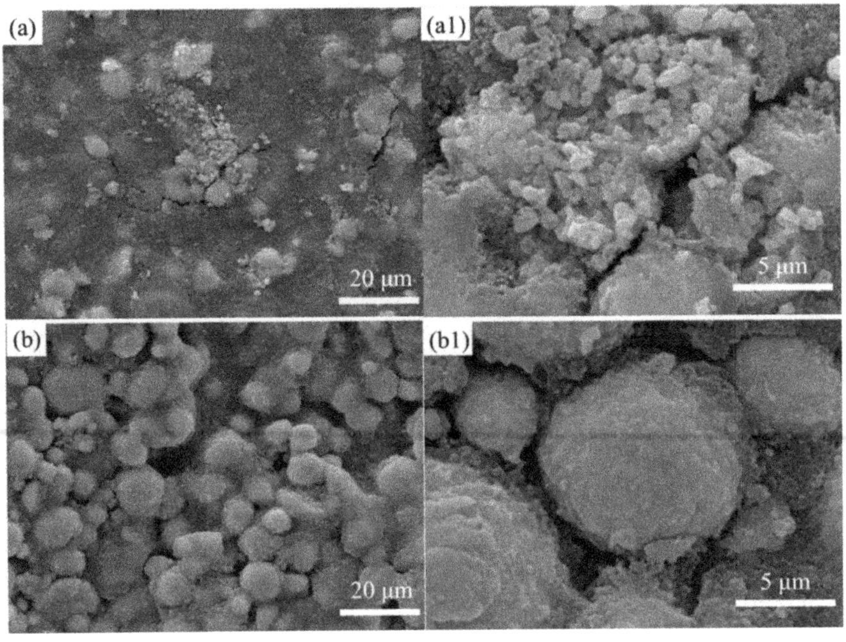


**Figure 6.** SEM images of NCM811 and $Y_2O_3$-modified NCM811 samples after 200 charge/discharge cycles; (**a**) bare NCM811; (**b**) NCM@YO700; and (**a1**,**b1**) the enlarged images of (**a**,**b**).

To further understand the electrochemical performance of the electrode, cyclic voltammetry (CV) tests were carried out, as shown in Figure 7a,b. The CV curves of the NCM811 and NCM811@YO700 samples are similar. It shows that there is no significant change in the electrochemical reaction in the case of $Y_2O_3$-coated NCM811. In the second cycle, two redox peaks are observed for NCM811, with the lower peak at 3.945 V/3.624 V corresponding to $Ni^{2+}/Ni^{3+}$ and the higher peak at 4.265 V/4.097 V corresponding to $Ni^{3+}/Ni^{4+}$. For NCM811@YO700, the redox peak is observed at 3.859 V/3.653 V and 4.253 V/4.128 V. The difference of potential difference ΔE becomes smaller during circulation. The decrease in the potential difference between the anodic and cathodic peaks in NCM811@YO700 (0.47 V) compared to NCM811 (0.66 V) indicates that the $Y_2O_3$ coating effectively reduces the polarization of the original sample. This reduction in polarization can lead to improved electrochemical kinetics and enhanced overall performance of the material [45]. In addition, compared with the exposed NCM811, NCM811@YO700 shows highly overlapping reduction peaks in the first three cycles, indicating that NCM811@YO700 has a smaller irreversible capacity.

The electrochemical impedance spectra (EIS) of the original NCM811 and NCM811@Y O700 were analyzed to investigate the electrochemical kinetic properties. The EIS measurements were conducted after 150 charge–discharge cycles, as shown in Figure 7c, and the equivalent circuit fitting results were obtained with Z-View software. The EIS consists of a concave semicircle and a diagonal line. At high frequency, the intercept with the semicircular x-axis corresponds to the total resistance (Rs), which includes contact resistance, material resistance, and electrolyte ohmic resistance. The Rsf and Rct represent the resistance at the electrode–electrolyte interface and the resistance associated with charge transfer processes, respectively. The Warburg impedance (Zw) is a characteristic impedance resulting from the diffusion limitation of ions. At low frequencies, the EIS can be represented by a diagonal line; the non-ideal capacitance of the surface layer is represented by a constant phase angle element (CPE) [32,46]. The values of NCM811 and NCM811@YO700 are calculated with the given equivalent circuit. For NCM811 and the coated samples, Rsf is much smaller than Rct, which indicates that Rct is the main factor affecting the electrochemical reaction (Seen in Table 3 for details). The Rct value of NCM811@YO700 is smaller than the bare NCM811, which indicates that the coating of $Y_2O_3$ improves the conductivity of lithium ions. Similar results have appeared in previous studies [46,47]. The $Li^+$ ion diffusion coefficient is estimated by the following formula [48].

$$D_{Li^+} = \frac{R^2T^2}{2A^2n^2F^2C^2\sigma^2} \quad (1)$$

**Table 3.** Fitted values of impedance parameters of NCM811 and NCM811@YO700.

| Electrode | *Rs* (Ω) | *Rsf* (Ω) | *Rct* (Ω) | $D_{Li}{}^+$ ($cm^2\ s^{-1}$) |
|---|---|---|---|---|
| NCM811 | 14.5 | 179.6 | 535.1 | $1.12 \times 10^{-14}$ |
| NCM811@YO700 | 9.5 | 85.3 | 322.8 | $2.28 \times 10^{-14}$ |

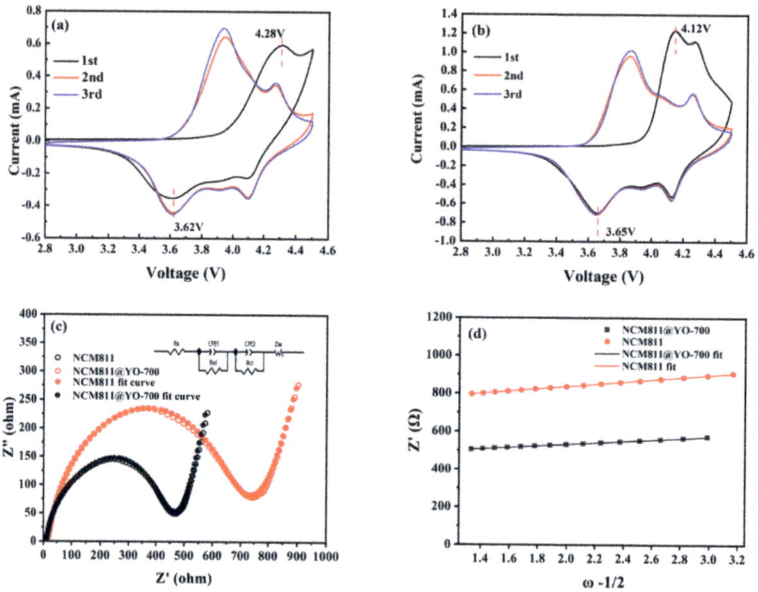


**Figure 7.** The CV curves in the voltage range of 2.8-4.3 V for (**a**) NCM811 and (**b**) NCM811@YO700. (**c**) The EIS and equivalent circuit-fitted graph after 150 charge/discharge cycles. (**d**) The relationship between Z′ and $\omega^{-1/2}$.

The gas constant (8.314 J $K^{-1}$ $mol^{-1}$), the absolute temperature (298 K), the electrode's surface (1.654 $cm^2$), the number of electrons involved in the reaction ($n$ = 1), the Faraday constant, and the Warburg coefficient are represented by the variables R, T, A, n, F, and σ, respectively [49]. C in this study is 0.00718 mol $cm^{-3}$, which is the concentration of lithium ions. The value of σ can be obtained by linearly fitting the slope of Z′ to the frequency $\omega^{-1/2}$ (Figure 7d). The $D_{Li}{}^+$ value of NCM811 increases from $1.12 \times 10^{-14}$ $cm^2$ $S^{-1}$ to $2.28 \times 10^{-14}$ $cm^2$ $S^{-1}$ after $Y_2O_3$ modification. Considering that NCM811 and NCM811@YO700 have similar structures and particle sizes, the enhancement of $Li^+$ ion mobility can be attributed to the $Y_2O_3$ coating on the NCM811 interface. There have been similar reports previously [32,50,51].

## 4. Conclusions

$Y_2O_3$-modified NCM811 samples were made by dry milling and heat treating at sintering temperatures of 400 °C, 700 °C, and 750 °C in order to enhance the electrochemical performance of NCM811. The discharge capacity, rate performance, and cycle stability are all significantly enhanced by the $Y_2O_3$ alteration. Particularly, the NCM811@YO700 (sintered at 700 °C) demonstrates excellent cyclic stability with a capacity retention efficiency of 96.58% after 150 cycles at 2.7 V–4.3 V, which is much higher than the 76.28% of the bare NCM811, and still retains the discharge capacity of 183.6 mAh $g^{-1}$ after cycling. Also, the rate capacity is greatly enhanced at high current density for $Y_2O_3$-modified NCM811. The discharge capacity of NCM811 and NCM811@YO700 is 107.54 mAh $g^{-1}$ and 138.88 mAh $g^{-1}$ at 5 C. The results show that dual coating and doping effects can be achieved through $Y_2O_3$ modification. $Y_2O_3$ coating can reduce the direct contact between the surface of NCM811 and the electrolyte, which is conducive to the inhibition of interfacial side reactions and the enhancement of electrochemical reactions. The other portion of Y diffuses into the layer structure, which can stabilize the crystal structure, promote lithium-ion transport, and enhance the cycling performance. The $Y_2O_3$ dry milling method makes them have great application potential as high-energy-density cathode material.

**Supplementary Materials:** The following supporting information can be downloaded at https://www.mdpi.com/article/10.3390/coatings14040443/s1: Figure S1: The positive electrode materials used in the experiment were pure NCM811 and Y2O3-modified NCM811 samples, (a) bare NCM811, (b) NCM@YO400, (c) NCM@YO700, (d) NCM@YO750.; Figure S2: Pictures of CR2032 button half-coin batteries in this experiment.

**Author Contributions:** Conceptualization, J.W., F.W. and X.W.; methodology, validation, X.W. and M.Z.; formal analysis, J.D., P.L., D.L. and X.W.; investigation, X.W. and M.Z.; writing—original draft preparation, X.W.; writing—review and editing, F.W., M.R. and J.W.; project admin-istration, J.D., M.R., P.L. and D.L.; funding acquisition, M.R. All authors have read and agreed to the published version of the manuscript.

**Funding:** This research was funded by the National Natural Science Foundation of China (grant numbers 52261003 and 52361037), the Guangxi Key Laboratory of Information Materials and the Guilin University of Electronic Technology, China (221023-Z), and the Innovation Project of GUET Graduate Education (2023 YCXS153).

**Institutional Review Board Statement:** Not applicable.

**Informed Consent Statement:** Not applicable.

**Data Availability Statement:** Data are contained within the article and Supplementary Materials.

**Conflicts of Interest:** The authors declare no conflicts of interest.

## References

1. Li, X.; Han, X.; Li, G.; Du, J.; Cao, Y.; Gong, H.; Wang, H.; Zhang, Y.; Liu, S.; Zhang, B.; et al. Nonsacrificial Nitrile Additive for Armoring High-Voltage $LiNi_{(0.83)}Co_{(0.07)}Mn_{(0.1)}O_{(2)}$ Cathode with Reliable Electrode-Electrolyte Interface toward Durable Battery. *Small* **2022**, *18*, e2202989. [CrossRef]
2. Yee, M.; An, K.; Nguyen, D.-T.; Yun, H.W.; Park, J.; Suk, J.; Song, S.-W. Confining nonflammable liquid in solid polymer electrolyte to enable nickel-rich cathode-based 4.2 V high-energy solid-state lithium-metal and lithium-ion batteries. *Mater. Today Energy* **2022**, *24*, 100950. [CrossRef]
3. Xu, G.L.; Liu, X.; Daali, A.; Amine, R.; Chen, Z.; Amine, K. Challenges and Strategies to Advance High-Energy Nickel-Rich Layered Lithium Transition Metal Oxide Cathodes for Harsh Operation. *Adv. Funct. Mater.* **2020**, *30*, 2004748. [CrossRef]
4. Goodenough, J.B.; Kim, Y. Challenges for Rechargeable Li Batteries. *Chem. Mater.* **2009**, *22*, 587–603. [CrossRef]
5. Zhai, Y.; Zhang, J.; Zhao, H.; Hu, Z.; Liu, X. Research progress of ternary layered oxide cathode materials for lithium ion batteries. *J. Eng. Stud.* **2017**, *9*, 523. [CrossRef]
6. Zhao, R.; Liang, J.; Huang, J.; Zeng, R.; Zhang, J.; Chen, H.; Shi, G. Improving the Ni-rich $LiNi_{0.5}Co_{0.2}Mn_{0.3}O_2$ cathode properties at high operating voltage by double coating layer of $Al_2O_3$ and $AlPO_4$. *J. Alloys Compd.* **2017**, *724*, 1109–1116. [CrossRef]
7. Li, J.; Xiang, J.; Yi, G.; Tang, Y.; Shao, H.; Liu, X.; Shan, B.; Chen, R. Reduction of Surface Residual Lithium Compounds for Single-Crystal $LiNi_{0.6}Mn_{0.2}Co_{0.2}O_2$ via $Al_2O_3$ Atomic Layer Deposition and Post-Annealing. *Coatings* **2022**, *12*, 84. [CrossRef]

8. Bian, X.; Fu, Q.; Bie, X.; Yang, P.; Qiu, H.; Pang, Q.; Chen, G.; Du, F.; Wei, Y. Improved electrochemical performance and thermal stability of Li-excess $Li_{1.18}Co_{0.15}Ni_{0.15}Mn_{0.52}O_2$ cathode material by $Li_3PO_4$ surface coating. *Electrochim. Acta* **2015**, *174*, 875–884. [CrossRef]
9. Ma, F.; Wu, Y.; Wei, G.; Qiu, S.; Qu, J. Enhanced electrochemical performance of $LiNi_{0.8}Co_{0.1}Mn_{0.1}O_2$ cathode via wet-chemical coating of MgO. *J. Solid State Electrochem.* **2019**, *23*, 2213–2224. [CrossRef]
10. Yu, Z.; Tong, Q.; Zhao, G.; Zhu, G.; Tian, B.; Cheng, Y. Combining Surface Holistic Ge Coating and Subsurface Mg Doping to Enhance the Electrochemical Performance of $LiNi_{0.8}Co_{0.1}Mn_{0.1}O_2$ Cathodes. *ACS Appl. Mater. Interfaces* **2022**, *14*, 25490–25500. [CrossRef]
11. Feng, Y.; Zhang, X.; Lin, C.; Wang, Q.; Zhang, Y. Improving the electrochemical performance of $LiNi_{0.8}Co_{0.1}Mn_{0.1}O_2$ cathodes using a simple $Ce^{4+}$-doping and $CeO_2$-coating technique. *N. J. Chem.* **2021**, *45*, 21617–21623. [CrossRef]
12. Li, D.; Guo, H.; Jiang, S.; Zeng, G.; Zhou, W.; Li, Z. Microstructures and electrochemical performances of $TiO_2$-coated Mg–Zr co-doped NCM as a cathode material for lithium-ion batteries with high power and long circular life. *N. J. Chem.* **2021**, *45*, 19446–19455. [CrossRef]
13. Tang, W.; Peng, Z.; Shi, Y.; Xu, S.; Shuai, H.; Zhou, S.; Kong, Y.; Yan, K.; Lu, T.; Wang, G. Enhanced cyclability and safety performance of $LiNi_{0.6}Co_{0.2}Mn_{0.2}O_2$ at elevated temperature by $AlPO_4$ modification. *J. Alloys Compd.* **2019**, *810*, 151834. [CrossRef]
14. Yang, C.; Li, Y.; Zhang, X.; Xiao, J.; Xiong, H.; Li, W.; Guo, P.; Yang, Z.; Xie, M. Enhanced cyclic stability of $LiNi_{0.8}Co_{0.1}Mn_{0.1}O_2$ (NCM811) by $AlF_3$ coating via atomic layer deposition. *Ionics* **2022**, *28*, 4547–4554. [CrossRef]
15. Li, M.; Wang, H.; Zhao, L.; Zhang, F.; He, D. The combined effect of $CaF_2$ and graphite two-layer coatings on improving the electrochemical performance of Li-rich layer oxide material. *J. Solid State Chem.* **2019**, *272*, 38–46. [CrossRef]
16. Zhou, S.; Zhang, X.; Zhang, Z.; Liu, S.; Wang, R. Improving the Electrochemical Performance of $LiNi_{1/3}Co_{1/3}Mn_{1/3}O_2$ Cathode Material by LiF Modification. *Coatings* **2023**, *13*, 727. [CrossRef]
17. Dai, S.; Yan, G.; Wang, L.; Luo, L.; Li, Y.; Yang, Y.; Liu, H.; Liu, Y.; Yuan, M. Enhanced electrochemical performance and thermal properties of Ni-rich $LiNi_{0.8}Co_{0.1}Mn_{0.1}O_2$ cathode material via $CaF_2$ coating. *J. Electroanal. Chem.* **2019**, *847*, 113197. [CrossRef]
18. Dong, H.; Xie, M.; Cai, M.; Liu, H.; Zhang, Z.; Ye, B.; Zhao, P.; Dong, W.; Huang, F. Amorphous fluorine glaze for crack-free nickel-rich layered cathode grains under electrochemical cycling. *Chem. Eng. J.* **2022**, *436*, 135227. [CrossRef]
19. Breddemann, U.; Krossing, I. Review on Synthesis, Characterization, and Electrochemical Properties of Fluorinated Nickel-Cobalt-Manganese Cathode Active Materials for Lithium-Ion Batteries. *ChemElectroChem* **2020**, *7*, 1389–1430. [CrossRef]
20. Yang, S.; Meng, T.; Wang, Z.; Hu, X. Unveiling decaying mechanism of non-flammable all-fluorinated carbonate electrolytes in lithium metal batteries with 4.6-V $LiCoO_2$ cathodes at elevated temperatures. *Energy Storage Mater.* **2024**, *65*, 103177. [CrossRef]
21. Ren, Q.; Yuan, Y.; Wang, S. Interfacial Strategies for Suppression of Mn Dissolution in Rechargeable Battery Cathode Materials. *ACS Appl. Mater. Interfaces* **2021**, *14*, 23022–23032. [CrossRef]
22. Zhang, W.; Tang, X.; Xiao, L.; Xiao, Z. Influence of $Al^{3+}$ and $PO_4{}^{3-}$ co-doping on structure and electrochemical performance of $LiNi_{0.8}Co_{0.1}Mn_{0.1}O_2$ cathode materials. *J. Solid State Electrochem.* **2023**, *27*, 1185–1194. [CrossRef]
23. Li, Y.; Wang, Z.-T.; Liu, G.; Wang, J.; Wang, J. Boosting the electrochemical performance of $LiNi_{0.8}Co_{0.1}Mn_{0.1}O_2$ cathode materials with $Zn_3(PO_4)_2$ surface coating. *Adv. Powder Technol.* **2021**, *32*, 4651–4657. [CrossRef]
24. Liu, L.; Li, M.; Chu, L.; Jiang, B.; Lin, R.; Zhu, X.; Cao, G. Layered ternary metal oxides: Performance degradation mechanisms as cathodes, and design strategies for high-performance batteries. *Prog. Mater. Sci.* **2020**, *111*, 100655. [CrossRef]
25. Zhang, X.; Gao, S.; Gui, W.; Zeng, Q. First-principles study of structure, mechanical and optical properties of La- and Sc-doped $Y_2O_3$. *J. Rare Earths* **2019**, *37*, 879–885. [CrossRef]
26. Liu, X.-H.; Kou, L.-Q.; Shi, T.; Liu, K.; Chen, L. Excellent high rate capability and high voltage cycling stability of $Y_2O_3$-coated $LiNi_{0.5}Co_{0.2}Mn_{0.3}O_2$. *J. Power Sources* **2014**, *267*, 874–880. [CrossRef]
27. Zou, P.; Lin, Z.; Fan, M.; Wang, F.; Liu, Y.; Xiong, X. Facile and efficient fabrication of $Li_3PO_4$-coated Ni-rich cathode for high-performance lithium-ion battery. *Appl. Surf. Sci.* **2020**, *504*, 144506. [CrossRef]
28. Jo, C.-H.; Cho, D.-H.; Noh, H.-J.; Yashiro, H.; Sun, Y.-K.; Myung, S.T. An effective method to reduce residual lithium compounds on Ni-rich $Li[Ni_{0.6}Co_{0.2}Mn_{0.2}]O_2$ active material using a phosphoric acid derived $Li_3PO_4$ nanolayer. *Nano Res.* **2014**, *8*, 1464–1479. [CrossRef]
29. Kitsche, D.; Tang, Y.; Ma, Y.; Goonetilleke, D.; Sann, J.; Walther, F.; Bianchini, M.; Janek, J.; Brezesinski, T. High Performance All-Solid-State Batteries with a Ni-Rich NCM Cathode Coated by Atomic Layer Deposition and Lithium Thiophosphate Solid Electrolyte. *ACS Appl. Energy Mater.* **2021**, *4*, 7338–7345. [CrossRef]
30. Wang, X.; Yushin, G. Chemical vapor deposition and atomic layer deposition for advanced lithium ion batteries and supercapacitors. *Energy Environ. Sci.* **2015**, *8*, 1889–1904. [CrossRef]
31. Nisar, U.; Muralidharan, N.; Essehli, R.; Amin, R.; Belharouak, I. Valuation of Surface Coatings in High-Energy Density Lithium-ion Battery Cathode Materials. *Energy Storage Mater.* **2021**, *38*, 309–328. [CrossRef]
32. Wang, F.; Luo, Y.; Liu, P.; Balogun, M.-S.; Deng, J.; Wang, Z. Improved Cycling Performance and High Rate Capacity of $LiNi_{0.8}Co_{0.1}Mn_{0.1}O_2$ Cathode Achieved by $Al(PO_3)_3$ Modification via Dry Coating Ball Milling. *Coatings* **2022**, *12*, 319. [CrossRef]
33. Ran, A.; Chen, S.; Cheng, M.; Liang, Z.; Li, B.; Zhou, G.; Kang, F.; Zhang, X.; Wei, G. A single-crystal nickel-rich material as a highly stable cathode for lithium-ion batteries. *J. Mater. Chem. A* **2022**, *10*, 19680–19689. [CrossRef]

34. Zhang, M.; Lv, M.; Zhang, D.; Yan, Y.; Wang, Y.; Li, J.; Li, Z. Enhanced electrochemical properties of NCM811 cathode material due to synergistic modification with Sm as doping and coating agent. *J. Alloys Compd.* **2022**, *909*, 164712. [CrossRef]
35. Xu, Y.-D.; Xiang, W.; Wu, Z.-G.; Xu, C.-L.; Li, Y.-C.; Guo, X.-D.; Lv, G.-P.; Peng, X.; Zhong, B.-H. Improving cycling performance and rate capability of Ni-rich $LiNi_{0.8}Co_{0.1}Mn_{0.1}O_2$ cathode materials by $Li_4Ti_5O_{12}$ coating. *Electrochim. Acta* **2018**, *268*, 358–365. [CrossRef]
36. Zhao, S.; Zhu, Y.; Qian, Y.; Wang, N.; Zhao, M.; Yao, J.; Xu, Y. Annealing effects of $TiO_2$ coating on cycling performance of Ni-rich cathode material $LiNi_{0.8}Co_{0.1}Mn_{0.1}O_2$ for lithium-ion battery. *Mater. Lett.* **2020**, *265*, 127418. [CrossRef]
37. Chen, S.; He, T.; Su, Y.; Lu, Y.; Bao, L.; Chen, L.; Zhang, Q.; Wang, J.; Chen, R.; Wu, F. Ni-Rich $LiNi_{(0.8)}Co_{(0.1)}Mn_{(0.1)}O_{(2)}$ Oxide Coated by Dual-Conductive Layers as High Performance Cathode Material for Lithium-Ion Batteries. *ACS Appl. Mater. Interfaces* **2017**, *9*, 29732–29743. [CrossRef]
38. Zhang, C.; Liu, S.; Su, J.; Chen, C.; Liu, M.; Chen, X.; Wu, J.; Huang, T.; Yu, A. Revealing the role of $NH_{(4)}VO_{(3)}$ treatment in Ni-rich cathode materials with improved electrochemical performance for rechargeable lithium-ion batteries. *Nanoscale* **2018**, *10*, 8820–8831. [CrossRef]
39. Ding, Y.; Deng, B.; Wang, H.; Li, X.; Chen, T.; Yan, X.; Wan, Q.; Qu, M.; Peng, G. Improved electrochemical performances of $LiNi_{0.\ 6}Co_{0.\ 2}Mn_{0.\ 2}O_2$ cathode material by reducing lithium residues with the coating of Prussian blue. *J. Alloys Compd.* **2019**, *774*, 451–460. [CrossRef]
40. Li, L.; Zhang, Z.; Fu, S.; Liu, Z. F127-assisted synthesis of $LiNi_{0.\ 5}Co_{0.\ 2}Mn_{0.\ 3}O_{1.\ 99}F_{0.\ 01}$ as a high rate and long lifespan cathode material for lithium-ion batteries. *Appl. Surf. Sci.* **2019**, *476*, 1061–1071. [CrossRef]
41. House, R.A.; Marie, J.-J.; Pérez-Osorio, M.A.; Rees, G.J.; Boivin, E.; Bruce, P.G. The role of $O_2$ in O-redox cathodes for Li-ion batteries. *Nat. Energy* **2021**, *6*, 781–789. [CrossRef]
42. He, R.; Bai, X.; Wei, A.; Zhang, L.; Liu, P.; Liu, Z. $Y_2O_3$ modification on nickel-rich $LiNi_{0.8}Co_{0.1}Mn_{0.1}O_2$ with improved electrochemical performance in lithium-ion batteries. *J. Rare Earths* **2022**, *40*, 309–317. [CrossRef]
43. Lin, F.; Markus, I.M.; Nordlund, D.; Weng, T.-C.; Asta, M.D.; Xin, H.L.; Doeff, M.M. Surface reconstruction and chemical evolution of stoichiometric layered cathode materials for lithium-ion batteries. *Nat. Commun.* **2014**, *5*, 1–9. [CrossRef] [PubMed]
44. Du, M.; Yang, P.; He, W.; Bie, S.; Zhao, H.; Yin, J.; Zou, Z.; Liu, J. Enhanced high-voltage cycling stability of Ni-rich $LiNi_{0.8}Co_{0.1}Mn_{0.1}O_2$ cathode coated with $Li_2O–2B_2O_3$. *J. Alloys Compd.* **2019**, *805*, 991–998. [CrossRef]
45. Peng, Z.; Li, T.; Zhang, Z.; Du, K.; Hu, G.; Cao, Y. Surface architecture decoration on enhancing properties of $LiNi_{0.8}Co_{0.1}Mn_{0.1}O_2$ with building bi-phase $Li_3PO_4$ and $AlPO_4$ by $Al(H_2PO_4)_3$ treatment. *Electrochim. Acta* **2020**, *338*, 135870. [CrossRef]
46. Liu, Y.; Wang, Q.; Zhang, Z.; Dou, A.; Pan, J.; Su, M. Investigation the electrochemical performance of layered cathode material $Li_{1.\ 2}Ni_{0.\ 2}Mn_{0.\ 6}O_2$ coated with $Li_4Ti_5O_{12}$. *Adv. Powder Technol.* **2016**, *27*, 1481–1487. [CrossRef]
47. Tan, X.; Zhang, M.; Zhang, D.; Yan, Y.; Wang, Y.; Li, Z. Inhibited intracrystalline cracks and enhanced electrochemical properties of NCM811 cathode materials coated by EPS. *Ceram. Int.* **2021**, *47*, 32710–32719. [CrossRef]
48. Liu, X.; Chen, Q.; Li, Y.; Chen, C.; Zeng, W.; Yuan, M.; Wang, R.; Xiao, S. Synergistic Modification of Magnesium Fluoride/Sodium for Improving the Electrochemical Performances of High-Nickel Ternary (NCM811) Cathode Materials. *J. Electrochem. Soc.* **2019**, *166*, A3480–A3486. [CrossRef]
49. Zhao, M.; Chen, Q. Improving the electrochemical performance of $LiNi_{0.\ 7}Mn_{0.\ 3}O_2$ cathode material by $CeO_2$ coating via Ce doping. *Solid State Ion.* **2023**, *399*, 116296. [CrossRef]
50. Wu, Y.S.; Pham, Q.-T.; Yang, C.-C.; Chern, C.-S.; Babulal, L.M.; Seenivasan, M.; Jeyakumar, J.; Mengesha, T.H.; Placke, T.; Brunklaus, G.; et al. Coating of a Novel Lithium-Containing Hybrid Oligomer Additive on Nickel-Rich $LiNi_{0.8}Co_{0.1}Mn_{0.1}O_2$ Cathode Materials for High-Stability and High-Safety Lithium-Ion Batteries. *ACS Sustain. Chem. Eng.* **2022**, *10*, 7394–7408. [CrossRef]
51. Cheng, W.; Li, L.; Hao, S.; Liu, L.; Wu, Y.; Huo, J.; Ji, Y.; Liu, X. Face-lifting the surface of $LiNi_{0.8}Co_{0.15}Al_{0.05}O_2$ cathode via $Y(PO_3)_3$ to form an in situ triple composite Li-ion conductor coating layer with the enhanced electrochemical performance. *Nanotechnology* **2022**, *33*, 375701. [CrossRef]

coatings

MDPI

Article

# Controlling the Polymer Ink's Rheological Properties to Form Single and Stable Droplet

**Zhonghui Du [1], Lu Zhang [2], Yushuang Du [3], Xiaoqing Wei [1], Xiang Du [1], Xinyan Lin [1], Jiajun Liu [1], Yani Huang [1], Yan Xue [1,*], Ning Zhao [1,*] and Hongbo Liu [1,*]**

[1] School of Materials & Environmental Engineering, Shenzhen Polytechnic University, Shenzhen 518055, China; duzhonghui@szpu.edu.cn (Z.D.); aprilwei@szpu.edu.cn (X.W.); df2003110@szpu.edu.cn (X.D.); lin2618@szpu.edu.cn (X.L.); lzp790310@szpu.edu.cn (J.L.); huangguanlan@szpu.edu.cn (Y.H.);

[2] Department of Power Mechanical Engineering, Beijing Jiaotong University, Beijing 100044, China; 22110424@bjtu.edu.cn

[3] School of Chemistry Engineering, East China University of Science and Technology, Shanghai 200237, China; dyushuang@163.com

* Correspondence: xueyan@szpu.edu.cn (Y.X.); zhaoning@szpu.edu.cn (N.Z.); liuhongbo@szpu.edu.cn (H.L.)

**Abstract:** The formation of single and stable ink droplets is crucial for producing high-quality functional films in drop-on-demand (DOD) inkjet printing. The stability and singularity of droplet formation are significantly influenced by filament breakup behavior, governed by the rheological parameters of the ink formula. This study explores the droplet formation behavior of Poly3-hexylthiophene (P3HT) ink across various Weber numbers (*We*) and assesses the impact of the $Z$ value on the formation of single ink droplets. Observations reveal that as the *We* number increases, droplet morphology transitions from single to double, and eventually to sputtered droplets. Results demonstrate that stable, single droplets form when the *We* number $\leq 13$ and $12 < Z < 34$, with a pulse duration of approximately 340 μs. When the *We* number exceeds 13, the molecular chains of P3HT stretch due to high hydrodynamic forces, resulting in the formation of unwanted satellite droplets.

**Keywords:** inkjet printing; *We* number; single droplet; satellite droplets

## 1. Introduction

Because of its unique advantages, inkjet printing technology has been widely used in fine pattern processing, such as biochemical for cell guidance [1–3], organic light-emitting diodes [4–6], solar cells [7–9], sensors [10–13] and so on. In drop-on-demand (DOD) inkjet printer, with the mechanical deformation of the piezoelectric transducer in the ink chamber, the ink droplet is squeezed out when an electric voltage signal is applied to the piezoelectric transducer. Forming single and stable ink droplet is the key to printing high-quality functional films in drop-on-demand (DOD) inkjet printing [14–16].

In the early research, inks with no solute were often used in DOD inkjet printing [17–22]. Morris, Jeffrey F. et al. used the glycerin–water and glycerin–water–isopropanol mixtures as the inks and the main stages of DOD drop formation were analyzed [23]. The main stages of DOD drop formation include the ejection and stretching of liquid, the pinch-off of liquid thread from the nozzle exit, the contraction of liquid thread, the breakup of liquid thread into primary drop and satellites, and the recombination of primary drop and satellites. Daehwan Jang et al. have investigated the inter-relationship between inkjet printability and physical fluid properties by monitoring droplet formation dynamics using a mixture of ethyl alcohol and ethylene glycol [24]. Berend-Jan de Gans et al. have investigated the influence of architecture on the inkjet printability of polymer solutions by comparing linear and 6-arm star PMMA inks [25].

For certain inks, their physical characteristic parameters intricately affect the droplet formation process [26–28]. To simplify the influence process, dimensionless parameters were proposed to elucidate the dynamics of droplet formation [29–34]. One such parameter is the Weber number (*We*), representing the ratio of inertial and capillary forces, expressed as follows:

$$We = \frac{v^2 \alpha \rho}{\gamma} \quad (1)$$

The Reynolds number (*Re*) is utilized to characterize fluid flow, representing the ratio of inertial to viscous forces, and is expressed as follows:

$$\mathrm{Re} = \frac{v \alpha \rho}{\eta} \quad (2)$$

Z is defined as the ratio of the Reynolds number to the square root of the Weber number.

$$Z = \frac{Re}{We^{1/2}} = \frac{(\gamma \rho \alpha)^{1/2}}{\eta} \quad (3)$$

In the above formulas, $v$ is the droplet velocity, $\alpha$ is the characteristic length (e.g., the diameter of the jetting nozzle), $\rho$ is the fluid density, $\eta$ is the fluid viscosity, and $\gamma$ is the surface tension. Proper parameters are crucial for achieving stable injection. Fromm originally proposed that inkjet printing is possible if $Z > 4$ [35]. These limits have been experimentally studied, and a slightly wider range has been proposed with $1 < Z < 14$ [36]. To generate a small radius droplet, the surface tension and associated Laplace pressure must be overcome before a drop can be ejected from a printer. Duineveld proposed that this can be described by a minimum value of the Weber number, $We > 4$, below which there is insufficient fluid flow to overcome surface tension [37]. When $We < 4$, the droplet cannot be injected from the nozzle.

Ink properties substantially influence the quality of droplet formation in inkjet printing, with optimal performance parameters varying across different ink formulations [38–40]. This study investigates how the physical properties of inks influence droplet formation, focusing specifically on conditions enabling a stable and single droplet ejection of Poly3-hexylthiophene (P3HT) ink. We have comprehensively examined the roles of the dimensionless Weber number (*We*) and the Z value in the dynamics of droplet formation. This analysis not only enhances our understanding of P3HT ink behavior but also yields valuable insights applicable to optimizing printing processes for other functional materials. By exploring the interplay between these parameters, our research aims to delineate the 'printable window' for P3HT inks, defined as the specific range of conditions under which the ink exhibits optimal jetting performance without forming satellite droplets or other anomalies. Understanding these conditions is expected to facilitate precise material deposition in applications ranging from electronics to biomedical devices, where the exact placement and structure of materials are critical [41,42]. The findings of this study could serve as a foundational reference for further experimental and theoretical research on the fluid dynamics of inkjet printing.

## 2. Materials and Methods

The P3HT ink solution was meticulously prepared by solving Poly3-hexylthiophene (P3HT) with chlorobenzene (CB) solvent at ambient temperature. The P3HT powder was synthesized to have a molecular weight (MW) of 102 kDa and a polydispersity index (PDI) of 2.19. The CB solvent was procured from Sigma-Aldrich Co., Ltd. (St. Louis, Missouri, USA) without any additional purification of the solvent. The solution was concentrated at 5 mg/mL using the specified 102 kDa molecular weight P3HT material. To ensure the complete dissolution of the P3HT, the mixture was gently heated for 10 min and then allowed to rest at room temperature for 24 h before use. This procedure ensures that the ink has the appropriate viscosity and chemical consistency for optimal printing performance.

The inkjet printing experiments were conducted using the AD-P-8000 Printing System (Cambridge, UK) with a high-resolution built-in CCD shown in Figure 1, provided by

Microdrop Technologies GmbH. Its nozzle diameter is 50 μm. Its viscosity and surface tension range are 0.4–100 mPa·s and 28–72 mN/m. And its max printing frequency is 2000 Hz. To prevent the clogging of the printing head—an issue that can impede consistent droplet formation—the ink solutions were filtered through a 0.2 μm filter prior to use. This filtering step ensures a clean, smooth flow of ink through the printing system.

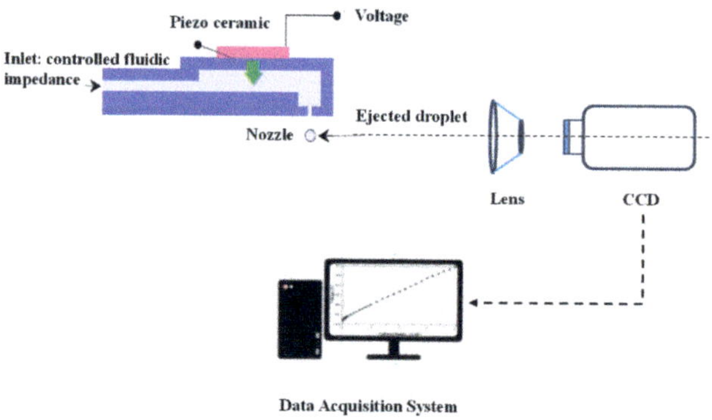


**Figure 1.** Schematic diagram of the experimental setup.

During the experiment, the formation of ink droplets under various driving voltages was meticulously captured using a high-speed CCD camera. This advanced imaging setup allowed for the detailed observation of the droplet dynamics with an interframe time of just 2 microseconds between each image. Such high temporal resolution is crucial for accurately analyzing the rapid processes involved in droplet formation and detachment from the nozzle.

The viscosity of the inks was meticulously characterized using the LVDV-III+ Programmable Control Rheometer from Brookfield Ltd. (Toronto, Canada) This measurement was conducted at room temperature, ensuring the ink's rheological properties were assessed under standard conditions to maintain consistency and reliability in data.

Additionally, the surface tension of the P3HT inks was accurately determined using the pendant drop method with the assistance of the DSA10 liquid drop shape analyzer from KRUSS GmbH, Hamburg, Germany. This method involves meticulously forming a single ink droplet at the tip of a needle or nozzle, which is then analyzed to determine its shape. The shape of the pendant droplet serves as a crucial indicator of surface tension. To ensure the highest accuracy of these measurements, a complete droplet of ink was extruded during the testing process to minimize potential errors caused by partial droplet formation or interference from environmental factors. The physical parameters of the Poly3-hexylthiophene (P3HT) ink are shown in Table 1.

**Table 1.** Physical parameters of the Poly3-hexylthiophene (P3HT) ink.

| Parameters | Value |
|---|---|
| Density, $\rho$ | 1068.40 Kg/m$^3$ |
| Surface tension, $\gamma$ | 31.0 mN/m |
| Viscosity, $\eta$ | 1.20 mPa·s |
| Conductivity, $G$ | 450 S/m |
| $Z$ value | 33.91 |

As previously stated, in drop-on-demand (DOD) inkjet printing, the production of uniform and stable ink droplets is crucial for fabricating high-quality functional films. The morphology of these droplets is chiefly influenced by ink viscosity, surface tension, and ejection velocity, with the Weber number (*We*), a dimensionless quantity representing the relative importance of inertial forces to surface tension, commonly employed to comprehend the droplet shape and stability. This study, by regulating the ink ejection velocity, examines the droplet ejection characteristics of P3HT ink under various Weber number conditions as depicted in Table 2, thereby enhancing the understanding and controlling of its application in DOD inkjet printing processes.

**Table 2.** The value of *We* for the Poly3-hexylthiophene (P3HT) at different dropping velocities.

| Velocity (m/s) | *Re* | *We* |
|---|---|---|
| 1.1 | 50.8 | 2.2 |
| 1.8 | 81.1 | 5.7 |
| 2.3 | 103.1 | 9.3 |
| 2.6 | 115.8 | 11.7 |

The concentration of the ink directly affects the Z value, and the change in the Z value will influence the inkjet printing dynamic process of the ink. Inks with different Z values were also investigated in Table 3.

**Table 3.** Z value of different concentrations of Poly3-hexylthiophene (P3HT).

| Concentration (mg/mL) | Z |
|---|---|
| 5 | 33.91 |
| 10 | 24.31 |
| 15 | 18.30 |
| 20 | 12.75 |

## 3. Results and Discussion

Firstly, we investigated the influence of *We* number on the ink droplet formation using the P3HT ink solution. The ink solution was prepared with a concentration of 5 mg/mL, dissolving 102 kDa molecular weight P3HT material. Our analysis was visually documented through a series of high-resolution jetting images, collectively referred to as Figure 2. These images captured the dynamic evolution of droplet formation at varying Weber numbers: 2.2, 5.7, 9.3, and 11.7. The imaging interval was maintained at 20 μs, covering a range from 80 to 500 μs to observe the droplet formation and fusion process in real-time. In Figure 2a, the initial formation of a liquid column is evident at 80 μs. A mere 20 μs later, a ligament extends from the nozzle, indicating the beginning of droplet separation. By 140 μs, a secondary, smaller droplet is visible trailing the primary droplet. This secondary droplet gains kinetic energy, progressively closing the distance to the primary droplet. By 300 μs, the two droplets merge into a single, larger droplet due to the kinetic energy surplus. This phenomenon of droplet fusion, driven by the Weber number, recurs in subsequent images (Figure 2b–d), illustrating a consistent pattern across different *We* values.

To provide a quantitative perspective on these observations, we meticulously measured the distances between the primary and secondary droplets during the jetting process. These measurements are graphically represented in Figure 3, which plots the distance of each droplet from the nozzle against time. The trajectory of the primary droplet is depicted with a black line, while the trajectory leading to the formation of the secondary droplet is marked in red. Our analysis highlights a notable trend for the sample with *We* = 2.2, where the gap between the primary and secondary droplets decreases steadily over time until the primary and the second droplets merge. The same phenomenon occurs in Figure 3b–d. This trend underscores the substantial influence of the Weber number on the dynamics of droplet formation and fusion.

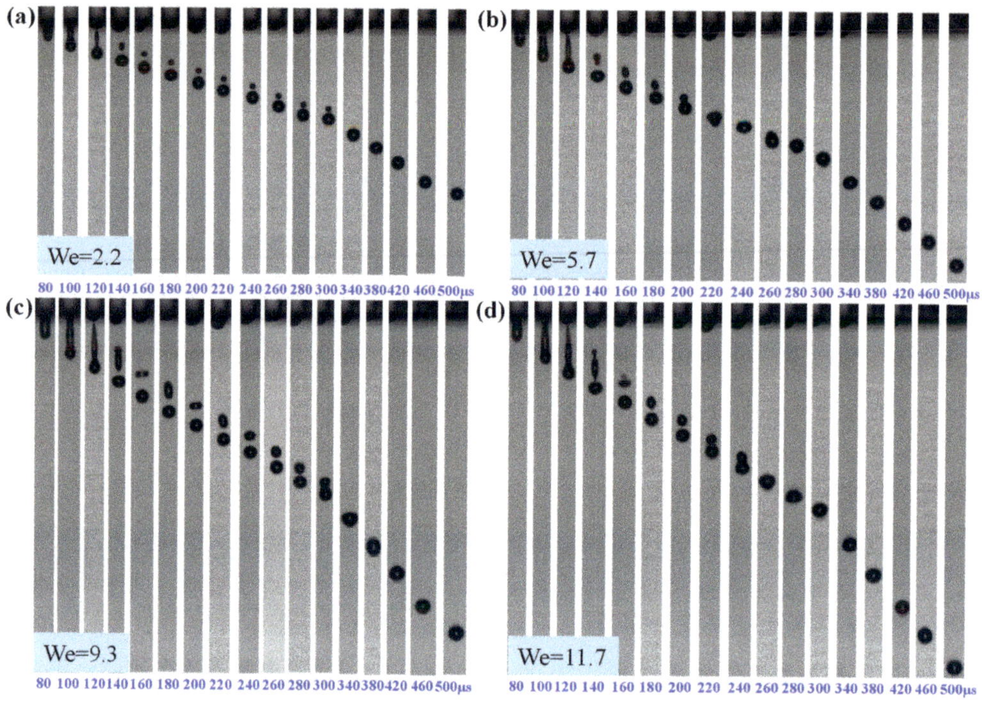


**Figure 2.** Droplet formation behavior at different *We* numbers, (**a**) *We* = 2.2, (**b**) *We* = 5.7, (**c**) *We* = 9.3, and (**d**) *We* = 11.7.

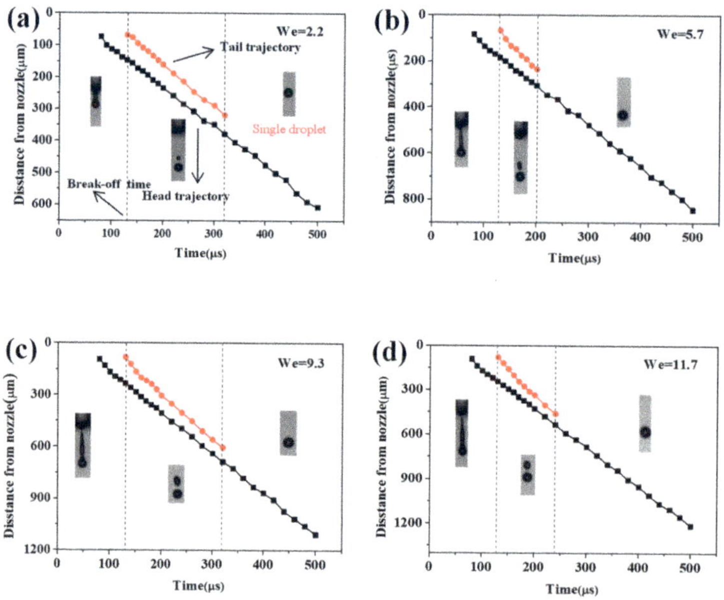


**Figure 3.** Trajectories (the black lines: head trajectory, the red lines: tail trajectory) of ejected droplets versus inkjetting time for inks with different Weber numbers, (**a**) We = 2.2, (**b**) *We* = 5.7, (**c**) *We* = 9.3, and (**d**) *We* = 11.7.

Figure 4 details the formation dynamics of P3HT ink droplets at *We* numbers of 13.4 and 36.5. Initially, a primary droplet forms, connected to the nozzle by a ligament, observed at 80 μs and 100 μs. By 160 μs, a smaller secondary droplet emerges, trailing behind the primary droplet. Unlike the scenarios depicted in Figure 1, these primary and secondary droplets do not merge into a single droplet, even after 500 μs. This persistent separation leads to the formation of an additional, third droplet at durations of 180 μs and 200 μs in Figure 4b.

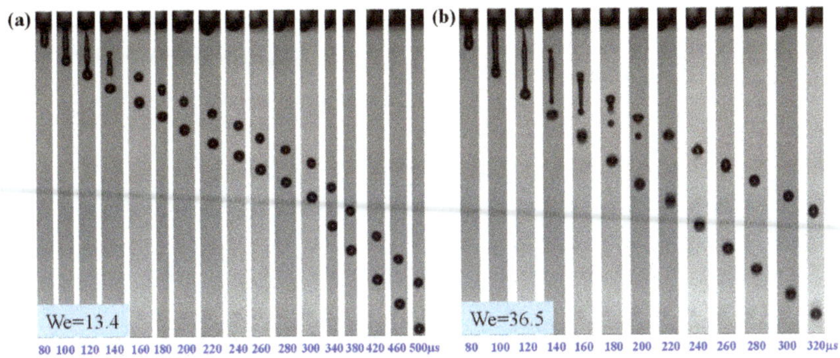


**Figure 4.** Droplet formation behavior at higher Weber numbers, specifically (**a**) *We* = 13.4 and (**b**) *We* = 36.5.

Figure 5 further explores the distances these droplets travel from the nozzle. The behavior of the droplets is classified into two distinct regions: the filament and the double droplet regions. In the filament region, the droplet forms a lengthy filament extending from the nozzle, significantly longer than those depicted in Figure 2. In the double droplet region, the separation between the primary droplet's leading edge and the secondary droplet's trailing edge significantly widens as time progresses. This increasing gap primarily prevents the droplets from merging, ultimately resulting in the formation of two distinct droplets. This behavior underscores the complex interplay between the kinetic energy and fluid dynamics at higher Weber numbers. The failure to merge at higher *We* values illustrates the limitations of the cohesive forces within the ink, exacerbated by the increased kinetic energy that propels the droplets apart rather than allowing them to coalesce. These observations are crucial for understanding the limitations and challenges in tuning ink formulations and printing parameters to achieve the desired outcomes, particularly in high-resolution printing applications where droplet precision is paramount.

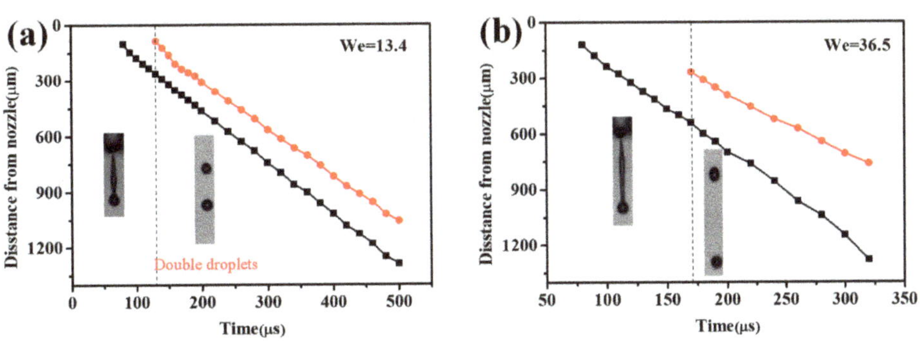


**Figure 5.** Trajectories (the black lines: head trajectory, the red lines: tail trajectory) of ejected droplets versus inkjetting time for inks with Weber numbers of (**a**) 13.4 and (**b**) 36.5.

To investigate the influence of kinetic energy on droplet volume, we analyzed the correlation between the droplet radius and the *We* number. As illustrated in Figure 6, a clear trend emerges: the droplet radius increases gradually with the *We* number. For example, at a *We* number of 3, the droplet radius measures 44 μm, which increases to 54 μm when the *We* number reaches 11.7, corresponding to a 50% increase in droplet volume. This progression suggests that higher Weber numbers correspond to increased kinetic energy affecting the droplet, thereby leading to an expansion in droplet volume.

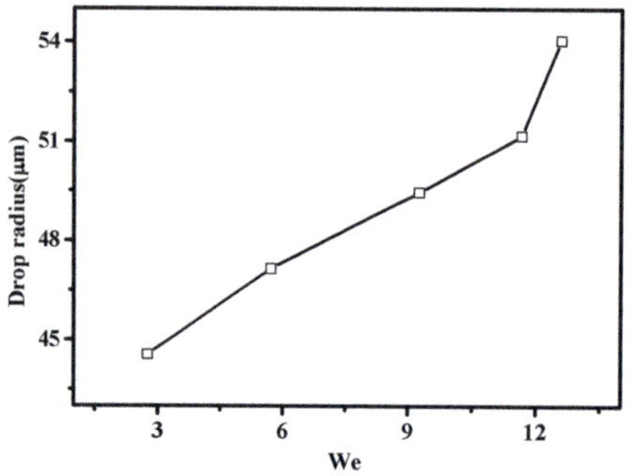


**Figure 6.** The relationship between the droplet radius and the *We* number.

Figure 7 presents a comprehensive summary of the droplet behavior under varying *Z* values and *We* numbers for P3HT ink, illustrating the impact of these parameters on inkjet printing dynamics. The observed droplet behaviors can be classified into three distinct scenarios: the formation of a single droplet, the formation of double droplets, and splashing events. Notably, when the Weber number exceeds 13, depicted by blue dots in the figure, a splashing phenomenon emerges. This phenomenon is characterized by numerous small, scattered droplets, indicative of excessive kinetic energy that disrupts droplet cohesion and results in erratic deposition patterns.

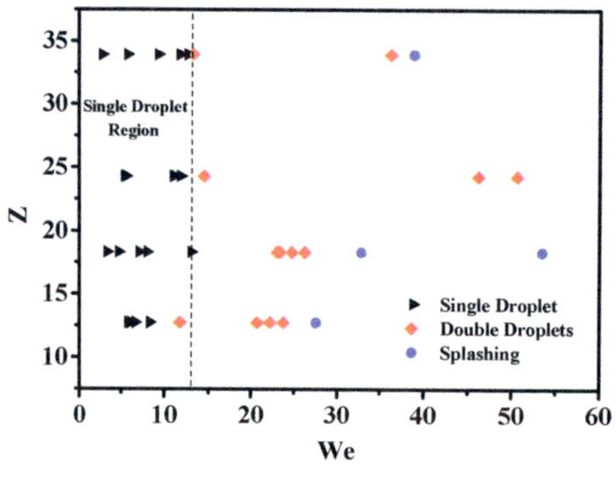


**Figure 7.** The printable region as a function of the *We* number relationship with respect to *Z*.

Conversely, when the Weber number is below 13 and $12 < Z < 34$, the ink demonstrates the formation of single, stable droplets. This situation represents the optimal conditions for controlled droplet deposition, crucial for achieving precision in applications such as electronic circuitry fabrication or the layering of sensitive biochemical substrates. The stability of these droplets ensures that the deposited materials adhere to precise locations and configurations, which are critical for the functionality and reliability of the final products.

This differentiation in droplet behaviors underscores the critical importance of fine-tuning the printing parameters, especially the Weber number, to align with the specific requirements of the printing task. By selecting the appropriate Weber number, manufacturers can manipulate droplet formation to mitigate unwanted splashing and ensure high-quality printing with minimal material waste.

## 4. Conclusions

In this study, we systematically investigated the influence of the Weber number (*We*) on droplet formation in inkjet printing processes. As the *We* number increases, droplet morphology transitions from single to double, and eventually to sputtered droplets. As for single droplet formation, the droplet radius increases gradually with the *We* number. Our comprehensive analysis across various driving voltages revealed that a Weber number (*We*) below 13 and $12 < Z < 34$ provides an optimal range for achieving the consistent formation of single droplets. Specifically, a pulse duration of 38.5 μs, employing a single waveform, is adequate for forming the stable droplets of P3HT ink without the presence of satellite droplets or other instabilities.

These findings not only confirm the crucial role of the Weber number in determining the printing capabilities of P3HT inks but also establish clear guidelines for its application in precision inkjet printing. They offer a framework for predicting and controlling the fidelity of droplet formation, which is especially valuable for the development of printing protocols for advanced materials, where the fidelity of droplet formation directly impacts the quality and functionality of the final product.

**Author Contributions:** Z.D., L.Z., Y.D., X.W., X.D., X.L., J.L. and Y.H. contributed to the experiment design, data analysis, and writing. The material preparation and data collection were performed by Y.X., N.Z. and H.L. All authors have read and agreed to the published version of the manuscript.

**Funding:** This work was supported by the Shenzhen Polytechnic Research Fund (6023310025K) and Post-doctoral Later-stage Foundation Project of Shenzhen Polytechnic (6023271017K), the funding from the Basic and Applied Basic Research Funding of Guangdong (No.2020A1515110551), Provincial Education Department of Guangdong (No.2021KTSCX277), Shenzhen Science and Technology Program (No.RCBS20210609103648040), and Provincial High Vocational Education Teaching Reform Research and Practice Project (GDJG2021422). Key funded projects of special funds for scientific and technological innovation strategy of Guangdong Province (pdjh2022a0973, pdjh2022001, pdjh2023a0967), Shenzhen Science and Technology Program under grant (KJZD20230923114412026), Post-doctoral Later-stage Foundation Project of Shenzhen Polytechnic under grant (6021271013K), and the Founding of Shenzhen polytechnic (No. 6021310007K, 6022312061K, 6022310034K, 6021271011K, 6021230059K, 6021210078K, LHRC20220402, 1019-8122310001C0, 6022310030K) provided funding for this work.

**Institutional Review Board Statement:** Not applicable.

**Informed Consent Statement:** Not applicable.

**Data Availability Statement:** All data generated or analyzed during this study are included in this article.

**Conflicts of Interest:** The authors declare that they have no known competing financial interests or personal relationships that could have appeared to influence the work reported in this paper.

## References

1. Saunders, R.E.; Derby, B. Inkjet printing biomaterials for tissue engineering: Bioprinting. *Int. Mater. Rev.* **2014**, *59*, 430–448. [CrossRef]
2. Zhang, G.; Qu, Z.; Tao, W.-Q.; Wang, X.; Wu, L.; Wu, S.; Xie, X.; Tongsh, C.; Huo, W.; Bao, Z.; et al. Porous Flow Field for Next-Generation Proton Exchange Membrane Fuel Cells: Materials, Characterization, Design, and Challenges. *Chem. Rev.* **2022**, *123*, 989–1039. [CrossRef] [PubMed]
3. Xu, C.; Zhang, M.; Huang, Y.; Ogale, A.; Fu, J.; Markwald, R.R. Study of Droplet Formation Process during Drop-on-Demand Inkjetting of Living Cell-Laden Bioink. *Langmuir* **2014**, *30*, 9130–9138. [CrossRef] [PubMed]
4. Wang, J.J.; Li, D.Y.; Mu, L.; Li, M.Z.; Luo, Y.; Zhang, B.B.; Mai, C.H.; Guo, B.A.; Lan, L.F.; Wang, J.; et al. Inkjet-Printed Full-Color Matrix Quasi-Two-Dimensional Perovskite Light-Emitting Diodes. *ACS Appl. Mater. Interfaces* **2021**, *13*, 41773–41781. [CrossRef] [PubMed]

5. Xue, Y.; Wang, L.; Zhang, Y.; Liang, G.; Chu, J.; Han, B.; Cao, W.; Liao, C.; Zhang, S. 31-Inch 4K Flexible Display Employing Gate Driver With Metal Oxide Thin-Film Transistors. *IEEE Electron Device Lett.* **2021**, *42*, 188–191. [CrossRef]
6. Cai, X.; Su, S.-J. Marching Toward Highly Efficient, Pure-Blue, and Stable Thermally Activated Delayed Fluorescent Organic Light-Emitting Diodes. *Adv. Funct. Mater.* **2018**, *28*, 1802558. [CrossRef]
7. Aernouts, T.; Aleksandrov, T.; Girotto, C.; Genoe, J.; Poortmans, J. Polymer based organic solar cells using ink-jet printed active layers. *Appl. Phys. Lett.* **2008**, *92*, 033306. [CrossRef]
8. Hoth, C.N.; Schilinsky, P.; Choulis, S.A.; Brabec, C.J. Printing Highly Efficient Organic Solar Cells. *Nano Lett.* **2008**, *8*, 2806–2813. [CrossRef]
9. Krebs, F.C. Fabrication and processing of polymer solar cells: A review of printing and coating techniques. *Sol. Energy Mater. Sol. Cells* **2009**, *93*, 394–412. [CrossRef]
10. Tu, N.; Lo, J.C.C.; Lee, S.W.R. Development of Uniform Polydimethylsiloxane Arrays through Inkjet Printing. *Polymers* **2023**, *15*, 462. [CrossRef]
11. Gupta, S.; Thombare, M.R.; Patil, N.D. Pinning and Depinning Dynamics of an Evaporating Sessile Droplet Containing Mono- and Bidispersed Colloidal Particles on a Nonheated/Heated Hydrophobic Substrate. *Langmuir* **2023**, *39*, 3102–3117. [CrossRef] [PubMed]
12. Bai, J.Y.; Hu, H.L.; Yu, Y.S.; Zhu, Y.B.; Xu, Z.W.; Zheng, W.C.; Zhao, H.B.; Lin, L.H.; Guo, T.L.; Li, F.S. Achieving high performance InP quantum dot light-emitting devices by using inkjet printing. *Org. Electron.* **2023**, *113*, 106705. [CrossRef]
13. Huang, Q.; Al-Milaji, K.N.; Zhao, H. Inkjet Printing of Silver Nanowires for Stretchable Heaters. *ACS Appl. Nano Mater.* **2018**, *1*, 4528–4536. [CrossRef]
14. Driessen, T.; Jeurissen, R. Drop Formation in Inkjet Printing. In *Fundamentals of Inkjet Printing*; Wiley-VCH Verlag GmbH & Co. KGaA: Weinheim, Germany, 2016; pp. 93–116.
15. Notz, P.K.; Basaran, O.A. Dynamics and breakup of a contracting liquid filament. *J. Fluid Mech.* **2004**, *512*, 223–256. [CrossRef]
16. Schulkes, R. The contraction of liquid filaments. *J. Fluid Mech.* **1996**, *309*, 277–300. [CrossRef]
17. Bernasconi, R.; Brovelli, S.; Viviani, P.; Soldo, M.; Giusti, D.; Magagnin, L. Piezoelectric Drop-On-Demand Inkjet Printing of High-Viscosity Inks. *Adv. Eng. Mater.* **2022**, *24*, 2100733. [CrossRef]
18. Satpathi, N.S.; Malik, L.; Ramasamy, A.S.; Sen, A.K. Drop Impact on a Superhydrophilic Spot Surrounded by a Superhydrophobic Surface. *Langmuir* **2021**, *37*, 14195–14204. [CrossRef] [PubMed]
19. Jung, S.J.; Hoath, S.D.; Hutchings, I.M. The role of viscoelasticity in drop impact and spreading for inkjet printing of polymer solution on a wettable surface. *Microfluid. Nanofluidics* **2013**, *14*, 163–169. [CrossRef]
20. Hoath, S.D.; Jung, S.; Hutchings, I.M. A simple criterion for filament break-up in drop-on-demand inkjet printing. *Phys. Fluids* **2013**, *25*, 5. [CrossRef]
21. Hoath, S.D.; Hutchings, I.M.; Harlen, O.G.; McLlroy, C.; Morrison, N.F. *Regimes of Polymer Behaviour in Drop-on-Demand Ink-Jetting*; Society for Imaging Science & Technology: Springfield, VA, USA, 2012; pp. 408–411.
22. Hoath, S.; Martin, G.; Castrejon-Pita, R.; Hutchings, I. *Satellite Formation in Drop-on-Demand Printing of Polymer Solutions*; Society for Imaging Science & Technology: Springfield, VA, USA, 2007; pp. 331–335.
23. Dong, H.; Carr, W.W.; Morris, J.F. An experimental study of drop-on-demand drop formation. *Phys. Fluids* **2006**, *18*, 072102. [CrossRef]
24. Jang, D.; Kim, D.; Moon, J. Influence of Fluid Physical Properties on Ink-Jet Printability. *Langmuir* **2009**, *25*, 2629–2635. [CrossRef]
25. de Gans, B.J.; Xue, L.J.; Agarwal, U.S.; Schubert, U.S. Ink-jet printing of linear and star polymers. *Macromol. Rapid Commun.* **2005**, *26*, 310–314. [CrossRef]
26. Suryo, R.; Doshi, P.; Basaran, O.A. Nonlinear dynamics and breakup of compound jets. *Phys. Fluids* **2006**, *18*, 082107. [CrossRef]
27. Aytouna, M.; Paredes, J.; Shahidzadeh-Bonn, N.; Moulinet, S.; Wagner, C.; Amarouchene, Y.; Eggers, J.; Bonn, D. Drop Formation in Non-Newtonian Fluids. *Phys. Rev. Lett.* **2013**, *110*, 5. [CrossRef]
28. Du, Z.; Zhou, H.; Cao, W.; Yu, X.; Han, Y. The effects of coil–stretch transition behavior of polyfluorene inks on single droplet formation during inkjet printing. *Chin. Chem. Lett.* **2020**, *31*, 3216–3220. [CrossRef]
29. Kang, D.J.; Gonzalez-Garcia, L.; Kraus, T. Soft electronics by inkjet printing metal inks on porous substrates. *Flex. Print. Electron.* **2022**, *7*, 033001. [CrossRef]
30. Yan, K.; Li, J.A.; Pan, L.J.; Shi, Y. Inkjet printing for flexible and wearable electronics. *APL Mater.* **2020**, *8*, 120705. [CrossRef]
31. Rivadeneyra, A.; Lopez-Villanueva, J.A. Recent Advances in Printed Capacitive Sensors. *Micromachines* **2020**, *11*, 367. [CrossRef] [PubMed]
32. Smith, P.J.; Stringer, J. Applications in Inkjet Printing. In *Fundamentals of Inkjet Printing*; Wiley-VCH Verlag GmbH & Co. KGaA: Weinheim, Germany, 2016; pp. 397–418.
33. Shi, Y.; Liu, J.; Yang, Y. Device performance and polymer morphology in polymer light emitting diodes: The control of thin film morphology and device quantum efficiency. *J. Appl. Phys.* **2000**, *87*, 4254–4263. [CrossRef]
34. Nguyen, T.Q.; Doan, V.; Schwartz, B.J. Conjugated polymer aggregates in solution: Control of interchain interactions. *J. Chem. Phys.* **1999**, *110*, 4068–4078. [CrossRef]
35. Fromm, J.E. Numerical-Calculation of the Fluid-Dynamics of Drop-On-Demand Jets. *IBM J. Res. Dev.* **1984**, *28*, 322–333. [CrossRef]
36. Derby, B.; Reis, N. Inkjet printing of highly loaded particulate suspensions. *MRS Bull.* **2003**, *28*, 815–818. [CrossRef]
37. Derby, B. Inkjet printing ceramics: From drops to solid. *J. Eur. Ceram. Soc.* **2011**, *31*, 2543–2550. [CrossRef]

38. Du, Z.; Zhou, H.; Yu, X.; Han, Y. Controlling the polarity and viscosity of small molecule ink to suppress the contact line receding and coffee ring effect during inkjet printing. *Colloids Surf. A Physicochem. Eng. Asp.* **2020**, *602*, 125111. [CrossRef]
39. Hoath, S.D.; Harlen, O.G.; Hutchings, I.M. Jetting behavior of polymer solutions in drop-on-demand inkjet printing. *J. Rheol.* **2012**, *56*, 1109–1127. [CrossRef]
40. Du, Z.; Yu, X.; Han, Y. Inkjet printing of viscoelastic polymer inks. *Chin. Chem. Lett.* **2018**, *29*, 399–404. [CrossRef]
41. Du, Z.; Lin, Y.; Xing, R.; Cao, X.; Yu, X.; Han, Y. Controlling the polymer ink's rheological properties and viscoelasticity to suppress satellite droplets. *Polymer* **2018**, *138*, 75–82. [CrossRef]
42. Jiang, C.; Zhong, Z.; Liu, B.; He, Z.; Zou, J.; Wang, L.; Wang, J.; Peng, J.; Cao, Y. Coffee-Ring-Free Quantum Dot Thin Film Using Inkjet Printing from a Mixed-Solvent System on Modified ZnO Transport Layer for Light-Emitting Devices. *ACS Appl. Mater. Interfaces* **2016**, *8*, 26162–26168. [CrossRef]

coatings

MDPI

Article

# Preparation of Highly Efficient and Stable All-Inorganic $CsPbBr_3$ Perovskite Solar Cells Using Pre-Crystallization Multi-Step Spin-Coating Method

**Yulong Zhang [1,†], Zhaoyi Jiang [2,*,†], Jincheng Li [1], Guanxiong Meng [3], Jiajun Guo [3,*] and Weijia Zhang [4,*]**

1 Department of Basic Courses, Officers College of PAP, Chengdu 610213, China; zyl881117@buaa.edu.cn (Y.Z.); jinchengli@alu.scu.edu.cn (J.L.)
2 School of Applied Science, Taiyuan University of Science and Technology, Taiyuan 610213, China
3 Hangzhou Acme Optoelectronics Technology Co., Ltd., Hangzhou 310020, China
4 Center of Condensed Matter and Material Physics, School of Physics, Beihang University, Beijing 100191, China
* Correspondence: jiangzhaoyi1991@buaa.edu.cn (Z.J.); jiangzhaoyan1991@163.com (J.G.); zwjghx@buaa.edu.cn (W.Z.)
† These authors contributed equally to this work.

**Abstract:** All-inorganic $CsPbBr_3$ perovskite solar cells have garnered extensive attention in the photovoltaic domain due to their remarkable environmental stability. Nevertheless, $CsPbBr_3$ prepared using the conventional sequential deposition method suffers from issues such as inferior crystallinity, low phase purity, and poor film morphology. Herein, we propose a pre-crystallization methodology by introducing a minute quantity of CsBr into the $PbBr_2$ precursor solution to generate a small amount of $CsPb_2Br_5$ crystals within the $PbBr_2$ film, leading to a porous $PbBr_2$ film with enhanced crystallinity. Under the influence of more pores and $CsPb_2Br_5$ crystals as nucleation sites for inducing growth, a $CsPbBr_3$ film with a larger crystal size, lower grain boundary density, stronger crystallinity, and higher phase purity is formed. Compared with untreated devices, photovoltaic devices prepared using the pre-crystallization method achieved a champion photovoltaic conversion efficiency (PCE) of 8.62%. Furthermore, pre-crystallized devices demonstrate higher stability than untreated ones and can still retain 94% of the original PCE after being exposed to air for 1000 h without encapsulating.

**Keywords:** perovskite solar cells; sequential deposition; $CsPbBr_3$; highly efficient and stability; porous $PbBr_2$

## 1. Introduction

Organic–inorganic hybrid perovskite possesses outstanding photovoltaic performance, rendering it the most popular photovoltaic material in recent years. Ever since its initial preparation in 2009, the power conversion efficiency (PCE) of organic–inorganic hybrid perovskite solar cells (PSCs) has witnessed a remarkable increase from 3.8% to the current maximum of 26.41% [1–4]. Nevertheless, on account of the hydrophobic nature and instability of organic cations ($MA^+$ and $FA^+$), organic–inorganic hybrid perovskites demonstrate chemical instability when exposed to harsh environments such as humidity, high temperature, light, and oxygen, thereby restricting the long-term stability of PSCs [5,6]. The method of using $Cs^+$ ions in lieu of organic cations to enhance stability and thermal stability has been verified as feasible [7]. Among them, $CsPbI_3$ PSC has attained the highest PCE of 21.86% [8]. However, $CsPbI_3$ perovskite demands a high temperature to sustain its ideal black cubic $\alpha$ phase, and it readily undergoes a transformation into the yellow orthorhombic $\delta$-non-perovskite phase at room temperature and in moist environments, boasting unfavorable photoelectric properties [9]. The employment of $Br^-$ as a halide anion leads to a more stable black phase at room temperature, thanks to the enhanced effective tolerance factor and lower phase-transition temperature [10]. In comparison to

$CsPbI_2Br$ and $CsPbIBr_2$, $CsPbBr_3$ perovskites that completely substitute $I^-$ with $Br^-$ enjoy the most robust environmental stability [11–13]. Nevertheless, the light-absorption capacity of $CsPbBr_3$ is compromised by its wider bandgap, posing a challenge to enhancing the efficiency of the single-layer $CsPbBr_3$ PSC [14,15].

Since no suitable solvent exists that exhibits excellent solubility for both $PbBr_2$ and CsBr, $CsPbBr_3$ is typically prepared through a multi-step method. Liang et al. proposed a solvent vapor-phase annealing technique to prepare porous $PbBr_2$ films by utilizing the N,N-dimethylformamide (DMF) gas phase environment during the annealing of $PbBr_2$ films. The porous $PbBr_2$ framework can facilitate a reaction with CsBr, thereby attaining dense, uniform, and high-quality large-grain $CsPbBr_3$ films. Consequently, a higher PCE can be obtained, and the unencapsulated device can still retain over 90% of its original PCE after being placed for 30 days [16]. Cao et al. employed a polymer of polyethylene oxide (PEO) as an additive to the $PbBr_2$ solution, transforming the small-grain $CsPbBr_3$ films into equiaxed large grains with a preferred crystal orientation, significantly enhancing the efficiency of the fabricated carbon hole transport layer (HTL) solar cells [17]. Zhao et al. proposed a two-in-one additive engineering strategy in which 4-tert butylpyridine (TBP) and dimethyl sulfoxide (DMSO) were added as additives to the $PbBr_2$ precursor solution and CsBr solution, respectively. The prepared perovskite layer exhibits better permeability, a lower carrier recombination rate, and high absorption capacity, thereby improving the power conversion efficiency. Moreover, the unencapsulated PSC maintains extremely high stability after being stored in ambient air for 1600 h [18]. Based on the two-step method, Che et al. prepared perovskite films using a total of four steps: spin coating low concentration CsBr solution, $PbBr_2$ solution, CsBr solution, and low concentration $PbBr_2$ solution. A $PbBr_2$ film with fewer pores, enhanced crystallinity, and a smoother surface was obtained, and the introduction of the derivative phase $CsPb_2Br_5$ promoted the formation of high-quality $CsPbBr_3$ films with larger grains. This film reduces interface defects, enhances grain growth, significantly improves power conversion efficiency, and can maintain stability for over 2400 h in an air environment [19]. Chi et al. introduced indium tribromide ($InBr_3$) into the precursor solution of $PbBr_2$ to form an $InBr_3(DMF)_3$ adduct, thereby accelerating the release of organic molecules in $PbBr_2(DMF)$ and preparing a porous $PbBr_2$ film. It not only optimizes the morphology of the $CsPbBr_3$ film but also reduces impurity phases. As a result, $InBr_3$:$CsPbBr_3$ perovskite solar cells achieved an efficiency of 7.28% [20]. In recent research, it can be observed that in the process of preparing $CsPbBr_3$ perovskite films via the multi-step method, optimizing the morphology of $PbBr_2$ films can further enhance the crystal quality of perovskite films, thereby improving the performance and stability of their photovoltaic devices [21–24].

In this work, the pre-crystallization method we previously explored for the preparation of organic–inorganic hybrid perovskite was adopted for the preparation of all-inorganic $CsPbBr_3$ perovskite solar cells [25]. By introducing a small quantity of CsBr into the precursor solution of $PbBr_2$, a minor amount of $CsPb_2Br_5$ crystals emerged in the $PbBr_2$ film, and a porous $PbBr_2$ film with higher crystallinity was acquired. Under the influence of the $C_sPb_2Br_5$ crystal as the nucleus induction and the porous $PbBr_2$ film, a uniform and pure-phase highly crystalline $CsPbBr_3$ film with large grains was formed. The device with the FTO/$SnO_2$/$CsPbBr_3$/carbon configuration prepared using the pre-crystallization method displayed a champion PCE of 8.62%, a short-circuit current density ($J_{sc}$) of 7.96 mA/$cm^2$, an open-circuit voltage ($V_{oc}$) of 1.351 V, as well as a fill factor (FF) of 0.802. Additionally, the unencapsulated devices were still capable of maintaining over 94% of their initial efficiency after being placed in an air environment for 1000 h.

## 2. Experimental Details

### 2.1. Materials

N,N-Dimethylformamide (DMF, 99.8%), methanol (99.8%), anhydrous isopropanol (IPA, 99.5%), and chlorobenzene (CB, 99.8%) were procured from Sigma-Aldrich, St. Louis, MO, USA. Lead bromide ($PbBr_2$, 99.99%) and cesium bromide (CsBr, 99.9%) were acquired from Xi'an Yuri Solar Co., Ltd., Xi'an, China, $SnCl_2 \cdot 2H_2O$ (98%) and $SnO_2$ ($SnO_2$ 15% in $H_2O$ colloidal dispersion) were purchased from Alfa Aesar. Carbon slurry (99%) was sourced from Shanghai MaterWin New Materials Co., Ltd., Shanghai, China.

### 2.2. Solution Preparation

A total of 367.0 mg of $PbBr_2$ was added to 1 mL of DMF solution and magnetically stirred at 90 °C for 30 min to obtain a 1 M $PbBr_2$/DMF precursor solution. A total of 367.0 mg of $PbBr_2$ and 20.4 mg of CsBr were added to 1 mL of DMF solution and magnetically stirred at 90 °C for 30 min to obtain a clear pre-crystalline precursor solution. A total of 14.3 mg of CsBr was added to 1 mL of methanol solution to obtain a 0.07 M CsBr/methanol solution. The above solutions were filtered using a 0.44 μm filter head after standing for 12 h.

### 2.3. Device Fabrication

A 25 mm × 25 mm FTO glass should be cleaned with deionized water, acetone, alcohol, and deionized water successively through ultrasonic cleaning (15 min for each) and then subjected to UV ozone treatment (30 min). An electron transport layer (ETL) is prepared by adopting a spin-coating water-bath method. The instructions are as follows: spin-coat 3% $SnO_2$ hydrogel at 3000 rpm for 30 s and anneal at 180 °C for 1 h. Place the sample in a water bath solution containing urea (1 g), hydrophobic acetic acid (20 μL), HCl (0.5 mL), $SnCl_2 \cdot 2H_2O$ (0.01 M), and deionized water (80 mL). Keep the water bath at 70 °C for 3 h, dry with $N_2$, and then anneal at 180 °C for 1 h. Preheat the $SnO_2$/FTO substrate and two precursor solutions to 80 °C. Spin-coat the 1 M $PbBr_2$/DMF precursor solution and pre-crystallized precursor solution at 3000 rpm for 30 s, add counter solvent CB during spin-coating, and anneal at 90 °C for 30 min to obtain the untreated and pre-crystallized $PbBr_2$ film. Spin-coat the 0.7 M CsBr/methanol solution at 2000 rpm for 30 s, anneal at 250 °C for 5 min, and repeat this process 5 times. Clean the excess CsBr with IPA solution to acquire the $CsPbBr_3$ film. Utilize screen-printed carbon electrodes and anneal at 100 °C for 30 min. Eventually, a photovoltaic device with a Glass/FTO/$SnO_2$/$CsPbBr_3$/Carbon structure is obtained.

### 2.4. Device Characterizations

Scanning electron microscope (SEM) images were acquired using Hitachi S-4800. The X-ray diffraction (XRD) patterns of the films within the 10°−50° region were obtained using a Bruker D8 X-ray diffractometer with CuKa radiation (1.5418 Å). The absorption spectra of the films were determined with a Hitachi U-4100 UV-Vis spectrophotometer. Photospectral response was evaluated by an external quantum efficiency (EQE) measurement system (QEX10, PV Measurement, Boulder, CO, USA). The photoluminescence (PL) spectra of the samples were recorded via the Edinburgh Instrument FLS920, and the excitation light source was a 450 W xenon lamp. The time-resolved photoluminescence (TRPL) spectra were measured through the Edinburgh Instrument FLS920, and the excitation light source was a 470 nm pulsed laser beam. The *J-V* curves under AM1.5G illumination (standard 100 mW/cm$^2$) of the devices with perovskite films were determined using Agilent B1500A Semiconductor parameter analysis instrument, and the light source was a 450 W xenon lamp.

## 3. Results and Discussion

To investigate the influence of the pre-crystallization method on the morphology of $PbBr_2$ precursor films and $CsPbBr_3$ films, top-down angle scanning electron microscopy was employed for examination. Figure 1a presents the SEM images of untreated $PbBr_2$ films as a control group, manifesting a relatively compact morphology along with a small number of pores. Figure 1b shows the SEM image of the $PbBr_2$ film prepared by the pre-crystallization method. Upon comparison, it could be discerned that the $PbBr_2$ film prepared using the pre-crystallization method displays more pores, featuring similar pore dimensions and uniform distribution. This porous configuration can leave diffusion channels for the second-step solution and enhance the contact area between the $PbBr_2$ film and CsBr, thereby accelerating the reaction rate and increasing the crystal size of $CsPbBr_3$ [26]. Figure 1c,d present the SEM images of untreated and pre-crystallized perovskite films. It can be observed that the pre-crystallization method renders the perovskite film more uniform, with a larger grain size and lower grain boundary density, with most crystal sizes exceeding 1 μm. Simultaneously, the pre-crystallization method enables $CsPbBr_3$ to achieve a single-layer crystal structure in the longitudinal direction, as shown in Figure S1. These demonstrate that the pre-crystallization method can enhance the quality of perovskite films, thereby reducing non-radiative recombination phenomena and increasing the migration efficiency of charge carriers, ultimately effectively enhancing the photoelectric performance of PSCs [27].

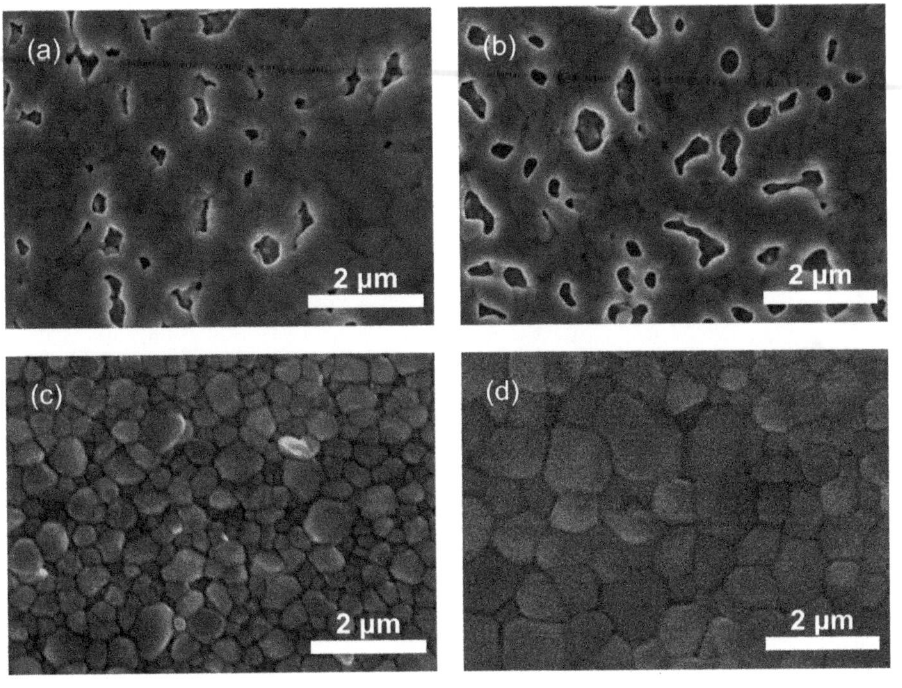


**Figure 1.** Top-view SEM images of (**a**) untreated and (**b**) pre-crystallized $PbBr_2$ films and (**c**) untreated and (**d**) pre-crystallized perovskite films.

Furthermore, the crystallization of $PbBr_2$ and $CsPbBr_3$ films was analyzed via X-ray diffraction spectroscopy testing. In Figure 2a, diffraction peaks situated at 18.7°, 21.7°, 30.3°, and 39.5° can be identified, which, respectively, correspond to the (020), (120), (130), and (140) crystal planes of $PbBr_2$. Meanwhile, diffraction peaks at 11.7°, 23.3°, and 29.4° can also be noticed in the pre-crystallized $PbBr_2$ film, corresponding to the (002), (210), and (213) crystal planes of $CsPb_2Br_5$, respectively. In the $PbBr_2$ film prepared using the pre-crystallization method, a minor amount of $CsPb_2Br_5$ crystals was generated through the chemical reaction between $PbBr_2$ and a small quantity of CsBr, as indicated in Equation (1). Additionally, in contrast to untreated perovskite films, the diffraction peak intensity of $PbBr_2$ in the pre-crystallized samples was enhanced, particularly on the (020) and (120) crystal planes. The phase transition during the preparation of $CsPbBr_3$ films using the multi-step spin-coating method can typically be depicted as the following chemical reactions [21]:

$$2PbBr_2 + CsBr \rightarrow CsPb_2Br_5 \text{ (with excess amount of } PbBr_2\text{)} \quad (1)$$

$$CsPb_2Br_5 + CsBr \rightarrow 2CsPbBr_3 \quad (2)$$

$$CsPbBr_3 + 3CsBr \rightarrow Cs_4PbBr_6 \text{ (with excess amount of CsBr)} \quad (3)$$

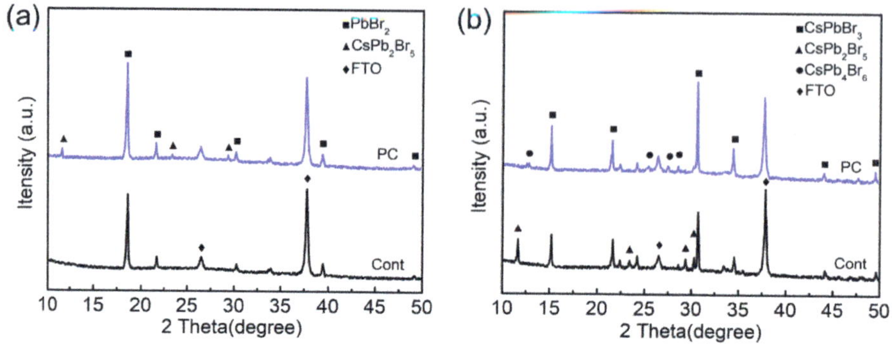


**Figure 2.** X-ray diffraction spectroscopy patterns of (**a**) $PbBr_2$ films and (**b**) $CsPbBr_3$ films.

Figure 2b shows the X-ray diffraction spectra of perovskite films prepared using untreated and pre-crystallized methods. Diffraction peaks situated at 15.2°, 21.6°, 30.7°, 34.5°, 44.1°, and 49.6° can be conspicuously perceived in both untreated and pre-crystallized films, respectively corresponding to the (100), (110), (200), (210), (220), and (310) crystal planes of $CsPbBr_3$. The $CsPb2Br_5$ phase can still be detected in the untreated perovskite film, which represents a common phenomenon in the multi-step spin-coating method when $PbBr_2$ is excessive [28]. After the employment of the pre-crystallization method, the $CsPb_2Br_5$ phase was mitigated, and diffraction peaks at 25.4°, 27.5°, and 28.6° were observed, respectively, corresponding to the (024), (131), and (214) crystal planes of $CsPb_4Br_6$. The phase of $CsPb_4Br_6$ is rather feeble, and the entire procedure is in accordance with Equations (1)–(3). Meanwhile, the pre-crystallization method also augmented the intensity of the $CsPbBr_3$ diffraction peak. In the course of preparing the $CsPbBr_3$ film using the pre-crystallization method, not only does the porous structure of the $PbBr_2$ film accelerate the reaction rate with the CsBr solution, but also a small quantity of $CsPb_2Br_5$ crystals within the $PbBr_2$ film function as nucleation sites to expedite the complete conversion of $PbBr_2$ to $CsPbBr_3$. Owing to the influence of the two factors, the pre-crystallized perovskite film exhibits better crystallinity and higher phase purity.

Figure 3a presents the UV-Vis absorption spectra of untreated and pre-crystallized $CsPbBr_3$ films. The pre-crystallized perovskite film exhibits a significantly enhanced light-absorption capacity relative to the control group within the 350–535 nm range, and the absorption band edge corresponds to the 2.32 eV band gap of $CsPbBr_3$, as shown in Figure S2. The enhancement of the light-absorption capability of the pre-crystallized sample mainly results from higher crystallinity, which is conducive to achieving a higher $J_{sc}$, in line with the measurement outcomes presented in Figure 2b.

Furthermore, the steady-state photoluminescence (PL) behavior of perovskite films was investigated. Figure 3b presents the steady-state PL spectrum of a perovskite film deposited directly onto an FTO substrate. Owing to the absence of ETL and HTL, excited state carriers cannot be promptly extracted, thereby leading to radiative recombination. Consequently, a higher fluorescence intensity suggests fewer traps or defects [29]. The pre-crystallized samples manifest a stronger PL emission peak intensity, indicating that the trap states in $CsPbBr_3$ films are mitigated. The PL emission peaks of the control group and pre-crystallization were 531 nm and 529 nm, respectively. The minor blue shift of the emission peak can be accounted for by the passivation of near-band edge defects or sub-band gap defects [30], further substantiating that the pre-crystallization method can effectively minimize the defect states of $CsPbBr_3$ films. Figure 3c shows the time-resolved

PL (TRPL) spectrum of perovskite films, which can be appropriately fitted using a double exponential equation:

$$I(t) = A_1 \exp(-t/\tau_1) + A_2 \exp(-t/\tau_2) \tag{4}$$

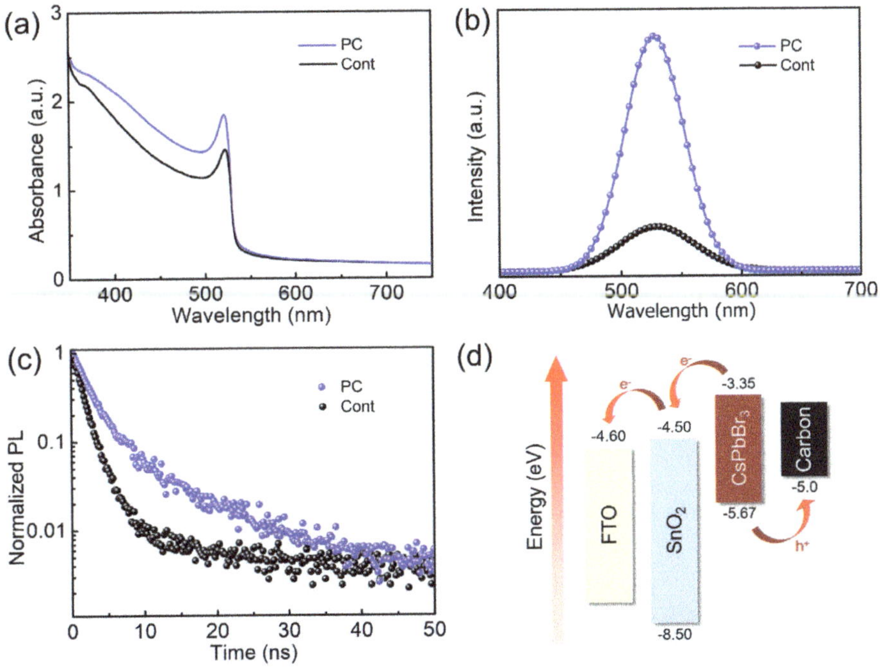


**Figure 3.** (**a**) Absorption spectrum, (**b**) steady-state PL spectrum, (**c**) TRPL spectrum, and (**d**) energy level diagram of photovoltaic devices for control and pre-crystallized perovskite films.

Here, $\tau_1$ and $\tau_2$ denote the slow decay time constants and fast decay time constants, respectively. $A_1$ and $A_2$ stand for the relative amplitudes. Commonly, the slow decay component mirrors the trap-assisted radiation-recombination process within the perovskite crystal, whereas the fast decay component is associated with the quenching of charge carriers at the interface [31]. Table S1 recapitulates the TRPL parameters of two $CsPbBr_3$ films. After the employment of the pre-crystallization method, the average carrier lifetime ($\tau_{ave}$) escalated from 3.57 ns to 7.32 ns, an increase of over 100%. The elongation of the carrier lifetime implies a reduction in defect density within perovskite films, attributable to higher crystallinity and a decrement in the grain boundary density engendered by larger grains [32]. This will enhance the efficiency of charge extraction and collection, facilitating the improvement in $J_{sc}$ and $V_{oc}$ in the device.

A PSC with FTO/$SnO_2$/$CsPbBr_3$/carbon structure was fabricated by employing untreated and pre-crystallized $CsPbBr_3$ films. The production process was conducted in an air environment, and the energy level diagram of the entire device is presented in Figure 3d. The cascaded conduction band arrangement between FTO, $SnO_2$, and $CsPbBr_3$ is beneficial for electron transfer and collection. Meanwhile, carbon has a higher valence band than $CsPbBr_3$, which is conducive to electron blocking and hole extraction. The *J-V* characteristics of untreated and pre-crystallized devices under AM 1.5 G single solar irradiation (100 mW/cm$^2$) are shown in Figure 4a. Table 1 summarizes the photovoltaic parameters of PSCs fabricated using untreated and pre-crystallized methods, including $J_{sc}$, $V_{oc}$, FF, and PCE, with 20 devices fabricated for each type. The device based on untreated

$CsPbBr_3$ merely attained an average PCE of 5.97%, with the champion device achieving a 6.77% PCE, $J_{sc}$ of 7.25 mA/cm$^2$, $V_{oc}$ of 1.260 V, and FF of 0.741. After pre-crystallization treatment, all photovoltaic parameters were significantly enhanced, and the average PCE of the device was increased to 7.93%. The champion device presented a $J_{sc}$ of 7.96 mA/cm$^2$, $V_{oc}$ of 1.351 V, and FF of 0.802, attaining a high PCE of 8.62%. The improvement in device performance is mainly ascribed to the increased crystallinity, reduced defects, and extended carrier lifetime of $CsPbBr_3$ films resulting from the pre-crystallization method. Additionally, due to the reduction in the grain boundary density, the pre-crystallization method can effectively mitigate hysteresis phenomena, as shown in Figure S3.

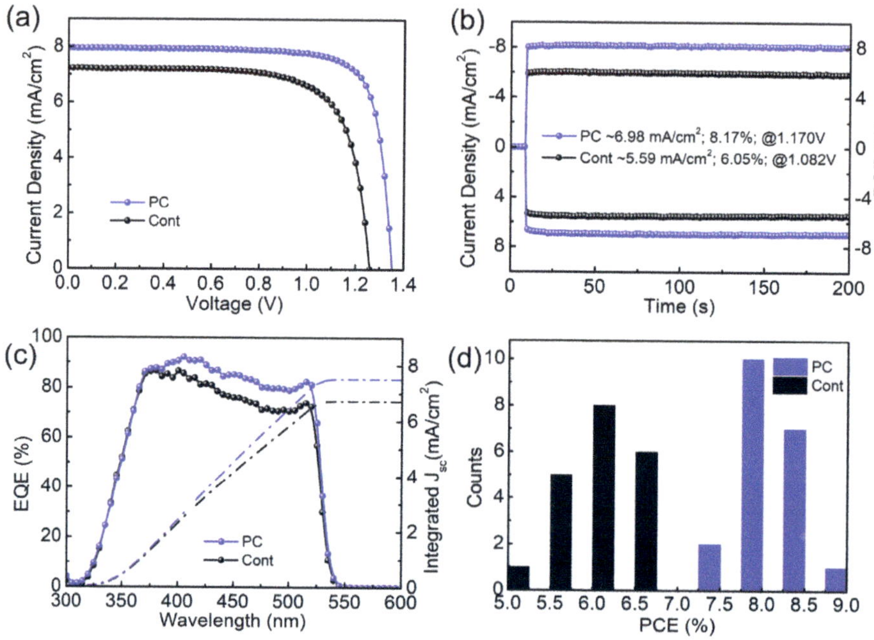


**Figure 4.** (**a**) *J-V* curves of $CsPbBr_3$ solar cells, (**b**) steady-state photocurrent and PCE at maximum output point voltage, (**c**) IPCE spectra and corresponding integrated $J_{sc}$, and (**d**) PCE distribution of 20 devices fabricated by control group and pre-crystallization method, respectively.

**Table 1.** Photovoltaic parameters of the untreated and pre-crystallized PSCs, with 20 prepared for each device.

| Device | | $J_{sc}$ (mA/cm$^2$) | $V_{oc}$ (V) | FF | PCE (%) |
|---|---|---|---|---|---|
| None | Average | 7.01 ± 0.72 | 1.210 ± 0.073 | 0.723 ± 0.041 | 5.97 ± 0.80 |
| | Champion | 7.25 | 1.260 | 0.741 | 6.77 |
| PC | Average | 7.66 ± 0.52 | 1.292 ± 0.063 | 0.787 ± 0.029 | 7.93 ± 0.69 |
| | Champion | 7.96 | 1.351 | 0.802 | 8.62 |

Figure 4b shows the steady-state output of the champion devices at their maximum power point. The photocurrent and PCE of both devices can rapidly attain their maximum values under illumination. The pre-crystallized device achieved a stable PCE of 8.17% at a bias voltage of 1.170 V, corresponding to a stable photocurrent density of 6.98 mA/cm$^2$. This was significantly higher than the 6.05% PCE obtained by the control group device at a bias voltage of 1.082 V, corresponding to a photocurrent density of 5.59 mA/cm$^2$. Additionally, after continuous illumination for 200 s, the PCE of the pre-crystallized device remained unchanged, demonstrating better stability. We also measured the incident photo-conversion efficiency (IPCE) spectrum of the champion device under short-circuit conditions, as presented in Figure 4c. Owing to the large bandgap of $CsPbBr_3$, the devices predominantly respond to wavelengths within the range of 300–550 nm, showing a cutoff wavelength similar to Figure 3a at 535 nm. The pre-crystallized device exhibits higher quantum yields within the response range, resulting in a higher integrated current density than the untreated device. The current densities of the untreated and pre-crystallized devices were 6.72 mA/cm$^2$ and 7.51 mA/cm$^2$, respectively, slightly lower than the *J-V* results, attributed to the inherent properties of the perovskite film [33]. Furthermore, the PCEs of 20 devices fabricated by the untreated and pre-crystallization methods were statistically analyzed, as shown in Figure 4d. Compared with untreated devices, the pre-crystallization method can prepare more efficient photovoltaic devices more stably, and other photovoltaic parameters are also enhanced, as shown in Figure S4.

To explore the causes by which the pre-crystallization method enhances the performance of PSCs, the dark *J-V* characteristics were determined through the utilization of the structure of FTO/$SnO_2$/$CsPbBr_3$/PCBM/Carbon. The trap state density (ntrap) of $CsPbBr_3$ films can be derived via the space charge limited current (SCLC) method, and the calculation equation is presented as follows [34]:

$$n_{trap} = 2\varepsilon_r\varepsilon_0 V_{TFL}/eL^2 \quad (5)$$

Here, $\varepsilon_r$ is the relative dielectric constant of $CsPbBr_3$ ($\varepsilon_r$ = 16.64), where $\varepsilon_0$ denotes the vacuum dielectric constant, $V_{TFL}$ refers to the limit voltage for trap filling, *e* symbolizes the fundamental charge of electrons, and *L* stands for the thickness of the perovskite film (*L* = 300 nm). As shown in Figure 5a, the $V_{TFL}$ of untreated and pre-crystallized devices are 1.16 V and 0.79 V, respectively. After calculation, following the utilization of the pre-crystallization method, the $n_{trap}$ of the device has significantly decreased from $7.08 \times 10^{15}$ cm$^{-3}$ to $4.82 \times 10^{15}$ cm$^{-3}$. This implies that the pre-crystallization method is beneficial for reducing defects and traps in $CsPbBr_3$ films.

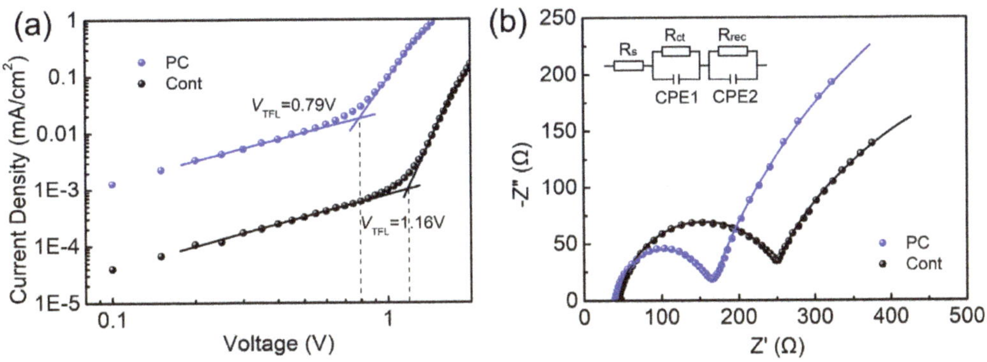


**Figure 5.** (**a**) SCLC plots and (**b**) Nyquist plot of untreated and pre-crystallized PSCs in dark environment.

Electrochemical impedance spectroscopy (EIS) measurements were also carried out to explore charge transfer and recombination in PSCs. Figure 5b presents the Nyquist plot of untreated and pre-crystallized devices under standard sunlight irradiation (100 mW/cm$^2$) at a bias voltage of 0.8 V and within a frequency range from 1 MHz to 0.01 Hz. Two primary semicircles can be perceived in the figure. The small semicircle at high frequencies is connected to hole transport at the interface of perovskite/HTL or HTL/electrode, corresponding to charge transport resistance ($R_{ct}$). The large semicircle at low frequencies is associated with the electron-recombination process at the $SnO_2$/perovskite interface, corresponding to the charge in recombination resistance ($R_{rec}$) [35]. Compared to untreated devices, pre-crystallization decreases the Rct of the device from 214 Ω to 129 Ω, leading to more rapid charge extraction, as shown in Table S2. At the same time, the $R_{rec}$ of the pre-crystallized device is 1887 Ω, higher than that of the untreated device at 1171 Ω, indicating a lower recombination rate. Faster charge extraction and lower carrier recombination lead to higher $J_{sc}$ and $V_{oc}$ for pre-crystallized devices. In addition, it can be deduced from the starting point in the real part of the Nyquist plot that the series resistance ($R_s$) of the pre-crystallized device is 37.2 Ω, which is lower than the 42.1 Ω of the untreated device, which means that the conductivity of the pre-crystallized perovskite film is higher and the charge-transport capacity is stronger, facilitating the attainment of higher FF [36].

Finally, the stability of the device was examined. In an environment with a temperature of 20 °C and a relative humidity (RH) of 20%, untreated and pre-crystallized devices demonstrate excellent stability, as presented in Figure 6a. This is ascribed to the outstanding environmental stability of $CsPbBr_3$, and the carbon electrode also contributes to enhancing the long-term stability of the device [37,38]. The pre-crystallized device can still retain an initial value of over 94% after 1000 h, pointing out that the pre-crystallization method can effectively enhance the stability of the $CsPbBr_3$ device. In a high-temperature environment (80 °C, 20% RH), the pre-crystallized device also manifested superior thermal stability without significant degradation, as shown in Figure 6b.

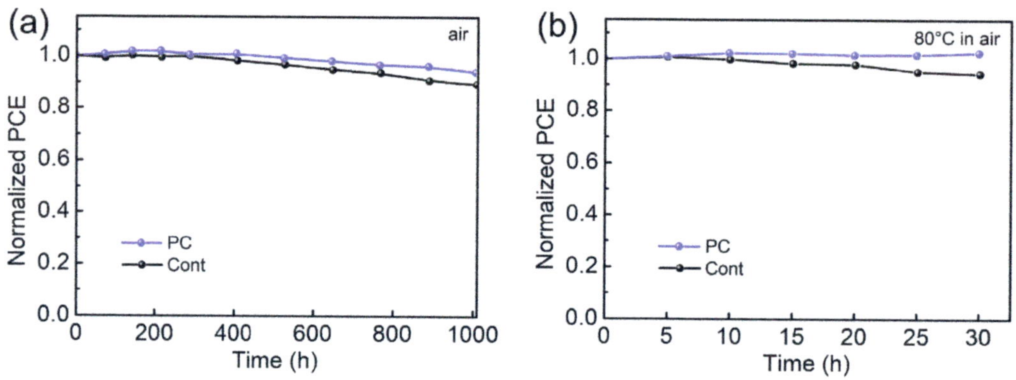


**Figure 6.** Long-term stability of the untreated and pre-crystallized devices in (**a**) 20 °C air (relative humidity of 20%) and (**b**) thermal stability in 80 °C air.

## 4. Conclusions

In summary, during the multi-step fabrication process of $CsPbBr_3$ perovskite solar cells, we employed a pre-crystallization method. By introducing a minor quantity of CsBr into the precursor solution of $PbBr_2$, mesoporous $PbBr_2$ films containing a small amount of $CsPb_2Br_5$ crystals were produced. A series of test results indicated that the $CsPbBr_3$ films prepared using the pre-crystallization method possess larger crystal sizes, lower grain boundary density, enhanced crystallinity, and higher phase purity. This reduces the trap state density of perovskite films, effectively facilitating charge transfer and suppressing carrier recombination. The device fabricated using the pre-crystallization method achieved a champion PCE of 8.62%, which is 27% higher than that of the untreated device. Additionally, the pre-crystallized devices demonstrate superior stability and thermal stability.

**Supplementary Materials:** The following supporting information can be downloaded at: https://www.mdpi.com/article/10.3390/coatings14070918/s1, Table S1: Parameters of the TR-PL spectroscopy for the $CsPbBr_3$ samples with varied CsBr spin-coating cycles. Table S2: EIS parameters of various $CsPbBr_3$ PSCs in the dark condition. Figure S1: Cross-sectional SEM image of the pre-crystallized $CsPbBr_3$ film. Figure S2: Tauc diagram of $CsPbBr_3$ films. Figure S3: Forward and reverse scanning $J$-$V$ curves of (a) untreated and (b) pre-crystallized devices. Figure S4: Box plots of (a) $J_{sc}$,

(b) $V_{oc}$, (c) FF, and (d) PCE of untreated and pre-crystallized PSCs, respectively. Each group is fabricated with 20 devices.

**Author Contributions:** Conceptualization, Y.Z., Z.J. and W.Z.; methodology, Y.Z.; software, J.L.; validation, J.L.; formal analysis, Y.Z.; investigation, Y.Z.; resources, G.M. and J.G.; data curation, Y.Z.; writing—original draft preparation, Y.Z.; writing—review and editing, Y.Z.; visualization, supervision, Z.J.; project administration, Z.J.; funding acquisition, Z.J. and W.Z. All authors have read and agreed to the published version of the manuscript.

**Funding:** This research was funded by TAIYUAN UNIVERSITY OF SCIENCE AND TECHNOLOGY SCIENTIFIC RESEARCH INITIAL FUNDING, grant number 20232091.

**Institutional Review Board Statement:** Not applicable.

**Informed Consent Statement:** Not applicable.

**Data Availability Statement:** Data are contained within the article/Supplementary Material; further inquiries can be directed to the corresponding author.

**Acknowledgments:** We gratefully acknowledge Beihang University and Hangzhou Acme Opto-electronics Technology Co., Ltd. for providing access to their experimental equipment to fabricate samples and their measurements.

**Conflicts of Interest:** Author Guanxiong Meng, Jiajun Guo were employed by Hangzhou Acme Optoelectronics Technology Co., Ltd. The remaining authors declare that the research was conducted in the absence of any commercial or financial relationships that could be construed as a potential conflict of interest.

## References


1. Kojima, A.; Teshima, K.; Shirai, Y.; Miyasaka, T. Organometal halide perovskites as visible-light sensitizers for photovoltaic cells. *J. Am. Chem. Soc.* **2009**, *131*, 6050–6051. [CrossRef] [PubMed]
2. Zhou, J.; Tan, L.; Liu, Y.; Li, H.; Liu, X.; Li, M.; Wang, S.; Zhang, Y.; Jiang, C.; Hua, R.; et al. Highly efficient and stable perovskite solar cells via a multifunctional hole transporting material. *Joule* **2014**, *8*, 1691–1706. [CrossRef]
3. Heo, D.Y.; Lee, T.H.; Iwan, A.; Kavan, L.; Omatova, M.; Majkova, E.; Kamarás, K.; Jang, H.W.; Kim, S.Y. Effect of lead thiocyanate ions on performance of tin-based perovskite solar cells. *J. Power Sources* **2022**, *458*, 228067. [CrossRef]
4. Tian, X.; Stranks, S.D.; You, F. Life cycle assessment of recycling strategies for perovskite photovoltaic modules. *Nat. Sustain.* **2021**, *4*, 821–829. [CrossRef]
5. Yang, Y.; You, J. Make perovskite solar cells stable. *Nature* **2017**, *544*, 155–156. [CrossRef] [PubMed]
6. Wang, L.; Shahiduzzaman, M.; Muslih, E.Y.; Nakano, M.; Karakawa, M.; Takahashi, K.; Tomita, K.; Nunzi, J.M.; Taima, T. Double-layer CsI intercalation into an $MAPbI_3$ framework for efficient and stable perovskite solar cells. *Nano Energy* **2021**, *86*, 106135. [CrossRef]
7. Liu, Q.; Qiu, J.; Yan, X.; Fei, Y.; Qiang, Y.; Chang, Q.; Wei, Y.; Zhang, X.; Tian, W.; Jin, S.; et al. Surface passivation and hole extraction: Bifunctional interfacial engineering toward high-performance all-inorganic $CsPbIBr_2$ perovskite solar cells with efficiency exceeding 12%. *J. Energy Chem.* **2022**, *74*, 387–393. [CrossRef]
8. Qiu, J.; Mei, X.; Zhang, M.; Wang, G.; Zou, S.; Wen, L.; Huang, L.; Hua, Y.; Zhang, X. Dipolar Chemical Bridge Induced $CsPbI_3$ Perovskite Solar Cells with 21.86 % Efficiency. *Angew. Chem. Int. Ed.* **2024**, *63*, e202401751. [CrossRef]
9. Wang, Z.; Tian, Q.; Zhang, H.; Xie, H.; Du, Y.; Liu, L.; Feng, X.; Najar, A.; Ren, X.; Liu, S. Managing Multiple Halide-Related Defects for Efficient and Stable Inorganic Perovskite Solar Cells. *Angew. Chem. Int. Ed.* **2023**, *62*, e202305815. [CrossRef]
10. Cheng, J.; Fan, Z.; Dong, J. Research Progress of Green Solvent in $CsPbBr_3$ Perovskite Solar Cells. *Nanomaterials* **2023**, *13*, 991. [CrossRef]
11. Sutton, R.J.; Eperon, G.E.; Miranda, L.; Parrott, E.S.; Kamino, B.A.; Patel, J.B.; Hörantner, M.T.; Johnston, M.B.; Haghighirad, A.A.; Moore, D.T.; et al. Bandgap-tunable cesium lead halide perovskites with high thermal stability for efficient solar cells. *Adv. Energy Mater.* **2016**, *6*, 1502458. [CrossRef]
12. Wei, L.; Rothmann, M.U.; Liu, A.; Wang, Z.; Zhang, Y.; Pascoe, A.R.; Lu, J.; Jiang, L.; Chen, Y.; Huang, F.; et al. Phase segregation enhanced ion movement in efficient inorganic $CsPbIBr_2$ solar cells. *Adv. Energy Mater.* **2017**, *7*, 1700946. [CrossRef]
13. Duan, J.; Zhao, Y.; He, B.; Tang, Q. High-purity inorganic perovskite films for solar cells with 9.72% efficiency. *Angew. Chem. Int. Ed.* **2018**, *57*, 3787–3791. [CrossRef] [PubMed]
14. Liang, J.; Wang, C.; Wang, Y.; Xu, Z.; Lu, Z.; Ma, Y.; Zhu, H.; Hu, Y.; Xiao, C.; Yi, X.; et al. All-inorganic perovskite solar cells. *J. Am. Chem. Soc.* **2016**, *138*, 15829–15832. [CrossRef]
15. Zhu, J.; Tang, M.; He, B.; Shen, K.; Zhang, W.; Sun, X.; Sun, M.; Chen, H.; Duan, Y.; Tang, Q. Ultraviolet filtration and defect passivation for efficient and photostable $CsPbBr_3$ perovskite solar cells by interface engineering with ultraviolet absorber. *Chem. Eng. J.* **2021**, *404*, 126548. [CrossRef]

16. Liang, K.; Wu, Y.; Zhen, Q.; Zou, Y.; Zhang, X.; Wang, C.; Shi, P.; Zhang, Y.; Sun, W.; Li, Y.; et al. Solvent vapor annealing-assisted mesoporous $PbBr_2$ frameworks for high-performance inorganic $CsPbBr_3$ perovskite solar cells. *Surf. Interfaces* **2023**, *37*, 102707. [CrossRef]
17. Cao, X.; Zhang, G.; Su, G.; Zhang, H.; Song, W.; Xie, X.; Wang, Y.; Zeng, Q.; He, X.; Wei, J. Fabrication of large columnar grains composed $CsPbBr_3$ films for efficient solar cells through ligand engineering. *Colloids Surf. A Physicochem. Eng. Asp.* **2023**, *674*, 131817. [CrossRef]
18. Zhao, Y.; Wang, S.; Ran, H.; Zhang, Y.; Tang, Y. Two-in-one additive engineering strategy induced high-quality $CsPbBr_3$ film in printable mesoscopic perovskite solar cells. *J. Alloys Compd.* **2023**, *965*, 171317. [CrossRef]
19. Che, G.; Wang, X.; Cui, C.; Pang, B.; Wang, X.; Dong, H.; Feng, J.; Yu, L.; Dong, L. Boosting the efficiency and stability of $CsPbBr_3$ perovskite solar cells through modified multi-step spin-coating method. *J. Alloys Compd.* **2023**, *969*, 172423. [CrossRef]
20. Chi, K.; Zhai, Y.; Wang, Y.; Zhang, X.; Liang, F.; Wang, S. Porous $PbBr_2$ films modulation by indium tribromide additive for the fabrication of $SnO_2$-based $CsPbBr_3$ perovskite solar cells. *Ceram. Int.* **2024**, *50*, 28386–28393. [CrossRef]
21. Liu, X.; Tan, X.; Liu, Z.; Ye, H.; Sun, B.; Shi, T.; Tang, Z.; Liao, G. Boosting the efficiency of carbon-based planar $CsPbBr_3$ perovskite solar cells by a modified multistep spin-coating technique and interface engineering. *Nano Energy* **2019**, *56*, 184–195. [CrossRef]
22. Liu, X.; Liu, Z.; Tan, X.; Ye, H.; Sun, B.; Xi, S.; Shi, T.; Tang, Z.; Liao, G. Novel antisolvent-washing strategy for highly efficient carbon-based planar $CsPbBr_3$ perovskite solar cells. *J. Power Sources* **2019**, *439*, 227092. [CrossRef]
23. Yan, J.; Hou, S.; Li, X.; Dong, J.; Zou, L.; Yang, M.; Xing, J.; Liu, H.; Hao, H. Preparation of highly efficient and stable $CsPbBr_3$ perovskite solar cells based on an anti-solvent rinsing strategy. *Sol. Energy Mater. Sol. Cells* **2022**, *234*, 111420. [CrossRef]
24. Chen, D.; He, Y.; Fan, G.; Zhang, Z.; Zhu, W.; Xi, H.; Zhou, L.; Zhang, C.; Zhang, J.; Hao, Y. Ultrahigh fill-factor all-inorganic $CsPbBr_3$ perovskite solar cells processed from two-step solution method and solvent additive strategy. *J. Mater.* **2023**, *9*, 717–724. [CrossRef]
25. Zhang, Y.; Jiang, Z.; Zhang, W.; Yan, L.; Lu, C.; Ni, C. Pre-crystallization applied in two-step spin-coating method for high-quality perovskite films in photovoltaic devices. *J. Alloys Compd.* **2020**, *832*, 153616. [CrossRef]
26. Zhao, Y.; Duan, J.; Wang, Y.; Yang, X.; Tang, Q. Precise stress control of inorganic perovskite films for carbon-based solar cells with an ultrahigh voltage of 1.622 V. *Nano Energy* **2020**, *67*, 104286. [CrossRef]
27. Liu, J.; Zhu, L.; Xiang, S.; Wei, Y.; Xie, M.; Liu, H.; Li, W.; Chen, H. Growing high-quality $CsPbBr_3$ by using porous $CsPb_2Br_5$ as an intermediate: A promising light absorber in carbon-based perovskite solar cells. *Sustain. Energy Fuels* **2019**, *3*, 184–194. [CrossRef]
28. Teng, P.; Han, X.; Li, J.; Xu, Y.; Kang, L.; Wang, Y.; Yang, Y.; Yu, T. Elegant Face-Down Liquid-Space-Restricted Deposition of $CsPbBr_3$ Films for Efficient Carbon-Based All-Inorganic Planar Perovskite Solar Cells. *ACS Appl. Mater. Interfaces* **2018**, *10*, 9541. [CrossRef] [PubMed]
29. Luo, J.; Jia, C.; Wan, Z.; Han, F.; Zhao, B.; Wang, R. The novel dopant for hole-transporting material opens a new processing route to efficiently reduce hysteresis and improve stability of planar perovskite solar cells. *J. Power Sources* **2017**, *342*, 886–895. [CrossRef]
30. Cao, X.; Zhang, G.; Cai, Y.; Jiang, L.; Chen, Y.; He, X.; Zeng, Q.; Jia, Y.; Xing, G.; Wei, J. Enhanced performance of $CsPbBr_3$ perovskite solar cells by reducing the conduction band offsets via a Sr-modified $TiO_2$ layer. *Appl. Surf. Sci.* **2020**, *529*, 147119. [CrossRef]
31. Liu, X.; Liu, Z.; Ye, H.; Tu, Y.; Sun, B.; Tan, X.; Shi, T.; Tang, Z.; Liao, G. Novel efficient C60-based inverted perovskite solar cells with negligible hysteresis. *Electrochim. Acta* **2018**, *288*, 115–125. [CrossRef]
32. Duan, J.; Zhao, Y.; Yang, X.; Wang, Y.; He, B.; Tang, Q. Lanthanide ions doped $CsPbBr_3$ halides for HTM-free 10.14%-efficiency inorganic perovskite solar cell with an ultrahigh open-circuit voltage of 1.594 V. *Adv. Energy Mater.* **2018**, *8*, 1802346. [CrossRef]
33. Saliba, M.; Etgar, L. Current density mismatch in perovskite solar cells. *ACS Energy Lett.* **2020**, *5*, 2886–2888. [CrossRef]
34. Wang, S.; Wang, P.; Chen, B.; Li, R.; Ren, N.; Li, Y.; Shi, B.; Huang, Q.; Zhao, Y.; Grätzel, M.; et al. Suppressed recombination for monolithic inorganic perovskite/silicon tandem solar cells with an approximate efficiency of 23%. *eScience* **2022**, *2*, 339–346. [CrossRef]
35. Wang, D.; Li, W.; Du, Z.; Li, G.; Sun, W.; Wu, J.; Lan, Z. Highly efficient $CsPbBr_3$ planar perovskite solar cells via additive engineering with NH4SCN. *ACS Appl. Mater. Interfaces* **2020**, *12*, 10579–10587. [CrossRef] [PubMed]
36. Hauff, E.; Klotz, D. Impedance spectroscopy for perovskite solar cells: Characterisation, analysis, and diagnosis. *J. Mater. Chem. C* **2022**, *10*, 742–761. [CrossRef]
37. Cao, X.; Zhang, G.; Jiang, L.; Cai, Y.; Wang, Y.; He, X.; Zeng, Q.; Chen, J.; Jia, L.; Wei, J. Achieving environment-friendly production of $CsPbBr_3$ films for efficient solar cellsviaprecursor engineering. *Green Chem.* **2021**, *23*, 2104–2112. [CrossRef]
38. Fu, Z.; Xu, M.; Sheng, Y.; Yan, Z.; Meng, J.; Tong, C.; Li, D.; Wan, Z.; Ming, Y.; Mei, A.; et al. Encapsulation of Printable Mesoscopic Perovskite Solar Cells Enables High Temperature and Long-Term Outdoor Stability. *Adv. Funct. Mater.* **2019**, *29*, 1809129. [CrossRef]

*coatings*

MDPI

*Article*

# Controllable Crystallization of Perovskite Films during the Blade-Coating Fabrication Process for Efficient and Stable Solar Cells

**Zhaoyi Jiang [1,*,†], Jiaqi Li [1,†], Kun Li [1], Rui Zhang [1], Zhenxing Qin [1], Yufei Zhang [1], Boheng Wu [1], Qiang Ma [2], Yulong Zhang [3] and Weijia Zhang [4,*]**

1 School of Applied Science, Taiyuan University of Science and Technology, Taiyuan 030024, China; 202318060306@stu.tyust.edu.cn (J.L.); likun@tyust.edu.cn (K.L.); zrzx_0921@tyust.edu.cn (R.Z.); s202318210879@stu.tyust.edu.cn (B.W.)

2 School of Advanced Interdisciplinary Studies, Ningxia University, Zhongwei 755000, China; maq@nxu.edu.cn

3 Department of Basic Courses, Officers College of PAP, Chengdu 610213, China

4 Center of Condensed Matter and Material Physics, School of Physics, Beihang University, Beijing 100191, China

* Correspondence: jiangzhaoyi1991@buaa.edu.cn (Z.J.); zwjghx@buaa.edu.cn (W.Z.)

† These authors contributed equally to this work.

**Abstract:** The scalable production of high-quality perovskite thin films is pivotal for the industrialization of perovskite thin film solar cells. Consequently, the solvent system employed for the fabrication of large-area perovskite films via coating processes has attracted significant attention. In this study, a solvent system utilizing a volatile solvent as the primary reagent has been developed to facilitate the rapid nucleation of volatile compounds. While adding the liquid Lewis base dimethylformamide (DMF) can help to improve the microstructure of perovskite films, its slow volatilization renders the crystal growth process uncontrollable. Based on the solvent system containing DMF and ethanol (EtOH), introducing a small amount of $NH_4Cl$ increases the proportion of the intermediate phase in the precursor films. This not only results in a controllable growth process for the perovskite crystals but also contributes to the improvement of the film microstructure. Under the simulated illumination (AM1.5, 1000 W/m$^2$), the photoelectric conversion efficiency (PCE) of the inverted solar cells has been improved to 20.12%. Furthermore, after 500 hours of continuous illumination, the photovoltaic device can retain 95.6 % of the initial, indicating that the solvent system is suitable for the scalable fabrication of high-quality $FAPbI_3$ thin films.

**Keywords:** perovskite film; solar cells; scalable fabrication; vacuum-assisted deposition; $FAPbI_3$; stability

## 1. Introduction

Over the past decade, the laboratory efficiency record of perovskite solar cells (PSCs) has experienced a remarkable increase, advancing from an initial 3.8 % to a certified 26.1% [1–3]. This substantial improvement has attracted considerable interest from both the academic community and industry stakeholders. However, the deployment of this innovative photovoltaic technology is impeded by several critical challenges that necessitate further investigation and resolution [4–6]. Moreover, a significant disparity exists in comparison to crystalline silicon solar cells. The principal limitation is attributed to the quality of the perovskite thin film, which includes factors such as morphology, uniformity, crystalline structure, and defect density, all of which are inferior to those observed in spin-coated small-area perovskite thin films [7–10].

To date, the predominant methodologies for the fabrication of large-area perovskite thin films include solution-coating (such as blade- and slit-coating), inkjet printing, and vacuum evaporation. Among these techniques, solution-coating demonstrates significant advantages in terms of production efficiency, cost-effectiveness, and the ability to preserve the nonstoichiometric composition of the coated perovskite film. The recent scholarly literature has extensively documented the application of solution-coating methods for the preparation of large-area perovskite thin films, with the objective of improving the conversion efficiency of large area modules [11,12]. Substantial progress has been made in this field of research. Irrespective of whether a small-area solution spin-coating technique or a large-area solution-coating method is utilized, both conform to the traditional LaMer crystal growth model [13–16]. Prior to attaining the critical nucleation concentration, the perovskite solute within the film undergoes rapid supersaturation facilitated by techniques such as anti-solvent extraction, air knife blowing, and rapid vacuum pumping. These post-processing methods induce instantaneous nucleation, thereby ensuring comprehensive substrate coverage. The formation of a defect-free film is of paramount importance. In the context of various post-processing methodologies, the regulation of the physical and chemical properties of the precursor solution is crucial for influencing the nucleation and crystallization processes of perovskite as well as determining the crystalline quality of the resultant thin film [17,18]. Solvents are generally categorized and optimized according to their volatility, solubility, and coordination properties to attain the optimal coating formulation. Deng et al. proposed the use of a solvent system consisting of high-volatility, low-coordination solvents in conjunction with a minor proportion of high-boiling-point, strong-coordination solvents. Through the application of air-knife-assisted film formation, they successfully prepared a large area of MAPbI3 perovskite film at a high speed of up to 99 cm/s [8,12]. However, a substantial quantity of acetonitrile (ACN) is utilized in the precursor solution, which can volatilize rapidly to facilitate high-speed film formation. Furthermore, in comparison to small-area perovskite films prepared via spin-coating, the current large-area perovskite films exhibit several significant issues: 1. The crystal phase and grain size distribution of perovskite crystals exhibit significant non-uniformity across a large area. 2. The surface roughness of the films varies considerably over an extensive area. 3. There is substantial variation in the thickness of the film across a large area. The aforementioned issues are intricately linked to the nucleation and growth mechanisms of perovskite crystals within thin films [14,19,20]. A significant challenge involves modulating the growth kinetics of perovskite thin film crystals to enable controlled fabrication over extensive areas [21–23]. Overcoming this challenge is crucial for the scalable production of high-quality perovskite thin films and the advancement of high-efficiency photovoltaic devices. In this study, we successfully prepared high-quality perovskite thin films through the incorporation of both solid and liquid Lewis base additives [24–27]. This can be achieved by employing a range of solvent systems and design strategies suitable for large-area fabrication. The coating process was conducted in dry air, adhering to the proposed design scheme.

In this study, $FAPbI_3$ thin films were fabricated using a vacuum flashing-assisted coating process with EtOH as the main solvent. The effects of DMF and ammonium chloride ($NH_4Cl$) on the nucleation and growth of $FAPbI_3$ perovskite crystals were systematically investigated. The addition of appropriate amounts of these Lewis base additives increased the size of the perovskite crystals. Notably, compared to the EtOH/DMF solvent system, the incorporation of a small amount of $NH_4Cl$ not only facilitated an increase in grain size but also conferred advantages such as uniform grain size distribution and enhanced surface flatness. In conjunction with the analysis of the crystallization kinetics mechanism of perovskite crystals, the incorporation of a small amount of $NH_4Cl$ into the precursor solution has been shown to effectively promote the formation of an intermediate structure in the unannealed film. This intermediate structure gradually transitions into crystalline form during the subsequent annealing process, thereby enhancing the controllable growth of the crystals. This method is advantageous for the scalable fabrication of high-quality $FAPbI_3$ films. Through the substantial enhancement of the microstructure of the $FAPbI_3$ perovskite thin film, the carrier transport properties in the associated photovoltaic device are significantly optimized. Consequently, this leads to marked improvements in both the photoelectric conversion efficiency (PCE) and the stability of the $FAPbI_3$ perovskite film solar cell.

## 2. Experimental Details

### 2.1. Materials

N, N-dimethylformamide (DMF, 99.8%), EtOH (99.8%), and IPA (99.5%) were brought from Sigma-Aldrich. $NH_4Cl$, lead(II) iodide($PbI_2$) powers (99.8%), and formamidinium iodide (FAI) powers (99.8%) were acquired from Xi'an Yuri Solar Co., Ltd. (Xi'an, China). Nickel(II) acetylacetonate (99%) was purchased from Tokyo Chemical Industry. The metallic Ag particles were purchased from Sino-Platinum Metals Co., Ltd. (Kunming, China) Unless stated otherwise, all solvents and regents were stored and used without any purification. The $C_{60}$ and bathocuproine (BCP) powers ($\geq$99.9%) were purchased from Advanced Election Technology Co., Ltd., Shenzhen, China.

### 2.2. Solution Preparation

The indium tin oxide (ITO)-coated glass substrates (Nippon Sheet Glass Co., Ltd., Osaka, Kitahama, Japan) were cleaned by sequential ultrasonication for 15 min with deionized water and ethanol. Then, the ITO substrates were dried by a high-speed $N_2$ gas flow before being stored. Then, a nickel(II) oxide (NiO) layer (hole transport layer) was deposited on the clean substrates. Next, NiO solution was prepared with a mixed solution of acetonitrile and ethanol (with a 90:10 volume ratio, 150 mL) of nickel acetylacetonate. The perovskite precursor solution (1.2 M) was prepared by mixing FAI and $PbI_2$ in mixed solvents (EtOH: DMF/Volume: Volume = 9:1).

### 2.3. Device Fabrication

The ITO glass should be washed with sequential ultrasonication for 15 min with detergent, deionized water, and ethanol and then subjected to UV ozone treatment (10 min). First, a NiO layer (hole transport layer, HTL) was deposited on the clean substrates. The prepared nickel acetylacetonate solution was automatically sprayed by an air nozzle on the hot ITO glass (200 °C) for 10 min, which was transferred into the glove box filled with $N_2$. The films were blade-coated with a precursor solution (coating speed of 20 mm/s), and then the wet films were transferred in a vacuum chamber to accelerate the evaporation of most solvents. Undergoing an annealing process (150 °C for 10 min), the $FAPbI_3$ films were prepared. Subsequently, the 20 nm-$C_{60}$ and 2 nm-BCP were sequentially deposited on the prepared $FAPbI_3$ films by thermal evaporation in a vacuum atmosphere ($\leq 1 \times 10^{-4}$ Pa), and the deposition rate was controlled below 0.2 nm/s. Finally, the Ag (120 nm) was deposited on the surface of the devices in a vacuum chamber ($\leq 3 \times 10^{-4}$ Pa) for the electrodes.

### 2.4. Characterizations

Morphologies of the prepared films and devices were carried out by scanning electron microscope (SEM) images acquired using Hitachi S-4800 (Tokyo, Japan). The XRD patterns of annealed and unannealed films were obtained by an Empyrean X-ray diffractometer with Cu Ka radiation (PANalytical B.V. Co., Almelo, The Netherlands). The absorption spectra of the perovskite films were carried out by a Hitachi U-4100 UV–Vis spectrophotometer (Tokyo, Japan). The photoluminescence (PL) spectra of the samples were recorded using the Edinburgh Instrument FLS920 (the excitation light source was a 450 W xenon lamp) (Livingston, Scotland). The electrochemical impedance spectroscopy (EIS) results of the devices were collected with a Zennium electrochemistry workstation (Zahner, Berlin, Germany). The photoluminescence (PL) spectra of the perovskite films prepared on glasses were recorded via the Edinburgh Instrument FLS920, and the excitation light source was a 450 W xenon lamp. Then the current density–voltage (J–V) curves and corresponding photovoltaic parameters of perovskite solar cells (PSCs, the active area was 0.09 $cm^2$) were achieved with a black mask under a simulated illumination (standard AM1.5G,

1000 W/m$^2$), and a silicon photodiode was used for calibrating the simulated illumination. The operational stability of the prepared PSCs was performed under continuous white-light LED array illumination in a dry atmosphere (RH $\leq$ 10%), and the test temperature was controlled at around 30 $\pm$ 5 °C.

## 3. Results and Discussion

Currently, the solvent system employed for the preparation of large-area perovskite films via vacuum/nitrogen knife-assisted crystallization blade-coating necessitates the use of volatile solvents at ambient temperature and pressure, such as EtOH and ACN. However, the nucleation and growth processes of perovskite crystals cannot be effectively regulated solely through the use of highly volatile solvents, thereby constraining their applicability in the scalable production of high-quality perovskite films. In this study, we employ EtOH with high volatility as the primary solvent, incrementally incorporating small quantities of DMF and $NH_4Cl$ to maintain an adequate volatilization rate. The rapid evaporation of a substantial volume of solvent facilitates the swift and uniform nucleation of perovskite crystals. Concurrently, the formation of a minor DMF/$NH_4Cl$-related intermediate phase within the precursor films enhances the controllability of perovskite crystal growth during the post-annealing process. Based on previous research [4], three kinds of precursor solutions were prepared for this study: EtOH, EtOH + DMF (with a molar ratio of DMF/$Pb^{2+}$ = 0.8), and EtOH + DMF + $NH_4Cl$ (with a molar ratio of $NH_4Cl/Pb^{2+}$ = 0.2). As illustrated in Figure 1a, wet films were initially formed by coating these precursor solutions onto a clean substrate at a coating speed of 20 mm/s in a controlled dry environment (relative humidity RH $\leq$ 10%). Subsequently, the coated substrates were placed in a vacuum chamber for a brief period of 5 seconds. Then, the samples were removed onto the heating table for a subsequent annealing treatment at 100 °C for a duration of 10 minutes. Figure 1b–d illustrate the surface morphologies of the perovskite thin films synthesized from three distinct precursor solutions.

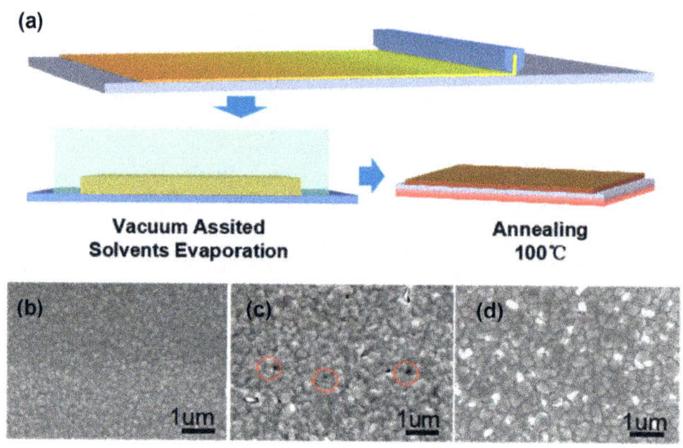


**Figure 1.** (**a**) Schematic illustration of the coating process for perovskite films. Top-view SEM images of the annealed perovskite films prepared (**b**) without the additives and with (**c**) DMF and (**d**) DMF + $NH_4Cl$.

The perovskite film synthesized from the precursor solution containing solely EtOH exhibits a flat surface and a dense microstructure [28,29]. The grain sizes within the film are uniformly distributed, albeit relatively small, ranging from 50 to 100 nm. This limited grain size is primarily attributed to the rapid volatilization of the solvent and the consequent increase in solution concentration, conditions under which nucleation predominantly occurs within the precursor films. Within a brief duration, a dense microstructure develops in the film, thereby inhibiting the crystal growth process. Concurrently, crystal growth ceases entirely once the solvent has fully volatilized. These factors collectively result in a smooth and compact microstructure of the film, albeit with a small grain size. During this process, the introduction of a minor quantity of DMF facilitates the gradual transformation of the mesophase structure present in the precursor film into a perovskite crystal structure during the post-annealing stage. Concurrently, DMF progressively volatilizes and dissolves the crystal boundaries, leading to the formation of larger crystal particles. Nevertheless, due to the low volatility of DMF, a limited number of voids are observed in the prepared film, as illustrated in Figure 1c. Subsequently, a minor quantity of $NH_4Cl$ was introduced into the precursor solution, resulting in an enhancement of the crystal particles within the film. Concurrently, the surface morphology of the film exhibited significant improvement. To further investigate the dynamic process of crystal growth in the perovskite thin film, this study analyzed the crystal structure of three types of perovskite thin films both before and after annealing, utilizing X-ray diffraction spectra, as illustrated in Figure 2a,b.

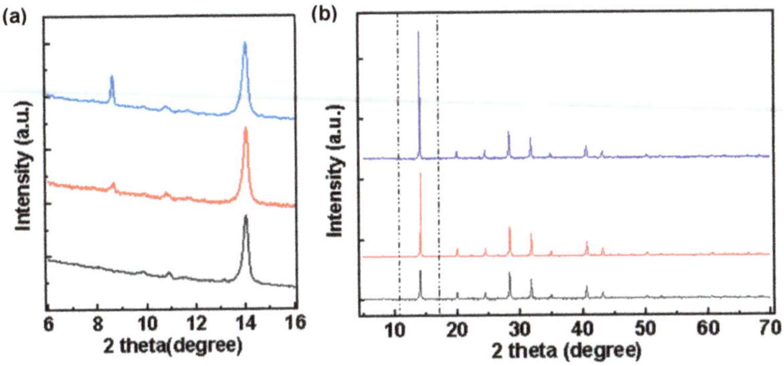


**Figure 2.** (**a**) X-ray diffraction (XRD) patterns of unannealed perovskite films without the additives (black curve) and with DMF (red curve) and DMF + NH4Cl (blue curve); (**b**) X-ray diffraction (XRD) patterns of annealed perovskite films without the additives (black curve) and with DMF (red curve) and DMF + $NH_4Cl$ (blue curve).

Figure 2a presents the X-ray diffraction (XRD) spectra of various perovskite films, both before and after annealing at 100 °C for 10 minutes. In the sample devoid of DMF and $NH_4Cl$, the precursor film exhibits only a weak diffraction peak corresponding to the $\alpha$-phase $FAPbI_3$ perovskite. This indicates the presence of a substantial number of fine $\alpha$-phase perovskite crystal nuclei within the film. The full width at half maximum (FWHM) of the primary XRD peak associated with the $\alpha$-phase crystal nuclei is 14.1°, and the crystal nucleus size, as calculated from the FWHM, is approximately 50 nm. With the addition of DMF to the precursor film, the relative integral area of the $\alpha$-phase perovskite nucleus (100) correlated XRD diffraction peak (approximately 14.1°) progressively decreases. This observation indicates a gradual reduction in the proportion of the $\alpha$-phase crystal (100) component within the film [4,30–32]. Concurrently, a secondary phase diffraction peak emerges between 8° and 9°, which, according to literature reports, is likely attributable to the formation of an intermediate phase resulting from the coordination of DMF with $FAPbI_3$. Upon the addition of a small quantity of $NH_4Cl$ to the precursor solution, there is a progressive enhancement in the diffraction peak intensity of the secondary phase relative to the $\alpha$-phase perovskite crystal nucleus. This indicates an increased relative content of the intermediate phase within the surface precursor film, corroborating the aforementioned analytical results. Post-annealing, the diffraction peaks corresponding to DMF and the intermediate phase are no longer observed, indicating a complete transformation of all products into $\alpha$-phase perovskite crystals. After adding the $NH_4Cl$, the intensity of the corresponding 14.1° diffraction peak progressively increases, while the half-height width correspondingly decreases. This observation suggests that the size of the perovskite crystals within the film enlarges, corroborating the SEM test results. Consequently, the presence of the intermediate phase appears to facilitate the crystal growth of the perovskite film during the subsequent annealing process.

Subsequently, in light of the aforementioned conclusions, an analysis of the optical properties of perovskite thin films synthesized from the three precursor solutions was conducted. Figure 3a presents the absorption characteristic curve of the perovskite thin film. The absorption spectra of the perovskite thin films, fabricated using a precursor solution with pure EtOH as the primary solvent, predominantly span the range of 450 nm to 800 nm. The intersection of the absorption curve with the X-axis occurs at approximately 780 nm. This observation aligns closely with the absorption characteristic curve of perovskite thin films reported in the literature, suggesting that the thin film primarily comprises α-phase perovskite crystals [31]. Simultaneously, the absorption intensity of the film sample is significantly enhanced by the addition of DMF, primarily due to the increased crystal particle size within the film. Additionally, it was observed that the slope of the absorption edge (approximately 780 nm) in the film's absorption curve is steeper compared to the aforementioned samples. This phenomenon is predominantly attributed to the effective improvement in the microstructure of the perovskite film following the incorporation of DMF. Furthermore, the absorption intensity of the thin film was significantly enhanced, and the slope of the absorption edge of the absorption curve was also improved following the addition of an appropriate amount of $NH_4Cl$ to the precursor solution. This suggests that the incorporation of $NH_4Cl$ further optimized the microstructure of the perovskite thin film [33]. Figure 3b presents the photoluminescence curves of the film samples prepared from the three different precursor solutions. The addition of a small amount of DMF and $NH_4Cl$ to the precursor solution results in a gradual increase in the luminescence intensity of the thin film sample, suggesting an enhancement in the crystal structure of the synthesized $FAPbI_3$ thin film. This observation aligns with the findings from XRD and SEM analyses, which further corroborate that the incorporation of DMF and $NH_4Cl$ in the precursor solution facilitates the improvement of the quality and controllability of $FAPbI_3$ thin films.

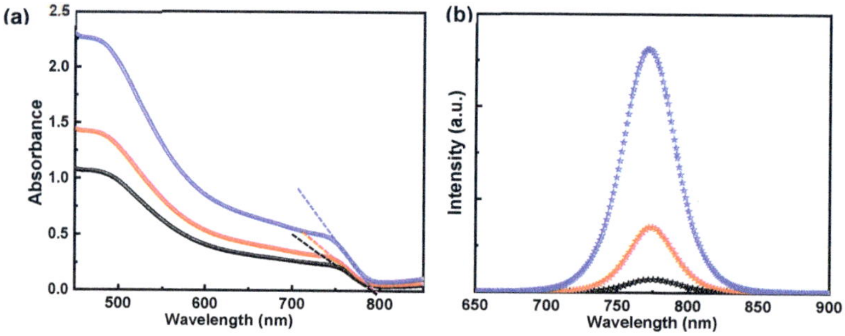


**Figure 3.** (**a**) Absorption spectra of the annealed perovskite films on glasses without the additives (black curve) and with DMF (red curve) and DMF + $NH_4Cl$ (blue curve). (**b**) Photoluminescence (PL) spectroscopy of annealed perovskite films on glasses without the additives (black curve) and with DMF (red curve) and DMF + $NH_4Cl$ (blue curve).

To further analyze the effects of DMF and $NH_4Cl$ on the microstructure of $FAPbI_3$ films, Figure 4a presents the grain size distribution statistics obtained from SEM images of the thin film samples. It is evident that compared to the $FAPbI_3$ film prepared using a DMF precursor solution, the grain sizes of the sample prepared with an $NH_4Cl$ precursor solution are generally larger and exhibit a more concentrated distribution. These findings indicate that the incorporation of DMF and $NH_4Cl$ enhances the controllability of perovskite crystal formation, which is advantageous for subsequent scalability. Simultaneously, Atomic Force Microscopy (AFM) images of perovskite films (Figure 4b,c) reveal that the surface roughness of the $FAPbI_3$ film prepared using the $NH_4Cl$ precursor solution is lower compared to that of the $FAPbI_3$ film prepared with the DMF precursor solution [34]. Consequently, the flatness of the film surface is enhanced with the addition of $NH_4Cl$, which establishes a foundation for further improving the carrier transport performance of photovoltaic devices. Additionally, Figure 4d presents the cross-sectional Scanning Electron Microscopy (SEM) images of the film samples. The presence of numerous voids at the interface between the $FAPbI_3$ film prepared using a DMF precursor solution and the substrate is evident, which is likely to result in suboptimal carrier transport within photovoltaic devices. Conversely, the incorporation of $NH_4Cl$ significantly reduces the voids at the interface of the as-prepared perovskite thin film samples and the substrate. Numerous studies indicate that this phenomenon is primarily attributed to the volatilization of DMF film samples during the annealing process, which leads to the formation of numerous voids at the film's base. Consequently, the introduction of a small quantity of $NH_4Cl$ enhances the interface morphology between the film and the substrate. This improvement is crucial for optimizing carrier transport.

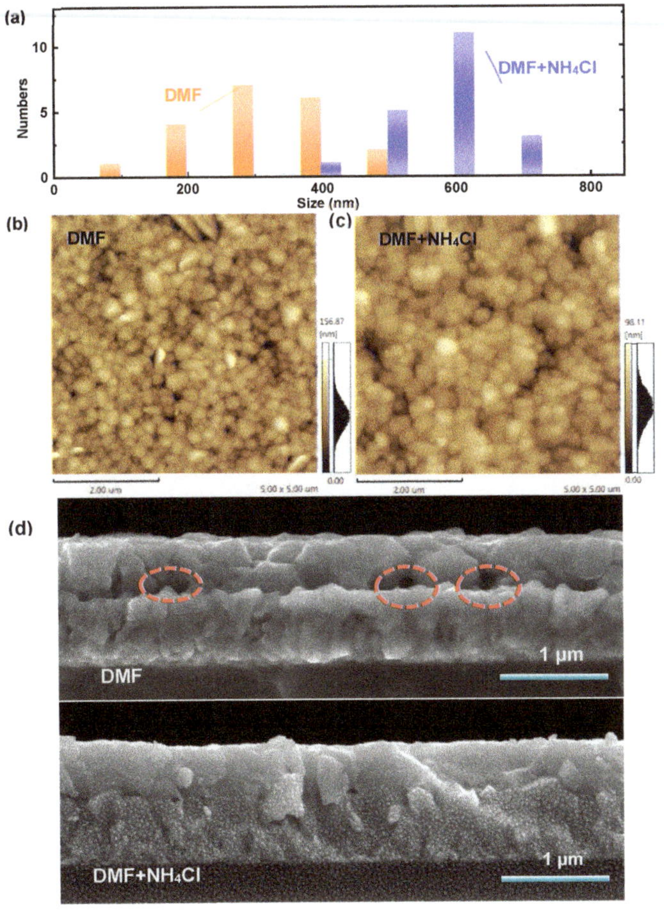


**Figure 4.** (**a**) Statistical grain size of perovskite crystals in the annealed films prepared with DMF and DMF + $NH_4Cl$. Atomic force microscope (AFM) images of the annealed perovskite films with (**b**) DMF and (**c**) DMF + $NH_4Cl$. (**d**) Cross-SEM images of the annealed perovskite films with DMF and DMF + NH4Cl.

To further investigate the transport properties of carriers in photovoltaic devices, reverse-structure thin film photovoltaic devices were fabricated utilizing the aforementioned perovskite thin films. The architecture of these thin film photovoltaic devices comprises ITO glass/NiO layer (hole transport layer, HTL)/$FAPbI_3$/$C_{60}$ (electron transport layer, ETL)/BCP/Ag. Cross-sectional SEM images of the photovoltaic device are presented in Figure 5a, with an active area of 0.09 $cm^2$. Figure 5 illustrates the voltage–current characteristics of the photovoltaic device under dark conditions. The data indicate a decrease in the reverse saturation current density with the incorporation of $NH_4Cl$. This observation, when integrated with the analysis of the thin film microstructure, suggests that the addition of $NH_4Cl$ mitigates defects within the perovskite thin film. Consequently, this reduction in defects diminishes carrier recombination within the device, thereby enhancing carrier transport properties. These improvements are anticipated to significantly elevate the photovoltaic performance of the perovskite thin film solar cell. Simultaneously, electrochemical impedance spectroscopy (EIS) measurements were conducted (Figure 5c) to evaluate the charge transfer and recombination processes in perovskite solar cells (PSCs) fabricated using various precursor solutions. The charge recombination resistance (Rrec) of the PSC prepared with DMF + NH4Cl was found to be higher than that of the sample prepared with DMF alone, suggesting that the interface characteristics of the thin film photovoltaic devices are substantially improved upon the incorporation of $NH_4Cl$. This result is consistent with the analytical results obtained from the cross-sectional SEM images of the aforementioned thin film samples. Moreover, enhanced charge extraction can contribute to improvements in both the short-circuit current density ($J_{sc}$) and the open-circuit voltage ($V_{oc}$) of photovoltaic devices.

Furthermore, the J–V characteristic curve of the photovoltaic device under AM1.5 illumination is presented in Figure 5d,e, with detailed parameters provided in Table 1. The J–V characteristic curve of the photovoltaic device sample incorporating DMF exhibits a pronounced S-shaped distortion, suggesting inferior photo-generated carrier transportation under illuminated conditions. Under illumination, the photovoltaic device exhibits an open-circuit voltage ($V_{OC}$) of 1.09 V, a short-circuit current density ($J_{SC}$) of 18.76 mA/$cm^2$, a fill factor (*FF*) of 64.80%, and a PCE of 13.25%. After the addition of $NH_4Cl$, the current–voltage (J–V) characteristic curve of the thin film photovoltaic device exhibits a progressive flattening. As a result, the open-circuit voltage increases to 1.1 V, the short-circuit current density rises to 22.86 mA/$cm^2$, and the PCE improves to 20.12%. Simultaneously, the relatively concentrated forward and reverse scanning results of photovoltaic devices with $NH_4Cl$ addition suggest a high degree of repeatability in the process method. The observed decrease in photoelectric conversion efficiency may be attributed to an increase in the internal resistance of series cells. In subsequent stages, the photovoltaic efficiency can be enhanced by further adjusting the parameters of the precursor solution and optimizing the interface contact to minimize performance losses. Simultaneously, Figure 6a,b illustrate the statistical power conversion efficiency (PCE) of solar cells fabricated with and without the incorporation of $NH_4Cl$ into the precursor solution, based on ten samples prepared under each set of conditions. The results indicate that the PCE of samples prepared without $NH_4Cl$ ranges from approximately 4% to 7%, whereas the PCE of samples prepared with $NH_4Cl$ ranges from approximately 16% to 21%, and the related parameters are shown in Table S1. These results further demonstrate that the addition of $NH_4Cl$ significantly enhances the PCE and repeatability of perovskite thin film solar cells, thereby facilitating the scalable fabrication of PSCs.

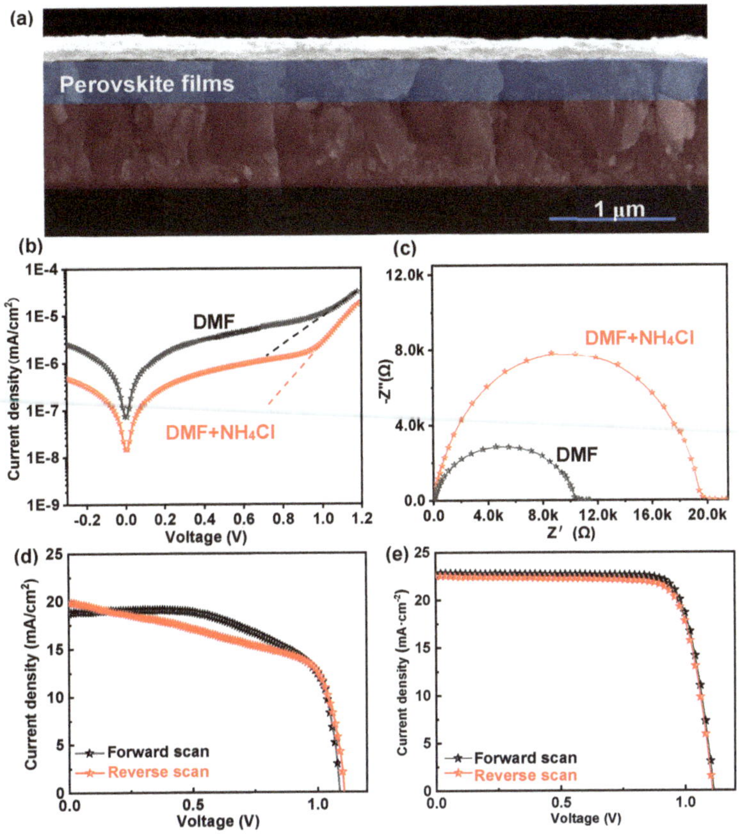


**Figure 5.** (**a**) Cross-SEM images of the prepared perovskite film solar cells. (**b**) Dark J–V curves of the PSCs prepared with DMF and DMF + $NH_4Cl$. (**c**) Nyquist plots of the PSCs prepared with DMF and DMF + $NH_4Cl$ in the frequency range of 20 kHz to 0.1 kHz under standard sunlight irradiation (100 mW/cm$^2$) at a bias voltage of 0.8 V. J–V curves for the champion PSCs prepared with (**d**) DMF and (**e**) DMF + $NH_4Cl$.

**Table 1.** Photovoltaic parameters of the champion PSCs fabricated with precursor solutions containing DMF and DMF + $NH_4Cl$.

| | | $J_{SC}$ (mA/cm$^2$) | $V_{OC}$ (V) | *FF* (%) | PCE (%) |
|---|---|---|---|---|---|
| DMF | Forward scan | 18.76 | 1.09 | 64.8% | 13.25% |
| | Reverse scan | 19.90 | 1.10 | 59.6% | 13.05% |
| DMF + $NH_4Cl$ | Forward scan | 22.86 | 1.11 | 79.3% | 20.12% |
| | Reverse scan | 22.46 | 1.10 | 79.9% | 19.74% |

Furthermore, we evaluated the operational stability of the photovoltaic device under continuous illumination. To enhance stability, a 20 nm layer of bismuth (Bi) was applied as an inert buffer layer before the evaporation of the silver (Ag) electrode [35,36]. This buffer layer serves to mitigate iodine-induced corrosion of the Ag electrode, which can result from the decomposition and diffusion of perovskite. The maximum power point (MPP) of the photovoltaic device was continuously monitored and assessed using a white LED as the aging light source, with an equivalent light intensity of 1 sun. The module temperature was maintained at 30 ± 5 °C, as depicted in Figure 6. After subjecting the aforementioned small photovoltaic device, with an effective area of 0.09 $cm^2$, to 500 hours of continuous illumination, the photoelectric conversion efficiency of the device incorporating $NH_4Cl$ remains at 95.6% of its initial efficiency. In contrast, photovoltaic devices fabricated using DMF exhibit a more pronounced efficiency decay. Consequently, the incorporation of $NH_4Cl$ enhances the stability of thin film photovoltaic devices, primarily due to the high-quality, low-defect-density perovskite thin film and the improved carrier transport mechanism at the $NiO/C_{60}$ stable interface.

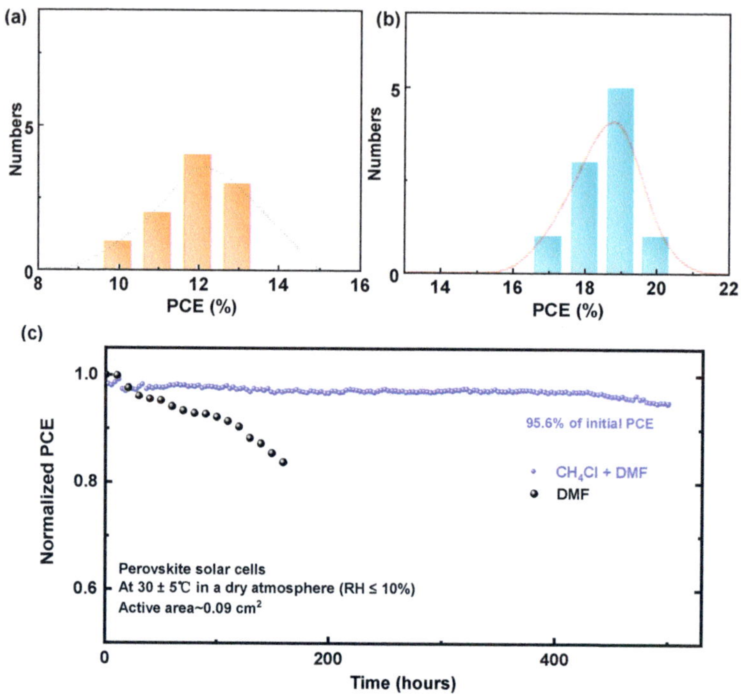


**Figure 6.** PCE distribution of PSCs (ten samples) fabricated precursor solutions contained (**a**) DMF and (**b**) DMF + $NH_4Cl$. (**c**) Operational stability of the PSCs exposed under continuous 1 sun illumination (the light source for the aging test was achieved by a 1-sun equivalent white-light LED array) with MPP tracking in dry air (RH $\leq$ 10%).

## 4. Conclusions

In this study, high-quality $FAPbI_3$ perovskite thin films were synthesized using a coating method. The incorporation of DMF and $NH_4Cl$ enhanced the stability and temporal window of the wet film and moderated the crystal growth rate within the film. This approach increased the content of the intermediate phase in the wet film, allowing perovskite grains to grow further during the annealing process. Consequently, the grain sizes were enlarged, the density of grain boundary defects was reduced, and the crystalline quality of the film was significantly improved.

Furthermore, the interfacial contact between the perovskite film and the substrate is significantly enhanced through the incorporation of liquid DMF and solid $NH_4Cl$. An analysis of the carrier transport mechanism under dark conditions reveals that the high-quality $FAPbI_3$ perovskite thin film, along with the improved interface performance, enhances the carrier transport properties of the corresponding photovoltaic device. Based on these findings, under simulated illumination conditions, the photovoltaic performance parameters of the thin film device were enhanced by the incorporation of a small amount of DMF and $NH_4Cl$, resulting in a photoelectric conversion efficiency of 20.12%. Furthermore, the decay curve of the photoelectric conversion efficiency under continuous irradiation with an equivalent 1-sun white-light source indicates that the improved microstructure of the perovskite thin film achieved through the addition of DMF and $NH_4Cl$ also enhances the operational stability of the photovoltaic device. The results indicate that solid Lewis base additives enable the control and scalability of perovskite crystal growth, which is crucial for advancing the industrialization of perovskite thin film photovoltaic devices.

**Supplementary Materials:** The following supporting information can be downloaded at https://www.mdpi.com/article/10.3390/coatings14091113/s1, Table S1. Related photovoltaic parameters of PSCs fabricated with precursor solutions containing DMF and DMF + CH4Cl.

**Author Contributions:** Conceptualization, Z.J. and Q.M.; methodology, J.L.; software, Y.Z. (Yufei Zhang) and Y.Z. (Yulong Zhang); validation, B.W.; formal analysis, Z.Q.; investigation, Z.Q.; resources, W.Z.; data curation, R.Z.; writing—original draft preparation, Z.J.; writing—review and editing, Z.J.; visualization, supervision, Z.J.; project administration, K.L. and Q.M. All authors have read and agreed to the published version of the manuscript.

**Funding:** This research was funded by Taiyuanuniversity of Science and Technology Scientific Research Initial Funding, grant number 20232091, the Fundamental Research Program of Shanxi Provinces (No. 0210302123221), the Fundamental Research Program of Shanxi Provinces (No. 202203021222207), the Fundamental Research Program of Shanxi Province (202103021224268), the Special Fund for Science and Technology Innovation Teams of Shanxi Province (No. 02204051001002), and the Fundamental Research Program of Shanxi Province (No. 202203021221146).

**Institutional Review Board Statement:** Not applicable.

**Informed Consent Statement:** Not applicable.

**Data Availability Statement:** Data are contained within the article/Supplementary Materials; further inquiries can be directed to the corresponding author.

**Acknowledgments:** We gratefully acknowledge Beihang University and Hangzhou Acme Optoelectronics Technology Co., Ltd. for providing access to their experimental equipment to fabricate samples and their measurements.

**Conflicts of Interest:** The authors declare that they have no known competing financial interests or personal relationships that could have appeared to influence the work reported in this paper.

## References

1. Kojima, A.; Teshima, K.; Shirai, Y.; Miyasaka, T. Organometal halide perovskites as visible-light sensitizers for photovoltaic cells. *J. Am. Chem. Soc.* **2009**, *131*, 6050–6051. [CrossRef] [PubMed]
2. Green, M.A.; Ho-Baillie, A.; Snaith, H.J. The emergence of perovskite solar cells. *Nat. Photonics* **2014**, *8*, 506–514. [CrossRef]
3. Green, M.A. Commercial progress and challenges for photovoltaics. *Nat. Energy* **2016**, *1*, 15015. [CrossRef]
4. Jiang, Z.; Wang, B.; Zhang, W.; Yang, Z.; Li, M.; Ren, F.; Imran, T.; Sun, Z.; Zhang, S.; Zhang, Y. Solvent engineering towards scalable fabrication of high-quality perovskite films for efficient solar modules. *J. Energy Chem.* **2023**, *80*, 689–710. [CrossRef]
5. Chen, R.; Zhang, W.; Guan, X.; Raza, H.; Zhang, S.; Zhang, Y.; Troshin, P.A.; Kuklin, S.A.; Liu, Z.; Chen, W. Rear electrode materials for perovskite solar cells. *Adv. Funct. Mater.* **2022**, *32*, 2200651. [CrossRef]
6. Yang, Z.; Zhang, W.; Wu, S.; Zhu, H.; Liu, Z.; Liu, Z.; Jiang, Z.; Chen, R.; Zhou, J.; Lu, Q. Slot-die coating large-area formamidinium-cesium perovskite film for efficient and stable parallel solar module. *Sci. Adv.* **2021**, *7*, 3749. [CrossRef]
7. Kam, M.; Zhu, Y.; Zhang, D.; Gu, L.; Chen, J.; Fan, Z. Efficient mixed-cation mixed-halide perovskite solar cells by all-vacuum sequential deposition using metal oxide electron transport layer. *Sol. RRL* **2019**, *3*, 1900050. [CrossRef]
8. Deng, Y.; Xu, S.; Chen, S.; Xiao, X.; Zhao, J.J. Huang, Defect compensation in formamidinium–caesium perovskites for highly efficient solar mini-modules with improved photostability. *Nat. Energy* **2021**, *6*, 633–641. [CrossRef]
9. Bu, T.; Li, J.; Li, H.; Tian, C.; Su, J.; Tong, G.; Ono, L.K.; Wang, C.; Lin, Z.; Chai, N. Lead halide–templated crystallization of methylamine-free perovskite for efficient photovoltaic modules. *Science* **2021**, *372*, 1327–1332. [CrossRef]
10. Feng, Q.; Huang, X.; Tang, Z.; Hou, Y.; Chang, Q.; Nie, S.; Cao, F.; Niu, X.; Yin, J.; Li, J. Governing PbI6 octahedral frameworks for high-stability perovskite solar modules. *Energy Environ. Sci.* **2022**, *15*, 4404–4413. [CrossRef]
11. Si, H.; Liao, Q.; Kang, Z.; Ou, Y.; Meng, J.; Liu, Y.; Zhang, Z.; Zhang, Y. Deciphering the $NH_4PbI_3$ Intermediate Phase for Simultaneous Improvement on Nucleation and Crystal Growth of Perovskite. *Adv. Funct. Mater.* **2017**, *17*, 1701804. [CrossRef]
12. Deng, Y.; Zheng, X.; Bai, Y.; Wang, Q.; Zhao, J.; Huang, J. Surfactant-controlled ink drying enables high-speed deposition of perovskite films for efficient photovoltaic modules. *Nat. Energy* **2018**, *3*, 560–566. [CrossRef]

13. Guo, F.; Qiu, S.; Hu, J.; Wang, H.; Cai, B.; Li, J.; Yuan, X.; Liu, X.; Forberich, K.; BrabecC, J. A Generalized Crystallization Protocol for Scalable Deposition of High-Quality Perovskite Thin Films for Photovoltaic Applications. *Adv. Sci.* **2019**, *6*, 1901067. [CrossRef] [PubMed]
14. Deng, Y.; Van Brackle, C.H.; Dai, X.; Zhao, J.; Chen, B.; Huang, J. Tailoring solvent coordination for high-speed, room-temperature blading of perovskite photovoltaic films. *Sci. Adv.* **2019**, *5*, 7537. [CrossRef] [PubMed]
15. Gao, Y.; Wu, Y.; Lu, H.; Chen, C.; Liu, Y.; Bai, X.; Yang, L.; Yu, W.W.; Dai, Q.; Zhang, Y. CsPbBr3 perovskite nanoparticles as additive for environmentally stable perovskite solar cells with 20.46% efficiency. *Nano Energy* **2019**, *59*, 517–526. [CrossRef]
16. Lee, J.W.; Bae, S.H.; Marco, N.D.; Hsieh, Y.T.; Dai, Z.; Yang, Y. The role of grain boundaries in perovskite solar cells. *Mater. Today Energy* **2018**, *7*, 149–160. [CrossRef]
17. Wang, B.; Wong, K.Y.; Yang, S.; Chen, T. Crystallinity and defect state engineering in organo-lead halide perovskite for high-efficiency solar cells. *J. Mater. Chem. A* **2016**, *4*, 3806–3812. [CrossRef]
18. Huang, F.; Dkhissi, Y.; Huang, W.; Xiao, M.; Benesperi, L.; Rubanov, S.; Rubanov, S.; Zhu, Y.; Lin, X.; Jiang, L.; et al. Gas-assisted preparation of lead iodide perovskite films consisting of a monolayer of single crystalline grains for high efficiency planar solar cells. *Nano Energy* **2014**, *10*, 10–18. [CrossRef]
19. Wang, M.; Cao, F.; Deng, K.; Li, L. Adduct phases induced controlled crystallization for mixed-cation perovskite solar cells with efficiency over 21%. *Nano Energy* **2019**, *63*, 103867. [CrossRef]
20. Ma, Q.; Zhang, Y.L.; Lu, C.Q.; Zhang, R.; Wang, X.J.; Zhang, W.J.; Jiang, Z.Y. MACl-induced controlled crystallization in sequentially deposited perovskites for high-efficiency and stable perovskite solar cells. *Coatings* **2023**, *13*, 1885. [CrossRef]
21. Docampo, P.; HanuschF, C.; Stranks, S.D.; Doblinger, M.; Feckl, J.M.; Ehrensperger, M.; Minar, N.K.; Johnston, M.B.; Snaith, H.J.; Bein, T. Solution Deposition-Conversion for Planar Heterojunction Mixed Halide Perovskite Solar Cells. *Adv. Energy Mater.* **2014**, *4*, 1400355. [CrossRef]
22. Chen, J.; Zuo, L.; Zhang, Y.; Lian, X.; Fu, W.; Yan, J.; Li, J.; Wu, G.; Li, C.Z.; Chen, H. High-Performance Thickness Insensitive Perovskite Solar Cells with Enhanced Moisture Stability. *Adv. Energy Mater.* **2018**, *8*, 1800438. [CrossRef]
23. Zhang, Y.; Jiang, Z.; Zhang, W.; Yan, L.; Lu, C.; Ni, C. Pre-crystallization applied in two-step spin-coating method for high-quality perovskite films in photovoltaic devices. *J. Alloys Compd.* **2020**, *832*, 153616. [CrossRef]
24. Zhao, Y.; Duan, J.; Wang, Y.; Yang, X.; Tang, Q. Precise stress control of inorganic perovskite films for carbon-based solar cells with an ultrahigh voltage of 1.622 V. *Nano Energy* **2020**, *67*, 104286. [CrossRef]
25. Liu, J.; Zhu, L.; Xiang, S.; Wei, Y.; Xie, M.; Liu, H.; Li, W.; Chen, H. Growing high-quality CsPbBr3 by using porous CsPb2Br5 as an intermediate: A promising light absorber in carbon-based perovskite solar cells. *Sustain. Energy Fuels* **2019**, *3*, 184–194. [CrossRef]
26. Liu, X.; Liu, Z.; Ye, H.; Tu, Y.; Sun, B.; Tan, X.; Shi, T.; Tang, Z.; Liao, G. Novel efficient C60-based inverted perovskite solar cells with negligible hysteresis. *Electrochim. Acta* **2018**, *288*, 115–125. [CrossRef]
27. Teng, P.; Han, X.; Li, J.; Xu, Y.; Kang, L.; Wang, Y.; Yang, Y.; Yu, T. Elegant Face-Down Liquid-Space-Restricted Deposition of CsPbBr3 Films for Efficient Carbon-Based All-Inorganic Planar Perovskite Solar Cells. *ACS Appl. Mater. Inter.* **2018**, *10*, 9541. [CrossRef]
28. Luo, J.; Jia, C.; Wan, Z.; Han, F.; Zhao, B.; Wang, R. The novel dopant for hole-transporting material opens a new processing route to efficiently reduce hysteresis and improve stability of planar perovskite solar cells. *J. Power Sources* **2017**, *342*, 886–895. [CrossRef]
29. Shi, D.; Adinolfi, V.; Comin, R.; Yuan, M.; Alarousu, E.; Buin, A.; Chen, Y.; Hoogland, S.; Rothenberger, A.; Katsiev, K.; et al. Low trap-state density and long carrier diffusion in organolead trihalide perovskite single crystals. *Science* **2015**, *347*, 519–522. [CrossRef]
30. Xiao, Z.; Dong, Q.; Bi, C.; Shao, Y.; Yuan, Y.; Huang, J. Solvent Annealing of Perovskite-Induced Crystal Growth for Photovoltaic-Device Efficiency Enhancement. *Adv. Mater.* **2014**, *26*, 6503–6509. [CrossRef]
31. Li, L.; Chen, Y.; Liu, Z.; Chen, Q.; Wang, X.; Zhou, H. The additive coordination effect on hybrids perovskite crystallization and high-performance solar cell. *Adv. Mater.* **2016**, *28*, 9862–9868. [CrossRef] [PubMed]
32. Saliba, M.; Etgar, L. Current density mismatch in perovskite solar cells. *ACS Energy Lett.* **2020**, *5*, 2886–2888. [CrossRef]
33. Zhang, Y.L.; Jiang, Z.Y.; Li, J.C.; Meng, G.X.; Guo, J.J.; Zhang, W.J. Preparation of efficiency and stability all-inorganic CsPbBr3 perovskite solar cells by pre-crystallization multi-step spin coating method. *Coatings* **2024**, *14*, 918. [CrossRef]
34. Li, M.; Li, B.; Cao, G.; Tian, J. Monolithic MAPbI3 films for high-efficiency solar cells via coordination and a heat assisted process. *J. Mater. Chem. A* **2017**, *40*, 21313–21319. [CrossRef]
35. Qiu, J.J.; Yang, T.T.; Li, Y.F.; Qian, W.H.; Liu, X.Y. Au@Ag@Pt core–shell nanorods regulating Ag release behavior endow titanium antibacterial activity and biocompatibility. *Rare Met.* **2022**, *41*, 630–638. [CrossRef]
36. Aristidou, N.; Eames, C.; Sanchez-Molina, I.; Bu, X.; Kosco, J.; Islam, M.S.; Haque, S.A. Fast oxygen diffusion and iodide defects mediate oxygen-induced degradation of perovskite solar cells. *Nat. Commun.* **2017**, *8*, 15218. [CrossRef]

*coatings*

MDPI

*Article*

# Effect of the Atmosphere on the Properties of Aluminum Anodizing

**Gabriela Baltierra-Costeira [1], Jesús Emilio Camporredondo-Saucedo [2], Marco Arturo García-Rentería [3,*], Lázaro Abdiel Falcón-Franco [3], Laura Guadalupe Castruita-Ávila [2] and Adrián Moisés García-Lara [2,*]**

1 Facultad de Ciencias Químicas, Universidad Autónoma de Coahuila, Blvd. Venustiano Carranza, Saltillo 25000, Coahuila, Mexico; gabriela.baltierra@uadec.edu.mx

2 Facultad de Ingeniería Mecánica y Eléctrica, Universidad Autónoma de Coahuila, Avenida Barranquilla s/n, Colonia Guadalupe, Monclova 25760, Coahuila, Mexico; emiliocamporredondo@uadec.edu.mx (J.E.C.-S.); laura_castruita@uadec.edu.mx (L.G.C.-Á.)

3 Facultad de Metalurgia, Universidad Autónoma de Coahuila, Carr. 57, km 4.5, Monclova 25710, Coahuila, Mexico; lazarofalcon@uadec.edu.mx

* Correspondence: marcogarciarenteria@uadec.edu.mx (M.A.G.-R.); adrian.garcia@uadec.edu.mx (A.M.G.-L.); Tel.: +55-8661655327 (M.A.G.-R.); +55-8662125485 (A.M.G.-L.)

**Abstract:** This study aims to quantify the effect of process parameters on the anodizing of Al6061 aluminum. To achieve this, studies on layer thickness, the porosity of the anodized surface, electrochemical techniques, X-ray diffraction, grain size estimation, and statistical analysis were conducted for three different atmospheres (without air, air, and oxygen). Parameter levels were established as follows: temperature (30 °C, 45 °C, and 60 °C), time (20 min, 40 min, and 60 min), electrolyte concentration (0.5 M), voltage (9 V), and current intensity (0.600 A). A $3^3$ experimental design (three factors, three levels) was proposed, and mathematical models were obtained using general factorial design. The experimental design was used to determine the three most important variables in the optimal condition. A total of 27 tests were conducted using sulfuric acid electrolytic solutions, of which 12 samples were selected by the factorial design method, which simultaneously evaluates the effects of factors and their interactions in a single experiment. Measurement of porosity and oxide layer thickness was performed using scanning electron microscopy. The purity of the anodic layer formed was characterized using X-ray diffraction techniques with a vertical goniometer X-ray diffractometer. The electrochemical behavior is presented through potentiodynamic polarization curves for the anodic layer. A general factorial design and an analysis of variance (ANOVA) were conducted to establish the significant factors for layer thickness, grain size, and reaction rate. Finally, the best results and their parameters for each response are presented.

**Keywords:** anodizing; aluminum alloy; Al6061; sulfuric acid; corrosion protection

## 1. Introduction

Aluminum is one of the most important non-ferrous metals in engineering, used in many chemical and mechanical processes, and produced in both pure form and as alloys. It has a strong affinity for oxygen with the standard molar heat of the formation of $Al_2O_3$ of −1,675,700 J at 25 °C $\left(2Al_{(s)} + \frac{3}{2}O_{2(g)} \rightarrow Al_2O_{3(s)}\right)$ [1]. From a physicochemical perspective, the anodizing process explains that when current passes through the electrolyte containing an aqueous solution of sulfuric acid, oxygen is released on the surface of the aluminum connected to the anode. This oxygen allows for the accelerated oxidation of the aluminum. At the cathode (AISI 304 steel), hydrogen is released. In other words, it is a process of water decomposition that generates gases at both electrodes [2]. Ion exchange is the primary mechanism for oxide formation at the metal–oxide interface. When aluminum is polarized as an anode, the basic oxidation reaction that occurs at the interface with the electrolyte can be summarized as follows:

$$2Al + 3H_2O \rightarrow Al_2O_3 + 6H^+ + 6e^- \quad (1)$$

For anodizing in a sulfuric acid electrolyte, the reaction proceeds as follows:

$$4Al + 6(H_2SO_4) \rightarrow 2Al_2O_3 + 6(SO_3^-) + 3(H_2)_g + 6H^+ + 6e^- \quad (2)$$

The simultaneous oxidation of aluminum and the electrolysis of the aqueous sulfuric acid electrolyte allows the production of sulfite ions and hydrogen gas at the interface.

Due to this high activity, in the presence of oxygen, it reacts in the environment to spontaneously form a thin natural surface layer of alumina ($Al_2O_3$), which controls the corrosion rate and protects the metal, enabling the creation of durable aluminum and aluminum alloys [3]. However, it is necessary to increase the thickness of this thin layer to protect the aluminum surface from contact with corrosive agents, because the metallic surface acquires an electric potential that causes the migration of metallic ions into the solution as cations, making the pure aluminum or its alloys prone to corrosion in the electrolyte [4]. The amount of $Al_2O_3$ formed during anodizing is directly proportional to the current density and time, while the film's growth depends on the chemical composition, electrolyte concentration, and anodizing conditions. Some electrolytes have little or no solvent action on the oxide coating, forming very thin films usually known as barrier-type coatings, whose thickness depends only on the applied voltage. This type of coating is typically produced in borate and boric acid solutions. Sulfuric acid is an electrolyte that slightly dissolves the formed coating, achieving the formation of porous films as the oxidation process continues [5]. Sulfuric acid anodizing ($H_2SO_4$) was patented by Gower et al. in 1927 for surface finishing and corrosion protection of aluminum and its alloys [6]. Since then, sulfuric acid has become one of the most popular electrolytes for forming a porous oxide on aluminum films, from basic laboratory research to industrial applications, due to the low cost of anodizing electrolytes [6–9]. It is the most widely used in industry; some researchers have proposed it, obtaining high data on thickness, hardness, and reaction rate [10–12].

The corrosion phenomenon occurs because these materials tend to reach their natural state over time, which constitutes a lower potential energy state, allowing them to stabilize thermodynamically with their environment [13]. Most metallic materials tend to show surface changes when exposed to the environment, with the most common changes being color, texture, thickness, chemical composition, and physical properties. Due to this, efforts have been made to counteract the corrosion phenomenon using paints or some electrochemical methods that can delay or prevent its appearance, such as the anodizing process, which is used to improve the surface properties of aluminum by creating a hard oxide layer that is highly wear- and corrosion-resistant [11]. Generally, the microstructure and oxide layer of aluminum depend directly on the effect of the anodizing process variables such as electrolyte nature, concentration, pH, temperature, time, current density, and voltage [14].

Based on the above, this study focused on elucidating a mathematical model for the anodizing process of commercial 6061 aluminum by systematically varying the parameters of time and temperature, with oxygen injection, air injection, and without air injection to improve the mechanical properties and stability of the anodic layer formed on the aluminum, correlating the response through electrochemical techniques such as potentiodynamic polarization.

## 2. Materials and Methods

### 2.1. Materials and Pretreatment

In this study, plates of commercial 6061 aluminum 3 mm thick were used; the chemical composition is shown in Table 1. Chemical analysis was performed by optical emission spark spectroscopy Thermo Fisher Scientific, ARL iSpark Plus. Each sample was prepared as shown in Figure 1 for handling within the electrochemical cell of the same material and then sectioned into 2 × 1 $cm^2$ samples for microstructural analysis and corrosion tests.

**Table 1.** Elemental composition of 6061 aluminum alloy.

| Elemento | S | Fe | Cu | Mn | Mg | Cr | Ni | Zn | Ti | Co | Al |
|---|---|---|---|---|---|---|---|---|---|---|---|
| % by weight | 0.420 | 0.275 | 0.033 | 0.018 | 0.493 | 0.013 | 0.007 | 0.027 | 0.018 | 0.005 | Bal |

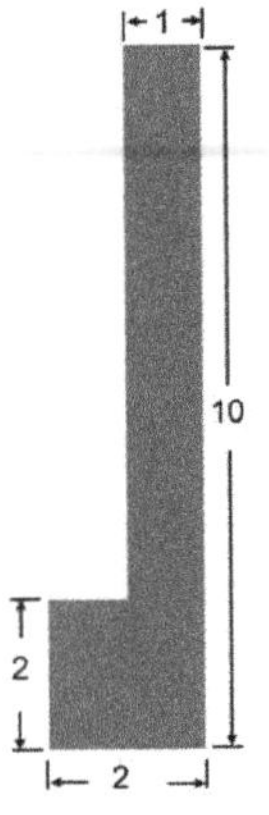


**Figure 1.** 6061 Aluminum plate geometry (cm).

Samples were subjected to mechanical preparation, cleaning in acetone for 5 min, rinsing with distilled water, then drying at room temperature and sanding with silicon carbide paper of grit sizes 320, 500, 800, 1200, and 1500 from lower to higher. Subsequently, they were polished for 3 min on cotton cloth with 1-micron white diamond paste (5 gr) to a mirror finish and stored in a plastic box. To remove impurities in the chemical preparation, they were immersed in a 10% $v/v$ sodium hydroxide (NaOH) solution for one minute, using $HNO_3$ to neutralize the caustic soda effect, and after neutralization, samples were rinsed with distilled water, then dried at room temperature and hermetically stored.

### 2.2. Anodizing

The anodizing process was carried out under direct energy supply using a two-electrode reactor with an AISI 304 stainless steel plate as the cathode and 6061 commercial aluminum sheets as the anode, maintaining a 2 cm distance between each electrode. During the entire experiment, 50 mL of 5% $v/v$ $H_2SO_4$ was used as the electrolyte in each test. The selection of operating variables during the anodizing process is shown in Table 2; fixed and manipulable according to the $3^2$ factorial design (three factors, two levels), and the mathematical model was obtained using the general factorial design.

**Table 2.** Classification of operating variables $3^3$.

| Operating Variables | Fixed | Manipulable | |
|---|---|---|---|
| Voltage (V) | ✓ | | |
| Current Density (*i*) | ✓ | | |
| Cell Type | ✓ | | |
| pH | ✓ | | |
| Interelectrode Distance (d) | ✓ | | |
| Electrolyte | ✓ | | |
| Electrolyte Concentration [ ] | ✓ | | |
| Treatment Time (*t*) | | ✓ | 20, 40, and 60 min |
| Electrolyte Temperature (T) | | ✓ | 30°, 45°, and 60 °C |
| Agitation Speed (v) | ✓ | | |
| Gas | | ✓ | Without air |
| | | ✓ | With air |
| | | ✓ | With oxygen |

The anodizing process was divided into three steps, using 5% pure sulfuric acid, 9 V voltage, and a current intensity of 0.600 A for each test. In the first step, times of 20, 40, and 60 min were used at temperatures of 30, 45, and 60 °C without air injection. In the second step, the same conditions of time and temperature were applied with air injection using an electric air compressor, HP 1.80, Max Pressure 135 psi, 120 VAC Voltage, and a specific flow rate of 5 lt/min was maintained through control valves. In the third step, oxygen was injected with the same specific flow rate of 5 lt/min and the same conditions of time and temperature. At the end of each step, the anodized electrodes were removed, the anode was washed with 1 mL of ethanol, and then stored in an airtight container, repeating the processes as necessary.

A selection of 12 samples was carried out using the general factorial design method for 3 factors, 2 levels, and high and low levels, as shown in Table 3. This method simultaneously evaluates the effects of the factors and their interactions in a single experiment, allowing for a comprehensive analysis of experimental results and identifying the best level for those factors that influence the mean of the response variable [15]. Table 4 shows the variables manipulated during the anodizing process.

**Table 3.** General factorial design, 3 factors, 2 levels ($3^2$).

| Table | A Low | B Low | C Low |
|---|---|---|---|
| 1 | A Low | B Low | C Low |
| 2 | A Low | B High | C High |
| 3 | A High | B Low | C High |

**Table 4.** Values of the variables manipulated during the anodizing process $3^3$.

| Variable | Low (−1) | Intermediate | High (+1) |
|---|---|---|---|
| Temperature (T) | 30 °C | 45 °C | 60 °C |
| Treatment time (*t*) | 30 min | 40 min | 60 min |
| | No air | No air | No air |
| Gas | No air | With air | With air |
| | With oxygen | With oxygen | With oxygen |

### 2.3. Structure and Composition of Anodic Films

In this experiment, an X-ray diffractometer with a vertical goniometer (PANalytical, model EMPYREAM) was used to analyze the structure of the anodic films using the grazing incidence technique. The 2θ range was between 20° and 90°, with a scanning speed of 1°/min. The crystallite size was estimated using the Scherrer formula $\tau = 0.9\lambda/B\cos(B\_max)$ [16]. The chemical composition of the anodic films was obtained through the integrated analytical platform with Tescan Essence, which combines SEM images with elemental composition analysis using the Tescan Mira 3 LMU scanning electron microscope (SEM). It was performed by energy dispersive spectroscopy (EDS).

### 2.4. Morphology

The thickness measurement of the anodic layer was performed using a Tescan Mira 3 LMU SEM with a backscattered electron detector (BSE) [17]. The current intensity was standardized at 15 kV with a magnification of 3500×.

### 2.5. Electrochemical Techniques

For the electrochemical characterization of the anodized aluminum surface, potentiodynamic polarization curves were carried out to determine the current density and corrosion potential. The samples were cleaned with an approximate area of 1 $cm^2$ in a CJ-008 high-frequency ultrasonic cleaner, characterized in a three-electrode cell using anodized aluminum as the working electrode, platinum as the counter electrode, and a saturated KCl Ag/AgCl reference electrode. The tests were performed with a Gamry Instruments Interface 1010E potentiostat, using a 3.5% NaCl solution as electrolyte for one hour, with pH measurements taken before and after each test using a basic Ohaus ST2200-F pH meter.

The open circuit time was 600 s, with a sweep speed of 1.66 mV/s. Cyclic polarization was performed from a 1 V cathode arm to a −1 V anode overpotential with a sweep of 1.66 mV/s.

The adjustment of the Tafel curves obtained makes it possible to determine the cathodic and anodic slopes (βa and βc), and thus determine the corrosion current density (Icorr) and the corrosion rate, calculating the localized corrosion rate (r) in accordance with ASTM G46.

## 3. Results and Discussion

Micrographs of the SEM show the morphology of the anodic aluminum oxide (AAO) films of aluminum 6061 at times of 30 and 60 min and temperatures of 30 and 60 °C are shown in Figure 2 without air injection. It was used to ensure there was no formation of sulfides and other agents that react with oxygen. It is not for quantifying the film thickness, but to identify that there is not a significant amount of $SO_4^{2-}$ on the surface, contaminant phases in the aluminum oxide film, and to search for phases that may be on the surface of the material, because the formation is thin, but impurities may be present. The same conditions with air injection are shown in Figure 3, and with oxygen injection in Figure 4, following the nomenclature T, t, condition. As shown in Figure 2a, the AAO surface formed at 30-20-SA has an irregular and non-uniform surface. Figure 2b–d show uniform and stable layers formed at 60-20-SA, 30-60-SA, and 60-60-SA, respectively. The thickness of the layer increases with shorter times, with an average thickness of 5.47 µm in the cross-section of samples without air injection.

Figure 3a, 30-20-CA y 3b, 60-20-CA show uniform formation of AAO in both micrographs, with lower film thicknesses due to acid dissolution of the anodic layer under temperature conditions of 20 °C and 60 °C with averages of 2.7 and 3.2 µm. Figure 3c (30-60-CA) has an average of 5.5 µm, while Figure 3d (60-60-CA) shows a decrease of 2.7 and 1.4 µm, respectively, with a continuous uniform and stable layer.

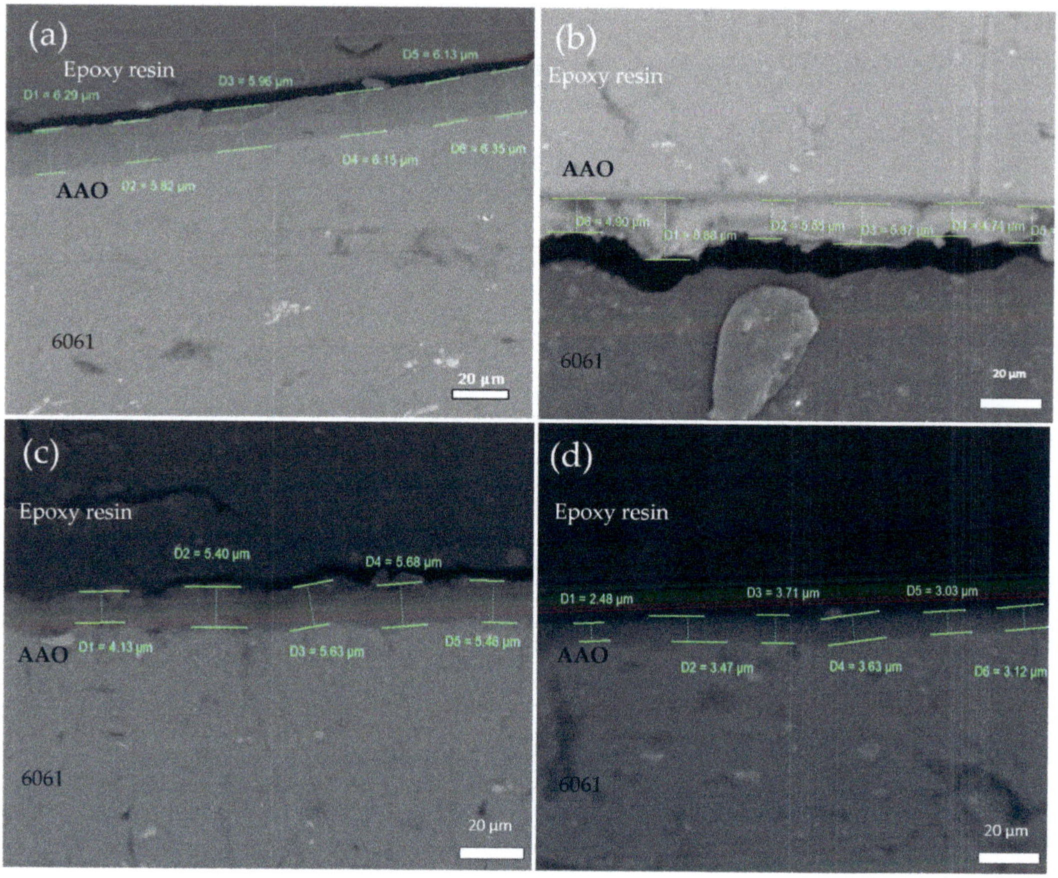


**Figure 2.** SEM cross-sectional micrograph of AAO of aluminum alloy 6061 at different times and temperatures, using sulfuric acid as the electrolyte: (**a**) 30-20-SA, (**b**) 60-20-SA, (**c**) 30-60-SA, and (**d**) 60-60-SA, without air injection.

Figure 4a (30-20-CO) shows an anodic layer formation of 3.5 µm. In Figure 4b (30-60-CO), the average thickness of the anodized layer is 3.15 µm with uniform morphology. In Figure 4c (60-20-CO), the average thickness visibly decreases to 1.4 µm, maintaining uniformity but with deformation. In Figure 4d (60-60-CO), the film formation is 1.45 µm, which is six times smaller than without air injection, as seen in Figure 2a. The growth rate of the oxide film at high temperatures was much faster than in the low-temperature electrolyte, suggesting that as temperature increased, the driving force to promote longitudinal growth of the oxide film increased with the oxygen bubble overflow and acceleration of the electrochemical reaction.

The relative percentage of the constituent elements of the AAO is shown in Table 5. EDS is a semiquantitative analysis, showing an approach to the chemical composition of the alumina films. A high percentage of sulfur is observed without air injection and low content with air and oxygen injection, due to the formation of sulfates during the anodization process.

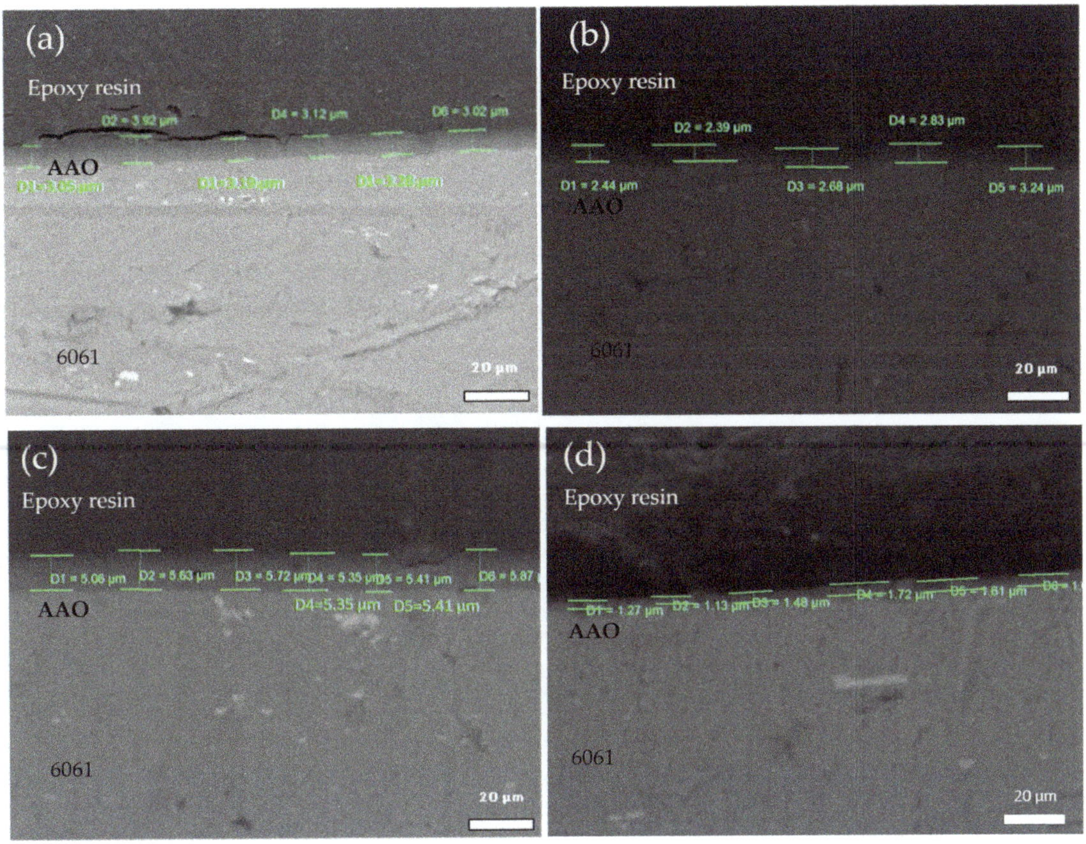


**Figure 3.** SEM cross-sectional micrograph of AAO of aluminum alloy 6061 at different times and temperatures, using sulfuric acid as the electrolyte: (**a**) 30-20-CA, (**b**) 60-20-CA, (**c**) 30-60-CA, and (**d**) 60-60-CA, with air injection.

**Table 5.** Relative percentage of elemental energy spectrum of AAO of aluminum alloy 6061.

| Element | | Atomic % | |
|---|---|---|---|
| | 30-60-SA | 30-20-CA | 30-20-CO |
| Carbon | 61.60 | 57.05 | 54.86 |
| Oxygen | 27.57 | 32.67 | 34.30 |
| Silicon | 0.01 | 0.25 | 0.83 |
| Sulfur | 10.17 | 0.52 | 0.33 |
| Aluminum | 0.64 | 9.51 | 9.69 |
| | 100 | 100 | 100 |

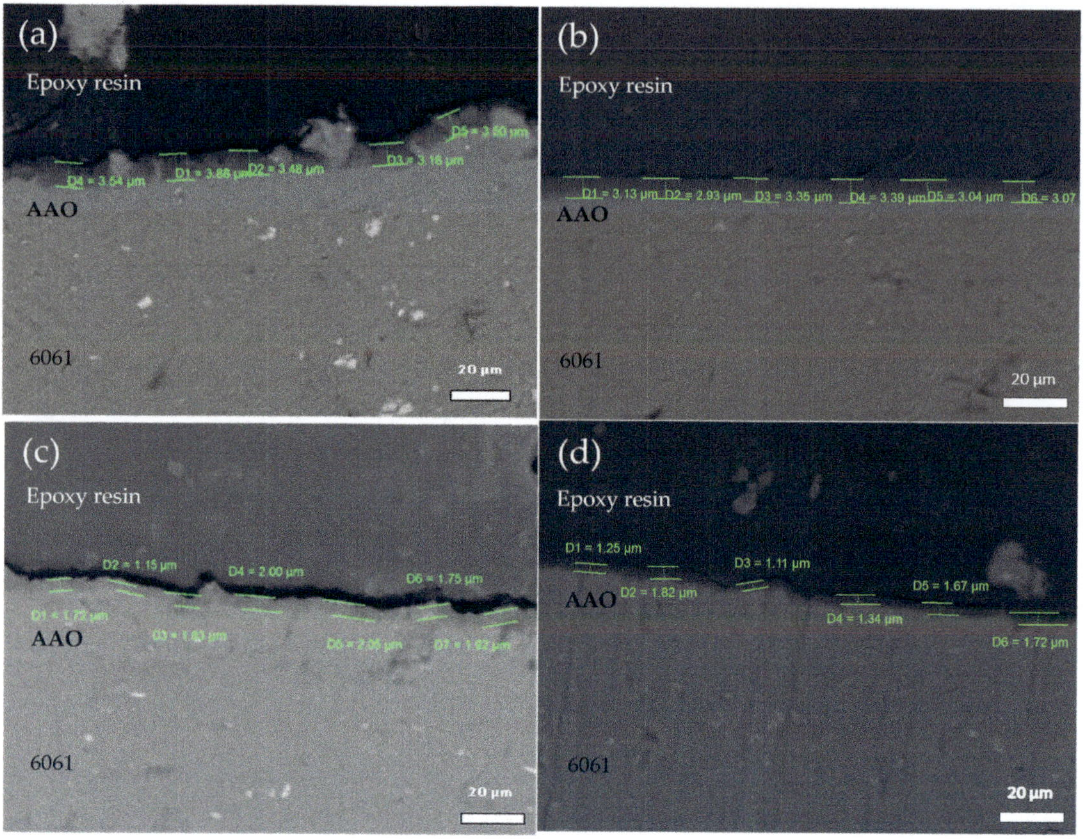


**Figure 4.** SEM cross-sectional micrograph of AAO of aluminum alloy 6061 at different times and temperatures, using sulfuric acid as the electrolyte: (**a**) 30-20-CO, (**b**) 60-20-CO, (**c**) 30-60-CO, and (**d**) 60-60-CO, with oxygen injection.

From a molecular perspective, there may be a better arrangement; the one with better electrochemical behavior might be more compact. When it is observed that the film is neither compact nor porous, it is assumed that surface defects in the growth of the oxide film may be generated due to the presence of nitrogen from the air, which can be absorbed on the surface of the aluminum or trapped during the film's growth. Sulfur ions might also be absorbed on the aluminum surface. If air is present and the temperature is increased, the growth is limited, and over time the layer may dissolve. With air, increasing the temperature favors growth, but over time it becomes more compact. With oxygen at lower temperatures, the film becomes more uniform.

The AAO formation thicknesses are shown in Figure 5 according to the sample identification, the main conditions under which the anodization process was carried out, and the anodized film thickness obtained under each established parameter. The maximum layer growth limit was 7.4 µm on the base substrate in the sample, corresponding to Figure 1a, 30-20-SA.

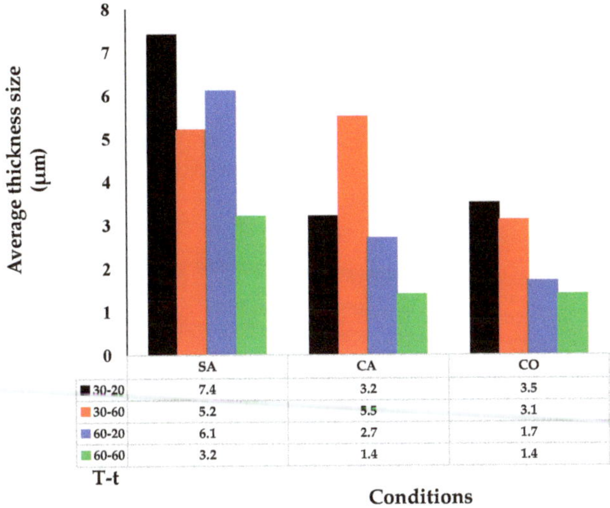


**Figure 5.** Average formation thickness of AAO by anodization process.

X-ray diffraction patterns (XRD) are shown in Figure 6a without air injection at temperatures of 30 °C and 60 °C, and 20 and 60 min using the grazing incidence technique. It indicates the presence of a broad diffraction peak near 22 °C of the XRD curves, explaining that AAO was not crystalline. The diffraction peaks show the presence of the aluminum phase, indicating that the X-rays penetrated the non-crystalline oxide layer. The diffraction intensities of AAO formation at 30 °C and 60 °C, at 20 and 60 s without air injection, were $2\theta = 40°$, 47°, 80°, and 85°, corresponding to the Miller indices of the cubic $Al_2O_3$ phase (111), (200), (311), and (222), according to the crystallographic chart 01-75-0921. The diffraction intensity of AAO at 60 °C was higher, indicating that AAO in the high-temperature electrolyte could transmit more X-rays than the film prepared at low temperature, further confirming that the oxide layer density decreased as the electrolyte temperature increased. In Figure 6b, with air injection at temperatures of 30 °C and 60 °C, and 20 and 60 min using the grazing incidence technique, the presence of a broad diffraction peak at 30° of the XRD curves is observed. The diffraction peaks $2\theta$ at 40°, 47°, 68°, 80°, and 85° correspond to the Miller indices of the cubic $Al_2O_3$ phase (111), (200), (311), (222), respectively, with the crystallographic chart 01-75-0921. The most notable diffraction intensity of AAO at 60 °C was higher, indicating that AAO in the high-temperature electrolyte could transmit more X-rays.

In Figure 7a, the potentiodynamic polarization results of the AAO films at temperatures of 30 °C and 60 °C, and 20 and 60 min without air injection are shown. At 30 °C—20 min, activation polarization is present, with the cathodic arm showing stable behavior until reaching the corrosion potential (Ecorr), indicating major susceptibility to corrosion due to an increase in current density, resulting from the low resistance of the sample. At 30 °C—60 min, the corrosion potential (Ecorr) changed to a nobler value with a current density almost a decade lower logarithmically, indicating improved corrosion resistance. In samples 60 °C—20 min and 60 °C—60 min, there is activation polarization, and the current density is an order of magnitude lower than the previous ones. In Figure 6b, the behavior of the anodic film with air injection at 30 °C and 60 °C, and 20 and 60 min is observed. At 30 °C—20 min, a passive layer formation was achieved, showing the cathodic and anodic arms behavior, with a negative Ecorr of −900 mV, indicating active dissolution of aluminum. At 30 °C—60 min, the passive layer formation achieved was stable with a corrosion current density of 2.48 mA/cm$^2$, with the Ecorr moving to more negative values at −1200 mV. At 60 °C—20 min and 60 °C—60 min, passive layer formation with stable behavior was observed, with the Ecorr at −800 mV and −1200 mV, indicating anodic dissolution. Figure 6c shows the anodic film behavior with oxygen injection at 30 °C and 60 °C, and 20 and 60 min. At 30 °C—20 min, the passive layer formation was achieved, with the Ecorr at −1000 mV, showing low corrosion current density (Icorr) behavior at 1.2 mA/cm$^2$, indicating anodic dissolution. At 30 °C—60 min, the passive layer formation was stable, with the Ecorr moving to −1100 mV. At 60 °C—20 min and 60 °C—60 min, stable passive layer formation was observed, with the Ecorr at −1100 mV and −1250 mV, indicating anodic dissolution.

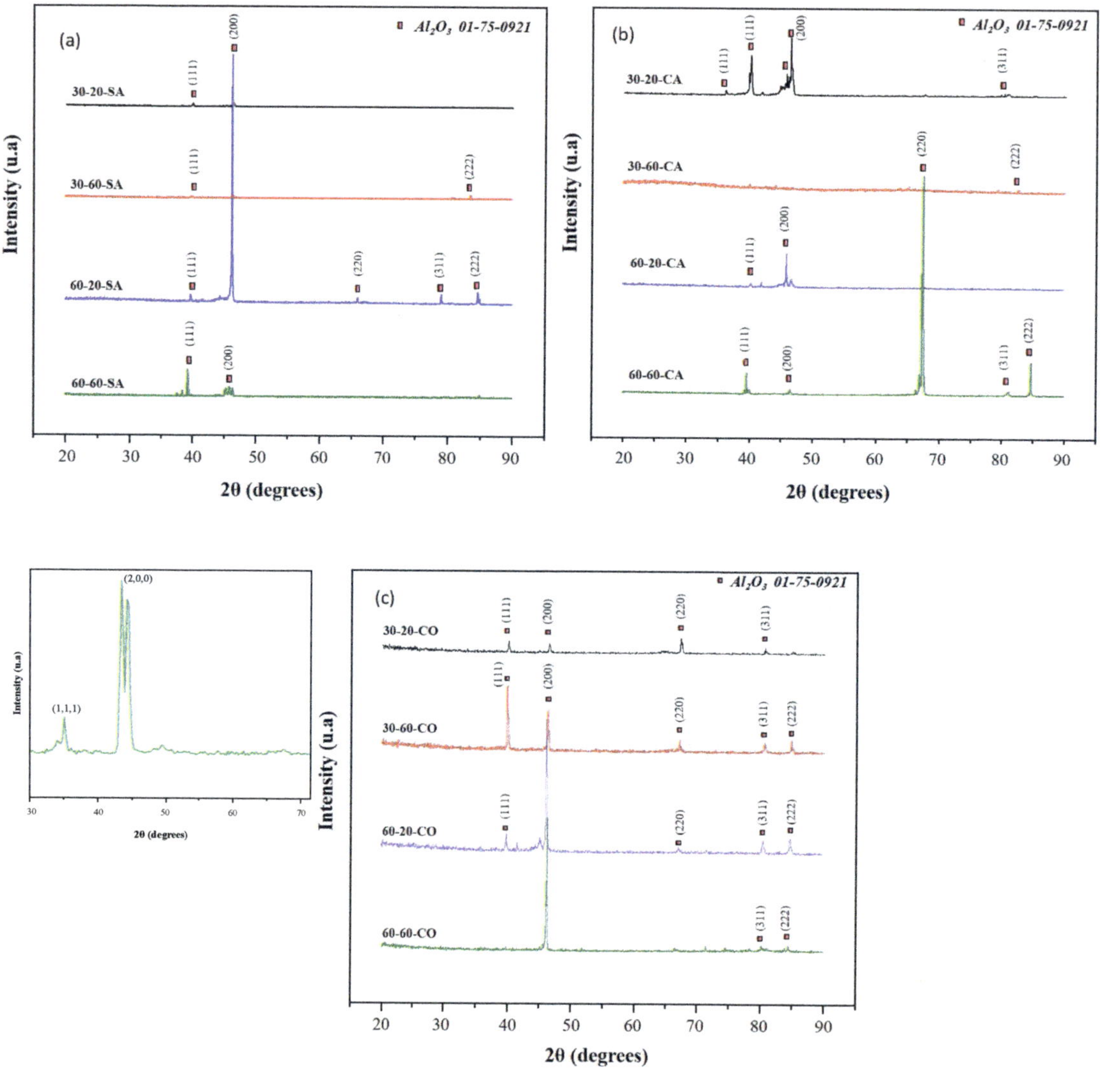


**Figure 6.** X-ray diffraction pattern of AAO in different temperature and time environments: (**a**) without air injection, (**b**) with air injection, and (**c**) with oxygen injection, using the grazing incidence technique.

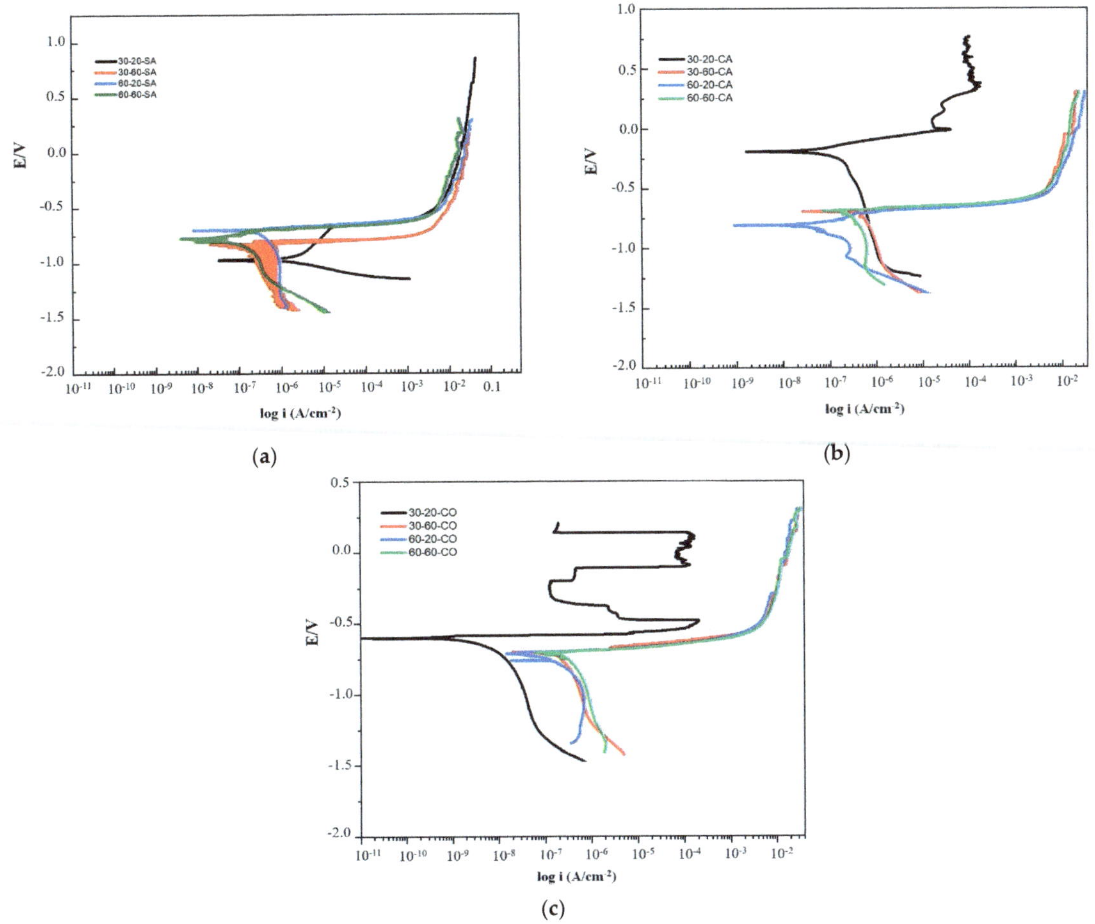


**Figure 7.** (**a**) Potentiodynamic polarization curves of aluminum alloy 1061 as a function of processing parameters without air injection; *y*-axis scale potential vs. current density *x*-axis scale. (**b**) Potentiodynamic polarization curves of aluminum alloy 1061 as a function of processing parameters with air injection; *y*-axis scale potential vs. current density *x*-axis scale. (**c**) Potentiodynamic polarization curves of aluminum alloy 1061 as a function of processing parameters with oxygen injection; *y*-axis scale potential vs. current density *x*-axis scale.

Therefore, at 30 ° and 60 °C at a time of 20 min, increasing the temperature limits the growth of the layer, produces malformations, and accelerates the kinetics of the reactions.

At 30 °C, from 20 to 60 min, increasing the time degrades the growth of the layer due to the reaction products.

By increasing the residence time and temperature, there is a degradation or dissolution of the layer that forms in the first minutes, which is attributed to the kinetic behavior of the system that is located when increasing the temperature or increasing the residence time.

In Figure 7b, the results of the potentiodynamic polarization of AAO films at temperatures of 30 °C and 60 °C, and durations of 20 and 60 min with air injection are shown. At 30 °C—20 min, a mixed behavior is observed with passivity between $5 \times 10^{-5}$ and $6 \times 10^{-4}$ A/cm$^2$ in terms of current density. This phenomenon occurs due to the growth of a stable, compact, and impermeable oxide film, which is the result of the environment to which the sample is exposed. During this process, an additional oxide layer forms that protects the aluminum (passive layer). At 30 °C–60 min, a concentration polarization with similar limiting current values to the previous sample is presented, but with more active values. Regarding 60 °C—20 min, it shows concentration polarization with lower limiting currents by approximately one logarithmic decade, and at 60 °C—60 min, an increase in the limiting current is again observed, showing less corrosion resistance. It is also necessary to discuss this in terms of the structural properties of the anodized film. Therefore, 30° and 60° at 20 min favor the growth of the layer in both cases, obtaining uniform growth, without malformations, which indicates that there is no presence of pores in the film, 30° and 60° at 60 min the size of the layer decreases without malformations, where the electrolyte acts by detaching part of the layer and subsequently becoming passivated again.

In Figure 7c, the results of the potentiodynamic polarization of AAO films at temperatures of 30 °C and 60 °C, and durations of 20 and 60 min with air injection, are shown. At 30 °C—20 min, the test shows the lowest (Icorr) and highest (Ecorr) in this group, demonstrating a passivation phenomenon at $1 \times 10^{-4}$ A/cm$^2$ and 0.5 V vs. Ag/AgCl over a range of 300 mV, indicating better corrosion resistance. Subsequently, the passive film breaks and re-passivates, possibly due to the formation of anodic sites that are later protected by the passive $Al_2O_3$ film, or a pitting corrosion phenomenon might occur. After this, the passive layer attempts to form but is not sufficiently resistant, leading to the generation of pits on the sample as the corrosion potential increases. This is attributed to the oxygen injection during the anodizing process. The curves for 30 °C—60 min, 60 °C—20 min, and 60 °C—60 min show concentration polarization with increments in the limiting current of more than one logarithmic decade, reflecting a sudden decrease in corrosion resistance.

In Table 6, the results of the corrosion rate in mpy are shown. The sample 30-20-CO presents the best corrosion rate, which is due to the fact that materials like aluminum are mainly known for their corrosion resistance. This explains why most samples exhibit a low corrosion rate.

**Table 6.** Potentiodynamic polarization parameters for the corrosion of anodized aluminum at different pH Values with 3.5 wt.% NaCl, with and without oxygen injection.

| Sample | $I_{corr}$ (A/cm$^2$) | $E_{corr}$ (V) | $\beta_a$ (mV) | $\beta_c$ (mV) | Corrosion Rate MPY (Mils per Year) |
|---|---|---|---|---|---|
| 30-20-SA | $2.0327 \times 10^{-6}$ | −0.9891 | 342.65 | −78.84 | $3.7941 \times 10^{-7}$ |
| 30-60-SA | $9.8137 \times 10^{-8}$ | −0.90115 | −231.53 | 45.911 | $1.0380 \times 10^{-7}$ |
| 60-20-SA | $2.3762 \times 10^{-7}$ | −0.69415 | 23.214 | −284 | $4.4352 \times 10^{-8}$ |
| 60-60-SA | $2.2916 \times 10^{-8}$ | −0.77737 | 70.197 | −95.838 | $4.2773 \times 10^{-9}$ |
| 30-20-CA | $2.9719 \times 10^{-8}$ | −0.1886 | 62.925 | −79.81 | $5.5471 \times 10^{-9}$ |
| 30-60-CA | $4.4063 \times 10^{-7}$ | −0.69703 | 22.375 | −563.12 | $8.2245 \times 10^{-8}$ |
| 60-20-CA | $2.9470 \times 10^{-8}$ | −0.80927 | 70.07 | −139.93 | $5.5006 \times 10^{-9}$ |
| 60-60-CA | $2.1897 \times 10^{-7}$ | −0.68639 | 24.807 | −401.06 | $4.0871 \times 10^{-8}$ |
| 30-20-CO | $6.3584 \times 10^{-13}$ | −0.24103 | −178.71 | 528.16 | $6.7253 \times 10^{-13}$ |
| 30-60-CO | $3.0776 \times 10^{-7}$ | −0.72403 | 43.62 | −953.58 | $5.7444 \times 10^{-8}$ |
| 60-20-CO | $2.3397 \times 10^{-7}$ | −0.74397 | 44.787 | −466.87 | $4.3671 \times 10^{-8}$ |
| 60-60-CO | $3.5176 \times 10^{-7}$ | −0.72688 | 37.626 | −594.76 | $6.5657 \times 10^{-8}$ |

The results of all the potentiodynamic polarization tests highlight a more corrosion-resistant behavior for the 30 °C—20 min specimen with oxygen injection. This is because it has the least material loss per year and the lowest corrosion current with a higher corrosion potential based on the presented graphs.

There are no changes in the mechanisms, but there is a variation in the activation energy. Nonetheless, when the temperature increases and the time decreases without air, the mechanism changes to a cathodic polarization by concentration. The controls are the same in all the curves under different atmospheres. In the cathodic part, concentration occurs first, and then very close to the corrosion potential, we have mixed polarization. At

30 °C—60 min, the corrosion potential (Ecorr) changed to a more noble value because the layer is more compact as concentration polarization is due to the hydrogen evolution.

## 4. Discussion of Statistical Analysis

Table 7 shows the real variables of (T) temperature, (t) time, and conditions with air (CO), without air (SA), and with oxygen (CO) scaled, and the response to determine the best anodizing conditions. For the analysis of variance, variables were assigned to designate the response variables: thickness size with Q, grain size with U, and reaction rate with V. For the real variables, $i_1$ and $i_2$ are for low and high time, $t_1$ and $t_2$ are for low and high temperature, and $v_1$, $v_2$, and $v_3$ are for each condition, considering the effects of significant variables and their respective calculated coefficients.

**Table 7.** General factorial design for aluminum anodizing treatments.

| | Variables | | | | | | Answer | | |
|---|---|---|---|---|---|---|---|---|---|
| | Real | | | Staggered | | | | | |
| Sample | Temperature (°C) | Time (min) | Condition | $X_1$ | $X_2$ | $X_3$ | Thickness Size (μm) | Crystallite Size (τ) | Reaction Rate (MPY) |
| 1 | 60 | 60 | SA | 2 | 2 | 1 | 3.2 | 0.08 | $4.28 \times 10^{-9}$ |
| 2 | 60 | 60 | CA | 2 | 2 | 2 | 1.4 | 0.16 | $4.09 \times 10^{-8}$ |
| 3 | 30 | 60 | CO | 1 | 2 | 3 | 3.1 | 0.35 | $5.74 \times 10^{-8}$ |
| 4 | 30 | 20 | CA | 1 | 1 | 2 | 3.2 | 0.04 | $5.55 \times 10^{-9}$ |
| 5 | 60 | 20 | CA | 2 | 1 | 2 | 2.7 | 1.11 | $5.50 \times 10^{-9}$ |
| 6 | 60 | 20 | CO | 2 | 1 | 3 | 1.7 | 1.83 | $4.37 \times 10^{-8}$ |
| 7 | 30 | 60 | CA | 1 | 2 | 2 | 5.5 | 0.63 | $8.22 \times 10^{-8}$ |
| 8 | 30 | 60 | SA | 1 | 2 | 1 | 5.2 | 1.88 | $1.04 \times 10^{-7}$ |
| 9 | 60 | 20 | SA | 2 | 1 | 1 | 6.1 | 0.17 | $4.44 \times 10^{-8}$ |
| 10 | 30 | 20 | CO | 1 | 1 | 3 | 3.5 | 0.94 | $6.73 \times 10^{-13}$ |
| 11 | 30 | 20 | SA | 1 | 1 | 1 | 7.4 | 0.72 | $3.79 \times 10^{-7}$ |
| 12 | 60 | 60 | CO | 2 | 2 | 3 | 1.4 | 1.83 | $6.57 \times 10^{-8}$ |

The resultant was used to predict grain size (3), thickness size (4), and reaction rate (5) by performing a general linear model. The experimental fit shows a certainty of 95.07% ($R^2$ = 0.9507), providing evidence that the mathematical model adequately fits the data.

$$U = 0.812 - 0.052\ t_1 + 0.52t_2 - 0.010\ i_2 + 0.010\ i_2 - 0.099\ v_1 - 0.037\ v_2 - 0.426\ v_3 \quad (3)$$

$$Q = 3.700 + 0.950\ t_1 - 0.950\ t_2 + 0.400\ i_1 - 0.400\ i_2 + 1.775\ v_1 - 0.500\ v_2 - 1.275\ v_3 \quad (4)$$

$$V = 5.026\ -\ 0.402\ t_2 + 0.402\ t_2 + 0.335 i_1 - 0.035\ i_2 + 0.062 v_1 - 0.592 v_2 + 0.658 v_3 \quad (5)$$

The results obtained by ANOVA for the response variables Q, U, and V are shown in Table 8. According to the results of this study, it is confirmed that there is significance in the interaction Q—iQ-iQ—i ($p > 0.05$, $p > 0.05$, $p > 0.05$) for grain size and thickness, but for the reaction rate, this significance is not met. For the model (individual effects and main effects), the effect is significant because the $p$-value is much smaller ($p = 0.001$, $p = 0.001$, $p = 0.001$) than the fixed value of α\alphaα.

**Table 8.** Analysis of variance (ANOVA) for the variables investigated during the aluminum anodizing process, adjusted to a quadratic model.

| Response Variable | Source of Variation | Sum of Squares | Degrees of Freedom | Mean Squares | F | *p* |
|---|---|---|---|---|---|---|
| Q | Temperature | 10.830 | 1 | 10.830 | 8.85 | 0.021 |
| | Time | 1.920 | 1 | 1.920 | 1.57 | 0.521 |
| | Condition | 20.105 | 2 | 10.053 | 8.22 | 0.015 |
| | Error | 8.565 | 7 | 1.224 | | |
| | Total | 41.420 | 11 | | | |
| U | Temperature | 0.03203 | 1 | 0.03203 | 0.05 | 0.826 |
| | Time | 0.00120 | 1 | 0.00120 | 0.00 | 0.966 |
| | Condition | 1.19152 | 2 | 0.595758 | 0.97 | 0.426 |
| | Error | 4.31782 | 7 | 0.616831 | | |
| | Total | 5.54257 | 11 | | | |
| V | Temperature | 1.944 | 1 | 1.944 | 0.46 | 0.520 |
| | Time | 1.345 | 1 | 1.345 | 0.32 | 0.591 |
| | Condition | 3.164 | 2 | 1.582 | 0.37 | 0.702 |
| | Error | 29.695 | 7 | 4.242 | | |
| | Total | 36.148 | 11 | | | |

The $p$-value of Q as a function of temperature and condition does have an effect on the variable grain size, in contrast to time, which has a value greater than $p = 0.005$. Regarding the response variable thickness size U, there is no effect as the value is above the estimated $p$-value; similarly, the response variable reaction rate V presents the same effect with a value greater than $p$. The levels assigned to the factors were very narrow, so the difference between what happens at one level and another is practically imperceptible for the response variable.

## 5. Conclusions

From the results of this work, the conclusions regarding the mathematical model for the anodizing process under the variations in the atmosphere and time of expositions and its response to corrosion resistance are as follows.

1. The results from ANOVA show the response variables for the grain size $U$, thickness $Q$, and reaction rate $V$. These response variables are associated with the time, temperature, and the different atmospheres in good agreement. Thus, the thickness of the layer is correlated to the kinetics of the reaction.
2. The experiments show a similar linear kinetic behavior in the three systems or atmospheres: they react in the same way, forming the layer uniformly on the surface of the aluminum, but as time passes, there are areas where the aluminum is already occupied, and the reaction passes from linear to parabolic; therefore, the growth of the layer is no longer uniform, and the optimal value for grain size growth was at 30 °C for 20 min with air injection.
3. Samples obtained through different atmospheres are completely different on the microstructure; this fact causes the anodized layers to present different characteristics and protective properties.
4. With respect to the corrosion properties, samples present different behavior depending on their structural characteristics. Thus, in the potentiodynamic polarization tests, a more corrosion-resistant behavior can be highlighted for the sample 30-20-CO because it has the lowest material loss per year. It also contains the lowest corrosion current in its process.
5. At a low temperature of 30 °C, for a short time of 20 min and without air injection, are the optimal values for the linear growth of the anodic layer, where the thickness grows by 7.40 µm. In the potentiodynamic polarization tests, a more corrosion-resistant behavior can be highlighted for the sample 30-20-CO because it has the lowest material

loss per year. It also contains the lowest corrosion current in its process, with a higher corrosion potential.

**Author Contributions:** Conceptualization, A.M.G.-L.; Formal Analysis, L.G.C.-Á.; Funding Acquisition, A.M.G.-L.; Project Administration, L.A.F.-F.; Supervision, M.A.G.-R.; Validation, J.E.C.-S.; Writing—Original Draft, G.B.-C.; Writing—Review and Editing, M.A.G.-R. All authors have read and agreed to the published version of the manuscript.

**Funding:** Universidad Autónoma de Coahuila.

**Institutional Review Board Statement:** Not applicable.

**Informed Consent Statement:** Not applicable.

**Data Availability Statement:** The data presented in this study are available on request from the corresponding author.

**Conflicts of Interest:** The authors declare no conflict of interest.

## References

1. Gaskell, D.R.; Laughlin, D.E. *Introduction to the Thermodynamics of Materials*, 6th ed.; CRC Press: Boca Raton, FL, USA, 2017. [CrossRef]
2. Runge, J.M. Anodizing as an Industrial Process. In *The Metallurgy of Anodizing Aluminum*; Springer International Publishing: Cham, Switzerland, 2018; pp. 149–190. [CrossRef]
3. Rashid, K.H.; Khadom, A.A.; Mahood, H.B. Aluminum ASA 6061 Anodizing Process by Chromic Acid Using Box–Wilson Central Composite Design: Optimization and Corrosion Tendency. *Met. Mater. Int.* **2021**, *27*, 4059–4073. [CrossRef]
4. Benmohamed, M.; Benmounah, A.; Haddad, A.; Yahi, S. The effect of inhibiting molybdate used in anodizing-conversion treatment to improve corrosion protection of AA2030 aluminum alloy in different steps. *J. Eng. Appl. Sci.* **2022**, *69*, 40. [CrossRef]
5. Kikuchi, T.; Suzuki, Y.; Iwai, M.; Suzuki, R.O. Anodizing Aluminum and Its Alloys in Etidronic Acid to Enhance Their Corrosion Resistance in a Sodium Chloride Solution. *J. Electrochem. Soc.* **2020**, *167*, 121502. [CrossRef]
6. Gower, C.H.R.; O'Brien, S. Electrolyte Containing Sulphuric Acid. UK Patent 290901, 25 October 1927.
7. Ono, S.; Saito, M.; Asoh, H. Self-ordering of anodic porous alumina formed in organic acid electrolytes. *Electrochim. Acta* **2005**, *51*, 827–833. [CrossRef]
8. Parkhutik, V.P. Study of Aluminium Anodization in Sulphuric and Chromic Acid Solutions-II. Oxide Morphology and Structure. *Electrochim. Acta* **1990**, *35*, 961–966. [CrossRef]
9. Sulka, G.D.; Stroobants, S.; Moshchalkov, V.; Borghs, G.; Celis, J.-P. Synthesis of Well-Ordered Nanopores by Anodizing Aluminum Foils in Sulfuric Acid. *J. Electrochem. Soc.* **2002**, *149*, D97. [CrossRef]
10. Ono, S.; Asoh, H. A new perspective on pore growth in anodic alumina films. *Electrochem. Commun.* **2021**, *124*, 106972. [CrossRef]
11. Roshani, M.; Rouhaghdam, A.S.; Aliofkhazraei, M.; Astaraee, A.H. Optimization of mechanical properties for pulsed anodizing of aluminum. *Surf. Coat. Technol.* **2017**, *310*, 17–24. [CrossRef]
12. Saffari, H.; Sohrabi, B.; Noori, M.R.; Bahrami, H.R.T. Optimal condition for fabricating superhydrophobic Aluminum surfaces with controlled anodizing processes. *Appl. Surf. Sci.* **2018**, *435*, 1322–1328. [CrossRef]
13. Javaherdashti, R. *Microbiologically Influenced Corrosion—An Engineering Insight*; Springer: London, UK, 2008.
14. Kim, M.; Choi, E.; So, J.; Shin, J.-S.; Chung, C.-W.; Maeng, S.-J.; Yun, J.-Y. Improvement of corrosion properties of plasma in an aluminum alloy 6061-T6 by phytic acid anodization temperature. *J. Mater. Res. Technol.* **2020**, *11*, 219–226. [CrossRef]
15. Hernández, A.; de la Paz Guillón, M.; García, L. La metodología de Taguchi en el control estadístico de la Calidad. *Filosofía* **2015**, *23*, 65–83.
16. Cullity, B.D. *Elements of X-ray Diffraction*, 3rd ed.; Addison-Wesley: Reading, MA, USA, 1956.
17. Jones, D.A. *Principles and Prevention of Corrosion*, 2nd ed.; Prentice Hall: Upper Saddle River, NJ, USA, 1996.

coatings

MDPI

Article

# Effect of Low-Energy Implantation of $In^+$ Ions on the Composition and Electronic Structure of Single-Crystal GaP(111)

**Sardor Donaev [1,2,*], Ganjimurod Shirinov [1], Baltokhodja Umirzakov [1,3], Burkhan Donayev [4] and Shenghao Wang [5,*]**

1 Department of Electronic Equipment Production Technology, Faculty of Electronics and Automation, Tashkent State Technical University Named after Islam Karimov, Tashkent 100095, Uzbekistan; ganjimurod777@gmail.com (G.S.); be.umirzakov@gmail.com (B.U.)
2 Department of Theoretical and Experimental Physics, Faculty of Physics, Karshi State University, Karshi 180119, Uzbekistan
3 Institute of Ion-Plasma and Laser Technologies, Uzbekistan Academy of Sciences, Tashkent 100125, Uzbekistan
4 Karshi Engineering-Economical Institute, Karshi 180100, Uzbekistan; sb.donaev@gmail.com
5 Materials Genome Institute, Shanghai University, Shanghai 200444, China
* Correspondence: sardor.donaev@gmail.com (S.D.); shenghaowang@shu.edu.cn (S.W.)

**Abstract:** Using a complex of secondary and photoelectron spectroscopy methods, the effects of the implantation of $In^+$ ions with an energy of $E_0$ = 1 keV at different doses and subsequent annealing on the composition, electronic, and crystal structure of the GaP(111) surface were studied. It is shown that in the dose range $D \approx 5 \times 10^{14}$–$5 \times 10^{15}$ cm$^{-2}$ after annealing, nanocrystalline phases $Ga_{0.6}In_{0.4}P$ are formed with surface dimensions d ≈ 10–30 nm, and at $D \geq 6 \times 10^{16}$ cm$^{-2}$ nanofilm–$Ga_{0.6}In_{0.4}P$ with a thickness of 30–35 nm. It has been found that the band gap of nanophases ($E_g \approx$ 2–2.3 eV) is much larger than Eg of the film (~1.85 eV). For the first time, information was obtained on the density of state of electrons in the valence band of nanophases and nanofilm GaInP.

**Keywords:** ion implantation; distribution profile; Auger spectrum; photoelectrons; nanophase; nanofilm; quantum size effect; band gap

## 1. Introduction

Interest in nanosized structures grown on the basis of GaP is due to their wide use and their promise in the creation of a monolithic optoelectronic integrated circuit (MOEIC), photoelectric generators (solar cells), optical and electronic pumping lasers, and microwave technology devices [1–6]. Therefore, electronic, optical, emission, and electrophysical properties and the influence of various effects on these properties are currently being intensively studied. In particular, to create an MOEIC on a Si substrate, it will be necessary to grow an epitaxial film with a direct gap transition on its surface [1,7]. GaP lattice constants and Si practically coincide with each other, so it is possible to obtain a heteroepitaxial GaP/Si system.

Methods of molecular beam epitaxy [1–9], liquid phase and vapor phase epitaxy [10–16], MOCVD [5,17–19], and high-energy ion implantation [20–22] are widely used to obtain three- and four-component layers on the GaP surface. In recent years, low-energy ion implantation is often used to obtain nanosized structures on the surface and in the near-surface region of materials [23–32].

In [20], GaAlA nanophases and nanofilms were obtained by implanting Al ions into GaP(111) and GaAs(111). It has been established that in the case of GaAlP nanophases with surface sizes less than 35–40 nm and a thickness of 3.5–4 nm, quantum size effects appear. However, to date, nanoscale structures of the GaInP type have not been obtained by low-energy ion implantation.

The results of such studies are very important for elucidating the basic mechanisms of the formation of nanofilm heteroepitaxial systems based on $A^3B^5$ binary semiconductors, as well as for the development of new opto- and nanoelectronic devices.

In this work, the composition, electronic, and crystal structure of GaInP nanostructures obtained by implanting $In^+$ ions into GaP(111) with $E_0$ = 1 keV with different doses of D were studied for the first time. For comparison, the surface morphology of the GaInP/GaP(111) MBE film was also studied.

## 2. Experimental Method

Standard single-crystal samples of n-type GaP(111) (n = 3·1017 $cm^{-3}$) grown by the Czochralski method with a diameter of ~10 mm and a thickness of 1 mm were chosen as objects of study (samples were obtained from the Institute of Crystallography, Russia). Before ion implantation, GaP(111) was degassed under ultra-high vacuum conditions (P = $10^{-7}$ Pa) at T = 900 K for ~4 h. An ultra-high vacuum in the device was created by an oil-free pumping system: two zeolite sorption pumps and a cooled titanium getter pump. The experimental device consists of two chambers. In the first chamber, all technological operations (thermal heating, ion implantation) were carried out, and in the second chamber, the composition and electronic structure of the materials under research were studied using the methods of Auger electron spectroscopy (AES) (A.F.Ioffe Physical-Technical Institute, Russia) and ultraviolet photoelectron spectroscopy (UVES) (UV-1280, Shimadzu, Japan). The spectra of Auger electrons and photoelectrons were recorded using a four-grid electron analyzer. To determine the depth distribution profile of the atoms, a layer-by-layer Auger analysis was carried out by sputtering the sample surface with $Ar^+$ ions with an energy of 1 keV at an incidence angle of ~80–85° relative to the normal; the etching rate was ~(5 ± 1) Å/min. Ultraviolet photoelectron spectra were recorded at photon energy hν ≈ 10.8 eV. The photon source was a standard gas-discharge hydrogen lamp. Studies of the morphology of the crystalline structure of the surface of the samples were carried out in standard installations of scanning electron microscopy (SEM) (JSM-7200F), atomic force microscopy AFM (XE-100), and high-energy electron diffraction RHEED (EMR-2). The width of the forbidden zone was determined by a UV-1280 spectrophotometer.

## 3. Experimental Results

Figure 1 shows the initial part of the Auger spectra of pure GaP(111) and GaP implanted with $In^+$ ions with an energy of 1 keV at D = $D_{sat}$ ≈ $6 \cdot 10^{16}$ $cm^{-2}$ before and after heating at a temperature T = 950 K (where Dsat—In ion saturation dose). Warm-up time ~30 min. It can be seen that after implantation of $In^+$ ions, the intensity of the Ga and P Auger peaks in the spectrum decreases significantly and intense peaks characteristic of In appear. Ion-implanted layers are highly disordered up to amorphization. Calculations based on the full spectrum of Auger electrons of an ion-implanted sample using the formula [25] $C_x = \frac{I_x/S_x}{\sum I_i/S_i}$ showed that the surface concentrations of In, Ga, and P are, respectively, 40–45 at.%, 30–35 at.%, and 25–30 at.%. After heating this system at T = 950 K, the surface concentration of these atoms changes sharply: $C_{Ga}$ ≈ 28–33 at.%, $C_{In}$ ≈ 20–22 at.%, $C_P$ ≈ 48–52 at.%, i.e., a single-crystal film with an approximate composition of $Ga_{0.6}In_{0.4}P$ with a thickness of ~30–35 Å was formed.

Figure 2 shows the concentration profiles of the distribution of In atoms over the depth h′ for GaP(111) implanted with $In^+$ ions with $E_0$ = 1 keV at D = $6 \times 10^{16}$ $cm^{-2}$, obtained before and after heating. It can be seen that after ion implantation, the surface concentration of In atoms is ~40–45 at.%. With an increase in h′ to ~8–10 Å, the $C_{In}$ value sharply decreases by ~8–10 at.%. In the range h ≈ 10–25 nm, the In concentration does not change noticeably, and in the range h′ ≈ 25–80 Å, it decreases to a bullet. After heating at T = 950 K, the value of $C_{In}$ on the surface decreases to 20–22 at.%, and it practically does not change until a depth of 35–40 Å. In the range h′ ≈ 40–100 Å, $C_{In}$ monotonically decreases to 0, i.e., after heating, the concentration of atoms in the transition layer increases slightly.

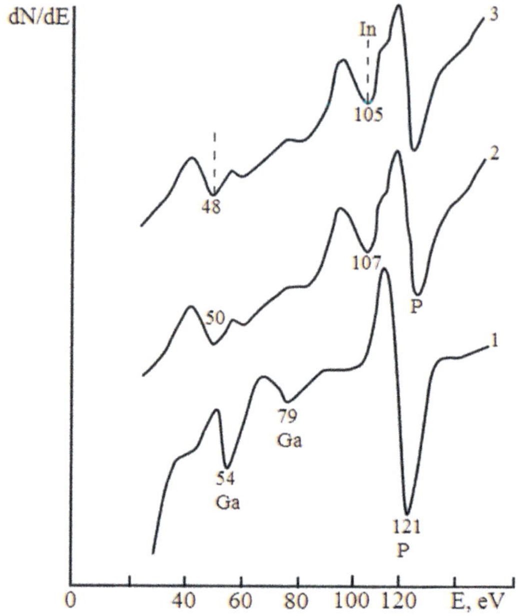


**Figure 1.** Auger spectra: 1—pure GaP(111); 2—GaP(111), implanted with $In^+$ ions with $E_0$ = 1 keV at D = 6 × $10^{16}$ $cm^{-2}$; 3—after annealing at T = 950 K. Annealing time 40 min.

A comparison of the areas under the $C_{In}(h')$ curves shows that after heating it noticeably decreases. Apparently, during the heating process, along with some diffusion of In atoms into the depths of GaP, their evaporation from the surface layers occurs.

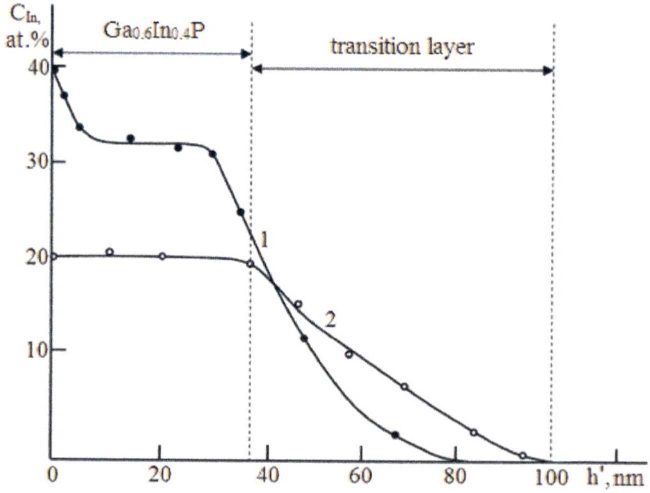


**Figure 2.** Profiles of the distribution of In atoms over the depth h′ for GaP(111) implanted with $In^+$ ions with $E_0$ = 1 keV at D = 6 × $10^{16}$ $cm^{-2}$ before (curve 1) and after heating at T = 950 K (curve 2).

Figure 3 shows SEM and AFM images of the surface of pure GaP(111) and GaP(111) with $Ga_{0.6}In_{0.4}P$ surface films, obtained by MBE (h = 50 nm) and ion implantation (h = 0.35–0.4 nm) methods. From Figure 3, it is clear that the surface of single-crystal

GaP(111) is atomically smooth. The estimated value of surface roughness, which was determined by AFM, is ~1 nm. Analysis of picture 3b shows that the MBE of the $Ga_{0.6}In_{0.4}P$ film consists of individual blocks. In this case, the surface roughness increases to 5–6 nm (Figure 3d). The surface of $Ga_{0.6}In_{0.4}P$ obtained by ion implantation in combination with heating is highly smooth, and the roughness is ~2 nm (Figure 3e).

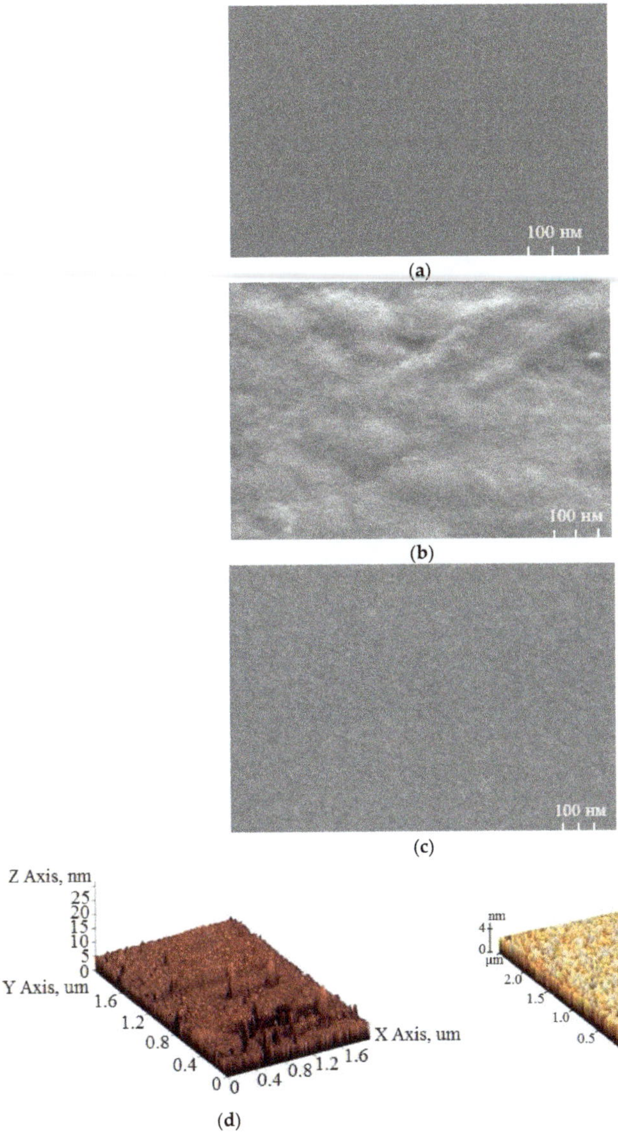


**Figure 3.** SEM (**a**–**c**) and AFM (**d**,**e**) images of the GaP surface with $Ga_{0.6}In_{0.4}P$ surface films obtained by MBE with d ≈ 50 nm (**b**,**d**) and ion implantation in combination with heating with d ≈ 3.5–4 nm (**c**,**e**).

Figure 4 shows HSED images of the surface of a pure GaP and GaP sample with a $Ga_{0.6}In_{0.4}P$ film with a thickness of 3–3.5 nm. The images clearly reveal reflections characteristic of cubic lattices. However, in the case of a three-component film, the reflections are bolder and blurrier. This is apparently due to some discrepancy between the lattice parameters of GaP and GaInP (see Table 1).

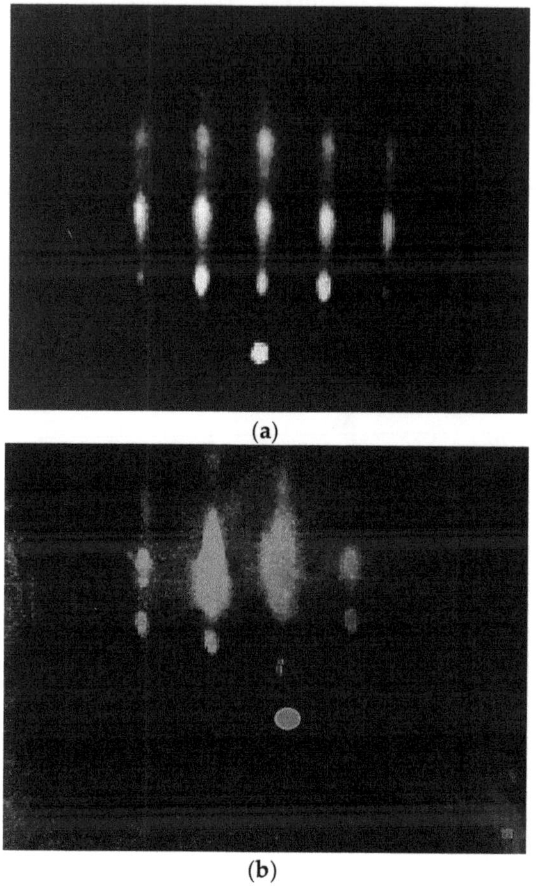

(a)

(b)

**Figure 4.** HSED images of the surface of GaP (**a**) and GaP with a $Ga_{0.6}In_{0.4}P$ film (**b**).

Figure 5 shows the photoelectron spectra obtained after heating at T = 950 K (heating time 40 min) of GaP implanted with $In^+$ ions with $E_0$ = 1 keV at doses of $10^{15}$ (curve 2) and $6 \times 10^{16}$ $cm^{-2}$ (curve 3) [29,32]. In the first case, nanocrystalline phases with surface diameters d = 15–20 nm were formed, and in the second case, a $Ga_{0.6}In_{0.4}P$ film was formed. These spectra well reflect the distribution of the density of the state of the valence band electrons. It can be seen that the spectrum of pure GaP exhibits maxima at binding energies of $E_{cb}$ ≈–0.8 eV; –2.2 eV; and 4 eV, probably due to the excitation of electrons from the 4p and 4p + 4s states of Ga, as well as the hybridization of the 4s states of Ga with the 3s states of P (Figure 5, curve 1) [29,32]. In the case of the $Ga_{0.6}In_{0.4}P$ film, the spectrum (Figure 5, curve 3) contains intense peaks with $E_{cb}$ = −1.2; −3.3; and −5.6 eV, apparently associated with the excitation of electrons from the hybridized states of electrons 4s(Ga) + 5p(In), 4s(Ga) + 5p(In) + 3d(P), and 4s (Ga) + 5s(In) + 3d(P). The spectrum of GaP with $Ga_{0.6}In_{0.4}P$ nanocrystals reveals some features characteristic of both GaP and $Ga_{0.6}In_{0.4}P$ (Figure 5, curve 2).

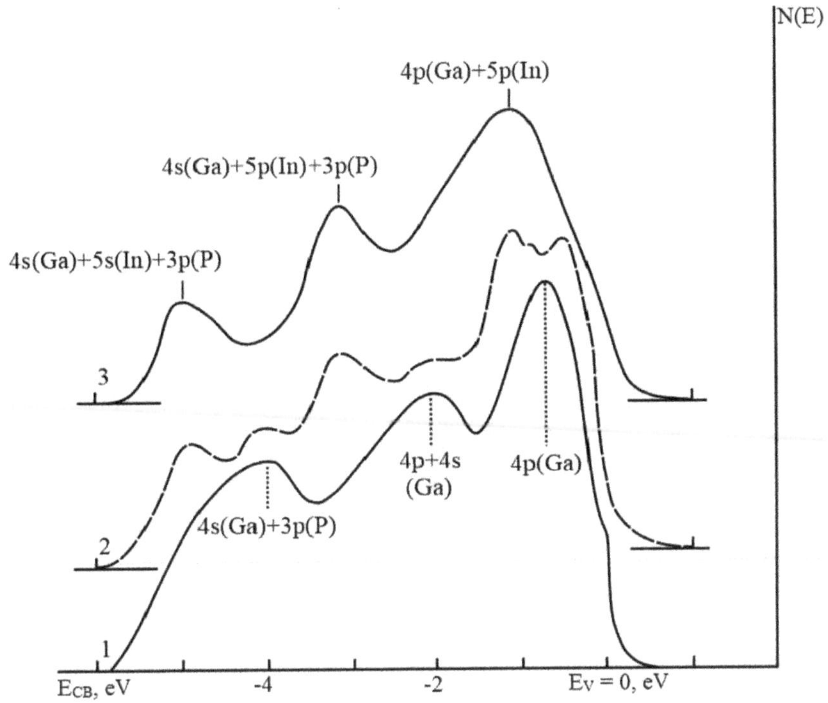


**Figure 5.** Photoelectron spectra obtained after heating at T = 950 K of GaP(111) implanted with $In^+$ ions with $E_0$ = 1 keV at doses D, $cm^{-2}$: 1—0; 2—$10^{15}$; 3—6 × $10^{16}$. Annealing time t ≈ 40 min [29,32].

To estimate the band gap of these samples, light absorption spectra were recorded (Figure 6), where K is the light absorption coefficient. It is known that at medium doses (5 × $10^{14}$–5·$10^{15}$ $cm^{-2}$), the ion-implanted surface consists of two phases: the ion-doped phase and the undoped portions of the semiconductor [29]. In our case, at D = $10^{15}$ $cm^{-2}$, the average surface diameters of the nanophases were d ~ 15–20 nm, and at D = 6 × $10^{16}$ $cm^{-2}$ they were 25-30 nm. The distance between the centers of the phases is ~50–60 nm. In the case of D ≈ 6 × $10^{16}$ $cm^{-2}$, after heating, a continuous homogeneous layer (film) of $Ga_{0.6}In_{0.4}P$ with h = 35–40 Å is formed. Figure 6 shows that in all dependences K(hν), first with increasing hν, the value of K practically does not change, and then K sharply decreases, where

$$K = I_0/I_n$$

$I_0$ is the intensity of light passing through GaP, and $I_n$ is the intensity of light passing through GaP with a GaInP nanostructure.

In the case of GaP and GaP with a $Ga_{0.6}In_{0.4}P$ film, a decrease in K is observed starting from hν ≈ 2.2 eV (curve 1) and hν ≈ 1.7 eV (curve 4), respectively, and in the case of GaP with $Ga_{0.6}In_{0.4}P$ nanocrystals, a sharp decrease in K is observed in two values of hν [33]. The first decrease is associated with light absorption in areas coated with GaInP nanocrystals, and the second-in areas of GaP are not coated with nanocrystals. The extrapolation of a sharply decreasing part of the curves to the hν axis gives a value of $E_g$ for a given material. It can be seen that $E_g$ for GaP(111) is approximately equal to 2.36 eV, for a $Ga_{0.6}In_{0.4}P$ film ~1.85 eV, and for nanocrystalline $Ga_{0.6}In_{0.4}P$ phases with d = 25 nm ~2.3 eV and d = 25–30 nm ~2.3 eV. With increasing surface diameters d nanotimes, the *K*(hν) curves shift towards lower energies, i.e., $E_g$ decreases (Table 1). Based on curve 2 (Figure 6), one can estimate the degree of coverage of the GaP(111) surface with GaInP Θ nanophases using the following formula [33]:

$$\Theta = 1 - K.$$

In the case of GaP with GaInP nanocrystals, the value of $\Theta \approx 0.40$. Table 1 shows the values of $E_g$ and Θ for the samples under study.

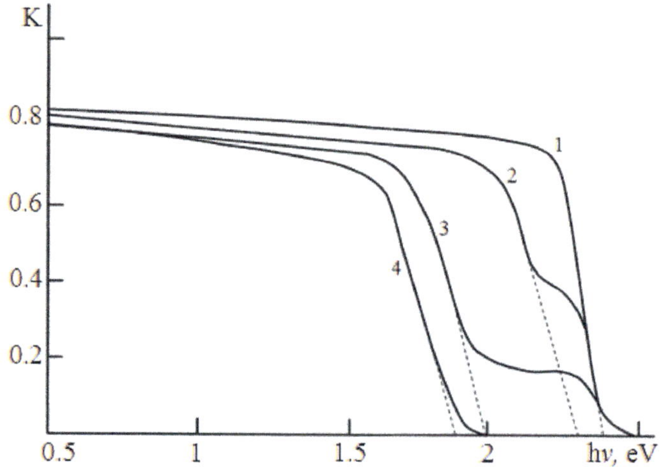


**Figure 6.** Spectra of light transmission through GaP(111) (curve 1) and GaP(111) with nanophases with surface diameters d ≈ 15–20 nm (2), d = 25–30 nm (3), and $Ga_{0.6}In_{0.4}P$ nanofilm (4).

**Table 1.** Dimensions of nanostructures (thickness h and surface diameter d) and values of $E_g$ and Θ of nanophases and GaInP films.

| Studied Samples | h, nm | d, nm | $E_g$, eV | Θ, % |
|---|---|---|---|---|
| GaP(111) | Monocrystal sample | | 2.36 | 0 |
| $Ga_{0.6}In_{0.4}P$/GaP(111) | 3–3.5 | 15–20 | 2.3 | 40 |
| | 3–3.5 | 25–30 | 2.0 | 70 |
| | 3–3.5 | continuous film | 1.85 | 100 |

From Figure 5 and the table, it can be seen that, at the same thicknesses, the values of $E_g$ for nanocrystalline phases are greater than for the film. These results show that in nanocrystalline phases, starting from a diameter of ≈ 25–30 nm, the value of $E_g$ increases with decreasing d.

## 4. Conclusions

For the first time, nanocrystalline phases and $Ga_{0.6}In_{0.4}P$ films with a thickness of 30–35 Å were obtained by the implantation of $In^+$ ions in combination with annealing on the GaP(111) surface. Their parameters of energy bands and density of valence band electrons, as well as the degree of coverage of the GaP surface with $Ga_{0.6}In_{0.4}P$ nanophases, were determined. It has been shown that in GaInP nanocrystals, the $E_g$ value is greater than the $E_g$ of the nanofilm.

**Author Contributions:** Conceptualization, B.U.; Validation, B.U.; Formal analysis, S.W.; Investigation, S.D., G.S., B.U., B.D. and S.W.; Resources, G.S.; Writing—original draft, B.U.; Writing—review & editing, G.S. and B.U.; Supervision, S.D., B.U. and S.W.; Project administration, S.D. All authors have read and agreed to the published version of the manuscript.

**Funding:** This research received no external funding.

**Institutional Review Board Statement:** Not applicable.

**Informed Consent Statement:** Not applicable.

**Data Availability Statement:** The data that support the findings of this study are available from the corresponding author on reasonable request.

**Conflicts of Interest:** The authors declare no competing interests.

## References


1. Lazarenko, A.A.; Nikitina, E.V.; Pirogov, E.V.; Sobolev, M.S.; Egorov, A.Y. Molecular beam epitaxy of GaPN, GaPAsN, and InGaPN nitride solid solutions. *Semiconductors* **2014**, *48*, 392–396. [CrossRef]
2. Abramkin, D.S.; Petrushkov, M.O.; Emelyanov, E.A.; Nenashev, A.V.; Yesin, M.Y.; Vasev, A.V.; Putyato, M.A.; Bogomolov, D.B.; Gutakovskiy, A.K.; Preobrazhenskiy, V.V. Formation of InAs/GaP Quantum-Well Heterostructures on Silicon Substrates by Molecular-Beam Epitaxy. *Semiconductors* **2021**, *55*, 139–146. [CrossRef]
3. Sun, C.; Wade, M.T.; Lee, Y.; Orcutt, J.S.; Alloatti, L.; Georgas, M.S.; Waterman, A.S.; Shainline, J.M.; Avizienis, R.R.; Lin, S.; et al. Single-chip microprocessor that communicates directly using light. *Nature* **2015**, *528*, 534–538. [CrossRef] [PubMed]
4. Uvarova, A.V.; Baranova, A.I.; Vyacheslavovaa, E.A.; Kalyuzhnyic, N.A.; Kudryashova, D.A.; Maksimovaa, A.A.; Morozova, I.A.; Mintairovc, S.A.; Saliic, R.A.; Gudovskikha, A.S. Formation of Heterostructures of GaP/Si Photoconverters by the Combined Method of MOVPE and PEALD. *Tech. Phys. Lett.* **2021**, *47*, 730–733. [CrossRef]
5. Ladugin, M.A.; Andreev, A.Y.; Yarotskaya, I.V.; Ryaboshtan, Y.L.; Bagaev, T.A.; Padalitsa, A.A.; Marmalyuk, A.A.; Vasilev, M.G. Comparative Study of GaAs/GaInP and GaAs/AlGaAs Quantum Wells Grown by Metalorganic Vapor Phase Epitaxy. *Inorg. Mater.* **2019**, *55*, 315–319. [CrossRef]
6. Razeghi, M. *Technology of Quantum Devices*; Springer: Boston, MA, USA, 2010; 560p.
7. Gudovskikh, A.S.; Uvarov, A.V.; Morozov, I.A.; Bukatin, A.S.; Baranov, A.I.; Kudryashov, D.A.; Kalyuzhnyy, N.A.; Mintairov, S.A.; Zubkov, V.I. Study of GaP Nucleation Layers Grown on Si by Plasma-Enhanced Atomic Layer Deposition. *Phys. Status Solidi A* **2019**, *217*, 1900532. [CrossRef]
8. Kunert, B.; Reinhard, S.; Koch, J.; Lampalzer, M.; Volz, K.; Stolz, W. First demonstration of electrical injection lasing in the novel dilute nitride Ga (NAsP)/GaP-material system. *Phys. Status Solidi C* **2006**, *3*, 614. [CrossRef]
9. Sobolev, M.S.; Lazarenko, A.A.; Nikitina, E.V.; Pirogov, E.V.; Gudovskikh, A.S.; Egorov, A.Y. MBE Growth of GaP on a Si Substrate. *Semiconductors* **2015**, *49*, 559–562. [CrossRef]
10. Gudovskikh, A.S.; Kalyuzhnyy, N.A.; Lantratov, V.M.; Mintairov, S.A.; Shvarts, M.Z.; Andreev, V.M. Properties of interfaces in GaInP solar cells. *Semiconductors* **2009**, *43*, 1363–1368. [CrossRef]
11. Baranov, A.I.; Uvarov, A.V.; Maksimova, A.A.; Vyacheslavova, E.A.; Kalyuzhnyy, N.A.; Mintairov, S.A.; Salii, R.A.; Yakovlev, G.E.; Zubkov, V.I.; Gudovskikh, A.S. Study of InP/GaP quantum wells grown by vapor phase epitaxy. *Tech. Phys. Lett.* **2023**, *49*, 163–167.
12. Dvoretckaia, L.N.; Bolshakov, A.D.; Mozharov, A.M.; Sobolev, M.S.; Kirilenko, D.A.; Baranov, A.I.; Mikhailovskii, V.Y.; Neplokh, V.V.; Morozov, I.A.; Fedorov, V.V.; et al. GaNP-based photovoltaic device integrated on Si substrate. *Sol. Energy Mater. Sol. Cells* **2019**, *206*, 110282. [CrossRef]
13. Baranov, A.I.; Gudovskikh, A.S.; Egorov, A.Y.; LeGall, S.; Kudryashov, D.A.; Kleider, J.P. Defect properties of solar cells with layers of GaP based dilute nitrides grown by molecular beam epitaxy. *J. Appl. Phys.* **2020**, *128*, 023105. [CrossRef]
14. Chang, L.L.; Ploog, K. (Eds.) *Molecular Beam Epitaxy and Heterostructures*; Springer: Dordrecht, The Netherlands, 2012; 728p.
15. Fedorchenko, I.V.; Kushkov, A.R.; Gaev, D.S.; Rabinovich, O.I.; Marenkin, S.F.; Didenko, S.I.; Legotin, S.A.; Orlova, M.N.; Krasnov, A.A. Growth method for AIIIBV and AIVBVI heterostructures. *J. Cryst. Growth* **2018**, *483*, 245–250. [CrossRef]
16. Mori, M.J. Lattice Mismatched Epitaxy of Heterostructures for Non-Nitride Green Light Emitting Devices. Ph.D. Thesis, Massachusetts Institute of Technology, Cambridge, MA, USA, June, 2008.
17. Uvarova, A.V.; Sharova, V.A.; Kudryashova, D.A.; Gudovskikha, A.S. Impact of Silicon Wafer Surface Treatment on the Morphology of GaP Layers Produced by Plasma Enhanced Atomic Layer Deposition. *Semiconductors* **2023**, *57*, 415–422. [CrossRef]
18. Uvarov, A.V.; Gudovskikh, A.S.; Nevedomskiy, V.N.; Baranov, A.I.; Kudryashov, D.A.; Morozov, I.A. Low temperature epitaxial growth of GaP on Si by atomic-layer deposition with plasma activation. *J. Phys. D Appl. Phys.* **2020**, *53*, 345105. [CrossRef]
19. Zhang, X.B.; Ryou, J.H.; Dupius, R.D.; Walter, G.; Holonyak, N. Metalorganic Chemical Vapor Deposition Growth and Characterization of GaInP/GaAs Superlattices. *J. Electron. Mater.* **2006**, *35*, 705–710. [CrossRef]
20. Ushakov, V.V.; Dravin, V.A.; Melnik, N.N.; Zavaritskaya, T.V.; Loiko, N.N.; Karavanskii, V.A.; Timoshenko, V.Y. Ion implantation of porous gallium phosphide. *Semiconductors* **1998**, *32*, 886–890. [CrossRef]
21. Bhattachar-Yaa, D.; Vinokurov, D.A.; Gusinskii, G.M.; Elyukhin, V.A.; Kovalenkov, O.V.; Kyutt, R.N.; Marsh, J.H.; Naidenov, V.O.; Portnoi, E.L. Low-temperature photoluminescence of heavy-ion-implanted InGaP solid solutions. *Tech. Phys. Lett.* **1998**, *24*, 690–691. [CrossRef]
22. Evtushenko, N.G.; Stukalov, S.A. Defect formation in GaP during laser implantation of impurities of groups I and III. *Bull. Odesnik State Univ.* **2000**, *5*, 196–201.

23. Donaev, S.B.; Djurabekova, F.; Tashmukhamedova, D.A.; Umirzakov, B.E. Formation of nanodimensional structures on surfaces of GaAs and Si by means of ion implantation. *Phys. Status Solidi C* **2015**, *12*, 89–93. [CrossRef]
24. Ergashov, Y.S.; Tashmukhamedova, D.A.; Umirzakov, B.E. On the Synthesis of Nanoscale Phases of Metal Silicides in the Near-Surface Region of Silicon and the Study of Their Electronic Structures by Passing Light. *J. Surf. Investig. X-ray Synchrotron Neutron Tech.* **2017**, *11*, 480–484. [CrossRef]
25. Donaev, S.B.; Umirzakov, B.E. Effect of the Implantation of Al+ Ions on the Composition, Electronic and Crystalline Structure of the GaP(111) Surface. *Semiconductors* **2020**, *54*, 860–862. [CrossRef]
26. Karimov, M.K.; Kutliev, U.O.; Bobojonova, S.B.; Otabaeva, K.U. Investigation of Angular Spectrum of Scattered Inert Gas Ions from the InGaP (001) Surface. *Phys. Chem. Solid State* **2021**, *22*, 742–745. [CrossRef]
27. Kutliev, U.; Karimov, M.; Sadullaeva, B.; Otaboev, M. Investigation of the ion scattering process from the A3B5 semiconductors by the computer simulation method. *Compusoft* **2018**, *7*, 2749–2751.
28. Chanda, A.; Lenka, H.P.; Jacob, C. Study of high energy Mn+1 ion implantation in GaAs. *Appl. Phys. A* **2009**, *94*, 89–94. [CrossRef]
29. Shirinov, G.M.; Donaev, S.B.; Umirzakov, B.Y.; Loboda, V.V. Emission, optical and electrical properties of GaInP/GaP nanofilms. *St. Petersburg State Polytech. Univ. J. Phys. Math.* **2023**, *16*, 89–97.
30. Normurodov, M.T.; Rysbaev, A.S.; Bekpulatov, I.R.; Normurodov, D.A.; Tursunmetova, Z.A. Formation and electronic structure of barium-monosilicide and barium-disilicide films. *J. Surf. Investig.* **2021**, *15*, S211–S215. [CrossRef]
31. Rysbaev, A.S.; Khuzhaniyazov, Z.B.; Normuradov, M.T.; Rakhimov, A.M.; Bekpulatov, I.R. Peculiarities of the electron structure of nanosized ion-implanted layers in silicon. *Tech. Phys.* **2014**, *59*, 1705–1710. [CrossRef]
32. Donaev, S.B.; Karimova, V.N.; Azimov, A.T.; Boltaboyev, K.; Yakubova, M.M. Using of ion implantation for obtaining nanostructures with the wide band GaP based on GaP. *IOP Conf. Ser. Earth Environ. Sci.* **2020**, *614*, 012002. [CrossRef]
33. Altukhov, A.A.; Zhirnov, V.V. Analysis of the morphology and stocsiometry of CoSi2/Si(100) films obtained by SPE and RE methods. In Proceedings of the II All-Union Conference, Moscow, Izhevsk, Russia, 23–25 October 1991; pp. 15–22.

coatings

MDPI

Article

# Inorganic Salts as Chemical Foam Suppressors

Niravkumar Raykundaliya [1], Stoyan I. Karakashev [2,*], Vyomesh M. Parsana [3], Nikolay A. Grozev [4], Kristina Mircheva [4] and Dilyana Ivanova-Stancheva [2]

1 Research Scholar, Gujarat Technological University, Ahmedabad 382424, Gujarat, India; nirav.ray@gmail.com
2 Faculty of Natural Sciences, University of Shumen, 9700 Shumen, Bulgaria; d.ivanova@shu.bg
3 Chemical Engineering Department, V.V.P. Engineering College, Gujarat Technological University, Rajkot 360005, Gujarat, India; vm_parsana@yahoo.com
4 Department of Physical Chemistry, Sofia University, 1 James Bourchier Avenue, 1164 Sofia, Bulgaria; fhng@chem.uni-sofia.bg (N.A.G.); fhkm@chem.uni-sofia.bg (K.M.)
* Correspondence: s.karakashev@shu.bg; Tel.: +359-899720050

**Abstract:** Can the foamability of surfactant aqueous solutions be controlled chemically? Well-known antifoams can prevent foaming by inducing the coalescence of the bubbles, but can the surfactants be deactivated chemically? If yes, how does this affect the surface tension of their aqueous solutions and their foaming capacity? To shed a light on these fundamental questions, we chose a well-known surfactant containing in its molecule a sulfate group (Sodium dodecyl sulfate, SDS) and mixed it with $BaCl_2$, (the solubility of $BaSO_4$ is 0.245 mg/100 mL water, T = 20 °C), $Pb(NO_3)_2$ (the solubility of $PbSO_4$ is 40.4 mg/100 mL water, T = 25 °C) and $FeCl_3$ (the solubility of $Fe_2(SO_4)_3$ is 25.6 g/100 mL water, T = 20 °C) at different molar ratios ($MX_n$/SDS): 1/2, 1/1, 2/1, 4/1. The results were surprising: in the case of $BaCl_2$, despite being in stoichiometric molar ratio with SDS ($BaCl_2$ + 2SDS -> $Ba(DS)_2$ + 2 NaCl), or in excess of $BaCl_2$, which should convert the whole amount of SDS into a sediment, the surface tension value remained significantly lower than that of the single surfactant. At the same time, foamability was either low or absent. It therefore appears that all of the surfactants should be converted into a sediment with very small solubility, but the low surface tension indicates the opposite. The lack of foamability indicated the opposite of that opposite. With $Pb(NO_3)_2$ and $FeCl_3$, the results are even stranger. The surface tension values are substantially smaller than those of the single surfactants, and at the same time, low foamability or lack of foamability was observed. It appears that the surfactant exists and at the same time does not exist in the aqueous solution. Where is the truth? Future studies will shed a light.

**Keywords:** foam; surfactants; salts; surface tension; foamability

## 1. Introduction

The effect of salts on amphiphilic molecules, interface boundaries, and dispersed systems has been widely studied since the time of Hofmeister [1–5], who was the first to study the effect of anions and cations on the precipitation of blood proteins, thus ordering them in a series of increasing precipitation capacity:

$$Li^+ < Na^+ < NH_4^+ < K^+ < Cs^+ < OH^- < F^- < CH_3COO^- < Cl^- < Br^- < NO_3^- < I^- < ClO_4^-$$

He established that anions are stronger precipitators than cations. It was established later [6,7] that salts can precipitate ionic surfactants, thus affecting their foaming capacity. In some of cases, the precipitation, caused by a certain salt, coincided with a drastic decrease in the foamability of the surfactant, while in other cases, the precipitation of the same surfactant, caused by another salt, practically did not affect the foaming capacity of the surfactant, but required a significantly larger salt concentration to completely suppress the foaming. Evidently, the overall picture of the ion-specific salting out and the subsequent suppressing of the surfactant's foamability is a complex phenomenon. Generally, the effect of the ions of a certain salt on the organic matter in the water could be: (i) their direct interaction with the organic molecules/ions; (ii) their interaction with the molecules of the water, just affecting its activity and ability to form hydrogen bonds with the polar groups of the organic molecules/ions; (iii) a combination of the first two effects. Koga [8,9] introduced a recent experimental thermodynamic method called 1-propanol probing methodology, thus studying the effect of many solutes (inorganic salts and amphiphilic and organic substances) on the 3D network of the water molecules. He concluded that the precipitation of solutes is a tendency of the nature of the avoid harsh effects on the water, i.e., on the hydrogen bonding network. Christov et al. [10–13] developed a thermodynamic model based on Pitzer theory [14–17] and established that the salt precipitates when a maximum in the osmotic coefficient is reached at a certain salt concentration. The new phase forms beyond this maximum. Evidently, the solute affects the properties of the bulk water, but this is only part of the whole picture. The salts specifically increase the surface tension of the water [18]. They decrease the surface tension of the aqueous solutions of ionic surfactants, depending on the type of the counter-ion of the added salt [19–23], which evidently shows that they interact with the air/water interface, thus affecting the state of the ionic surfactant adsorption layer. Ninham was the first to suggest the van der Waals interaction between the ions and the other entities of the aqueous solution as the main reason for the emergence of the ion-specific effects [24–26]. Tavares [27] followed his approach to study the van der Waals interaction of the salts with the proteins. Ivanov et al. [28] followed Ninham's approach as well and developed a theory for calculating the specific adsorption energy of the ions on the air/water interface. Hence, when the counter-ions adsorb on the air/water interface, they affect its surface potential and the very state of the surfactant adsorption layer. And here comes the puzzle: the reduction in surface tension caused by the added salt's counter-ion should boost the foamability of the surfactant's aqueous solutions. Pandey et al. [29] studied the surface tension, foamability, and foam durability of lithium dodecyl sulfate (LiDS), sodium dodecyl sulfate (NaDS), cesium dodecyl sulfate (CsDS), and magnesium dodecyl sulfate ($Mg(DS)_2$). Hence, they were able to study how the type of counter-ion affects surface tension, foamablity, and foam durability at identical concentrations of the surfactant solutions. They have established a correlation between the size of the counter-ion and the surface tension value. The surface tension values at identical surfactant concentrations diminish in the following order of counter-ions: $Li^+ < Na^+ < Cs^+ < Mg^{2+}$. The larger the counter-ion, the larger their adsoprtion on the air/water interface is. Larger adsorption of the counter-ion causes a larger drop in the surface potential value of the air/water interface and consequently a bigger adsorption of the surface-active co-ions ($DS^-$). The foamability of the above-mentioned surfactant solutions decreases in an opposite series to the counter-ions regarding surface tension: $Li^+ > Na^+ > Cs^+ > Mg^{2+}$. This looks, at first glance, strange, because the solution with the largest value of surface tension (LiDS) has the largest foamability. Its bubbles during the generation of foam are the most charged out of the whole series of surfactant solutions and consequently undergo the strongest repulsion between themselves even at high ionic strengths (50 mmol/L), thus ending in largest initial foam volume. This is an indication that the generation of foam occurs under the regime of electrostatic stabilization of the bubbles. The effect of the type of counter-ions on foam durability can be expressed in the following series: $Li^+ < Na^+ < Cs^+ < Mg^{2+}$. One can see that it is opposite to the series regarding foamability and the same as the series regarding surface tension. Once the foam is generated, the surface-active co-ions start adsorbing on the bubbles, tending to reach equilibrium. The most condensed adsorption layer will be in the case of the largest counter-ions, causing the smallest value of surface potential. This is an indication that the decay of the foam occurs under the regime of electrostatically controlled surfactant adsorption. So far, so good. Sett et al. [6], Amani et al. [7], and Petkova et al. [30] studied the foamability and foam durability of sodium dodecyl sulfate (SDS) in the presence of added excess of lithium chloride (LiCl), sodium chloride (NaCl), and potassium chloride (KCl). They confirmed the establishments of Pandey et al. [29]. The foam durability increases in the following order of added counter-ions: $Li^+ < Na^+ < K^+$, but beyond a certain critical concentration of KCl, precipitation occurs, and at the same time the foam durability drops substantially, thus reaching complete lack of foaming at a certain concentration of KCl. Hence, KCl can boost or suppress the foam depending on its concentration. Evidently, the formation of sediment of KDS exhausts the solution of $DS^-$ co-ions, thus causing a drop in both foamability and foam durability. The formation of sediment is a result of successful ion pairing, causing the beginning of precipitation. Collins [31] introduced a criterion about successful ion pairing. The ion–counter-ion pair should have similar affinity to water to realize coupling. So, ions with similar sizes should pair, while ions with significant differences in size should not pair. Unfortunately, this is valid only for inorganic ions. Meanwhile, Vlachy et al. [32] suggest a Hofmeister-like ordering of surfactant head groups able to make ion pairs with counter-ions. This approach is based on the Collins approach for determining the affinity of the ions to water. Unfortunately, only a few surfactant head groups were studied in this particular work. Independently, Karakashev [33] extended the scope of their series by studying the adsorption energy of a number of surfactant head groups (anionic, cationic, and nonionic) on the air/water interface. This is directly related to their affinity to water. Menhringer et al. [34] report on the effect of the hydrophobic tail on the affinity of the surfactant head group to water. Zakharova et al. [35] studied the salting in and out effect of hydrotropes and their relation to ion-specific effects. They stress that a linear dependence between the logarithm of the critical aggregation concentration of the ionic surfactants and the number of their carbon atom in the chain exists.

All these dependencies are based on the affinity of either surfactant head groups or certain inorganic ions and its counter-ions to water, following the simple rule derived by Collins [31] about the similarities of their radii. Yet, this rule cannot explain why $BaSO_4$ has ultra-low solubility while $K_2SO_4$ is very soluble in water if the radii of $Ba^{2+}$ and $K^+$ are similar ($R_{Ba2+}$ = 1.36 Å, $R_{K+}$ = 1.38 Å, $R_{SO4}{}^{2-}$ = 2.3 Å [36]). There is even about a 1 Å difference between the radii of the cation and anion. Yet, this theory can explain the difference in solubility between $BaSO_4$ and $PbSO_4$ ($R_{Ba2+}$ = 1.36 Å, $R_{Pb}{}^{2+}$ = 1.18 Å, $R_{SO4}{}^{2-}$ = 2.3 Å [36]).

To shed a light on this controversial topic, we decided to test the effect of three cations ($Ba^{2+}$, $Pb^{2+}$, and $Fe^{3+}$) on the foaming capacity of sodium dodecyl sulfate (SDS) and its surface tension, keeping in mind the experimental data on the solubilities of their sulfate salts ($BaSO_4$, $PbSO_4$, and $Fe_2(SO_4)_3$).

In the present study, we mixed sodium dodecyl sulfate (SDS) with salts at different ratios, containing cations interacting with the sulfate group of the SDS molecule, thus producing sulfate sediments with wide range of solubility. We measured the surface tension and the foamability of these precipitates, thus making correlations between surface tension and foamability.

## 2. Materials and Methods

All the chemicals (SDS, $BaCl_2$, $Pb(NO_3)_2$, and $FeCl_3$) were purchased from Sigma-Aldrich (Darmstadt, Germany). SDS was purified by means of three-time recrystallization with ethanol. The deionized water was produced by a Millipore water purification system. SDS was mixed with each one of the above-mentioned salts in the following molar ratios of salt/SDS: 1/2, 1/1, 2/1 and 4/1. The SDS concentration range studied in the present work is 1 mmol/L–10 mmol/L. At each particular case, a sediment was formed. A droplet from this newly formed suspension was positioned between two glass slides and observed on an optical microscope, allowing photos to be taken. This allowed us to take photos of the sediment for each particular case. The surface tension was measured by means of an automatic tensiometer/goniometer (Model 290, Ramé–Hart Instrument Co., Ltd., Succasunna, NJ, USA). Prior to the measurement of the surface tension, each particular solution was centrifuged for 20 min with 3000 r. p. m. to remove the sediment. The foam was generated by means of the shaking method [37]. A two hundred and fifty milliliter graduated cylinder was filled with fifty milliliters of each particular solution. After this, it was vigorously shaken tenfold. The volume of the initial foam was measured and recorded for each particular case.

### 3. Results

Both foamability and foam durability are related to the surface tension and the Gibbs elasticity of the surfactant solution. One of the simplest adsorption models (Langmuir-Szyskowski) provides us with the following relations between the surface tension, surfactant adsorption, and the Gibbs elasticity [38]:

$$\sigma = \sigma_0 + \Gamma_\infty RT \ln\left(1 - \frac{\Gamma}{\Gamma_\infty}\right) + \sigma_d \quad (1)$$

$$E_g = RT\Gamma \frac{\Gamma_\infty}{\Gamma_\infty - \Gamma} \quad (2)$$

where $\sigma_0$ is the surface tension of water at temperature $T$, $R$ is the gas constant, $\Gamma$ is the adsorption of the surfactant, $\Gamma_\infty$ is the maximal surfactant adsorption, $\sigma_d$ is the contribution of the diffuse electrical layer to the surface tension, and $E_g$ is the Gibbs elasticity.

Figure 1 is an illustration about how both the Gibbs elasticity and the surface tension depend on the surfactant adsorption. One can see that when the surfactant adsorption approaches its maximal value, the surface tension drops significantly (at the critical micelle concentration, CMC), while the Gibbs elasticity value increases substantially as well. So, we can pose the following questions:

1. How will the addition of salts, interacting chemically with SDS, affect the state of the surfactant adsorption layer and consequently the surface tension and the Gibbs elasticity?
2. How will the foamability of the surfactant solution depend on both the concentrations of the surfactant and the added salt?

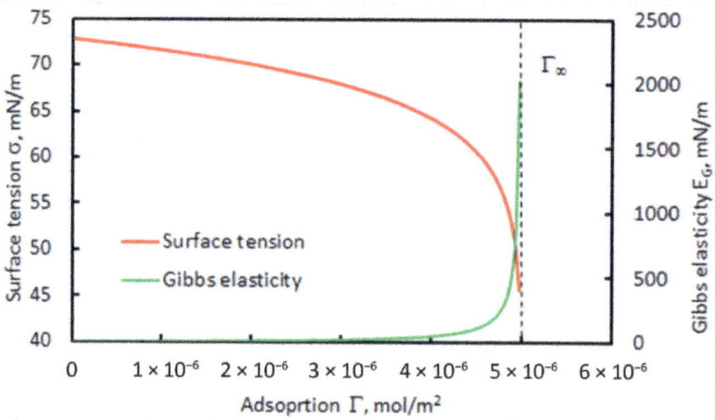


**Figure 1.** Surface tension and Gibbs elasticity versus the surfactant adsorption.

To shed a light on these questions, we conducted a series of experiments testing the foamability and surface tension of SDS solutions in the concentration range of 0.5 mmol/L–10 mmol/L in the presence of $BaCl_2$, $Pb(NO_3)_2$, and $FeCl_3$ at different salt/SDS molar ratios.

Shown in Figure 2 are the surface tension and the initial foam volume of aqueous solutions of SDS in the concentration range of 0.5 mmol/L–10 mmol/L in the absence of added salt.

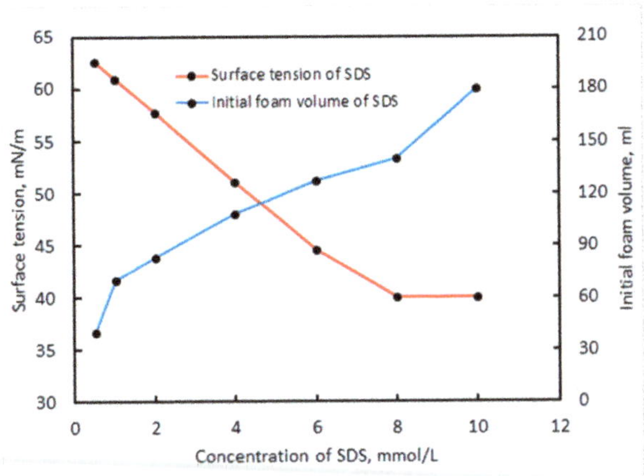


**Figure 2.** Surface tension and initial foam volume of vs. concentration of SDS in the concentration range of 0.5 mmol/L–10 mmol/L.

One can see that with the increase in the SDS concentration the surface tension decreases, while the initial foam increases. The CMC value of SDS is about 8 mmol/L, thus reaching maximal values of Gibbs elasticity. At concentrations larger than the CMC, the foamability is even larger. Another important question which can be raised is how the addition of salt interacting with SDS should affect both the surface tension and the foamability. Theoretically, two effects could be expected (see Figure 3): (i) the increase in the ionic strength due to the addition of salt should increase the adsorption of $DS^-$ co-ions, thus decreasing the surface tension [39] and increase the foamability; (ii) the chemical interaction of the salt with SDS will form a sediment/precipitate, thus exhausting the surfactant solution, causing an increase in the surface tension value and a decrease in foamability.

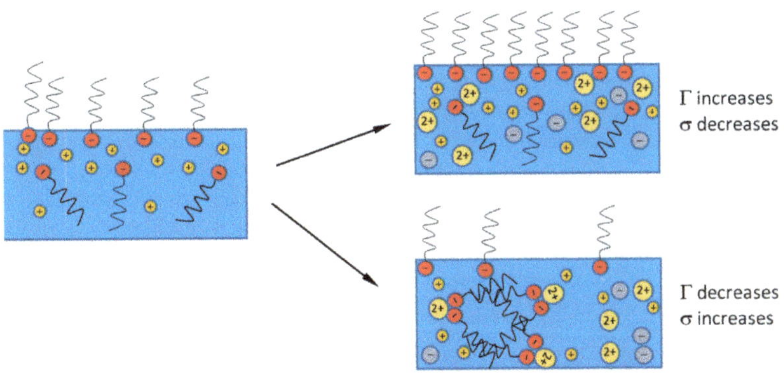


**Figure 3.** Illustration of the dual effect of the electrolyte on the surface tension of air/water interface.

The formation of sediment in the form of particles/aggregates should affect the foamability as well. Which one of all these effects will dominate? In all of the cases, metal sulfates are formed. $BaCl_2$ and $Pb(NO_3)_2$ react chemically with SDS via the following equations:

$$\begin{gathered} BaCl_2 + 2SDS \rightarrow Ba(DS)_2 + 2NaCl \\ Pb(NO_3)_2 + 2SDS \rightarrow Pb(DS)_2 + 2NaNO_3 \end{gathered} \tag{3}$$

Table 1 shows the solubilities of $BaSO_4$, $PbSO_4$, and $Fe_2(SO_4)_3$ per 100 mL water at T = 20 °C. One can see that $BaSO_4$ and $PbSO_4$ have very small solubilities in water and $BaSO_4$ is significantly less soluble than $PbSO_4$. The third salt ($Fe_2(SO_4)_3$) has significant solubility.

**Table 1.** Solubility of $BaSO_4$, $PbSO_4$ and $Fe_2(SO_4)_3$ in water at T = 20 °C.

| Salt | $BaSO_4$ | $PbSO_4$ | $Fe_2(SO_4)_3$ |
|---|---|---|---|
| Solubility, mg/100 mL water at T = 20 °C | 0.245 | 40.400 | 25,600 |

The sediment, which is formed in our experiments is $Ba(C_{12}H_{25}SO_4)_2$, $Pb(C_{12}H_{25}SO_4)_2$, and $Fe(C_{12}H_{25}SO_4)_3$. The formation of the third sediment has more of a physicochemical nature than a chemical one. We assumed at first approximation that the solubilities of these three organic salts are close to their inorganic analogs, shown in Table 1. One can see that solubility increases in the series $BaSO_4 < PbSO_4 < Fe_2(SO_4)_3$. So, we were curious to know how this significant difference in their solubility will affect both surface tension and foamability.

Figure 4 shows the surface tension and the initial foam volume of the aqueous solution of SDS in the concentration range of 0.5 mmol/L–10 mmol/L + added salt with a salt/SDS molar ratio of ½ (Figure 4A–C), which is the stoichiometric ratio for the cases of $Ba(DS)_2$ and $Pb(DS)_2$ and for $Fe(DS)_3$, even in excess, for the complete interaction of SDS with each one of the salts. The surface tension of only SDS is presented as well for comparison. Figure 4D shows the surface tension values of the same aqueous solutions as well in a separate graph. Photos of the sediments are presented in this Figure as well. One can see that in the case of $BaCl_2$, the surface tension dropped to the level of CMC even at 1 and 2 mmol/L SDS ($\sigma_{CMC} \approx 40$ mN/m). This corresponds to a saturated adsorption layer of SDS and the maximal value of Gibbs elasticity. Contrary to our expectations, these solutions produced no foam. The further increase in the SDS concentration caused an increase in surface tension, which is an indication of the dilution of the surfactant's adsorption layer, but at the same time, this gave a boost to the foamability of the surfactant solution. A sediment was formed as well, and the amount of sediment increases with the increase in the SDS concentration. The addition of $Pb(NO_3)_2$ at 1 mmol/L SDS dropped the surface tension to the CMC value of the single SDS solution ($\sigma_{CMC} \approx 40$ mN/m), which indicates a saturated adsorption surfactant layer on the air/water interface, but the further increase in the SDS concentration dropped the surface tension to $\sigma \approx 29$ mN/m at $C_{SDS} = 6$ mmol/L and this value did not change until reaching the final, largest concentration of SDS ($C_{SDS} = 10$ mmol/L). The foamability was not suppressed completely but kept very small values, in the range of 2 mL to 10 mL initial foam volumes. One can see in Figures 1 and 2 how Gibbs elasticity is related to surface tension and how the initial foam volume is related to Gibbs elasticity. At these values of surface tension, the surfactant solution should produce an enormous initial volume of foam, but contrary to our expectations, the foamability maintained very small values. In the case of $FeCl_3$, the surface tension decreased upon the increase in the SDS concentration until about 6 mmol/L. At this point, the surface tension reached the value $\sigma = 33.5$ mN/m and did not drop anymore. The foamability was reduced significantly compared to that of SDS alone, but was still larger than that of $Pb(NO_3)_2$. One can see as well that the sediment in the case of $FeCl_3$ is significantly better aggregated than in the two other cases.

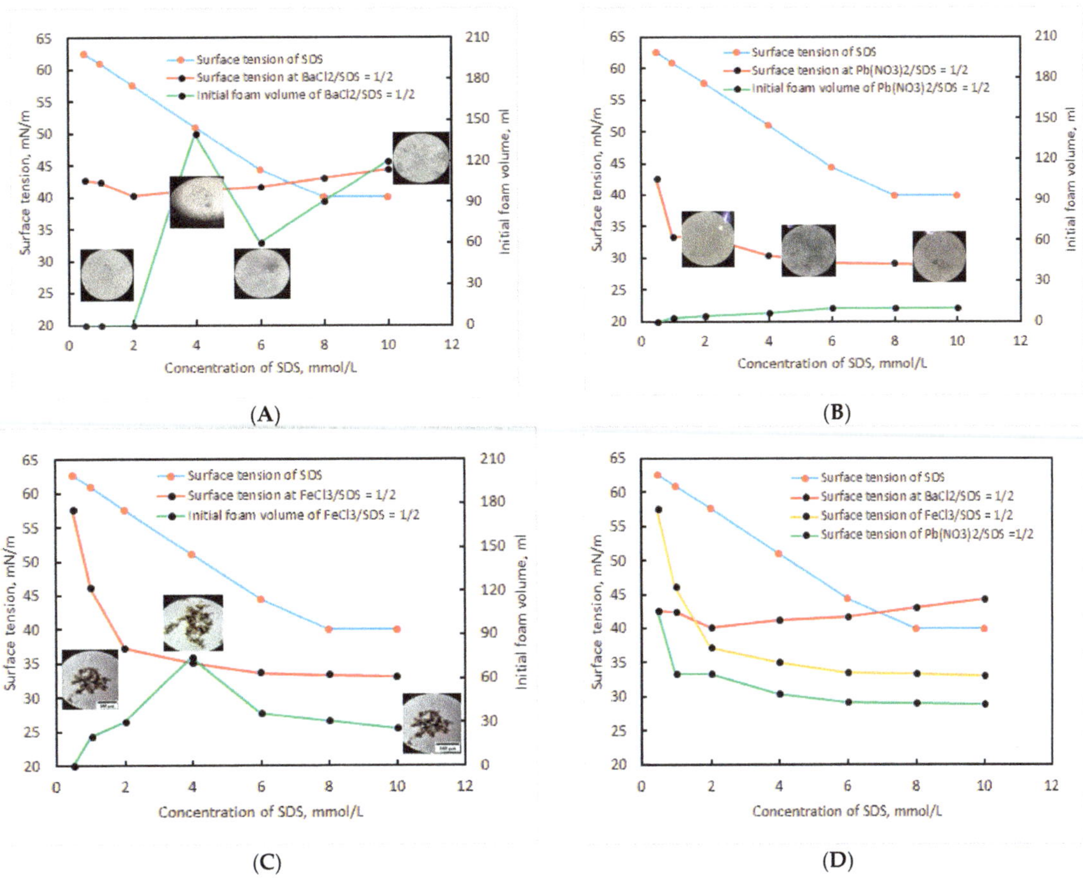


**Figure 4.** Surface tension and initial foam volume of vs. concentration of SDS in the concentration range of 0.5 mmol/L–10 mmol/L in the absence of added salt and a molar salt/SDS ratio of ½ and in presence of: (**A**) $BaCl_2$; (**B**) $Pb(NO_3)_2$, and (**C**) $FeCl_3$. (**D**) Surface tension isotherms of the above-mentioned aqueous solutions.

One can conclude in Figure 4 that, despite being in stoichiometric ratio between salt/SDS, only part of the SDS reacts with the salts. The surfactant adsorption layer looks saturated in the case of $BaCl_2$ and supersaturated in the cases of $Pb(NO_3)_2$ and $FeCl_3$. Evidently the mechanism of interaction between the salt and SDS is significantly different from the inorganic analog of the surfactant. Figure 5 shows the surface tension and the initial foam volume of the aqueous solution of SDS in the same concentration range + added salt with salt/SDS ratio = 1/1 (Figure 5A–C). The surface tension of only SDS is presented as well for comparison. Figure 5D shows the surface tension values of the same aqueous solutions in a separate graph. Photos of the sediments are presented in this Figure as well. In contrast to the case presented in Figure 4, the systems in Figure 5 are in excess of added salt. The surface tension of SDS + $BaCl_2$ as shown in Figure 5A keeps $\sigma \approx 45$ mN/m value in the whole concentration range of SDS, which is an indication of the saturated adsorption layer with $DS^-$ surface active co-ions. One can see the sediment as well on the photos in Figures 4A and 5A, and that its quantity increases upon the increase in the SDS concentration. One can see in Figure 1 that such a surface tension value ($\sigma \approx 45$ mN/m) in the case of SDS only produces about 90 mL foam, while the same surface tension in presence of $Ba(DS)_2$ sediment practically produces no foam. The situations with $Pb(NO_3)_2$ and $FeCl_3$ are even more strange, because the surface tension values in these cases are lower than those in the case of SDS only, which should produce a large amount of foam. Instead, they produce quite a small volume of foam: practically 10 mL in the case of $Pb(NO_3)_2$ (same as the case in Figure 4B, molar ratio = $Pb(NO_3)_2$/SDS =1/2) and less than 30 mL foam in the case of $FeCl_3$, which is less than in the case of molar ratio $FeCl_3$/SDS = 1/2.

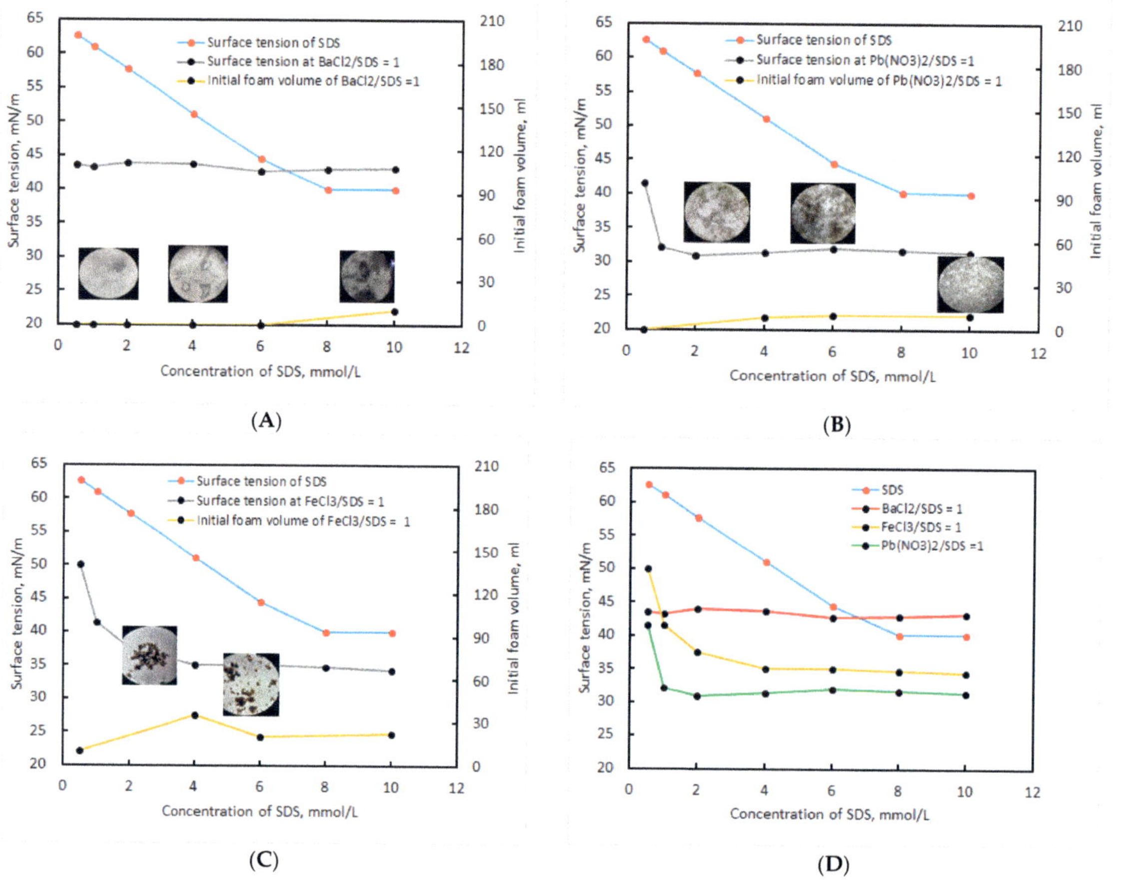


**Figure 5.** Surface tension and initial foam volume of vs. concentration of SDS in the concentration range of 0.5 mmol/L–10 mmol/L in absence of added salt and molar salt/SDS ratio 1/1 and in presence of: (**A**) $BaCl_2$; (**B**) $Pb(NO_3)_2$, and (**C**) $FeCl_3$. (**D**) Surface tension isotherms of the above-mentioned aqueous solutions.

Figure 6 shows the surface tension and the initial foam volume of the aqueous solution of SDS in the same concentration range + added salt with salt/SDS ratio = 2/1 (Figure 6A–C). The surface tension of only SDS is presented as well for comparison. Figure 6D shows the surface tension values of the same aqueous solutions in a separate graph. Photos of the sediments are presented in this Figure as well. There is no significant difference between the two systems $BaCl_2$/SDS = 1/1 and $BaCl_2$/SDS = 2/1. The surface tension values are close to $\sigma$ = 45 mN/m and there is almost no foamability. One can conclude that even at this significant stoichiometric excess of $BaCl_2$ and very small solubility of the sediment, there is still a certain amount of SDS, which keeps the value of the surface tension low. At the same time, there is no foamability. The case with molar ratio $Pb(NO_3)_2$ /SDS = 2/1 is similar to ratio 1/1, identical to the case of $BaCl_2$. The situation is similar in the case of molar ratio $FeCl_3$/SDS = 2/1. In both cases (molar ratios salt/SDS = 1/1 and salt/SDS = 2/1), the salts are in stoichiometric excess, whose value does not significantly affect the surface tension and the foamability of the systems. Yet, it is curious to note that the level of foamability in each particular case follows the solubility of the newly formed salts.

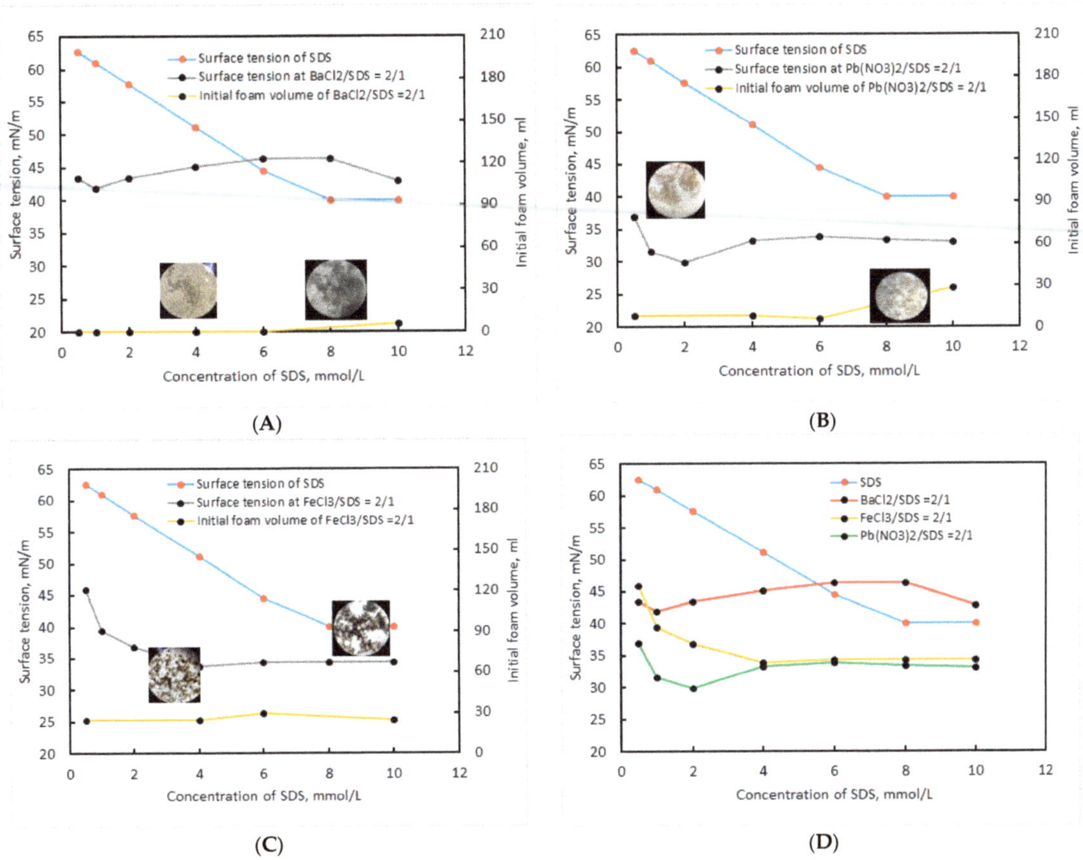


**Figure 6.** Surface tension and initial foam volume of vs. concentration of SDS in the concentration range of 0.5 mmol/L–10 mmol/L in absence of added salt and molar salt/SDS ratio 2/1 and in presence of: (**A**) $BaCl_2$; (**B**) $Pb(NO_3)_2$, and (**C**) $FeCl_3$. (**D**) Surface tension isotherms of the above-mentioned aqueous solutions.

Figure 7 shows the surface tension and the initial foam volume of the aqueous solution of SDS in the same concentration range + added salt with salt/SDS ratio = 4/1 (Figure 7A–C). The surface tension of only SDS is presented as well for comparison. Figure 7D shows the surface tension values of the same aqueous solutions in a separate graph. Photos of the sediments are presented in this Figure as well. The increase in the salt's stoichiometric excess to molar ratio salt/SDS = 4/1 significantly affects surface tension and foamability in the cases of $BaCl_2$ and $Pb(NO_3)_2$. The surface tension values increased roughly from $\sigma \approx 45$ mN/m (molar ratio $BaCl_2$/SDS = 2/1) to about $\sigma \approx 50$ mN/m at molar ratio $BaCl_2$/SDS = 4/1, which is an indication of exhausting the solution of SDS, but at the same time, there is no foamability. The surface tension values in the case of $Pb(NO_3)_2$ increase as well, roughly from $\sigma \approx 35$ mN/m (molar ratio $Pb(NO_3)_2$/SDS = 2/1) to about $\sigma \approx 40$ mN/m at molar ratio $Pb(NO_3)_2$/SDS = 4/1. The larger excess of $Pb(NO_3)_2$ exhausted the solution of SDS, which caused an increase in the surface tension and a complete lack of foamability. The increase in the stoichiometric excess in the case of $FeCl_3$ did not practically affect surface tension or foamability. It is interesting to note that the foamability of the systems are in line with the solubility of the sediments: **(i)** $BaSO_4$ has the lowest solubility (0.245 mg per 100 mL water at T = 20 °C) among the three salts and hence the system SDS + $BaCl_2$ has the lowest foamability; **(ii)** $PbSO_4$ is the next most soluble salt (40.4 mg per 100 mL water at T = 20 °C ) and hence the system SDS + $Pb(NO_3)_2$ has larger foamability; **(iii)** $Fe_2(SO_4)_3$ (91.2 g per 100 mL water at T = 20 °C) has the largest foamability. This corresponds to roughly 2 mol/L. At the same time, the sediment $Fe(DS)_3$ forms sediment as seen in Figure 4 even at 1 mmol/L SDS and 0.5 mmol/L $FeCl_3$. Moreover, it appears that the level of stoichiometric excess of $FeCl_3$ does not practically affect either the surface tension value or the foamability. Hence, the overall trend in salt foam suppressing ability is: $BaCl_2$ > $Pb(NO_3)_2$ > $FeCl_3$.

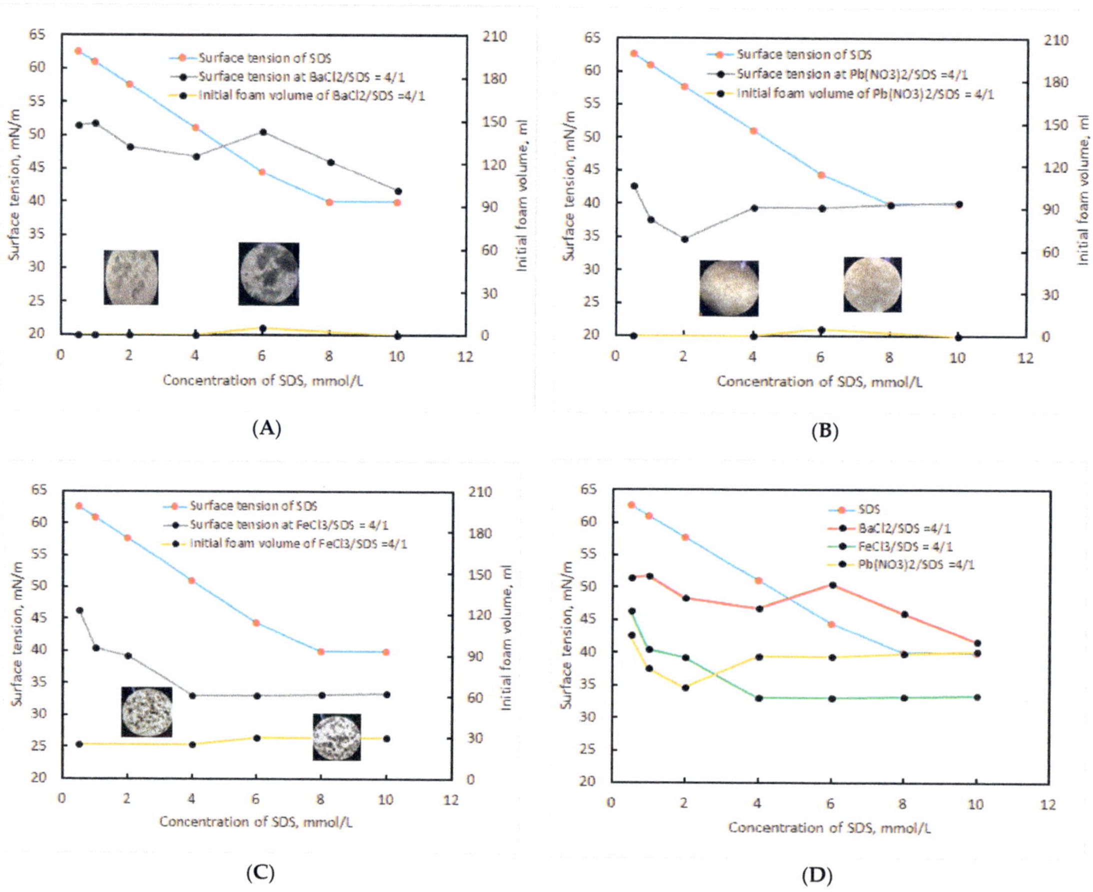


**Figure 7.** Surface tension and initial foam volume of vs. concentration of SDS in the concentration range of 0.5 mmol/L–10 mmol/L in absence of added salt and molar salt/SDS ratio 4/1 and in presence of: **(A)** $BaCl_2$; **(B)** $Pb(NO_3)_2$, and **(C)** $FeCl_3$. **(D)** Surface tension isotherms of the above-mentioned aqueous solutions.

## 4. Discussion

The foamability of SDS has been studied in the literature at high concentrations and in the presence of LiCl, NaCl, KCl, and $Mg(NO_3)_2$ [40,41], thus forming sediments, stabilizing the foam. We have studied the same surfactant at significantly lower concentrations in the presence of LiCl, NaCl, and KCl in preceding works of ours [6,7] as well. The idea of the present research is different. We mixed SDS with salts, which certainly make sediments with very low solubility, ($Ba(DS)_2$ and $Pb(DS)_2$), similarly to their inorganic analogs ($BaSO_4$ and $PbSO_4$). We used additional salt with cation ($Fe^{3+}$), which has a soluble inorganic analog ($Fe_2(SO_4)_3$) in water, just to compare the results with the other two salts. Such a study, using this idea, to our knowledge has not been conducted in the literature. So, we cannot compare our results with any in the literature.

Converting the surfactant into a sediment should increase its surface tension value and decrease its foamability. Ref. [6] reports about the foam suppressing ability of KCl when added to 0.5 mmol/L SDS. The latter forms a sediment of KDS, which, above 40 mmol/L KCl substantially suppresses the formation of foam, while the surface tension reaches $\sigma \approx 41$ mN/m at 25 mmol/L KCl. At 80 mmol/L added KCl, the foamability is suppressed completely. The mean precipitation concentration of the system SDS + KCl is about 6.36 mmol/L, while the solubility of $K_2SO_4$ is about 0.64 mol/L at T = 20 °C. So, these experimental results pose the following questions:

1. If the addition of KCl to SDS forms a sediment of KDS, thus suppressing foamability, keeping in mind that $K_2SO_4$ is very soluble, what will be the effect of added salt with counter-ions, whose sulfate has ultra-low solubility?
2. Why is the foamability suppressed at such a low value of surface tension upon the formation of the sediment?

To our knowledge, the experimental data in the literature about the effects of different salts on foamability is still poor. For this reason, we selected three salts, forming sulfates with: ultra-low solubility–(i) $BaCl_2$ (the solubility of $BaSO_4$ is 0.245 mg per 100 mL water at T = 20 °C); (ii) low solubility–$Pb(NO_3)_2$ (the solubility of $PbSO_4$ is 40.4 mg per 100 mL water at T = 20 °C); (iii) and medium solubility –$FeCl_3$ (the solubility of $Fe_2(SO_4)_3$ is 25.6 g per 100 mL water at T = 20 °C). The addition of an electrolyte to an aqueous solution of ionic surfactants should have basically two major effects:

1. A drop in the surface tension value due to screening of the electrostatic repulsion between the surfactant adsorption layer and the surface-active co-ions;
2. A precipitation of the surfactant thus forming sediment, exhausting the solution of surfactant species. This should increase the value of the surface tension and suppress the foamability of the system.

How are these two effects related to the solubility of the sulfates, the surface tension values, and the foamability in our particular case of SDS + added salts? We conducted this experimental research to shed light on these questions. We found the results intriguing, bringing about more questions.

To better understand the experimental results, we should begin from the observation of the surface tension and the foamability of SDS only. One can see in Figure 2 and Table 2 that the drop in surface tension is related to a significant increase in the initial foam volume. This is due to the increase in the Gibbs elasticity, as seen in Figure 1. The CMC value of SDS is about 7.5 mmol/L, at which it reaches its minimal value of surface tension of about $\sigma$ = 40.1 mN/m. So, the SDS concentration range of 0.5–10 mmol/L corresponds to a drop in the surface tension from $\sigma$ = 62.7 mN/m to $\sigma$ = 40.1 mN/m. Thus, the initial foam volume increases from V = 40 mL to V = 180 mL. The addition of $BaCl_2$ in molar ratio salt/SDS =1/2 (stoichiometric ratio) causes the formation of the least soluble sediment ($Ba(DS)_2$) of the three salts. In addition, it causes a substantial decrease in the surface tension value, even at 0.5 mmol/L SDS–from $\sigma$ = 62.7 mN/m to $\sigma$ = 43 mN/m, which is close to but above the minimal value of $\sigma$ = 40.1 mN/m at the CMC of SDS. This is due to electrostatic screening causing lower repulsion between the negatively charged surfactant adsorption layer and $DS^-$ co-ions. Evidently, not all of the $BaCl_2$ reacted with SDS in spite of their correct ½ molar stoichiometric ratio. Despite the low value of the surface tension ($\sigma \approx 43$ mN/), the foamability is completely suppressed up to a concentration of 2 mmol/L SDS. The increase in the concentration of SDS above 2 mmol/L, keeping the same molar ratio ($BaCl_2$/SDS = 1/2), gives rise to foamability, which is an indication of an increase in the excess of SDS in the aqueous solution. Evidently, only part of the $BaCl_2$ interacts with SDS. The further increase in the $BaCl_2$/SDS molar ratio (1/1, 2/1) increases the amount of $Ba^{2+}$ cations and promotes their interaction with SDS. The surface tension values practically did not change, while the foamability was suppressed. The low value of the surface tension indicates free species of SDS being adsorbed on the air/water interface, but the lack of foamability makes the puzzle of this phenomenon more difficult to understand. At molar ratio $BaCl_2$/SDS = 4/1, the value of the surface tension increases significantly, especially at lower concentrations of SDS, which shows that the solution is exhausted of SDS and the foamability is completely suppressed. On the contrary, the same value of surface tension at the single SDS system produces a substantial amount of foam. The situation of the system containing SDS + $Pb(NO_3)_2$ is not far different. At molar ratio $Pb(NO_3)_2$/SDS = 1/2 (stoichiometric ratio), the surface tension value drops from $\sigma \approx 42$ mN/m at 0.5 mmol/L SDS to $\sigma \approx 29$ mN/m at 6 mmol/L SDS, above which the surface tension does not change. The foamability in this SDS concentration range is low: between 0 and 10 mL initial foam volume. Indeed, the drop in surface tension significantly below the minimal one of the single SDS system was unexpected. As far as $PbSO_4$ is more soluble than $BaSO_4$, it forms less sediment and exhausts the aqueous solution less, but $Pb^{2+}$ counter-ions electrostatically screen the repulsion between the surfactant adsorption layer and $DS^-$ co-ions, causing significantly denser surfactant adsorption later on the air/water interface than in the single SDS case. Even at these small values of surface tension ($\sigma \approx 29$ mN/m), the foamability is low (between 0 and 10 mL initial foam volume). The further increase in the molar ratio $Pb(NO_3)_2$/SDS still keeps its initial drop from 0.5 mmol/L to about 2 mmol/L, but shifts up a linear part of the surface tension from $\sigma \approx 29$ mN/m (at $Pb(NO_3)_2$/SDS = 1/2) to $\sigma \approx 39.5$ mN/m (at $Pb(NO_3)_2$/SDS = 4/1). This indicatesa decrease in the concentration of free species of SDS due to their converting into $Pb(DS)_2$ sediment. The foamability is low at molar ratios $Pb(NO_3)_2$/SDS = 1/2, 1/1, 2/1 and completely suppressed at $Pb(NO_3)_2$/SDS = 4/1. Evidently, the surfactant and the salt interact, thus forming a sediment with small solubility, but still, it remains $Pb^{2+}$ and $DS^-$ ions, which are responsible for the substantial drop in surface tension below the minimal one for the single SDS, but the low foamability at this small value of surface tension becomes a real puzzle for further investigation. The system containing $FeCl_3$ + SDS is intriguing as well. It forms the most soluble sediment. At molar ratio $FeCl_3$/SDS = ½, the surface tension value drops from $\sigma \approx 57.6$ mN/m at 0.5 mmol/L SDS to $\sigma \approx 33.1$ mN/m at 10 mmol/L SDS, which, again, is below the minimal value of the surface tension for the SDS-only system. The foamability is reduced as compared with the SDS single system, but it is still significant. Evidently, $Fe^{3+}$ counter-ions act mostly to screen the electrostatic repulsion species, thus causing a denser surfactant layer on the air/water interface. The further increase in the molar ratio of $FeCl_3$/SDS decreases the surface tension value in the SDS concentration range of 0.5 mmol/L–2 mmol/L, while the rest of the concentration range of SDS (2 mmol/L–10 mmol/L) practically keeps a constant surface tension value ($\sigma \approx 35$ mN/m), not affected by the increase in the molar ratio of $FeCl_3$/SDS. Evidently, the SDS adsorption layer cannot be compressed more, despite the complete screening of the electrostatic repulsion. The foamability is not significantly affected by the increase in the $FeCl_3$/SDS molar ratio, thus keeping about 30 mL initial foam volume. It is interesting to correlate the solubility of the sulfates (see Table 1) with the foamability of the systems. Hence, one can order the counter-ions in a series of increasing foam-suppressing capacity in aqueous solutions of SDS: $Fe^{3+} < Pb^{2+} < Ba^{2+}$. This is in line with the solubility values of the sulfate sediments, as shown in Table 1. Furthermore, the three salts act as electrostatic repulsion screening agents, thus reducing the repulsion between the surfactant adsorption layer and $DS^-$ co-ions. The latter makes the surfactant adsorption layer more populated, thus reducing the surface tension value.

**Table 2.** Surface tension values and initial foam volumes of SDS and SDS + added salt in the specified molar ratios in the SDS concentration range of 0.5–10 mmol/L.

| | **SDS** | **$BaCl_2$ + 2SDS → $Ba(DS)_2$** | **$Pb(NO_3)_2$ + 2SDS → $Pb(DS)_2$** | **$FeCl_3$ + 3 SDS → $Fe(DS)_3$** |
|---|---|---|---|---|
| Salt/SDS ratio | | Surface tension in range of SDS: 0.5 mmol/L–10 mmol/L, mN/m | | |
| 1/2 | 62.7–40.1 | 43.0–44.4 | 42.7–28.9 | 57.6–33.1 |
| 1/1 | | 43.5–43.2 | 41.4–31.4 | 50.0–34.3 |
| 2/1 | | 43.9–42.9 | 37.0–33.1 | 45.9–34.3 |
| 4/1 | | 51.5–41.7 | 42.7–40.2 | 46.3–33.4 |
| Salt/SDS ratio | | Initial foam volume in range of SDS: 0.5 mmol/L–10 mmol/L, mL | | |
| 1/2 | 40–180 | 0–120 | 0–10 | 0–26 |
| 1/1 | | 0–10 | 0–10 | 10–22 |
| 2/1 | | 0–6 | 8–28 | 25–25 |
| 4/1 | | 0–0 | 0–0 | 25–30 |

## 5. Conclusions

The experimental results raised more questions about the nature of foam suppressing than existed prior to the beginning of this experiment. In short, the following conclusions were drawn from this work:

1. The counter-ions of the added salt react with $DS^-$ co-ions, thus forming sediments with increasing solubility in the series $Ba(DS)_2$ < $Pb(DS)_2$ < $Fe(DS)_3$ (see Table 1); the mechanism of this reaction is not clear, because they do not react completely, thus leaving certain free amounts of counter-ions and SDS; the latter varies with the variation in salt/SDS molar ratio;
2. The three salts reduce the foamability of the aqueous solutions of SDS following the same order of increasing foam suppressing capacity: $FeCl_3$ < $Pb(NO_3)_2$ < $BaCl_2$; this correlates well with the solubility of the sediments, suggesting the idea of the mechanism of exhausting of the solution from SDS.
3. The surface tension value drops significantly in the three cases, even substantially below the minimal value at CMC of the single SDS system. Despite the low values of surface tension, the foamability is either completely suppressed or significantly reduced. The hypothesis of exhausting the solution of SDS, due to the formation of sediment, cannot explain the combination of low surface tension values and lack of foamability (or significantly reduced foamability). Evidently, more studies are needed to explain this puzzle.

**Author Contributions:** Conceptualization, S.I.K. and V.M.P.; methodology, S.I.K.; software, S.I.K.; validation, S.I.K., N.R., N.A.G., K.M. and D.I.-S.; formal analysis, S.I.K. and V.M.P.; investigation, N.R., N.A.G., K.M. and D.I.-S.; data curation, S.I.K.; writing—original draft preparation, S.I.K.; writing—review and editing, S.I.K.; visualization, S.I.K.; supervision, S.I.K. and V.M.P.; project administration, S.I.K. and D.I.-S.; funding acquisition, S.I.K. and D.I.-S. All authors have read and agreed to the published version of the manuscript.

**Funding:** This research was funded by Shumen University "Episkop Konstantin Preslavsky", grant number No. RD-08-79/29.01.2024.

**Institutional Review Board Statement:** Not applicable.

**Informed Consent Statement:** Not applicable.

**Data Availability Statement:** Data are contained within the article.

**Acknowledgments:** S. Karakashev and D. Ivanova-Stancheva thank to Shumen University "Episkop Konstantin Preslavsky", grant number No. RD-08-79/29.01.2024 for the financial support.

**Conflicts of Interest:** The authors declare no conflicts of interest.

## References

1. Hofmeister, F. About regularities in the protein precipitating effects of salts and the relation of these effects with the physiological behaviour of salts. *Arch. Fuer Exp. Pathol. Und Pharmakol.* **1887**, *XXIV*, 247–260.
2. Hofmeister, F. About the water withdrawing effect of the salts. *Arch. Fuer Exp. Pathol. Und Pharmakol.* **1888**, *XXV*, 1–30.
3. Hofmeister, F. To the theory of the effect of the salts (Second message). *Arch. Exp. Pathol. Pharmakol.* **1888**, *24*, 247–260. [CrossRef]
4. Hofmeister, F. To the theory of the effect of the salts (First message). *Arch. Exp. Pathol. Pharmakol.* **1888**, *25*, 1–30. [CrossRef]
5. Hofmeister, F. To the theory of the effect of the salts (Free and disabled swelling). *Arch. Exp. Pathol. Pharmakol.* **1890**, *27*, 395–413. [CrossRef]
6. Sett, S.; Karakashev, S.I.; Smoukov, S.K.; Yarin, A.L. Ion-specific effects in foams. *Adv. Colloid. Interface Sci.* **2015**, *225*, 98–113. [CrossRef]
7. Amani, P.; Karakashev, S.I.; Grozev, N.A.; Simeonova, S.S.; Miller, R.; Rudolph, V.; Firouzi, M. Effect of selected monovalent salts on surfactant stabilized foams. *Adv. Colloid. Interface Sci.* **2021**, *295*, 102490. [CrossRef]
8. Koga, Y. 1-Propanol probing methodology: Two-dimensional characterization of the effect of solute on $H_2O$. *Phys. Chem. Chem. Phys.* **2013**, *15*, 14548–14565. [CrossRef]
9. Koga, Y. Two-dimensional characterization of the effect of solute on $H_2O$: A thermodynamic probing methodology. *J. Mol. Liq.* **2015**, *205*, 31–36. [CrossRef]
10. Christov, C.; Donchev, S.; Tsenov, T. Thermodynamic models for solution behavior and solid-liquid equilibrium in acetate binary systems from low to very high concentration at 25 °C. *Bulg. Chem. Commun.* **2024**, *54*, 26–31. [CrossRef]
11. Donchev, S.; Tsenov, T.; Christov, C. Development of accurate chemical thermodynamic database for geochemical storage of nuclear waste. Part II: Models for predicting solution properties and solid-liquid equilibrium in binary nitrate systems. *BioRisk* **2022**, *2022*, 389–406. [CrossRef]
12. Donchev, S.A.; Christov, C.I. Development of Accurate Chemical Thermodynamic Database for Geochemical Storage of Nuclear Waste. Part I: Models for Predicting Solution Properties and Solid-Liquid Equilibrium in Binary Nitrate Systems of the Type 1-1. *Ecol. Balk.* **2020**, *12*, 195–210. [CrossRef]
13. Tsenov, T.; Donchev, S.; Christov, C. Development of accurate chemical thermodynamic database for geochemical storage of nuclear waste. Part III: Models for predicting solution properties and solid-liquid equilibrium in cesium binary and mixed systems. *BioRisk* **2022**, *2022*, 407–422. [CrossRef]
14. Pitzer, K.S. Thermodynamics of electrolytes. I. Theoretical basis and general equations. *J. Phys. Chem.* **1973**, *77*, 268–277. [CrossRef]
15. Pitzer, K.S.; Kim, J.J. Thermodynamics of electrolytes. IV. Activity and osmotic coefficients for mixed electrolytes. *J. Am. Chem. Soc.* **1974**, *96*, 5701–5707. [CrossRef]
16. Pitzer, K.S.; Mayorga, G. Thermodynamics of electrolytes. II. Activity and osmotic coefficients for strong electrolytes with one or both ions univalent. *J. Phys. Chem.* **1973**, *77*, 2300–2308. [CrossRef]
17. Pitzer, K.S.; Mayorga, G. Thermodynamics of electrolytes. III. Activity and osmotic coefficients for 2-2 electrolytes. *J. Solut. Chem.* **1974**, *3*, 539–546. [CrossRef]
18. Heydweiller, A. Concerning the physical characteristics of solutions in correlation. II. Surface tension and elcetronic conductivity of watery salt solutions. *Ann. Der Phys.* **1910**, *33*, 145–185. [CrossRef]
19. Warszynski, P.; Lunkenheimer, K.; Czichocki, G. Effect of counterions on the adsorption of ionic surfactants at fluid-fluid interfaces. *Langmuir* **2002**, *18*, 2506–2514. [CrossRef]
20. Para, G.; Jarek, E.; Warszynski, P. The Hofmeister series effect in adsorption of cationic surfactants—theoretical description and experimental results. *Adv. Colloid. Inreface Sci.* **2006**, *122*, 39–55. [CrossRef]
21. Para, G.; Jarek, E.; Warszynski, P. The surface tension of aqueous solutions of cetyltrimethylammonium cationic surfactants in presence of bromide and chloride counterions. *Coll. Surf. A* **2005**, *261*, 65–73. [CrossRef]
22. Li, H.H.; Imai, Y.; Yamanaka, M.; Hayami, Y.; Takiue, T.; Matsubara, H.; Aratono, M. Specific counterion effect on the adsorbed film of cationic surfactant mixtures at the air/water interface. *J. Colloid. Interface Sci.* **2011**, *359*, 189–193. [CrossRef] [PubMed]
23. Lu, J.R.; Marrocco, A.; Su, T.; Thomas, R.K.; Penfold, J. Adsorption of dodecyl sulfate surfactants with monovalent metal counterions at the air-water interface studied by neutron reflection and surface tension. *J. Colloid. Interface Sci.* **1993**, *158*, 303–316. [CrossRef]
24. Ninham, B.W.; Yaminsky, V. Ion Binding and Ion Specificity: The Hofmeister Effect and Onsager and Lifshitz Theories. *Langmuir* **1997**, *13*, 2097–2108. [CrossRef]
25. Bostroem, M.; Williams, D.R.M.; Ninham, B.W. Surface tension of electrolytes: Specific ion effects explained by dispersion forces. *Langmuir* **2001**, *17*, 4475–4478. [CrossRef]
26. Bostrom, M.; Williams, D.R.M.; Ninham, B.W. Specific Ion Effects: Why DLVO Theory Fails for Biology and Colloid Systems. *Phys. Rev. Lett.* **2001**, *87*, 168103. [CrossRef]

27. Tavares, F.W.; Bratko, D.; Blanch, H.W.; Prausnitz, J.M. Ion-specific effects in the colloid-colloid or protein-protein potential of mean force: Role of salt-macroion van der Waals interactions. *J. Phys. Chem. B* **2004**, *108*, 9228–9235. [CrossRef]
28. Ivanov, I.B.; Marinova, K.G.; Danov, K.D.; Dimitrova, D.; Ananthapadmanabhan, K.P.; Lips, A. Role of the counterions on the adsorption of ionic surfactants. *Adv. Colloid. Interface Sci.* **2007**, *134–135*, 105–124. [CrossRef]
29. Pandey, S.; Bagwe, R.P.; Shah, D.O. Effect of counterions on surface and foaming properties of dodecyl sulfate. *J. Colloid. Interface Sci.* **2003**, *267*, 160–166. [CrossRef]
30. Petkova, N.; Ivanova-Stancheva, D.; Grozev, N.A.; Mircheva, K.; Karakashev, S.I. Counter-Ion Effect on the Surface Potential of Foam Films and Foams Stabilized by 0.5 mmol/L Sodium Dodecyl Sulfate. *Coatings* **2024**, *14*, 51. [CrossRef]
31. Collins, K.D. Why continuum electrostatics theories cannot explain biological structure, polyelectrolytes or ionic strength effects in ion–protein interactions. *Biophys. Chem.* **2012**, *167*, 43–59. [CrossRef] [PubMed]
32. Vlachy, N.; Jagoda-Cwiklik, B.; Vácha, R.; Touraud, D.; Jungwirth, P.; Kunz, W. Hofmeister series and specific interactions of charged headgroups with aqueous ions. *Adv. Colloid. Interface Sci.* **2009**, *146*, 42–47. [CrossRef] [PubMed]
33. Karakashev, S.I. How to determine the adsorption energy of the surfactant's hydrophilic head? How to estimate easily the surface activity of every simple surfactant? *J. Colloid. Interface Sci.* **2014**, *432*, 98–104. [CrossRef] [PubMed]
34. Mehringer, J.; Hofmann, E.; Touraud, D.; Koltzenburg, S.; Kellermeier, M.; Kunz, W. Salting-in and salting-out effects of short amphiphilic molecules: A balance between specific ion effects and hydrophobicity. *Phys. Chem. Chem. Phys.* **2021**, *23*, 1381–1391. [CrossRef]
35. Zakharova, L.Y.; Vasilieva, E.A.; Mirgorodskaya, A.B.; Zakharov, S.V.; Pavlov, R.V.; Kashapova, N.E.; Gaynanova, G.A. Hydrotropes: Solubilization of nonpolar compounds and modification of surfactant solutions. *J. Mol. Liq.* **2023**, *370*, 120923. [CrossRef]
36. Marcus, Y. *Ion Properties*; Marcel Dekker: New York, NY, USA, 1997.
37. Bartsch, O. Concerning the formation of foam and surface tension. *Kolloid-Zeitschrift* **1926**, *38*, 177–179. [CrossRef]
38. Kralchevsky, P.A.; Danov, K.D.; Denkov, N.D. Chemical Physics of Colloid Systems and Interfaces. In *Handbook of Surface and Colloid Chemistry*; Birdi, K.S., Ed.; CRC Press: Boca Raton, FL, USA, 2008; pp. 179–379.
39. Ivanov, I.B.; Ananthapadmanabhan, K.P.; Lips, A. Adsorption and structure of the adsorbed layer of ionic surfactants. *Adv. Colloid. Interface Sci.* **2006**, *123–126*, 189–212. [CrossRef]
40. Zhang, L.; Wang, H.; Zheng, B.; Du, H.; Salonen, A. Surfactant Crystals as Stimulable Foam Stabilizers: Tuning Stability with Counterions. *J. Surfactants Deterg.* **2019**, *22*, 1237–1245. [CrossRef]
41. Binks, B.P.; Shi, H. Aqueous Foams in the Presence of Surfactant Crystals. *Langmuir* **2020**, *36*, 991–1002. [CrossRef]

coatings

MDPI

*Article*

# A Spray-Deposited Modified Silica Film on Selective Coatings for Low-Cost Solar Collectors

**Francisco Ivan Lizama-Tzec [1,*], Marco de Jesús Cetina-Dorantes [1], Dallely Melissa Herrera-Zamora [2], Juan José Alvarado-Gil [1], Geonel Rodríguez-Gattorno [1], Manuel Alejandro Estrella-Gutiérrez [3], Octavio García-Valladares [2], Caridad Vales-Pinzón [4] and Gerko Oskam [5]**

1 Department of Applied Physics, Center for Research and Advanced Studies, Campus Merida, Km 6 Carretera Antigua a Progreso, Cordemex, Mérida 97319, Yucatán, Mexico; marco.cetina@cinvestav.mx (M.d.J.C.-D.); juan.alvarado@cinvestav.mx (J.J.A.-G.); geonelr@cinvestav.mx (G.R.-G.)
2 Instituto de Energías Renovables de la Universidad Nacional Autónoma de México (IER-UNAM), Privada Xochicalco s/n, Col. Centro, Temixco 62580, Morles, Mexico; dmhz@ier.unam.mx (D.M.H.-Z.); ogv@ier.unam.mx (O.G.-V.)
3 Facultad de Ingeniería Química, Universidad Autónoma de Yucatán, Periférico Norte, Kilómetro 33.5, Chuburná de Hidalgo Inn, Mérida 97203, Yucatán, Mexico; alejandro.estrella@correo.uady.mx
4 Facultad de Ingeniería, Universidad Autónoma de Yucatán, Avenida Industrias no contaminantes Anillo periférico Norte s/n, Mérida 97203, Yucatán, Mexico; caridad.vales@correo.uady.mx
5 Center for Nanoscience and Sustainable Technologies (CNATS), Department of Physical, Chemical and Natural Systems, Universidad Pablo de Olavide, 41013 Seville, Spain; gosk@upo.es
* Correspondence: flizama@cinvestav.mx

**Abstract:** Solar collectors represent an attractive green technology for water heating, where sunlight is efficiently absorbed by a selective coating and the generated heat is transferred to water. In this work, the improvement and scale-up of an electrodeposited black nickel selective coating with a modified silica (MS) film deposited by spray pyrolysis are reported. The MS material was prepared by the sol–gel method using tetraethyl orthosilicate with the addition of n-propyl triethoxysilane to obtain a porous film with an adequate refractive index and enhanced flexibility. The reflectance of electrodeposited selective coatings was characterized with and without the MS film and compared to a commercially available coating of black paint. The MS film increased the solar absorptance from 89% to 93% while maintaining a much lower thermal emittance than the painted coating. The reflectance of the MS film remained unchanged after prolonged thermal treatment at 200 °C (200 h). The fabrication process was scaled up to 193 cm × 12 cm copper fins, which were incorporated in commercial-size flat-plate solar collectors. Three complete collectors of an area of 1.7 m$^2$ were fabricated and their performance was evaluated under outdoor conditions. The results show that the electrodeposited selective coating with the MS film outperformed both the commercial black paint system and the system without the modified silica film.

**Keywords:** thermosolar energy; electrodeposition; solar collector; porous silica

## 1. Introduction

Solar-to-thermal energy conversion systems have diverse applications and are crucial for the energy transition. They can provide temperatures ranging from 30 °C to 200 °C for domestic water heaters and industrial applications and high temperatures (400–1000 °C) for electricity generation using concentrated solar power [1–3]. The efficiency of a thermosolar collector is boosted by the effectiveness of the selective coating used [4,5]. A typical structure of a selective coating includes a glass or metallic substrate coated, if needed, with a metallic layer to prevent heat loss through infrared radiation. This is typically followed by the application of a solar absorber film consisting of metal oxides, cermets, or multilayered materials [6,7]. The absorber must permit the transmission of mid-long infrared (IR) radiation to reduce the thermal emittance [8]. For a selective coating to be industrially viable, it must be low-cost and thermally stable and the fabrication method must be scalable [9]. Electrodeposition and spray pyrolysis represent two attractive techniques because of their low cost, easy scalability, and low energy consumption [10]. In addition, both techniques are sufficiently robust such that small fluctuations in the experimental conditions do not affect the properties of the scaled-up coatings.

Electrodeposition is a method for preparing surface coatings that has made significant advances towards becoming more eco-friendly. A specifically designed black paint is generally used as the solar absorber for copper fin-based solar collectors, but its selectivity is low. An alternative solar absorber is black nickel (BN), deposited via an electroless method, electrodeposition, spray pyrolysis, or chemical conversion [11–14]. Electrodeposited BN obtained from a chloride-based plating bath presents good optical properties, which are also stable under exposure to moisture [15]. Important features of the BN electrodeposition process are the low cost of precursor salts and the easy maintenance of the bath.

A silica coating (SC) can be applied on top of the selective coating, serving purposes such as increasing solar absorptance and improving chemical and thermal stability [16]. The SC prevents or lowers the rate of oxidation of the metallic particles in the solar absorber in the case of cermet-type materials. BN is a cermet-type material, and a functional SC may increase the solar absorptance and collector performance. Hence, in this work, we integrated silica-based top films in our selective coating [17]. Silica is a well-known antireflective (AR) material with an adequate refractive index; films can be deposited using several methods, such as sputtering, spray pyrolysis, and spin-coating [18–25]. Several studies show that silica AR coatings increase the spectral selectivity of the coatings, while films obtained by sputtering have been reported to increase the solar absorptance specifically [26–33]. There are reports on the effect of a silica film overlayer on the thermal stability of black nickel-based coatings but only a few reports on scale-up techniques to obtain the AR coating. To our knowledge, there are no reports on full-size prototypes detailing the scale-up of electrodeposited BN films covered with SC by spray pyrolysis.

Silica is often obtained using tetraethyl orthosilicate (TEOS) as a sol–gel precursor but, often, the material is not sufficiently flexible [34]. To make silica more flexible, a propyl group can be incorporated into the structure; the resulting material is called modified silica (MS). It has been demonstrated for electrodes in super-capacitors and Li-ion intercalation batteries that the coating flexibility and hydrophobicity can be improved by introducing short-chain organic moieties into the silica sol–gel precursor [35,36]. The propyl group is attractive because of its good thermal stability and hydrophobicity; additionally, it has been reported that it helps the formation of mesopores in the synthesis of titanium silicalite [37]. In the alkoxide precursor TEOS, all four ethoxy groups can hydrolyze (considering complete hydrolysis) and then condense, creating a 3-dimensional oxide network with ethoxy and hydroxyl side groups in the structure if condensation is not complete. When one ethoxy group in the precursor is replaced by an organic group that cannot hydrolyze such as n-propyl, the process of condensation still occurs, but the process and the resulting structures are modified. The stability of MS with ethoxy and n-propyl side groups to UV is high, as reported by J. Feenstra [38], because the organic moieties are not in the backbone of polysiloxane chains. G. Zimmermann [39] demonstrated that UV radiation below 300 nm can rapidly break C-C or C-Si bonds for larger organic groups with more than three carbons, but, for the solar spectrum at sea level, the UV component below 300 nm is almost zero; hence, the organic bond cleavage is not possible. Following this approach, in this work, we used n-propyl triethoxysilane (PTES) as a short-chain organic co-precursor in the TEOS-based sol–gel solution for the preparation of the MS coating. The films were deposited using spray pyrolysis as a cost-effective and scalable method, in line with the requirements for the commercialization of low-temperature, flat-plate solar collectors. Considering that the cover glass used in solar collectors can absorb UV radiation of a wavelength up to 320 nm, thus helping the long-term stability of organic moieties, and that the BN/MS surface needs to be isolated from environmental conditions to extend its durability, we decided to deposit the MS directly on the fins and not on the covering glass plate of the solar collector.

In this work, we describe the deposition of the selective coating, consisting of a metallic nickel (Ni) underlayer and BN absorber layer, both with and without the MS overlayer, on 16 $cm^2$ substrates. We provide a detailed characterization of the optical properties and thermal stability. In addition, we scaled up the system and here report the fabrication of three complete solar collector prototypes with a total area of 1.72 $m^2$ for three different systems: (i) electrodeposited nickel/black nickel selective coating; (ii) electrodeposited nickel/black nickel selective coating with MS overlayer; and (iii) commercial, black paint coating (CBP). The collectors were evaluated under outdoor conditions and the performance of the three systems was compared, highlighting the effect of the MS film.

## 2. Materials and Methods

Material Fabrication: A potentiostat–galvanostat Gamry Reference 3000 coupled to a Reference 30k booster was employed for galvanostatic electrodeposition. The Cu sheets and fins were washed with detergent and sanded. After the sanding process, the samples were rinsed with deionized water and acetone in an ultrasonic bath. The bright Ni film was electrodeposited from a bath composed of (0.45 M) $NiSO_4{\cdot}6H_2O$, (0.32 M) KCl, and (0.32 M) $H_3BO_3$ at pH 4.5; the current density was $-5$ mA $cm^{-2}$ for 585 s. The average film thickness was 1.0 $\pm$ 0.2 mm (see Supporting Information, Figure S1, for thickness measurement details). Later, the samples were rinsed and dried and the BN layer was electrodeposited. The BN layer was electrodeposited from a plating bath with (0.31 M) $NiCl_2$ and (0.40 M) KCl at pH 6.5, using a two-pulse method of 60 s at $-2.6$ mA $cm^{-2}$, followed by 90 s at $-1.4$ mA $cm^{-2}$; the average film thickness was 280 $\pm$ 50 nm (see Supporting Information, Figure S2). A three-electrode configuration was employed, and rectangular Cu sheet sections of 4 $\times$ 6 $cm^2$ with a 1 mm thickness were used as substrates; the electrodeposition area was 16 $cm^2$. For Cu sheets, the experimental setup can be seen in Figure S3a). Note that both deposition steps corresponded to reduction processes; hence, by convention, a negative current was applied to the Cu substrate. For fins, a two-electrode configuration was employed. The working electrode (cathode) consisted of fins with an active area of 2316 $cm^2$ on the front face. For the counter electrode, 11 titanium baskets filled with high-purity nickel plating rounds (99.95%) were used; the baskets were interconnected in series using a copper tube. The plating cell was a 2 m $\times$ 36 cm $\times$ 30 cm metallic container covered with epoxy paint. The distance between the fin and the baskets was 2 cm and the system was not agitated. A schematic representation of the experimental setup for Cu fins can be seen in the Supporting Information, Figure S3b). To compare the coatings, we prepared samples with commercial black paint that was spray-deposited directly onto the Cu substrate at 10 cm and 3.51 kg $cm^{-2}$ to obtain a thickness of about 40.0 $\pm$ 3 µm (see Supporting Information, Figure S4). The silica and MS films were deposited on top of the selective coating from a solution prepared by adding 10 mL tetraethyl orthosilicate (TEOS) to 100 mL ethanol at room temperature with agitation at 300 rpm. Then, 2.7 mL of an aqueous HCl solution at pH 2.0 was added slowly to this precursor solution. The molar ratio $[H_2O]/[TEOS]$ was 3.4 to avoid complete TEOS hydrolysis. For preparing the MS, 1 mL n-propyl triethoxysilane (PTES) was added to 10 mL TEOS before adding ethanol. The silica and MS coatings were obtained using spray pyrolysis after 2 h of the solution preparation. The solution was sprayed onto the selective coating or glass slide substrates from a 20 cm distance at a pressure of 3.51 kg $cm^{-2}$. The coated substrate was then heated to 200 °C on a hot plate. Two layers were applied to obtain the desired thickness, with a waiting time of 5 min between applications.

The thermal stability of the samples was evaluated using thermal treatments in a furnace (Thermo Scientific, Waltham, MA, USA, model FB1315M) in air, which consisted of one heating cycle for Cu samples at 200 °C for 200 h; the applied heating ramp was 10 °C $min^{-1}$. The samples are represented by the following notations: Substrate/Metallic interlayer/Absorber layer/SC or MS coating; the thermal treatment is indicated by adding (TT).

Characterization: The electrodeposited films were studied by field-emission scanning electron microscopy (FE-SEM; JEOL JSM-7600F, Tokyo, Japan). X-ray diffraction on the selective coatings was performed using a Siemens D-5000, while a Bruker D8-Advance (Billerica, MA, USA) with Bragg–Brentano geometry was employed to characterize SC and MS powders; both systems used monochromatic Cu-K$\alpha$ radiation ($\lambda$ = 1.5418 Å). The film thickness was measured by profilometry using a KLA-Tencor D-120, Chandler, AZ, USA. Attenuated total reflectance Fourier-transform infrared spectroscopy (ATR-FTIR) was performed using a Thermo Scientific Nicolet iS5 FTIR spectrometer. Contact angle measurements were performed using 10 µL deionized water droplets, and the angle was calculated from the photograph using Image J software. Atomic force microscopy (AFM) images were obtained using WITec Alpha 300 equipment (Oxford Instruments, Abingdon, UK). X-ray photoelectron spectroscopy (XPS) was performed using K-Alpha Surface Analysis, Thermo Scientific, equipped with an Al K$\alpha$ X-ray source. Depth profile XPS analysis was performed by eroding the samples with Ar in 20 s steps before each XPS measurement.

Optical properties: The total reflectance spectra of the selective coatings were obtained using the spectrophotometer UV–Vis Avantes (Avaspec 2048) and the infrared spectrometer Avantes (Avaspec-Nir 256-2.5), both equipped with the integrating sphere Avantes 50-Ls-HAL. For the UV–Vis–NIR measurements, the Avantes WS-2 white reflective tile was used as the reference. The reflectance spectra obtained in the range of 2.5 to 15 µm were measured using the FTIR spectrometer Perkin Elmer, Waltham, MA, USA (frontier NIR/MIR) equipped with an integrating sphere (PICO, Integrat IR model). A gold standard was used as the reference. All reflectance spectra were obtained at room temperature.

The solar absorptance ($\alpha$) was defined as the fraction of the solar power absorbed by the selective coating and was calculated by weighting the measured wavelength-dependent reflectance [R($\lambda$)] spectrum of the coatings against the solar radiation spectrum [ISUN($\lambda$)] (ASTM G173-0329), according to Equation (1) [40]. The thermal emittance ($\varepsilon$) was defined by Equation (2) [41] as the fraction of emitted power, as given by the black body radiation spectrum, and was also calculated using the reflectance spectra:

$$\alpha = \frac{\int_{0.3}^{2.5} I_{SUN}(\lambda) \times [1 - R(\lambda)]d\lambda}{\int_{0.3}^{2.5} I_{SUN}(\lambda)d\lambda} \tag{1}$$

$$\varepsilon = \frac{\int_{2.5}^{15} I_{BB}(\lambda) \times [1 - R(\lambda)]d\lambda}{\int_{2.5}^{15} I_{BB}(\lambda)d\lambda} \tag{2}$$

where the black body radiation intensity as a function of wavelength and temperature is given by $I_{BB}(\lambda, T) = c_1/\left\{\lambda^5\left[e^{\left(\frac{c_2}{\lambda T}\right)} - 1\right]\right\}$ with $c_1 = 3.743 \times 10^{-16}$ W m$^2$ and $c_2 = 1.4387 \times 10^{-2}$ m K. In this work, the reflectance spectra were measured at room temperature, but the thermal emittance was calculated using T = 100 °C in Equation (2) in order to estimate the performance of the coating at a typical, maximum operating temperature of a solar collector for domestic water heating. The spectral selectivity ($h$) could be used as an indicator of the performance of the coatings and was calculated using $h = \alpha - 0.5\varepsilon$. The optical transmittance of SC and MS films deposited on soda lime glass substrates was examined using an Agilent 8453 spectrophotometer from 190 to 1100 nm. The transmittance of the soda lime glass substrate was used as the baseline.

Collector Performance Characterization: The collectors were experimentally evaluated according to the instantaneous thermal efficiency test described in the protocol ISO 9806:2013 [42]. The collectors connected in series had an aperture area of approximately 1.72 m$^2$ and were equipped with two 150 L hot water tanks.

The procedure for evaluating the instantaneous thermal efficiency of solar collectors was carried out by isolating the exposed pipes of the solar collector and placing temperature and flow sensors at its inlet and outlet; this was to measure the temperature differences in the system as well as the mass flow of the fluid at that instant. The flat plate solar collector was mounted on a support structure and the collector angle was set according to the day of the year of each test to obtain solar irradiance perpendicular to the collector plane. Mass flow rate and inlet temperature were kept constant with a recirculation pump and electrical heaters installed in the storage tank for cold water. The outlet water coming from the flat-plate solar collector was stored in the hot water storage tank. The water in the hot water storage tank was returned with the recirculation pump to the cold water storage tank to fix a new temperature and mass flow rate to repeat the process and save water consumption.

According to the protocol to obtain the thermal efficiency of a flat-plate solar collector, the tests were performed at solar noon, using water as the working fluid. During the test, measurements were taken of the inlet ($T_i$) and outlet ($T_0$) temperatures of the collector, ambient temperature ($T_a$), solar irradiance at the collector plane, and mass flow every 10 s. Four instantaneous thermal efficiency points were obtained in four inlet temperature ranges (approximately 28 °C, 38 °C, 55 °C, and 65 °C). With the information obtained from the instantaneous thermal efficiency as a function of temperature difference and solar irradiance, the heat loss transfer coefficients and the collector heat removal factor were obtained. An experimental setup scheme for the collector's characterization is provided in the Supporting Information, Figure S5, with details on the interconnection of the different elements of the characterization system.

## 3. Results

The crystal structure, chemical composition, and morphology of the Ni/BN selective coatings were characterized with and without MS coating. The stagnation temperature in a flat-plate collector can reach temperatures up to about 200 °C; therefore, thermal treatments were chosen to be in air at 200 °C for 200 h, exceeding realistic operating conditions. Figure 1a,b show SEM images of the BN film surface, illustrating the nanostructured and porous morphology of a BN film. Nanoflakes of 10 ± 2 nm thickness and 100 ± 5 nm long were vertically oriented with respect to the substrate, and, besides the nanoflakes, 10 ± 2 nm spherical nanoparticles were observed; the morphology did not change after the thermal treatment. This morphology aids light trapping and is in concordance with previous reports, and the existence of different nickel compounds was expected [43,44]. Figure 1c shows the surface of a BN with the MS coating after the 200 °C thermal treatment, showing a porous material, illustrating that the modified silica film completely covered the black nickel nanoflakes. Interestingly, related to the incorporated functional groups, the MS films exhibited increased hydrophobic character for the thermally treated films. Figure 1d shows that the contact angle was $\theta = 73°$ for the sample with PTES and $\theta = 28°$ for the sample without PTES. Figure 1e shows the surface AFM images of the SC and MS films treated at 200 °C, illustrating that the MS film presented ordered and periodic pores, like the morphology observed with SEM. The formation of the pores in the MS decreased the film refractive index, approaching the reported ideal value and resulting in the desired anti-reflective effect [45]. The AFM images show that the coatings were composed of silica structures with a sharpened texture, resulting in a low refractive index that defined the antireflective and hydrophobic properties of the coatings related to the propyl moiety in the MS film [46]. The surface average roughness Ra was calculated for the two films from the AFM images; the MS film roughness was 1.1 ± 0.1 nm while 1.4 ± 0.1 nm was obtained for SC. The smaller Ra value for MS indicates that the film had less light scattering, thus contributing to the higher transmittance than that of SC. It was also necessary to consider the thickness of both films, which needed to be adequate to generate the destructive interference effect that contributed to increasing the transmittance; the film thickness for both films can be seen in the Supporting Information, Figure S8. Based on the transmittance results and profilometry, we infer that the thickness of the MS was better, even though both films were deposited under the same conditions.

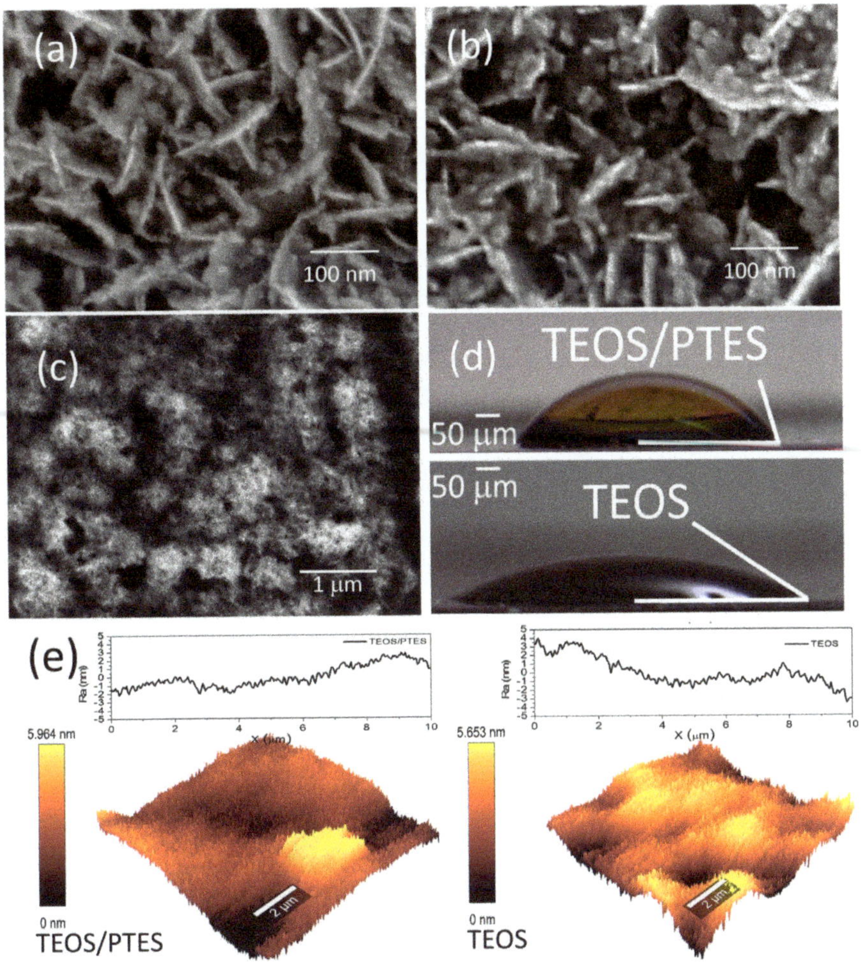


**Figure 1.** SEM images of the BN film (**a**) before and (**b**) after the thermal treatment at 200 °C. (**c**) BN/MS coating after the thermal treatment at 200 °C. (**d**) Image of the contact angle of a water drop on BN/S and BN/MS surfaces after thermal treatment at 200 °C (200 h). (**e**) AFM images of MS (Ra = 1.1 nm ± 0.1 nm) and silica (Ra = 1.4 nm ± 0.1 nm) surfaces; the corresponding film cross-sections are also shown.

Figure 2a shows the XRD pattern of the Cu/Ni/BN selective coating before and after thermal treatment at 200 °C for 200 h. Related to the nickel and black nickel films, both Cu from the substrate (ICDD PDF 004-0836) and Ni (ICDD PDF 004-0850) were detected. The electrodeposited films showed a face-centered cubic (fcc) structure for the metallic Ni, and the significant peak intensity corresponding to the (111) plane revealed the preferential growth of the film along this direction. It has been reported that chloride selectively suppresses the growth of certain planes, thus allowing preferential growth [47,48]. No peaks corresponding to nickel oxide were observed, indicating that the $NiO_x$ phases were very thin films and amorphous. The X-ray diffraction pattern for the thermally treated sample was identical to the as-deposited sample, illustrating that the oxides did not crystallize after the thermal treatment, which agrees with the report by Lizama-Tzec [43] related to the presence of amorphous $NiO_x$ on the black nickel surface and metallic nickel in the bottom of the film. In the Supporting Information, Figure S6, the XPS depth profile analysis is shown for samples of BN films without TT, showing the presence of $NiO_x$. As shown in Figure S6a for the 0 s analysis, a metallic Ni shoulder was detected at 852.8 eV, and the shoulders at 854.2 eV and 856 eV were assigned to $Ni^{2+}$ and $Ni^{3+}$ oxidation states, respectively [49]. After the increasing Ar etching time (in 20 s steps), the analysis showed an intensity increase of the metallic nickel peak, while the $Ni^{2+}$ and $Ni^{3+}$ intensities decreased; this was expected for a cermet material with graded metallic composition [50].

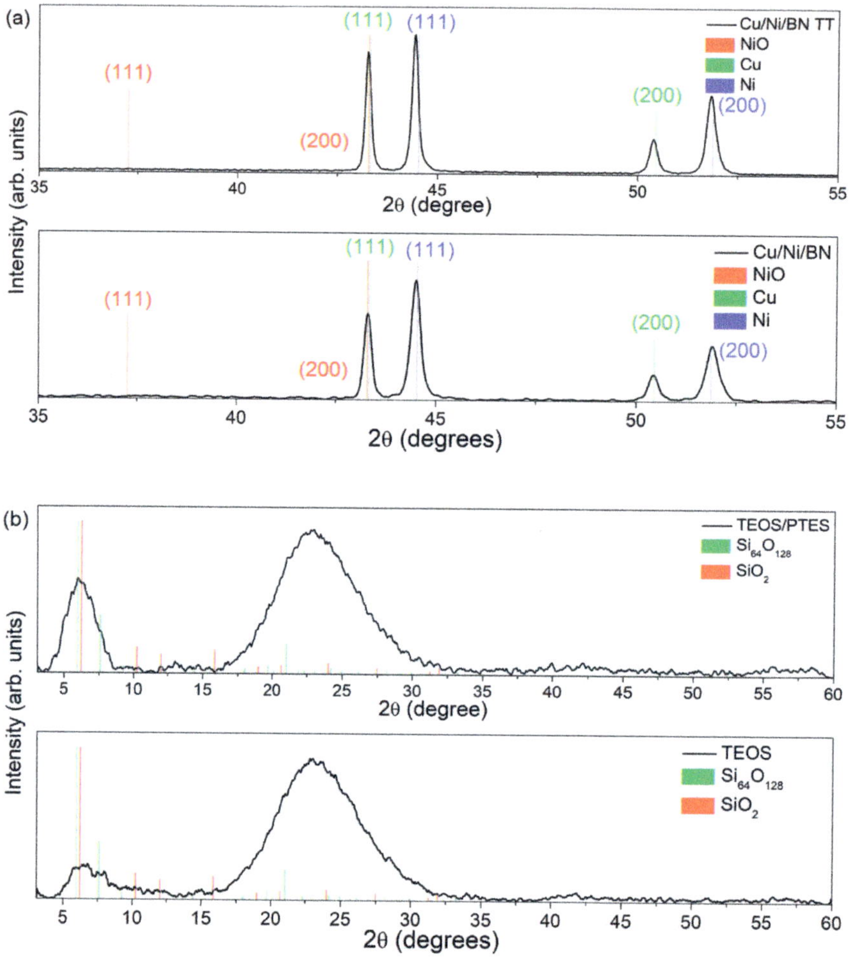


**Figure 2.** X-ray diffraction patterns for (**a**) Cu/Ni/BN films before and after thermal treatment (200 °C/200 h) and (**b**) MS and SC powders obtained from TEOS/PTES and TEOS solutions, respectively.

Figure S6b shows the XPS spectra for the oxygen window for the BN sample. The analysis at 0 s shows a wide and pronounced peak centered at 531.7 eV, associated with the oxygen from $Ni(OH)_2$. The peak covered the energies for the reported species at 532.1 eV (NiOOH) and 532.9 eV ($H_2O$) [49]. In addition, a less pronounced shoulder centered at 529.8 eV related to NiO was observed; as the successive Ar etching/XPS measurement steps were applied, it was found that the peak for $Ni(OH)_2$ decreased and the peak assigned to NiO increased. For the XPS analysis, after the etching step at 511 s, the spectra did not present peaks associated with $Ni(OH)_2$, $H_2O$, and NiOOH but only showed a weak peak associated with NiO. The previous results indicated that $NiO_x$ coexisted with other chemical species at the BN surface while, at the base of the film, metallic nickel dominated.

Figure 2b presents the X-Ray patterns for silica and modified silica powders. For the two powders, a broadened peak in the 2θ range of 15–30° was observed, corresponding to the amorphous silica matrix. The analysis matches the International Centre for Diffraction Data (ICDD) Powder Diffraction File (PDF) card numbers 050-0057 and 045-0112 assigned to $Si_{64}O_{128}$ and $SiO_2$, respectively; the patterns are inserted as visual references for the expected peaks. This pattern did not change after exposure to the deposition temperature (200 °C) used in spray pyrolysis, illustrating that the MS remained amorphous. A broadened shoulder over the 2θ range of 3° to 8° was observed for both powders, indicating ordered mesopore networks in the materials; this peak was more intense and sharper for the powder obtained with TEOS/PTES [51]. The ordered porous film could be attributed to the incorporation of a propyl functional group, which replaced some rigid Si-O-Si bonds, resulting in a more ordered material after the drying process. The larger intensity of the peak at low angles could be attributed to a larger number of ordered mesopore networks in the sample with PTES; these results are consistent with reports in the literature [52].

The optimum refractive index for the antireflection coating of the glass could be calculated using Equation (3):

$$n_f = (n_0\, n_s)^{1/2} \approx 1.22 \quad (3)$$

where $n_s \approx 1.5$ is the refractive index of the glass or other substrate material and $n_0$ is the refractive index of air. The lowest refractive indices known for dielectric materials are near 1.35; therefore, the value of $n_f$ is unreachable for conventional compact single AR coatings [53]. However, using mesoporous or microporous silica films for this objective with pore sizes smaller than the visible wavelengths, the refractive index of the medium is dependent on the film porosity (or film density). High porosity must be introduced in films to approach the refractive index of ~1.22 using silica with a refractive index of ~1.5. The expected effect of AR coating on the selective coating was to decrease the reflection in the solar range and increase the solar absorptance. Note that Equation (3) is only valid for non-absorbing film substrates, which was not the situation for the BN coating; however, the estimated refractive index of the BN absorbing film was about 1.7 [54].

The MS coatings were evaluated using transmittance spectroscopy. Figure 3 shows that the transmittance for an MS film in the 320–1100 nm range is about 101.2% on average, an increase of about 1.2% with respect to the glass slide without the AR coating. The transmittance exceeded 100%, demonstrating that the refractive index of the MS films was smaller than that of the glass substrate. As a result, the thin MS films presented an excellent antireflective effect, and films with similar properties were deposited on the selective coatings. The insert shows an image of a glass slide with the right half treated with the AR coating and the left half untreated. The difference in reflection is noticeable, in agreement with the lower transmittance of the glass (which was used as the 100% reference point). The results of transmittance spectroscopy and the image of the glass substrate exemplify the expected anti-reflective effect of the modified silica on selective coatings. Based on G. Helsch's [55] reported values, it was estimated that the porosity of the MS-coated films should be around 30% to exhibit an anti-reflective effect over a wide wavelength range, as observed for the TEOS/PTES film [56]. It is worth highlighting that in the range of 320–400 nm, the transmittance of the MS coating was around 2% above the reference value, showing that the film did not have a long-term problem due to cleavage of the Si-C or C-C bonds associated with solar spectrum UV radiation. On the other hand, MS showed a higher intensity for the X-ray shoulder at 6° and was less rough than SC; therefore, the refractive index of this film was closer to the ideal theoretical value of 1.22, resulting in a better anti-reflective effect.

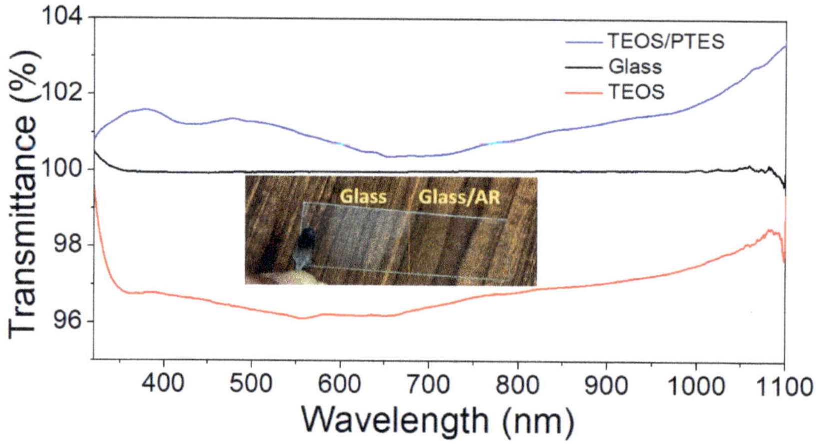


**Figure 3.** Transmittance spectra of MS (320–1100 nm) were applied using spray pyrolysis deposition of the TEOS/PTES precursor directly onto the glass slide: the thickness was about 113 ± 3 nm (see Supporting Information Figure S8 for the thickness measurement details).

Figure 4 shows the infrared absorbance spectra for films obtained with TEOS and TEOS/PTES precursor solutions with and without thermal treatment at 200 °C for 200 h. For graphic clarity, in Figure 4a the peaks are labeled and in Figure 4b they are only indicated with arrows. All peaks in Figure 4a,b were found at the same wavenumber values, except those indicated below. In both figures, a close-up is inserted in the range of 1200 cm$^{-1}$ to 3800 cm$^{-1}$. In Figure 4b, an additional close-up is inserted in the range of 2820 cm$^{-1}$ to 3050 cm$^{-1}$. The difference found between the spectra of films obtained with TEOS and TEOS/PTES, both as-deposited, was an increase in the relative intensity of the peaks that were assigned to Si-ethoxy bonds and siloxane groups. An increase in intensity for the peaks at 2873 cm$^{-1}$ and 2963 cm$^{-1}$ assigned to the vibrational modes of Si-$CH_2$-$CH_2$-$CH_3$ was found for the film obtained with PTES (see Table 1).

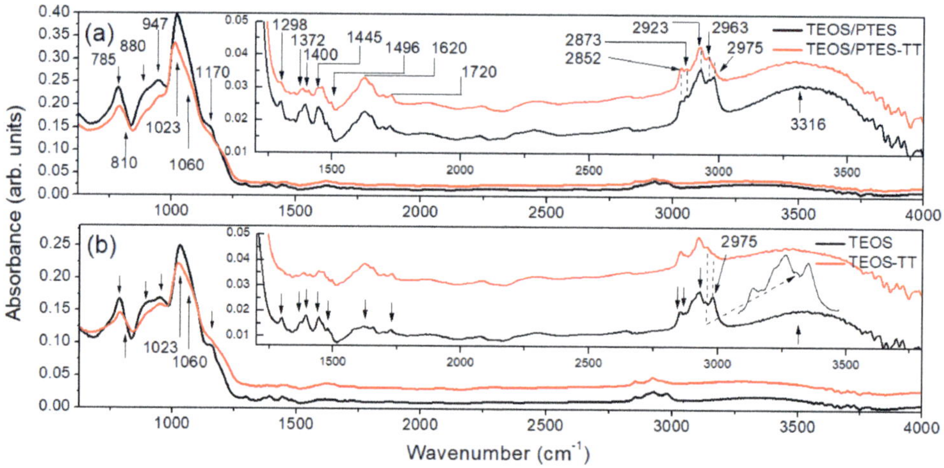


**Figure 4.** ATR-FTIR absorbance spectra of spray-pyrolysis deposited films: (**a**) silica film deposited on glass using a solution with TEOS/PTES before and after thermal treatment; (**b**) silica film deposited using a solution with TEOS before and after thermal treatment. In Figure (**b**), the arrows indicate the position of the wavenumbers assigned in (**a**).

In the comparison of the spectra between the TEOS and TEOS/PTES films, both thermally treated, a decrease in the intensity of the peaks associated with Si-ethoxy, ethyl, and siloxane groups was found. At 1720 $cm^{-1}$, C=O of acetaldehyde was observed, formed by the loss of the hydrogen of the carbon of the group -OH of ethanol; this band was not affected by the TT. For the film with PTES, the presence of the peak assigned to C-Si at 775 $cm^{-1}$ was detected overlapping with the peak at 785 $cm^{-1}$.

In the 2852–2975 $cm^{-1}$ range, peaks assigned to the symmetric and antisymmetric vibrations of the $-CH_2$ and $-CH_3$ groups were detected. After thermal treatment, the peak at 2873 $cm^{-1}$ did not disappear for samples prepared with PTES; this peak could be assigned to the $-CH_3$ group of n-propyl. The peak at 2963 $cm^{-1}$ was present in all four samples. The peak at 2975 $cm^{-1}$ was assigned to $-CH_3$ of ethanol and disappeared after the thermal treatment. In the 3000–3500 $cm^{-1}$ range, a wide band assigned to the -OH group was observed.

On the other hand, a slight decrease in the relative intensity of the 1023 $cm^{-1}$ and 947 $cm^{-1}$ bands was observed for the thermally treated samples, indicating a small degree of densification. The temperature of treatment at 200 °C was low, and the existence of remanent bands related to Si-ethoxy groups indicated that the TEOS and PTES did not hydrolyze completely. This was because to obtain dense, purely inorganic silica, it is necessary to increase the heat treatment temperature to about 300 °C, as was demonstrated by San Vicente [57].

As reported by Que [58], the presence of ethoxy groups after TT gives greater flexibility to the silica film, which is useful in the application of solar collectors due to the constant heating and cooling cycles. It has been reported that the presence of alkyl groups helps to modulate the hydrophobicity of the silica film due to the non-polarity of the alkyl groups [59].

For the samples, the infrared spectrum showed a shoulder at 1620 $cm^{-1}$ due to the formation of Si-OH groups at the removal of ethoxy groups during the condensation reaction. The addition of PTES increased the hydrophobicity of the silica film surface due to the presence of the n-propyl group ($Si-CH_2-CH_2-CH_3$). Figure 1d shows an example of the differences between the contact angle for a sample treated with PTES TT and a silica TT sample [60–64].

**Table 1.** Assignment of the FTIR-ATR peaks observed for the silica and MS films.

| Wavenumber ($cm^{-1}$) | Origin | Remark | Reference |
|---|---|---|---|
| 775 | Si-C | from $Si-CH_2-CH_2-CH_3$ | [65] |
| 785 | Si-O | overlapped, -Si-ethoxy | [61] |
| 810 | Si-O | siloxane groups: Si-O-Si, overlapped | [61,62] |
| 880 | C-O | from ethanol | [60] |
| 947 | Si-O | remnant shoulder Si-ethoxy, non-hydrolyzed | [60] |
| 1023 | Si-O | Si-ethoxy, non-hydrolyzed | [60] |
| ~1060 | O-Si-O | shoulder overlapped, stretching vibrations | [61] |
| 1170 | Si-O | remnant shoulder Si-ethoxy, non-hydrolyzed | [61] |
| 1298 | -CH | ethoxy, low intensity in TT samples | [63] |
| 1372–1496 | -CH | ethyl, low intensity in the TT samples | [63] |
| 1620 | Si-OH | condensation reaction of Si-OH | [64,66] |
| 1720 | C=O | all samples | [67] |
| 2852 | $-CH_2$ | symmetric stretching vibration | [61] |
| 2873 | $-CH_3$ | symmetric stretching vibration related to $Si-CH_2-CH_2-CH_3$ | [61,63] |
| 2923 | $-CH_2$ | asymmetric stretching vibration | [61] |
| 2963 | $-CH_3$ | asymmetric stretching vibration related to $Si-CH_2-CH_2-CH_3$ | [61,63] |
| 2975 | $-CH_3$ | from ethanol | [61,63] |
| ~3000–3500 (3316) | | hydroxyl group stretching; adsorbed water | [64] |

### 3.1. Optical Properties of the Selective Coatings

The optical properties of the selective coatings of Ni/BN on the Cu substrates, with and without the MS coating and before and after thermal treatment, were characterized using reflectance spectroscopy. Figure 5a displays the total reflectance spectra of Cu/Ni/BN and Cu/Ni/BN/MS coatings after thermal treatments at 200 °C. Table 2 shows the calculated values for solar absorptance and thermal emittance for these films. The as-deposited samples show the spectrum of a selective coating, with low reflectance in the visible and near-infrared and a large reflectance in the mid-long infrared. After thermal treatment at 200 °C, the reflectance spectrum was essentially the same, with increased solar absorptance. The thermally treated coating exhibited solar absorptance of 90.0 ± 0.1% and low thermal emittance of 11.0 ± 0.2%. The thermal treatment therefore improved the selectivity of the coating from 83.0 ± 0.4% to 85.0 ± 0.2%. The changes observed in the 300–2000 nm wavelength range were due to an increase in the BN film roughness from 27 ± 2 nm to 31 ± 2 nm, resulting in a slight increase in absorptance. Roughness profiles related to the BN films with and without TT are provided in Figure S7.

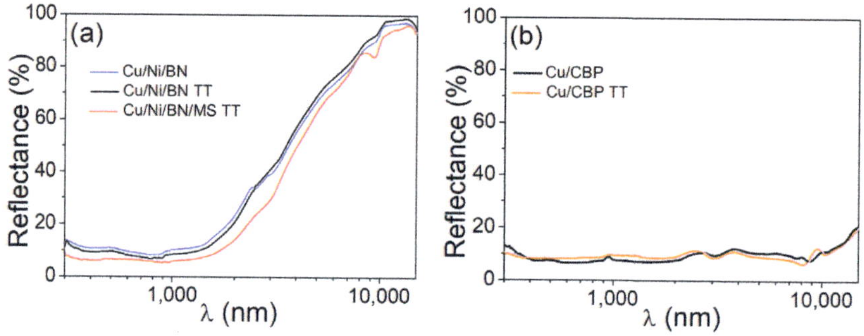


**Figure 5.** Total reflectance spectra for (**a**) Cu/Ni/BN and Cu/Ni/BN/MS before and after thermal treatment at 200 °C and (**b**) Cu/CBP before and after thermal treatment at 200 °C.

**Table 2.** Solar absorptance, thermal emittance, and spectral selectivity values were calculated for the selective coatings.

| | Solar Absorptance $\alpha$ (%) | Thermal Emittance $\varepsilon_{100\,^\circ C}$ (%) | Spectral Selectivity $h = \alpha - 0.5\,\varepsilon$ (%) [68] |
|---|---|---|---|
| Cu/Ni/BN | 89.0 ± 0.2 | 12.0 ± 0.4 | 83.0 ± 0.4 |
| Cu/Ni/BN TT | 90.0 ± 0.1 | 11.0 ± 0.2 | 85.0 ± 0.2 |
| Cu/Ni/BN/MS | 93.0 ± 0.1 | 13.0 ± 0.3 | 86.5 ± 0.3 |
| Cu/Ni/BN/MS TT | 93.0 ± 0.1 | 13.0 ± 0.2 | 86.5 ± 0.2 |
| Cu/CBP | 92.0 ± 0.1 | 88.0 ± 0.2 | 48.0 ± 0.2 |
| Cu/CBP TT | 91.0 ± 0.3 | 90.0 ± 0.1 | 46.0 ± 0.4 |

The reflectance spectra of the Cu/Ni/BN/MS sample after thermal treatment at 200 °C for 200 h compared to the spectrum for the Cu/Ni/BN TT coating show that there was an increase of 3% in solar absorptance in the spectral range of 300–2500 nm, while the increase in thermal emittance was slight at 2%. The spectra for the as-deposited MS samples and the thermally treated samples were identical, indicating no subsequent changes in the coating due to the prolonged thermal treatment time related to the decomposition of remaining functional groups or the densification of silica particles. In the samples with the MS film, the presence of the spray-deposited silica was signaled in the spectrum by the well-defined signal at 9430 (1060 cm$^{-1}$) nm. In fact, the thermal emittance was somewhat larger, mainly because of the absorption peak at 9430 nm. The absorption bands at about 9430 nm were assigned to the stretching vibration of Si-O-Si groups from the modified silica film, in accordance with the FTIR results.

Figure 5b shows the reflectance spectra for the Cu/CBP coating, characterized by low reflectance throughout the spectrum. The coating exhibited a large solar absorptance of 92.0 ± 0.1%. This was coupled with a lack of selectivity due to the calculated high thermal emittance of 88.0 ± 0.2%. The spectra changed slightly after the thermal treatment, resulting in lower solar absorptance and higher thermal emittance. The results show that the highest spectral selectivity of $h$ = 86.5 ± 0.3% was obtained for the full coating before the thermal treatment, related to the increase in solar absorptance to $\alpha$ = 93.0 ± 0.1% and combined with a minimal increase in thermal emittance to $\varepsilon$ = 13.0 ± 0.3%. In addition, the coatings exhibited excellent thermal stability at 200 °C. No peeling was detected in the MS coating, and the optimized transmittance of the MS coating helped to improve the solar absorptance of the selective coating.

### 3.2. *Scale-Up and Implementation in the Solar Collector*

The deposition process of the modified silica material was scaled up and implemented to provide the anti-reflective film on Ni/BN selective coatings electrodeposited onto Cu fins of 193 cm × 12 cm. We assembled solar collectors consisting of seven fins each, corresponding to a total aperture area of about 1.7 m$^2$.

Three solar collectors were fabricated to compare the performance of a commercial system with a black paint film (Cu/CBP) to collectors with the Ni/BN selective coating, both with and without the modified silica AR coating. The collectors were evaluated under outdoor conditions in Jiutepec, Morelos, Mexico; Figure 6 shows an image of the three collectors tested.

**Figure 6.** Picture of the flat-plate solar collectors under outdoor conditions in Jiutepec, Mexico. The maximum collector efficiency and heat loss transfer coefficient are shown.

The collectors were evaluated by measuring the temperature differences in the system for a certain mass flow rate of the fluid, which could be related to the efficiency, $\eta$, of the collector (see experimental section for details). The collector efficiency was determined by the radiant power absorbed and the heat losses related to conduction, convection, and thermal emittance. If there were no losses, which corresponded to the situation where the temperature difference was negligibly small, the maximum efficiency, $\eta_0$, was given by the solar absorptance, $\alpha$, corrected for the collector cover transmittance, $\tau$, and multiplied by a heat removal factor, $F_r$, which described the heat transfer to the collector fluid and which depended on the fluid properties, flow rate, and characteristics of the collector–fluid interface. For sufficiently small temperature differences, the total efficiency decreased linearly with increasing temperature differences according to the following relation [69–71]:

$$\eta = F_r \alpha \tau - F_r U_L \left( \frac{Ti - T_a}{G} \right) = \eta_0 - \eta_1 \left( \frac{T_i - T_a}{G} \right), \quad (4)$$

where $U_L$ is the heat loss transfer coefficient, G is the solar irradiance at the collector plane, and $(Ti - Ta)$ is the difference between the temperature of the fluid at the inlet of the collector and the ambient temperature. If the efficiency was graphed versus $(Ti - Ta)/G$, the intercept on the y-axis corresponded to the maximum collection efficiency, and the slope of the approximately straight line, $\eta_1 = -F_r U_L$, represented the total rate of heat loss from the collector to the environment [72,73]. Note that $F_r$ corresponds to the heat transfer to the collector fluid, while $U_L$ provides a quantification of the heat losses associated with conduction, convection, and thermal emittance. Figure 7 shows the instantaneous thermal efficiency curves obtained under outdoor conditions for the Cu/Ni/BN, Cu/Ni/BN/MS, and Cu/CBP collectors.

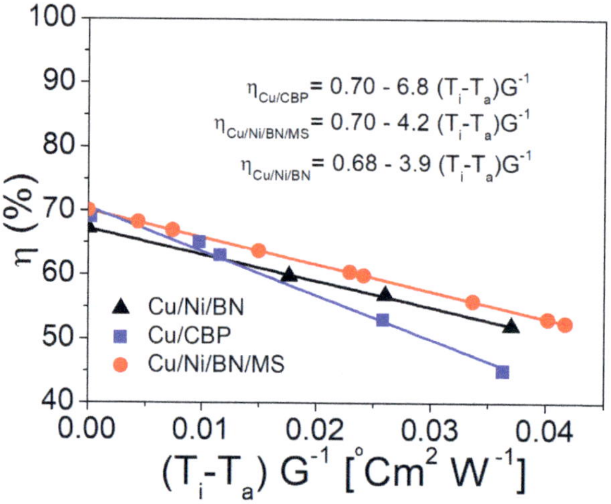


**Figure 7.** Instantaneous thermal efficiency measured under outdoor conditions in Jiutepec, Morelos, for the Cu/Ni/BN, Cu/Ni/BN/MS, and Cu/CBP solar collectors.

Figure 7 and Table 3 show that the maximum collector efficiency improved upon adding the MS film onto the Cu/Ni/BN selective coating, which was in accordance with the higher solar absorptance measured and was the same as for the commercial Cu/CBP collector. The efficiency decreased approximately linearly with the temperature difference, in agreement with Equation (4). An important observation was that the slope for the Cu/CBP collector was significantly larger than for the Cu/Ni/BN systems. Considering that the collector–fluid heat transfer was the same for all systems measured (i.e., $F_r$ was the same), this indicated a more significant value for $U_L$ describing the heat losses under operating conditions for the Cu/CBP collector. Since all other materials and interfaces were the same, this was most likely due to the large thermal emittance of the CBP film. Note that the slope for the Cu/Ni/BN/MS collector was slightly larger than for the system without the modified silica coating, which also correlated with a slightly larger thermal emittance. However, in the temperature difference range evaluated, the system with the modified silica coating had much better total performance due to the higher solar absorptance. These results also demonstrate that the characterization of the optical properties of the 4 cm × 4 cm samples can provide a good indication of the performance of scaled-up systems in a complete solar collector. On the other hand, based on the literature, it is thought that MS contributes to less heat loss since amorphous materials do not have a structural order, so there is no continuous crystalline network in which the atoms can vibrate, so that phonons propagate. As a result, for the phonon, the mean free paths are restricted to the interatomic spacing and the effective thermal conductivity of MS remains low.

**Table 3.** Performance characterization results for the evaluated solar collectors.

| Collector | $\alpha$ | $\varepsilon_{100\,^\circ C}$ | $\tau$ | $\eta_0$ | $\eta_1$ | $F_r$ | $(\alpha\tau)$ | $U_L$ |
|---|---|---|---|---|---|---|---|---|
| Cu/Ni/BN | 0.89 | 0.12 | 0.91 | 0.68 | 3.9 | 0.83 | 0.81 | 4.7 |
| Cu/Ni/BN/MS | 0.93 | 0.13 | 0.91 | 0.70 | 4.2 | 0.84 | 0.83 | 5.0 |
| Cu/CBP | 0.92 | 0.88 | 0.91 | 0.70 | 6.8 | 0.84 | 0.83 | 8.1 |

To compare the AR thermal stability results obtained on the 16 $cm^2$ samples with the results of the copper fins with the selective coating used in the solar collector, we sectioned one of the copper fins with the Ni/BN/MS coating, which was deposited by an identical process, and characterized the surface with ATR-FTIR. The measurement was performed in three different zones on the fin surface, and spectra were identical for the three zones, illustrating the reproducibility.

Figure 8 shows ATR-FTIR spectra obtained from sections of Ni/BN/MS TT-coated copper fins fabricated under equal conditions to those used for the evaluated collector. The spectra look like those obtained for glass slide substrates, as observed in Figure 4.

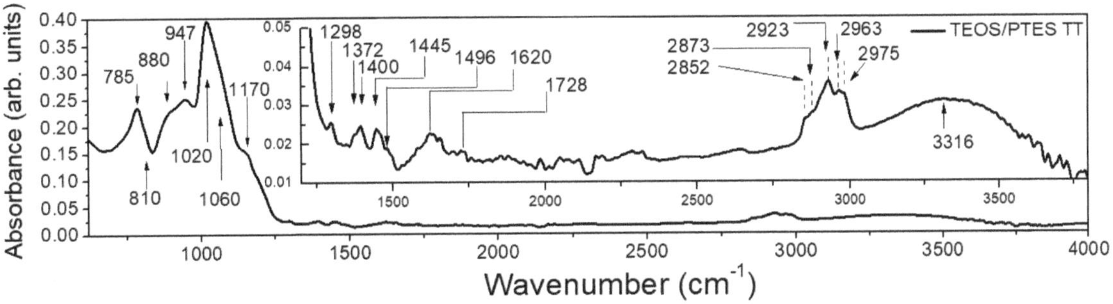


**Figure 8.** Infrared absorbance spectra of Cu/Ni/BN/MS obtained from a section of 193 cm Cu fin. The fin section was thermally treated at 200 °C for 200 h.

In comparison with Figure 4, the positions of some peaks were displaced as detailed below: as shown in Figure 8, for the sample, a band at 775 $cm^{-1}$ was detected, which was assigned to C-Si and was overlapping with the band at 785 $cm^{-1}$ assigned to Si-O of Si-ethoxy. The main peak associated with Si-ethoxy was detected at 1020 $cm^{-1}$. These peaks, compared to those detected in Figure 4 on glass, were close to each other. At 1170 $cm^{-1}$ [74].

The close-up inserted spectra in the 1200–3800 $cm^{-1}$ wavenumber range show that the spectra are essentially the same as those in Figure 4a for TEOS/PTES film TT, without significant shifts in the peak positions, except for peaks assigned to $H_2O$, C=O and the peak assigned to -CH at 1372 $cm^{-1}$. The peak at 1298 $cm^{-1}$ related to the -CH vibration of the ethoxy group was detected without changes. The changes in intensity observed were in the range of 2800–3000 $cm^{-1}$; however, the symmetric and asymmetric stretching vibrations related to the -$CH_3$ group from Si-$CH_2$-$CH_2$-$CH_3$ were detected. The peak assigned to $CH_3$ at 2975 $cm^{-1}$ of ethanol disappeared. In the 3000–3600 $cm^{-1}$ range, a widened and pronounced band was detected, indicating the existence of additional OH- groups on the BN surface; this was useful because it improved the adherence of MS at the black nickel surface. This result demonstrates the compatibility between MS and BN surfaces.

Based on these observations, we can infer that for the PTES-containing films, even after prolonged thermal treatment, peaks corresponding to the alkyl groups were present, indicating that the MS films were thermally stable. In addition, the surface morphology observed in the AFM image of the MS was regulated by steric effects [75]. This conclusion also correlates with the observation that the films showed increased hydrophobicity after thermal treatment. In addition, the films did not exhibit any cracking or peeling after the prolonged thermal treatment, which was related to the improved flexibility caused by the un-hydrolyzed ethoxy moieties and the intentionally incorporated propyl functional groups.

Finally, for a new system to be economically viable, it is essential to perform a cost–benefit analysis. Table 4 shows the costs of the different coatings used in the three collectors fabricated in this work in relation to the generated heat. We calculated the average useful heat, $Q_U$, using the following equation:

$$Q_U = \eta_{av}\, G\, A_c \qquad (5)$$

where $\eta_{av}$ corresponds to the average collector efficiency and $A_c$ is the collector aperture area (1.72 $m^2$). Table 4 shows that the solar collectors with the Ni/BN selective coating, both with and without MS film, produced more useful heat than the painted system. The fabrication of these more complex systems was somewhat more expensive, but the unit cost of the heat produced was lowest for the Cu/Ni/BN/MS system, providing an excellent motivation to consider the manufacture of this system. Detailed information about the cost estimate can be consulted in Tables S1 and S2 in the Supporting Information.

**Table 4.** Performance in terms of the average useful heat generated, $Q_U$, and cost analysis results for the three solar collectors fabricated and characterized in this work.

| Collector | $Q_U$ (J) | Cost (USD) | Unit Cost (USD/J) |
|---|---|---|---|
| Cu/CBP | 840 | 93.5 | 0.111 |
| Cu/Ni/BN | 894 | 95.6 | 0.107 |
| Cu/Ni/BN/MS | 990 | 105 | 0.106 |

Another benefit of the higher efficiency of the Cu/Ni/BN/MS collector was that it required less collector area than a commercial collector. This is an important aspect of sustainable manufacturing and reduces the collector's environmental impact. The area is an important parameter in solar collectors; just as in solar cells, efficiency is dependent on the device area. The preparation of selective coatings in small areas can give better properties than selective coatings for full-size solar collectors due to several factors implicit in the material preparation methodology and more optimal processes that can be applied in smaller areas. Therefore, being able to synthesize selective coatings in real-size prototypes and deposit a film with anti-reflective properties is a good achievement.

## 4. Conclusions

In this work, we evaluated the effects of implementing a modified silica (MS) film on top of electrodeposited Ni/BN selective coatings, both at the laboratory scale and in full-sized 1.72 $m^2$ solar collectors. We fabricated three collectors based on copper fins: one with commercial black paint (CBP) as an absorber and two with an electrodeposited black nickel selective coating, of which one was covered with an MS film. The MS film was deposited by spray pyrolysis from a [PTES]/[TEOS] molar ratio 0.09 precursor solution, resulting in a conformal mesoporous film that did not show change under thermal treatments up to 200 °C. The fabrication method was shown to be scalable up to collector size. The incorporation of the MS film increased the solar absorptance of the coating and improved the performance of a complete, 1.72 $m^2$ solar collector under real outdoor conditions. The Cu/Ni/BN/MS system maintained the advantage of electrodeposited Ni/BN selective coatings in terms of a much lower thermal emittance than coatings based on black paint. The novel system also demonstrated stability under heat treatments, illustrating the technical viability. A cost analysis indicated that the higher efficiency of the new collector resulted in a lower unit cost of heat produced compared to the commercial black paint system. Finally, higher efficiency is an important aspect of sustainable manufacturing as it decreases the amount of material needed for a given amount of heat and thus the carbon footprint.

**Supplementary Materials:** The following supporting information can be downloaded at: https://www.mdpi.com/article/10.3390/coatings14111368/s1, Figure S1. Bright nickel profile electrodeposited on copper sheets. The profiles were taken at three different points of the sheet (Cu/Bright Nickel 1, 2, and 3). The profile Cu/Bright nickel 4 is a representative profile was taken from a section of Cu fin. The average film thickness on the sheets was 1.0 ± 0.2 μm and the thickness on the fins was 1.0 ± 0.4 μm. Figure S2. Black nickel profile electrodeposited on Cu/Ni sheets. The profiles were taken from three different sections of the Cu fin. The average film thickness was 280 ± 50 nm. Figure S3. Schematic representation of electrodeposition experimental setup for (a) Cu sheets and (b) Cu fins. Figure S4. Paint profile on copper sheet. This shows three different profiles measured at different points of a sheet. The averaged film thickness was 40 ± 3 μm. Figure S5. Experimental setup used for the solar collector's characterization. Figure S6. XPS depth profile analysis of BN film samples: (a) window for nickel; (b) window for oxygen. Figure S7. Roughness profiles (red line) of (a) BN without thermal treatment and (b) BN with thermal treatment. The Ra values and thickness values were averaged from three measurements taken from different points. Figure S8. (a) Silica and (b) modified silica profiles measured from films deposited on glass slides. The thickness values were averaged from three measurements taken from different points. Table S1. Material and cost of black paint solar absorber solar collector. Table S2. Material and cost of black nickel solar absorber solar collector.

**Author Contributions:** Conceptualization, G.O., O.G.-V., and G.R.-G.; methodology, F.I.L.-T.; software, M.d.J.C.-D. and D.M.H.-Z.; validation, M.A.E.-G.; formal analysis, J.J.A.-G.; investigation, C.V.-P.; resources, M.A.E.-G.; data curation, M.d.J.C.-D.; writing—original draft preparation, F.I.L.-T.; writing—review and editing, F.I.L.-T.; project supervision, G.R.-G.; project administration, G.O.; funding acquisition, G.R.-G. and G.O. All authors have read and agreed to the published version of the manuscript.

**Funding:** This research was funded by Consejo Nacional de Humanidades, Ciencias y Tecnologías (CONAHCYT), and the Mexican Center for Innovation in Solar Energy (CeMIE-Sol), projects P-18, P-69 and P-81.

**Institutional Review Board Statement:** Not applicable.

**Informed Consent Statement:** Not applicable.

**Data Availability Statement:** The raw data supporting the conclusions of this article will be made available by the authors upon request.

**Acknowledgments:** The authors would like to thank Daniel Aguilar, William Cauich, Beatriz Heredia, José Bante-Guerra, and Santiago González Gómez for their technical help. The authors acknowledge the National Laboratory of Nano and Biomaterials (LANNBIO) funded by the projects FOMIX-Yucatan 2008-108160, CONACYT LAB-2009-01-123913, 292692, 294643, 188345, and 204822. Francisco Ivan Lizama-Tzec acknowledges CONAHCYT for the postdoc grant number 3095721 under the framework "Estancias posdoctorales por Mexico".

**Conflicts of Interest:** The authors declare no conflicts of interest.

## References

1. Ge, T.S.; Wang, R.Z.; Xu, Z.Y.; Pan, Q.W.; Du, S.; Chen, X.M.; Ma, T.; Wu, X.N.; Sun, X.L.; Chen, J.F. Solar Heating and Cooling: Present and Future Development. *Renew Energy* **2018**, *126*, 1126–1140. [CrossRef]
2. Singh, R.; Lazarus, I.J.; Souliotis, M. Recent Developments in Integrated Collector Storage (ICS) Solar Water Heaters: A Review. *Renew. Sustain. Energy Rev.* **2016**, *54*, 270–298. [CrossRef]
3. Marcos, J.D.; Izquierdo, M.; Parra, D. Solar Space Heating and Cooling for Spanish Housing: Potential Energy Savings and Emissions Reduction. *Sol. Energy* **2011**, *85*, 2622–2641. [CrossRef]
4. Spitz, J.; Maziere-Bezes, D. Selective Materials for Solar Energy Photothermal Conversion. *J. Opt.* **1984**, *15*, 325–332. [CrossRef]
5. Bogaerts, W.F.; Lampert, C.M. Review Materials for Photothermal Solar Energy Conversion. *J. Mater. Sci.* **1983**, *18*, 2847–2875. [CrossRef]
6. Cao, F.; McEnaney, K.; Chen, G.; Ren, Z. A Review of Cermet-Based Spectrally Selective Solar Absorbers. *Energy Environ. Sci.* **2014**, *7*, 1615–1627. [CrossRef]
7. Kennedy, C.E. *Review of Mid- to High-Temperature Solar Selective Absorber Materials*; National Renewable Energy Lab. (NREL): Golden, CO, USA, 2002.
8. Hutchins, M.G. Selective Thin Film Coatings For The Conversion of Solar Radiation. *Surf. Technol.* **1983**, *20*, 301–320. [CrossRef]

9. Selvakumar, N.; Karthik, G.; Jayaraj, S.; Barshilia, H.C. Sprayable PEDOT:PSS Based Spectrally Selective Coating for Solar Energy Harvesting. *Sol. Energy Mater. Sol. Cells* **2021**, *221*, 110906. [CrossRef]
10. Perednis, D.; Gauckler, L.J. Thin Film Deposition Using Spray Pyrolysis. *J. Electroceram.* **2005**, *14*, 103–111. [CrossRef]
11. Li, J.M.; Cai, C.; Song, L.X.; Li, J.F.; Zhang, Z.; Xue, M.Z.; Liu, Y.G. Electrodeposition and Characterization of Nano-Structured Black Nickel Thin Films. *Trans. Nonferrous Met. Soc. China (Engl. Ed.)* **2013**, *23*, 2300–2306. [CrossRef]
12. Abis, S.; Baudino Aluminia SpA, M.C. An Electroless Selective Black Nickel for Aluminium Substrates. *Mater. Sci. Eng.* **1986**, *77*, L1–L3. [CrossRef]
13. Madhusudana, M.; Sehgalt, H.K. Spray-Deposited Black Nickel Selective Absorber Surfaces For Solar Thermal Conversion. *Appl. Energy* **1982**, *10*, 65–74. [CrossRef]
14. Gogna, P.K.; Chopra, K.L. Selective Black Nickel Coatings on Zinc Surfaces by Chemical Conversion. *Sol. Energy* **1979**, *23*, 105–408. [CrossRef]
15. Koltun, M.; Gukhman, G.; Gavrilina, A. Stable Selective Coating "Black Nickel" for Solar Collector Surfaces. *Sol. Energy Mater. Sol. Cells* **1994**, *33*, 41–44. [CrossRef]
16. Khatibani, A.B.; Rozati, S.M. Spray Pyrolytically Grown NiAlOx Cermets for Solar Thermal Selective Absorbers: Spectral Properties and Thermal Stability. *Bull. Mater. Sci* **2016**, *39*, 97–107. [CrossRef]
17. Farchado, M.; San Vicente, G.; Germán, N.; Morales, Á. *Influence of Using Different $SiO_2$ Antireflective Coatings and Sintering Conditions on the Durability and Optical Performance of the Selective Solar Absorber*; International Solar Energy Society (ISES): Freiburg, Germany, 2019; pp. 1–11.
18. Li, D.; Huang, F.; Ding, S. Sol-Gel Preparation and Characterization of Nanoporous $ZnO/SiO_2$ Coatings with Broadband Antireflection Properties. *Appl. Surf. Sci.* **2011**, *257*, 9752–9756. [CrossRef]
19. Zhang, Y.; Zhang, X.; Ye, H.; Xiao, B.; Yan, L.; Jiang, B. A Simple Route to Prepare Crack-Free Thick Antireflective Silica Coatings with Improved Antireflective Stability. *Mater. Lett.* **2012**, *69*, 86–88. [CrossRef]
20. Xin, C.; Peng, C.; Xu, Y.; Wu, J. A Novel Route to Prepare Weather Resistant, Durable Antireflective Films for Solar Glass. *Sol. Energy* **2013**, *93*, 121–126. [CrossRef]
21. Zhu, L.W.; Feng, Y.D.; Wang, Y.; Wang, Z.M.; Zhao, K.; Su, X.M.; Qiu, J.W. A New Solar Selective Absorbing Structure of Al-$Al_2O_3$ Cermet Composite Films by Pulsed Direct Current Magnetron Sputtering. *Adv. Mat. Res.* **2008**, *53–54*, 349–353. [CrossRef]
22. Cao, F.; Kraemer, D.; Sun, T.; Lan, Y.; Chen, G.; Ren, Z. Enhanced Thermal Stability of W-Ni-$Al_2O_3$ Cermet-Based Spectrally Selective Solar Absorbers with Tungsten Infrared Reflectors. *Adv. Energy Mater.* **2015**, *5*, 1401042. [CrossRef]
23. Selvakumar, N.; Barshilia, H.C. Review of Physical Vapor Deposited (PVD) Spectrally Selective Coatings for Mid- and High-Temperature Solar Thermal Applications. *Sol. Energy Mater. Sol. Cells* **2012**, *98*, 1–23. [CrossRef]
24. Boström, T.; Westin, G.; Wäckelgård, E. Optimization of a Solution-Chemically Derived Solar Absorbing Spectrally Selective Surface. *Sol. Energy Mater. Sol. Cells* **2007**, *91*, 38–43. [CrossRef]
25. Becerril-Gonzalez, J.J.; Castro-Chong, A.M.; Oskam, G.; Arés-Muzio, O. Sputter Deposition of Mo-Alumina Cermet Solar Selective Coatings: Interrelation between Residual Oxygen Incorporation, Structure and Optical Properties. *Mater. Res. Express* **2021**, *8*, 105506. [CrossRef]
26. Farchado, M.; Rodríguez, J.M.; San Vicente, G.; Germán, N.; Morales, A. Optical Parameters of a Novel Competitive Selective Absorber for Low Temperature Solar Thermal Applications. *Sol. Energy Mater. Sol. Cells* **2018**, *178*, 234–239. [CrossRef]
27. Lisowski, W.; Van Den Berg, A.H.J.; Smithers, M.; Haanappep, V.A.C. Characterization of Thin Alumina Films Prepared by Metal-Organic Chemical Vapour Deposition (MOCVD) by High Resolution SEM, (AR)XPS and AES Depth Profiling. *Fresenius J. Anal. Chem.* **1995**, *353*, 707–712. [CrossRef]
28. Lira-Cantú, M.; Morales Sabio, A.; Brustenga, A.; Gómez-Romero, P. Electrochemical Deposition of Black Nickel Solar Absorber Coatings on Stainless Steel AISI316L for Thermal Solar Cells. *Sol. Energy Mater. Sol. Cells* **2005**, *87*, 685–694. [CrossRef]
29. Céspedes, E.; Rodríguez-Palomo, A.; Salas-Colera, E.; Fonda, E.; Jiménez-Villacorta, F.; Vila, M.; De Andrés, A.; Prieto, C. Role of $Al_2O_3$ Antireflective Layer on the Exceptional Durability of Mo-Si-N-Based Spectrally Selective Coatings in Air at High Temperature. *ACS Appl. Energy Mater.* **2018**, *1*, 6152–6160. [CrossRef]
30. Barshilia, H.C.; Selvakumar, N.; Vignesh, G.; Rajam, K.S.; Biswas, A. Optical Properties and Thermal Stability of Pulsed-Sputter-Deposited AlxOy/Al/AlxOy Multilayer Absorber Coatings. *Sol. Energy Mater. Sol. Cells* **2009**, *93*, 315–323. [CrossRef]
31. Cathro, K.J.; Constable, D.C.; Solaga, T. Durability Of Porous Silica Antireflection Coatings For Solar Collector Cover Plates. *Sol. Energy* **1981**, *27*, 491–495. [CrossRef]
32. Rebouta, L.; Sousa, A.; Capela, P.; Andritschky, M.; Santilli, P.; Matilainen, A.; Pischow, K.; Barradas, N.P.; Alves, E. Solar Selective Absorbers Based on $Al_2O_3$:W Cermets and AlSiN/AlSiON Layers. *Sol. Energy Mater. Sol. Cells* **2015**, *137*, 93–100. [CrossRef]
33. Boström, T.K.; Wäckelgård, E.; Westin, G. Anti-Reflection Coatings for Solution-Chemically Derived Nickel—Alumina Solar Absorbers. *Sol. Energy Mater. Sol. Cells* **2004**, *84*, 183–191. [CrossRef]
34. Katumba, G.; Lu, J.; Olumekor, L.; Westin, G.; Wäckelgard, E. Low Cost Selective Solar Absorber Coatings: Characteristics of Carbon-In-Silica Synthesized with Sol-Gel Technique. *J. Solgel. Sci. Technol.* **2005**, *36*, 33–43. [CrossRef]
35. Oskam, G.; Searson, P.C.; Jow, T.R. Sol-Gel Synthesis of Carbon/Silica Gel Electrodes for Lithium Intercalation. *Electrochem. Solid-State Lett.* **1999**, *2*, 610–612. [CrossRef]
36. Tsionsky, M.; Gun, G.; Glezer, V.; Lev, O. Sol-Gel Derived Ceramic-Carbon Composite Electrodes: Introduction and Scope of Applications. *Anal. Chem.* **1994**, *66*, 1747–1753. [CrossRef]

37. Liu, X.; Liu, X.; Shan, J.; Huai, J.; Yang, H.; Yan, X.; Zheng, Y.; Li, H. Synthesis of Amorphous Mesoporous $TiO_2$-$SiO_2$ and Its Excellent Catalytic Performance in Oxidative Desulfurization. *Inorg. Chem. Commun.* **2021**, *123*, 108336. [CrossRef]
38. Feenstra, J.; Van Leest, R.H.; Smeenk, N.J.; Oomen, G.; Bongers, E.; Mulder, P.; Vlieg, E.; Schermer, J.J. Flexible Shielding Layers for Solar Cells in Space Applications. *J. Appl. Polym. Sci.* **2016**, *133*, 43661. [CrossRef]
39. Zimmermann, C.G. On the Kinetics of Photodegradation in Transparent Silicones. *J. Appl. Phys.* **2008**, *103*, 083547. [CrossRef]
40. Duta, A.; Isac, L.; Milea, A.; Ienei, E.; Perniu, D. Coloured Solar-Thermal Absorbers—A Comparative Analysis of Cermet Structures. *Energy Procedia* **2014**, *48*, 543–553. [CrossRef]
41. Herrera-Zamora, D.M.; Lizama-Tzec, F.I.; Arés, O.; Oskam, G. Electrodeposition and Characterization of Selective Coatings Based on Black Cobalt for Solar-To-Thermal Energy Conversion. *ECS Trans.* **2015**, *69*, 7. [CrossRef]
42. *ISO 9806:2013*; Solar Energy-Solar Thermal Collectors-Test Methods. ISO: Geneva, Switzerland, 2013.
43. Lizama-Tzec, F.I.; Macías, J.D.; Estrella-Gutiérrez, M.A.; Cahue-López, A.C.; Arés, O.; de Coss, R.; Alvarado-Gil, J.J.; Oskam, G. Electrodeposition and Characterization of Nanostructured Black Nickel Selective Absorber Coatings for Solar–Thermal Energy Conversion. *J. Mater. Sci. Mater. Electron.* **2015**, *26*, 5553–5561. [CrossRef]
44. Ivan, L.-T.F.; Manterola-Villanueva, G.; García-Valladares, O.; Cetina-Dorantes, M.; Herrera-Zamora, D.M.; Pérez-Rábago, C.A.; Oskam, G.; Rodríguez-Gattorno, G. Nanostructured Black Nickel Electrodeposited Selective Coatings for Applications in Parabolic Trough Collectors. *J. Energy Eng. Optim. Sustain.* **2023**, *7*, 49–62. [CrossRef]
45. Lesaint, C.; Lebeau, B.; Marichal, C.; Patarin, J. Synthesis of Mesoporous Silica Materials Functionalized with n-Propyl Groups. *Microporous Mesoporous Mater.* **2005**, *83*, 76–84. [CrossRef]
46. Poddighe, M.; Innocenzi, P. Hydrophobic Thin Films from Sol–Gel Processing: A Critical Review. *Materials* **2021**, *14*, 6799. [CrossRef] [PubMed]
47. Zarębska, K.; Skompska, M. Electrodeposition of CdS from Acidic Aqueous Thiosulfate Solution-Invesitigation of the Mechanism by Electrochemical Quartz Microbalance Technique. *Electrochim. Acta* **2011**, *56*, 5731–5739. [CrossRef]
48. Lin, C.S.; Hsu, P.C.; Chang, L.; Chen, C.H. Properties and Microstructure of Nickel Electrodeposited from a Sulfamate Bath Containing Ammonium Ions. *J. Appl. Electrochem.* **2001**, *31*, 925–933. [CrossRef]
49. Potočnik, J.; Nenadović, M.; Bundaleski, N.; Popović, M.; Rakočević, Z. Effect of Thickness on Optical Properties of Nickel Vertical Posts Deposited by GLAD Technique. *Opt. Mater.* **2016**, *62*, 146–151. [CrossRef]
50. Lizama-Tzec, F.I.; Rodríguez-Pérez, M.; Vega-Poot, A.; Herrera-Zamora, D.M.; Estrella-Gutiérrez, M.A.; Canto-Aguilar, E.; Cetina-Dorantes, M.; Oskam, G. Electrochemistry for Solar Energy Conversion Systems: A Selection of Mexican Contributions. *J. Mex. Chem. Soc.* **2023**, *67*, 581–601. [CrossRef]
51. Yassin, F.M.; Fathy, M.M.; Fahmy, H.M.; Elshemey, W.M. Low-Angle X-ray Scattering for the Determination of the Size of Mesoporous Silica Nanoparticles. *Radiat. Phys. Chem.* **2021**, *179*, 109235. [CrossRef]
52. Wu, Z.; You, L.; Xiang, H.; Jiang, Y. Comparison of Dye Adsorption by Mesoporous Hybrid Gels: Understanding the Interactions between Dyes and Gel Surfaces. *J. Colloid Interface Sci.* **2006**, *303*, 346–352. [CrossRef]
53. Khoukhi, M.; Maruyama, S.; Komiya, A.; Behnia, M. Flat-Plate Solar Collector Performance with Coated and Uncoated Glass Cover. *Heat Transf. Eng.* **2006**, *27*, 46–53. [CrossRef]
54. Loshiku, E.M.; O'shea, K.R. Ellipsometry in the Study of Selective Radiation-Absorbing Surfaces. *Sol. Energy* **1977**, *19*, 271–276. [CrossRef]
55. Helsch, G.; Mös, A.; Deubener, J.; Höland, M. Thermal Resistance of Nanoporous Antireflective Coatings on Silica Glass for Solar Tower Receivers. *Sol. Energy Mater. Sol. Cells* **2010**, *94*, 2191–2196. [CrossRef]
56. Beganskiene, A.; Sakirzanovas, S.; Kazadojev, I.; Melninkaitis, A.; Sirutkaitis, V.; Kareiva, A. Sol-Gel Derived Antireflective Coating with Controlled Thickness and Reflective Index. *Mater. Sci.-Pol.* **2007**, *25*, 817–824.
57. San Vicente, G.; Bayón, R.; Germán, N.; Morales, A. Long-Term Durability of Sol-Gel Porous Coatings for Solar Glass Covers. *Thin Solid Films* **2009**, *517*, 3157–3160. [CrossRef]
58. Que, W.; Sun, Z.; Zhou, Y.; Lam, Y.L.; Chan, Y.C.; Kam, C.H. Optical and Mechanical Properties of $TiO_2/SiO_2$ /Organically Modified Silane Composite films Prepared by Solgel Processing. *Thin Solid Films* **2000**, *359*, 177–183. [CrossRef]
59. Arkles, B. Hydrophobicity-Hydrophilicity and Silane Surface Modification. *Gelest* **2011**, *215*, 547–1015.
60. Szczurek, A.; Paszkowski, M.; Lewandowski, D.; Gąsiorek, J.; Kaleta, J.; Krzak, J. Organically Functionalized Sol-Gel Silica Network Growth. *Ceram. Int.* **2020**, *46*, 13198–13204. [CrossRef]
61. Kamiya, K.; Yoko, T.; Tanaka, K.; Takeuchi, M. Thermal Evolution of Gels Derived from $CH_3Si(OC_2H_5)_3$ by the Sol-Gel Method. *J. Non-Cryst. Solids* **1990**, *121*, 182–187. [CrossRef]
62. Rovani, S.; Santos, J.J.; Corio, P.; Fungaro, D.A. An Alternative and Simple Method for the Preparation of Bare Silica Nanoparticles Using Sugarcane Waste Ash, an Abundant and Despised Residue in the Brazilian Industry. *J. Braz. Chem. Soc.* **2019**, *30*, 1524–1533. [CrossRef]
63. Pavia, D.; Lampman, G.; Kriz, G. *Introduction to Spectroscopy*, 3rd ed.; Vondelilng, J., Kiselica, S., Mula, P., Eds.; Thomson Learning: Belmont, CA, USA, 2001.
64. Fomina, P.S.; Proskurnin, M.A.; Mizaikoff, B.; Volkov, D.S. Infrared Spectroscopy in Aqueous Solutions: Capabilities and Challenges. *Crit. Rev. Anal. Chem.* **2023**, *53*, 1748–1765. [CrossRef]
65. Niu, Y.; Huang, G.; Chen, Y.; Dou, W.; Sun, D.; Li, J. Fabrication of Anti-Reflective Coatings on Solar Collector Tubes. *Adv. Mat. Res.* **2011**, *306–307*, 125–129. [CrossRef]

66. Yoshino, H.; Kamiya, K.; Nasu, H. IR Study on the Structural Evolution of Sol-Gel Derived $SiO_2$ Gels in the Early Stage of Conversion to Glasses. *J. Non Cryst. Solids* **1990**, *126*, 68–78. [CrossRef]
67. Launer, P.; Arkles, B. *Infrared Analysis of Organosilicon Compounds: Spectra-Structure Correlations*; Gelest: Morrisville, PA, USA, 2003; pp. 175–178.
68. Roberts, D.E. A Figure of Merit for Selective Absorbers in Flat Plate Solar Water Heaters. *Sol. Energy* **2013**, *98*, 503–510. [CrossRef]
69. Abdel Khalik, S.I. Heat Removal Factor For a Flat-Plate Solar Collector With a Serpentine Tube. *Sol. Energy* **1976**, *18*, 59–64. [CrossRef]
70. Montoya-Márquez, O.; Flores-Prieto, J.J. Heat Removal Factor in Flat Plate Solar Collectors: Indoor Test Method. *Energies* **2018**, *11*, 2783. [CrossRef]
71. García-Valladares, O.; Cesar-Munguia, A.L.; López-Vidaña, E.C.; Castillo-Téllez, B.; Ortiz-Sánchez, C.A.; Lizama-Tzec, F.I.; Domínguez-Niño, A. Effect by Using a Modified Solar Dryer on Physicochemical Properties of Carambola Fruit (*Averrhoa carambola* L.). *Rev. Mex. De Ing. Quim.* **2022**, *21*, Alim2650. [CrossRef]
72. Kong, W.; Wang, Z.; Fan, J.; Perers, B.; Chen, Z.; Furbo, S.; Andersen, E. Investigation of Thermal Performance of Flat Plate and Evacuated Tubular Solar Collectors According to a New Dynamic Test Method. *Energy Procedia* **2012**, *30*, 152–161. [CrossRef]
73. Malvi, C.S.; Gupta, A.; Gaur, M.K.; Crook, R.; Dixon-Hardy, D.W. Experimental Investigation of Heat Removal Factor in Solar Flat Plate Collector for Various Flow Configurations. *Int. J. Green Energy* **2017**, *14*, 442–448. [CrossRef]
74. Lopez, T.; Sanchez, E.; Bosch, P.; Meas, Y.; Gomez, R. FTIR and UV-Vis (Diffuse Reflectance) Spectroscopic Characterization of $TiO_2$, Sol-Gel. *Mater. Chem. Phys.* **1992**, *32*, 141–152. [CrossRef]
75. Li, S.; Zhang, L.; Li, J.; Wu, Z.; Yang, C. Silica Nanowires Reinforced Self-Hydrophobic Silica Aerogel Derived from Crosslinking of Propyltriethoxysilane and Tetraethoxysilane. *J. Solgel. Sci. Technol.* **2017**, *83*, 545–554. [CrossRef]

*coatings*

MDPI

*Article*

# Optimization of LPCVD Deposition Conditions of Silicon-Rich Silicon Nitride to Obtain Suitable Optical Properties for Photoluminescent Coating

**Francisco Uribe González [1,*], Karim Monfil Leyva [1,*], Mario Moreno Moreno [2], Alfredo Morales Sánchez [2], Ana L. Muñoz Zurita [3], José A. Luna López [1], Jesús Carrillo López [1], José A. D. Hernández de la Luz [1] and Alma S. L. Salazar Valdez [1]**

1 Researching Center on Semiconductor Devices, Meritorious University Autonomous of Puebla, Av. San Claudio s/n, Col. San Manuel, Puebla C.P. 72570, Mexico; jose.luna@correo.buap.mx (J.A.L.L.); jesus.carrillolopez@viep.com.mx (J.C.L.); jose.hernandez@correo.buap.mx (J.A.D.H.d.l.L.); alma.salazarva@alumno.buap.mx (A.S.L.S.V.)

2 National Institute of Astrophysics, Optics and Electronics, Luis Enrique Erro 1, Puebla C.P. 72840, Mexico; mmoreno@inaoep.mx (M.M.M.); alfredom@inaoep.mx (A.M.S.)

3 Electronics Faculty, Meritorious University Autonomous of Puebla, Av. San Claudio s/n, Col. San Manuel, Puebla C.P. 72570, Mexico; analuz.munoz@correo.buap.mx

* Correspondence: cross.ride1234@gmail.com (F.U.G.); karim.monfil@correo.buap.mx (K.M.L.)

**Abstract:** Silicon nitride is a commonly used material for ceramic applications and in the fabrication processes of integrated circuits (ICs). It has also increased in interest from the scientific community for use as a functional coating due to its physical, mechanical, electrical, and optoelectronic properties. In particular, silicon-rich silicon nitride (SRSN) has been considered in the photovoltaic industry as a down-conversion film for solar cells. In this work, SRSN films have been obtained by the Low-Pressure Chemical Vapor Deposition (LPCVD) technique at low to moderate deposition temperatures with a variation in the precursor gas pressure ratio. The SRSN films showed a wide photoluminescence (PL) in the visible region (without a high-deposition temperature or annealing process) and suitable optical properties (refractive index and absorption in the UV) to be used as photoluminescent coating on silicon solar cells. The absence of high-deposition temperatures could preserve the original structure of silicon solar cells, once the SRSN layer was applied. In addition, control of the reactive gas pressure ratio and deposition temperature showed an influence on the refractive index, the surface roughness, and the PL emission.

**Keywords:** silicon nitride; silicon-rich silicon nitride; refractive index; photoluminescent coating; down-conversion effect; optical properties

## 1. Introduction

Solar cells have become more prominent in the semiconductor market during the last decade; they have been continuously increasing in their relevance and impact on materials engineering and technology development. In the same way, the initial silicon solar cells with functional coatings have been investigated to improve efficiency and to reach higher competitive levels. In particular, the coatings used to create red-shift effects, which have shown important results, are among the recurrent add-ons [1,2]. Even different materials have been recently investigated to overcome optical loss by using them as antireflective coatings (ARCs), like silicon dioxide sol compositions, nanostructured $TiO_2$ or $SiO_2$/ITO, and ITO/TNO combinations [3–6]. However, silicon nitride ($Si_3N_4$) is a well-known material which has increased in interest from the industrial sector and the scientific community to be used as a functional coating due to its physical, mechanical, electrical, and optoelectronic properties [7–10]. $Si_3N_4$ was originally developed as a high-strength and high-toughness material, but recent reports have demonstrated that $Si_3N_4$

thin films are efficient and low-cost options to be used as an ARC [11–13]. Antireflective coatings should produce diffuse reflection and total internal reflection, but they should also preserve the properties of the materials on which they are deposited [14,15]. $Si_3N_4$ can be deposited using different techniques, such as Sputtering, Plasma-Enhanced Chemical Vapor Deposition (PECVD), Catalytic Chemical Vapor Deposition (Cat-CVD), Hot-Wire Chemical Vapor Deposition (HWCVD), etc. [16–19]. In particular, the physical properties of $Si_3N_4$ films can be varied by deposition temperature, growth time, and the ratio of the precursor gasses. When these $Si_3N_4$ films have been deposited at temperatures above 950 °C, they have shown interesting optical properties, but they have also faced some challenges when they are deposited on PN junctions. The deposition temperature of $Si_3N_4$ films is a very important parameter influencing their morphological, structural, and optical properties but the high-temperature deposition of these coatings on PN junctions can produce changes in the junction depth and consequently in the performance of solar cells [20–24]. On the other hand, these films have been deposited at unusually low temperatures, and their morphological, optical, and chemical characteristics have been clearly compromised [25,26]. In this way, the refractive index, root mean square roughness, and silicon excess of $Si_3N_4$ have shown an important dependence on deposition temperature, growth time, and the precursor gas ratio [10,16,27]. One possible alternative to preserve ARC characteristics in silicon nitride material and improve them with an effect of absorption on the UV range and emission in the visible region (down conversion) could be obtained with silicon-rich silicon nitride (SRSN or $SiN_x$) [28–30]. SRSN films can be obtained by LPCVD using a combination of $NH_3$ and $SiH_4$ as precursor gasses at temperatures usually above 725 °C. As-deposited SRSN films obtained by LPCVD have shown UV absorption and a wide photoluminescence (PL) spectrum in the visible range, where the maximum peak shifted from ~490 to ~590 nm as the silicon excess increased [31,32]. However, the behavior of these SRSN films not only depends on stoichiometry, but on all deposition conditions and thickness [21,33,34]. Information and interesting results of the SRSN films obtained by the PECVD technique have been reported by different authors [21,35,36], but SRSN films with interesting optical properties also can be obtained by a cheaper technique like LPCVD. In this way, an optimization on the deposition conditions of SRSN films obtained by LPCVD is required to obtain a suitable refractive index, roughness on the surface, and strong photoluminescence (PL) in the visible region even when SRSN films are deposited at moderate temperature ranges.

In this work, we provide a full description of the optimization of the deposition conditions like the deposition temperature and pressure gas ratio to obtain thin SRSN films by the LPCVD technique at temperatures $\leq$ 720 °C. No post-deposition thermal annealing was applied to the SRSN films. We provide an analysis of the optical, structural, and morphological characteristics from these SRSN films in order to control the thickness, refractive index, surface morphology, and PL emission. The optimization of the deposition conditions of the SRSN films allowed us to obtain suitable optical properties in this material to be applied as a photoluminescent coating.

## 2. Materials and Methods

SRSN films were deposited on p-type (100-oriented) silicon substrates with resistivity of 2–5 Ω cm by LPCVD. Before deposition, silicon wafers were cleaned using acetone, ethanol, and deionized water for 10 min followed by ultrasonic baths each, they were then cleaned with HF 10% (hydrofluoric acid) to remove native oxides, and finally, we used standard RCA solutions. The pressure ratio ($Ro_N$) of the reactive gasses ($NH_3$, $SiH_4$) was changed to increase the amount of silicon excess in the SRSN films in order to obtain optical characteristics like ARCs [12–15]. $Ro_N$ is defined by Equation (1):

$$Ro_N = \frac{P(NH_3)}{P(SiH_4)} \quad (1)$$

Two sets of samples were obtained: The first set was obtained by fixing the deposition temperature at 720 °C, and $Ro_N$ was changed in the range of 4 to 80. In the second set, we used a fixed $Ro_N = 4$, and different deposition temperatures were used in the range of 600 to 720 °C.

The thickness and refractive index of the SRSN films were calculated from measurements with a Gaertner L117 null ellipsometer (632.8 nm). Surface morphology was studied by atomic force microscopy measurements (AFM) using a Nanosurf EasyScan AFM system Version 2.3, operated in contact mode. A 450 µm long single-crystal Si cantilever operated at 12 kHz was used. AFM images were analyzed using the Scanning Probe Image Processor (SPIP) software 6.7.9. The SRSN vibrational modes were examined by Fourier Transform Infrared spectroscopy (FTIR) using a Bruker model Vector 22 (Coventry, UK) in the range of 400–4000 $cm^{-1}$ and a 1 $cm^{-1}/s$ resolution speed. Measurements of the chemical composition were obtained using Hitachi SU3500 SEM equipment (Tokyo, Japan) with an Electron Diffraction Spectroscopy (EDS) Bruker system operated at high vacuum. PL emission spectra were obtained with a Spectrofluorometer Fluoromax 3 Jobin Yvon (Horiba, NJ, USA) controlled with a computer. A set of filters were used to obtain only the UV excitation and visible emission. All SRSN films were excited using 270 nm radiation, and the PL emission signal was collected from 380 to 900 nm with a 1 nm/s resolution speed.

## 3. Results and Discussion

### 3.1. First Experiment

#### 3.1.1. Refractive Index, Deposition Rate, and Pressure Ratio

In this section, we describe the obtained results of the SRSN films in order to control refractive index using a stable deposition ratio. The ratio of precursor gasses ($Ro_N$) was changed in the range of 4 to 80 to obtain SRSN films with a different refractive index. Figure 1a shows the relationship between the refractive index and the $Ro_N$ at 720 °C. According to the refractive index value behavior, there is a limit on the atomic percentage of silicon excess introduced in SRSN films. In general, the thickness of the as-deposited SRSN films was calculated from ellipsometry measurements, and they showed a mean thickness of about 160 nm. The highest and lowest refractive indices were 2.21 and 1.98 when $Ro_N$ was 20 and 80, respectively.

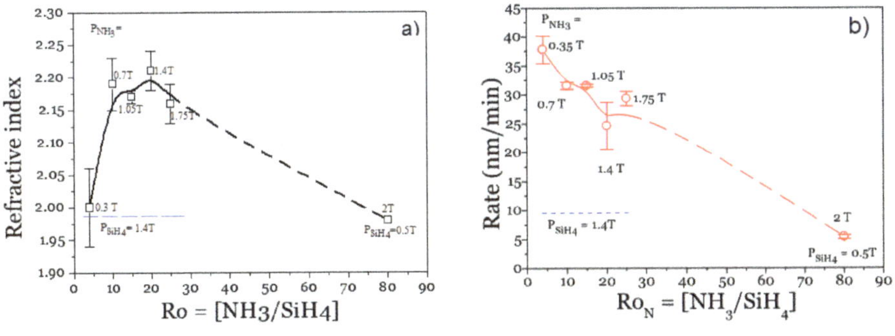


**Figure 1.** (**a**) Refractive index and (**b**) growth rate vs. pressure ratio $R_{oN}$ of SRSN LPCVD at 720 °C.

In Figure 1b, we can observe the deposition ratio vs. the pressure ratio of the SRSN films. The pressure ratio was changed by adjusting the precursor gas ratio as indicated. The refractive index showed an increase from 2.01 to 2.20 in the range of $Ro_N$ between 4 and 20 and a lineal decrease from 2.20 to 1.98 in the range of $Ro_N$ from 20 to 80. According to Cheng et al. [37] and our results in Figure 1a, there is a clear tendency for the refractive index to decrease as the pressure ratio is increased. The deposition ratio showed an almost lineal decay from 37.5 to 5 nm/min in the range of $4 \leq Ro_N \leq 80$. Table 1 shows the main deposition characteristics of each SRSN film and its corresponding refractive index.

**Table 1.** Main deposition characteristics and refractive index for the first experiment.

| Sample | $NH_3$ (Torr) | $SIH_4$ (Torr) | $Ro_N$ | RATE (nm/min) | REFRACTIVE INDEX |
|---|---|---|---|---|---|
| LPC01 | 0.35 | 1.40 | 4 | 37.5 | 2.00 |
| LPC02 | 0.70 | 1.40 | 10 | 32.5 | 2.19 |
| LPC03 | 1.05 | 1.40 | 15 | 31.9 | 2.16 |
| LPC04 | 1.40 | 1.40 | 20 | 24 | 2.21 |
| LPC05 | 1.75 | 1.40 | 25 | 30 | 2.15 |
| LPC06 | 2.00 | 0.50 | 80 | 5 | 1.98 |

According to these parameters, we obtained the lowest refractive index ($\leq$2.0) when the minimum and maxima $Ro_N$ were used during deposition; a very similar growth rate was reached at both limits, around 5 nm/min. The refractive index was changed in all the ranges of 4 < $Ro_N$ < 80, because the change in silane pressure produced a change in composition and film stoichiometry. This will be discussed in EDS Results and Analysis of Vibrational Modes sections.

3.1.2. Refractive Index Selection

In order to select the optimal refractive index of the SRSN films to work as an antireflective coating, we performed some calculations, considering the blue- to red-shift effect. By considering the lowest refraction to use these films as ARCs, the refractive index can be calculated as follows [33]:

$$n_{arc} = \sqrt{n_0 n_s} = \sqrt{1.0002926 \cdot 4.01} = 2.002791 \quad (2)$$

where $n_0$ is the silicon reflective index and $n_s$ is the antireflective coating reflective index. According to this calculation, we obtained only a refractive index ($n_{SRSN}$) of about 2.0 when $Ro_N$ = 4 but $n_{SRSN}$ was lower than 2 when $Ro_N$ = 80 was used for deposition. In this case, a value of $Ro_N$ = 4 produced an SRSN film with a refractive index closest to the calculated optimal value. This would be in agreement with a characteristic value for the reported refractive index of 2.01–2.05 for silicon nitride films obtained by CVD techniques [35–37].

The comparison between the refractive index of the SRSN films reported in Table 1 with the optimal refractive index calculation allowed us to select $R_{oN}$ = 4 for the second experiment. The second experiment was designedto obtain silicon-rich silicon nitride with a refractive index of 2 and down conversion effect (photoluminescence) at the lowest temperature as possible.

## *3.2. Second Experiment*

3.2.1. Thickness and Refractive Index Dependence on Temperature

We fixed the $NH_3$ and $SiH_4$ pressure at 0.35 and 1.4 Torr, respectively, to fix the pressure ratio at $Ro_N$ = 4 to reproduce the refractive index closest to the optimal $n_{arc}$, but we varied the deposition temperature from 600 to 720 °C to obtain another group of SRSN films with photoluminescence at the lowest deposition temperature as possible. Table 2 shows the main deposition characteristics of each SRSN film for the second experiment.

**Table 2.** Main deposition characteristics for the second experiment.

| Sample | Substrate | $Ro_N$ | Thickness (nm) | Deposition Temperature (°C) | Growth Time (min) |
|---|---|---|---|---|---|
| LP21 | 2″ N + (2.0 Ω·cm) | 4.00 | 75 | 600 | 5 |
| LP22 | 2″ N + (2.0 Ω·cm) | 4.00 | 95 | 650 | 5 |
| LP23 | 2″ N + (2.0 Ω·cm) | 4.00 | 112 | 700 | 5 |
| LP24 | 2″ N + (2.0 Ω·cm) | 4.00 | 250 | 720 | 5 |

We studied the temperature influence on thickness and refractive index by fixing the deposition time at 5 min. In Figure 2, we observed a clear dependence of thickness in temperature, where thickness was increased from 75 up to 250 nm according to the increase in deposition temperature from 600 up to 720 °C. According to Liu et al. [38], the growth rate of silicon nitride films obtained by LPCVD shows an Arrhenius behavior between 730 and 830 °C. In this case, the thickness of the silicon nitride samples should be clearly increased as the deposition temperature reaches 730 °C as with our SRSN sample obtained at 720 °C. We used the mean thickness values and the minimal squares theory in OriginPro 8.0 software to obtain a fitted curve that is directly proportional to the following equation:

$$\mathrm{Th(K)} = \mathrm{K}^3 - 0.91\mathrm{K}^2 + 587.71\mathrm{K} - 125955 \tag{3}$$

where Th = thickness and K = temperature.

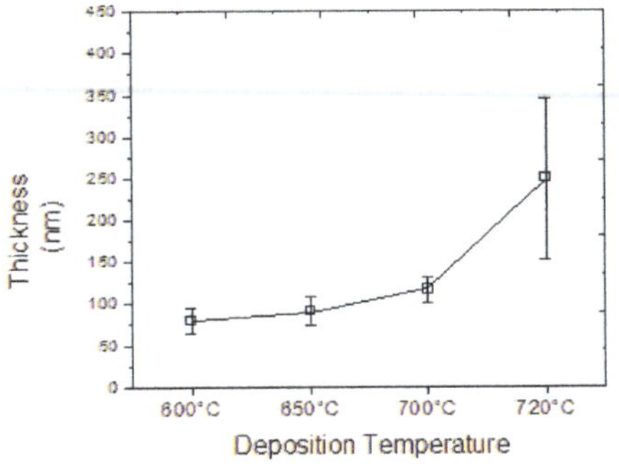


**Figure 2.** SRSN films thickness (nm) vs. deposition temperature (°C).

In the same way, we can observe in Figure 3 an inversely proportional dependence of the refractive index on deposition temperature, where the highest refractive index of 2.4 was obtained at a deposition temperature of 600 °C and decreased as temperature was increased. This would suggest that low-deposition temperature promotes non-stoichiometric silicon nitride content and high-deposition temperature enables an ordering of crystalline structure according to stoichiometric silicon nitride. Even when the same LPCVD equipment has been used to deposit the two series of samples, it is common that some optical, physical, or electrical properties are not exactly reproduced. However, according to Figure 3, the refractive index shows a decrease as deposition temperature is increased; this behavior is also related to an increase in crystallinity as a result of deposition temperature increments. In general, refractive index values are in a similar to that of silicon-rich nitride films deposited by different techniques and different deposition temperatures [30]. If we assumed $n$ = refractive index and $K$ = temperature, we could use the minimal squares approximation in the Origin Software again, and the obtained relationship is as follows:

$$n(K) = K^3 + 0.01K^2 - 3.72K + 805.15 \tag{4}$$

3.2.2. Refractive Index and Deposition Rate Dependence on Deposition Temperature

Figure 4 shows the deposition rate and refractive index dependence on temperature. In this case, we can observe that the growth rate was lower when 600 to 700 °C deposition temperatures were used, and it only increased when 720 °C was applied during deposition. The main values for the refractive index and growth rate can be seen at the intersection of both graphs; according to this, the best option to achieve the maximum growth rate with the best refractive index $n_{arc}$ would be using a deposition temperature of 703 °C.

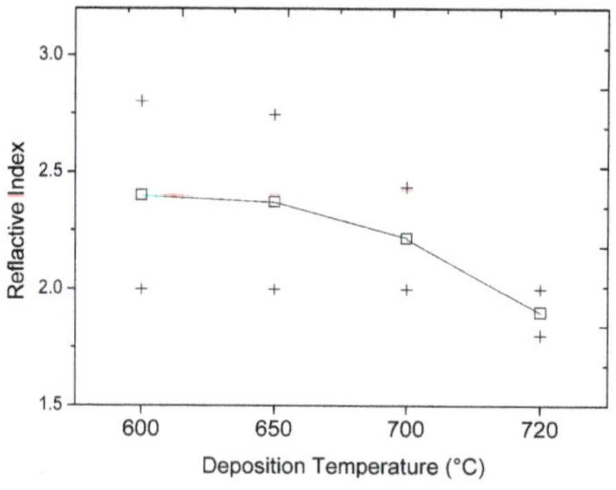


**Figure 3.** Refractive index vs. deposition temperature.

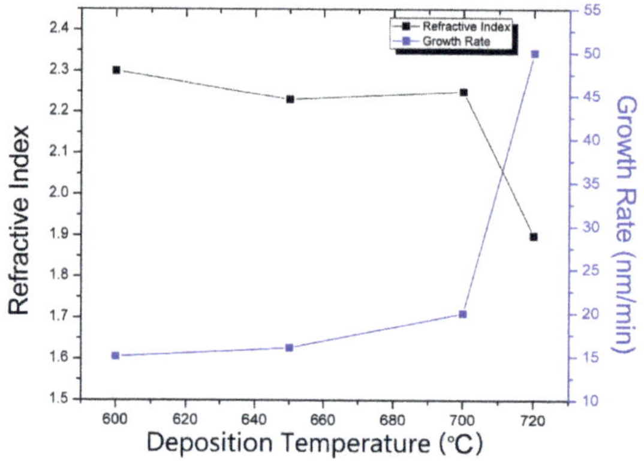


**Figure 4.** Relationship between refractive index and growth rate with deposition temperature.

### *3.3. EDS Results*

We conducted a qualitative analysis of the elemental composition in our SRNS samples through EDS measurements from three different regions on the films' surfaces. An accelerating voltage of 0.5 kV was used to minimize damage to the SRSN films. Silicon, nitrogen, and oxygen elements in the SRSN films were confirmed from their characteristic X-ray peaks. The presence of a hydrogen peak in SRSN films cannot be determined because EDS can only provide information on the chemical composition for elements with atomic number (Z) > 3. A quantitative analysis could not be conducted for all SRSN films due to the thin thickness of the samples. The electrons from the primary beam of the Bruker system can penetrate 0.5 μm and interact with the atoms in SRSN films but also in silicon substrate. In this case, only EDS values for the SRNS film deposited at 720 °C (250 nm thickness) are shown in Table 3. The silicon value was around 60; meanwhile, nitrogen was approximately 25 at.%. The silicon composition value is close to stoichiometric silicon nitride (60.1 at.% for Si and 39.9 at.% for Ni), but the lowest value of nitride in the SRSN film would suggest a probable replacement by hydrogen atoms and would confirm the non-stoichiometric nature of this film. A previously reported analysis on the SRSN composition by means of XPS has also shown the presence of oxygen at the outmost part of the layer through the film depth, being present mainly at the film surface [21,35,36]. This behavior is similar in every SRSN film deposited at different deposition temperatures where a small value of oxygen was registered.

**Table 3.** EDS results from SRSN film with $R_{oN}$ = 4 deposited at 720 °C.

| Sample | Deposition Temperature (°C) | Silicon (At.%) | Nitrogen (At.%) | Oxygen (At.%) |
|---|---|---|---|---|
| LP24 | 720 | 60 | 25 | 2.56 |

### 3.4. Surface Morphology Dependence on Temperature

Surface morphology was studied through different roughness measurements obtained by AFM from each SRSN film. The root mean square (RMS) height ($S_q$) was calculated using SPIP software to analyze Tapping Mode AFM images [30,39]. Figure 5 shows $S_q$ calculated results for the SRSN films deposited between 650 and 720 °C. The RMS values were around 2.5 and 6.8 nm on the SRSN films deposited in the range of 650 to 720 °C, respectively. A clear increase in RMS was observed as the deposition temperature was increased. In this case, an almost linear increasing dependance of $S_q$ on the deposition temperature was obtained. The increase in the obtained RMS values could be related to the incorporation of nitrogen and hydrogen atomic content, as observed in the reported EDS values in Table 3. The replacement of silicon by nitrogen or hydrogen atoms produces different defects in the film, one of which is random roughness of the upper boundaries, with a consequent increase in RMS values. The standard deviation for SRSN films deposited at 650, 700, and 720 °C was less than 1 nm, as observed on the box plots, which indicates the homogeneity of our films deposited by LPCVD. According to the deposition parameters, the highest surface roughness and RMS height ($S_q$) were obtained when a high deposition temperature was used; this would have an influence (increase) on the reflectivity of the surface that could be achieved [39,40].

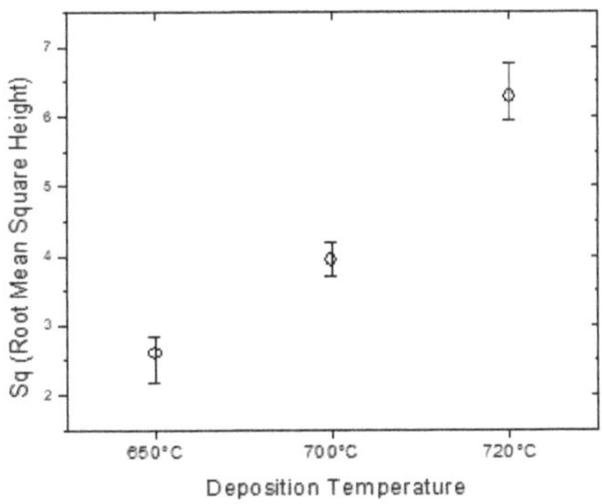


**Figure 5.** $S_q$ (mean square height) and temperature dependence.

### 3.5. Photoluminescence

PL from the samples was measured to confirm and determine the peak of emission in the visible range. In Figure 6, we show a comparison of normalized PL from the SRSN film of $R_{oN}$ = 80 deposited at 720 °C (SRSN80-720) with SRSN films of $R_{oN}$ = 4 deposited at 700 (SRSN4-700) and 720 °C (SRSN4-720); all samples showed a clear PL with a main emission band centered around the green region. These samples were compared due to the similar refractive index observed in Table 1. The main difference between the PL spectra of SRSN80-720 in the first experiment and SRSN4 films for the second, was the beginning of the PL emission band, where the PL of the SRSN80-720 started at the visible region but emission from SRSN4 films begins at the UV region. Moreover, the SRSN4-700 and SRSN4-720 films showed the same emission peaks (identified in Figure 7), and some shifts from the usual pattern of silicon nitride films were identified. The main difference in PL between these films was the PL intensity on the emission centered around 478.1 nm (2.59 eV). Previous works have reported that oxygen in SRSN films (like SRSN4-700 and SRSN4-720) will create a gap state of Si–O above the VBM, which will produce an electronic transition from the $K^0$ centers to ≡Si–O–Si, giving rise to the 478.1 nm (∼2.59 eV) emission [27,32].

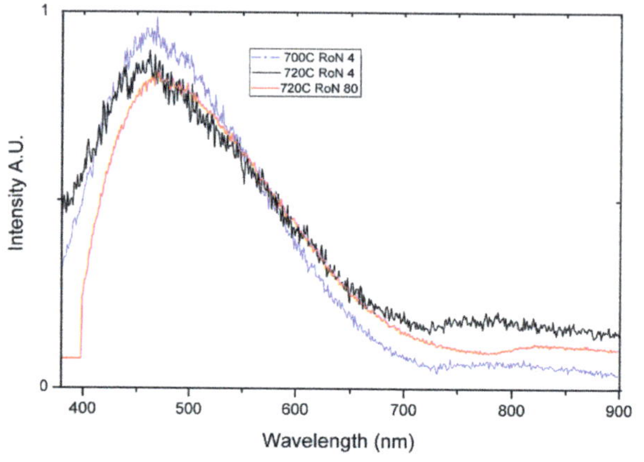


**Figure 6.** Photoluminescence spectra of SRSN films deposited at 700 and 720 °C.

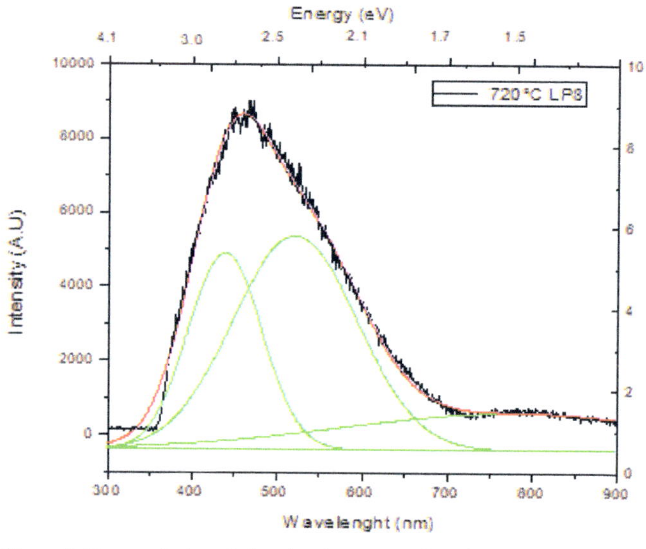


**Figure 7.** Deconvolution of photoluminescence of 720 °C sample.

According to the similar PL behavior for the SRSN4-700 and SRSN4-720 films, a detailed analysis was obtained only through the deconvolution of the PL from the SRSN film deposited at 720 °C, as shown in Figure 7. Three emission bands were identified, the first one in the near-infrared region (NIR), the second in the green region, and the last one in the blue region. In the first band, we obtained a low but non-negligible emission centered at 773.3 nm (1.6 eV); this could be considered in the range of reported photoluminescence of $SiN_x$ films. There are many radiative defects that could be responsible for the emission in this region, even though this peak in strong emission has been attributed to the effect of luminescence from Si nanocrystals (nCs) embedded in this kind of matrix by other authors [41,42].

The emission in the NIR also has been attributed to the presence of Silicon Quantum Dots (Si-QDs) that can promote the *QCE* effect (quantum confinement effect). A low excess of silicon in this sample (near or below stoichiometric silicon nitride), as shown in the EDS results, could be related to the main average size of the Si-QDs and, consequently, will be related to the observed PL emission [43–45]. However, further studies based on Transmittance Electron Microscopy will be required to confirm the presence of Si-nCs or Si-QDs.

The highest intensity of photoluminescence was found at 478.1 nm (2.59 eV) without symmetry in the emission band, which may be due to the size distribution of Si bonds and different species. This peak would be strongly influenced by the contribution of a radiative Si dangling bond $K^0$ center around 520 nm (2.39 eV) [46]. Finally, a strong emission peak is located in the blue region around 420 nm (2.81 eV), which may be due to a higher density of Si species than Si-N or an effect of surface passivation decreasing dangling bonds and increasing radiative recombination centers [46–49]. A correlation between the PL peak energy with the optical bandgap indicates that the luminescence is related to the band tail carrier recombination in the SRSN films.

Optimal Thickness Calculation

The optical thickness to produce down conversion using SRSN films needs to be calculated. The optimum optical thickness must be chosen like an antireflective coating (ARC) for an optimal energy conversion, so the optimum optical thickness was calculated based on the Fresnel equation for reflection theory [50] as follows:

$$R = |1 - T| = \frac{a_1 cos^2\delta_1 + a_2 sen^2\delta_1}{a_3 cos^2\delta_1 + a_4 sen^2\delta_1} \tag{5}$$

The following equation was obtained:

$$\frac{\delta_1 \lambda}{2\pi} = n_{arc} t_{arc} \tag{6}$$

$$\delta_1 = \frac{2\pi}{\lambda} n_{arc} t_{arc} \tag{7}$$

where $\delta_1$ is the phase difference introduced by the antireflective layer, $\lambda$ is the incident wavelength, $n_{arc}$ is the refractive index of the layer, and $t_{arc}$ is its thickness. According to antireflection coating theory and using $\delta_1 = 0.5\ \pi$, in most cases of interest, R vanishes when

$$t_{arc} = \frac{\lambda}{4n_{arc}}; \frac{3\lambda}{4n_{arc}}; \frac{5\lambda}{4n_{arc}} \ldots \tag{8}$$

Then, we used the calculated optimal refractive index ($n_{arc}$) and the Fresnel equation to obtain a suitable thickness with a red-shift effect (down conversion) [3,4,12,15,27,33] in the visible spectrum. In this way, looking for the film to absorb or transmit radiative energy with the expected down-conversion effect, we used the obtained wavelength of the highest PL peak from our SRSN films at $\lambda$ = 478.1 nm, and then, 478.1 nm/(4 $\times$ 2.002) = $t_{arc}$. The optimal thickness of the film for emission around 478.1 nm should be $\boldsymbol{t_{arc}} \approx 60$ nm.

### 3.6. Analysis of Vibrational Modes

Vibrational modes were identified by means of the obtained FTIR absorbance spectra of our SRSN films. Figure 8 shows a comparison of absorbance spectra of the SRSN films deposited at 700 °C and 720 °C; a deconvolution was carried out to locate the main peaks related to vibrational modes. In both samples, we could observe the conventional vibrations of silicon nitride corresponding to 960 and 1300 $cm^{-1}$ [51,52].

For the SRSN film deposited at 720 °C, we can observe a strong peak for the Si-H "Wagging" mode at 625 $cm^{-1}$ in (label A). In the same way, we can observe a clear and characteristic peak of the $Si_3N_4$ to $Si\text{-}N_3\text{-}H$ "Asymmetric Stretching" mode at 960 $cm^{-1}$ peak (label C) for both samples, which has been related to $Si_xN_y$ species; then, a non-stoichiometric behavior could be confirmed according to the PL emission peaks, atomic content in Table 3, and previous reported films obtained at similar conditions but different deposition temperatures [52–55]. The presence of different $Si_xN_y$ species confirms the relationship with the observed PL around 2.77 eV.

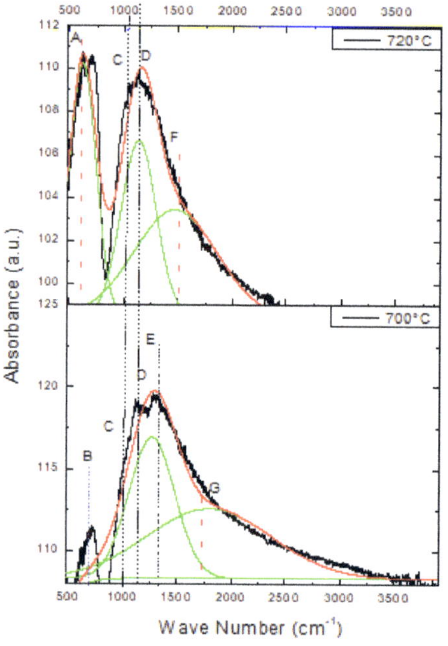


**Figure 8.** FTIR absorbance spectra of SRSN films deposited at 700 °C and 720 °C.

For the SRSN film obtained at 700 °C, we can observe a vibrational mode from 650 to 790 cm$^{-1}$ with a centered peak around 690 cm$^{-1}$ corresponding to Si-H and Si-N-H (label B) [52]. These species can be related to the deposition process without thermal annealing (as-deposited) and the off-stoichiometry nature in SRSN films with hydrogen enrichment [26,32,47] regarding the PL and EDS results. In the same way, there is a peak centered around 1010 cm$^{-1}$ (label C), where we observe a lower intensity peak in this film than that obtained from the 720 °C sample. This peak represents the Si-N "Asymmetric Stretching" mode but has also been related to H-Si-$N_3$ species [52,56]. The observed peak in both SRSN films around 1260 cm$^{-1}$ (label D-E) can be related to the H-$N_3$ and N-H Wagging mode. [48,49]. Also, 1550 cm$^{-1}$ (label F) can be related to the Wagging N-H vibration mode [55]. The Si-H Wagging mode [52,56], which has been related to the band centered at 1850 cm$^{-1}$ (label G) [57], could be due to the great abundance of hydrogen in our films produced by the silane precursor during the deposition process, as suggested by the EDS results.

Regarding the small difference in the intensity between the A and B labels, which correspond to Si-H, we can relate this variation to the small increase in deposition temperature from 700 to 720 °C. The hydrogen species predominating in the 720 °C sample can usually be removed by thermal treatments; however, our SRSN films did not have any annealing treatment to avoid N-H or Si-H species [51,52,57]. The absence of high-deposition temperatures or annealing treatments could preserve the original structure of silicon solar cells where the SRSN layer would be applied on top.

## 4. Conclusions

Silicon-rich silicon nitride films can be obtained by the LPCVD technique at temperatures under 720 °C in order to avoid physical changes in semiconductor junctions when they are applied as an antireflective coating on devices. SRSN films deposited at temperatures below 720 °C showed a refractive index in the range of 2.0 to 2.2 to be used as an antireflective coating. A mean square roughness from 2.5 to 6.5 nm with an almost linear increasing dependance of $S_q$ on deposition temperature was obtained due to an increase in nitrogen and hydrogen content. The SRSN films showed a strong PL in the blue to red region without any kind of post-deposition annealing. The vibrational modes related to $Si_3N_4$, Si-$N_3$-H, and Si-N-H corroborate the off-stoichiometry nature and nitrogen enrichment in our SRSN films. The radiative defects produced by $Si_xN_y$ species and QCE were related to the observed PL bands. According to the relationship of the growth rate and refractive index on the temperature and precursor gas ratio, the thickness and optical characteristics of the SRSN films can be controlled for application as a photoluminescent coating.

**Author Contributions:** Conceptualization, F.U.G., K.M.L. and A.L.M.Z.; methodology, F.U.G., K.M.L., and A.L.M.Z.; software, F.U.G.; validation, F.U.G., K.M.L., M.M.M., and A.L.M.Z.; formal analysis, F.U.G., K.M.L., M.M.M., A.M.S., A.L.M.Z., and J.A.D.H.d.l.L.; investigation, J.A.L.L., J.C.L., J.A.D.H.d.l.L., and A.S.L.S.V.; resources, K.M.L., M.M.M., A.M.S., J.A.L.L., J.C.L., and J.A.D.H.d.l.L.; writing—original draft preparation, F.U.G., K.M.L., M.M.M., and A.L.M.Z.; writing—review and editing, F.U.G., K.M.L., M.M.M., A.M.S., A.L.M.Z., J.A.L.L., J.C.L., J.A.D.H.d.l.L., and A.S.L.S.V.; supervision K.M.L., M.M.M., and A.L.M.Z. All authors have read and agreed to the published version of the manuscript.

**Funding:** This work has been partially supported by CONAHCyT [grant number: 217570727] and 100520056-VIEP2024 project.

**Institutional Review Board Statement:** Not applicable.

**Informed Consent Statement:** Not applicable.

**Data Availability Statement:** Data are contained within the article.

**Acknowledgments:** The authors acknowledge CIDS, INAOE, and IER laboratories for their help in the samples' obtention and characterization.

**Conflicts of Interest:** The authors declare no conflicts of interest.

## References

1. Luft, W. Effects of electron irradiation on N on P silicon solar cells. *Adv. Energy Convers.* **1965**, *5*, 21–41. [CrossRef]
2. Shah, A. Thin-Film Silicon Solar Cells. In *Practical Handbook of Photovoltaics*; Elsevier: Amsterdam, The Netherlands, 2012; pp. 209–281. [CrossRef]
3. Lokteva, A.A.; Kotelnikova, A.A.; Kovylin, R.S.; Konev, A.N.; Piskunov, A.V. Novel Antireflection Coatings Obtained by Low-Temperature Annealing in the Presence of Tetrabutylammonium Bromide and Gold Nanoparticles. *Materials* **2022**, *15*, 7658. [CrossRef] [PubMed]
4. Abu-Shamleh, A.; Alzubi, H.; Alajlouni, A. Optimization of antireflective coatings with nanostructured $TiO_2$ for GaAs solar cells. *Photonics Nanostruct. Fundam. Appl.* **2021**, *43*, 100862. [CrossRef]
5. Ho, W.-J.; Lin, J.-C.; Liu, J.-J.; Bai, W.-B.; Shiao, H.-P. Electrical and Optical Characterization of Sputtered Silicon Dioxide, Indium Tin Oxide, and Silicon Dioxide/Indium Tin Oxide Antireflection Coating on Single-Junction GaAs Solar Cells. *Materials* **2017**, *10*, 700. [CrossRef]
6. Lee, Y.S.; Chuang, L.Y.; Tang, C.J.; Yan, Z.Z.; Le, B.S.; Jaing, C.C. Investigation into the Characteristics of Double-Layer Transparent Conductive Oxide ITO/TNO Anti-Reflection Coating for Silicon Solar Cells. *Crystals* **2023**, *13*, 80. [CrossRef]
7. Bocanegra-Bernal, M.H.; Matovic, B. Mechanical properties of silicon nitride-based ceramics and its use in structural applications at high temperatures. *Mater. Sci. Eng. A* **2010**, 1314–1338. [CrossRef]
8. Nguyen, V.S.; Burton, S.; Pan, P. The Variation of Physical Properties of Plasma-Deposited Silicon Nitride and Oxynitride with Their Compositions. *J. Electrochem. Soc.* **1984**, *131*, 2348. [CrossRef]
9. De Brito Mota, F.; Justo, J.F.; Fazzio, A. Structural and electronic properties of silicon nitride materials. *Int. J. Quant. Chem.* **1998**, *70*, 973–980. [CrossRef]
10. Beliaev, L.Y.; Shkondin, E.; Lavrinenko, A.V.; Takayama, O. Optical, structural and composition properties of silicon nitride films deposited by reactive radio-frequency sputtering, low pressure and plasma-enhanced chemical vapor deposition. *Thin Solid Films* **2022**, *763*, 139568. [CrossRef]

11. Tahir, S.; Saeed, R.; Ashfaq, A.; Ali, A.; Mehmood, K.; Almousa, N.; Shokralla, E.A.; Macadangdang, R.R., Jr.; Soeriyadi, A.H.; Bonilla, R.S. Optical modeling and characterization of bifacial SiNx/AlOx dielectric layers for surface passivation and antireflection in PERC. *Prog. Photovolt. Res. Appl.* **2024**, *32*, 63–72. [CrossRef]
12. Lelièvre, J.-F.; Kafle, B.; Saint-Cast, P.; Brunet, P.; Magnan, R.; Hernandez, E.; Pouliquen, S.; Massines, F. Efficient silicon nitride SiNx:H antireflective and passivation layers deposited by atmospheric pressure PECVD for silicon solar cells. *Progress. Photovolt.* **2019**, *27*, 1007–1019. [CrossRef]
13. Yu, J.; Daoming, Y.; Yingchun, C.; Wentao, G.; Manqing, T. High-performance $SiO_2$-$SiN_X$ distributed Bragg reflectors fabricated by ion-assisted reactive magnetron sputtering. *Vacuum* **2024**, *220*, 112775. [CrossRef]
14. Lee, Y.; Gong, D.; Balaji, N.; Lee, Y.J.; Yi, J. Stability of $SiN_X$/$SiN_X$ double stack antireflection coating for single crystalline silicon solar cells. *Nanoscale Res. Lett.* **2012**, *7*, 50. Available online: http://www.nanoscalereslett.com/content/7/1/50 (accessed on 2 September 2024). [CrossRef]
15. Sharma, R. Silicon nitride as antireflection coating to enhance the conversion efficiency of silicon solar cells. *Turk. J. Phys.* **2018**, *42*, 350–355. [CrossRef]
16. Guler, I. Optical and structural characterization of silicon nitride thin films deposited by PECVD. *Mater. Sci. Eng. B* **2019**, *246*, 21–26. [CrossRef]
17. TuHuynh, T.C.; Keisuke, O. Long-term stability of low-temperature deposited Cat-CVD SiNx thin film against damp-heat stress. *Jpn. J. Appl. Phys.* **2024**, *63*, 01SP25. [CrossRef]
18. Verlaan, V.; Van Der Werf, C.H.M.; Houweling, Z.S.; Romijn, I.G.; Weeber, A.W.; Dekkers, H.F.W.; Goldbach, H.D.; Schropp, R.E.I. Multi-crystalline Si solar cells with very fast deposited (180 nm/min) passivating hot-wire CVD silicon nitride as antireflection coating. *Prog. Photovolt. Res. Appl.* **2007**, *15*, 563–573. [CrossRef]
19. Alain, E.K.; Jové, F.A.; Goff, J.; Arkles, B. Silicon Nitride and Silicon Nitride-Rich Thin Film Technologies: Trends in Deposition Techniques and Related Applications. *ECS J. Solid State Sci. Technol.* **2017**, *6*, P691. [CrossRef]
20. Jones, S.W. Difussion in Silicon. 2008 IC Knowledge LLCC. Available online: https://api.semanticscholar.org/CorpusID:14437811 (accessed on 5 September 2024).
21. Braña, A.F.; Gupta, H.; Bommali, R.K.; Srivastava, P.; Ghosh, S.; Casero, R.P. Enhancing efficiency of c-Si solar cell by coating nano structured silicon rich silicon nitride films. *Thin Solid Films* **2018**, *662*, 21–26. [CrossRef]
22. Ali, R.M.; Zahran, M.B.; Youssif, A.M.; Eliwa, A. Characterization of Monocrystalline Silicon Solar Cells based on the Phosphorus Diffusion Temperature. *Int. J. Eng. Sci. Invent.* **2021**, *10*, 01–07. [CrossRef]
23. de la Mora, M.B.; Amelines-Sarria, O.; Monroy, B.M.; Hernández-Pérez, C.D.; Lugo, J.E. Materials for down conversion in solar cells: Perspectives and challenges. *Sol. Energy Mater. Sol. Cells* **2017**, *165*, 59–71. [CrossRef]
24. El amrani, A.; Menous, I.; Mahiou, L.; Tadjine, R.; Touati, A.; Lefgoum, A. Silicon nitride film for solar cells. *Renew. Energy* **2008**, *33*, 2289–2293. [CrossRef]
25. Jana, T.; Mukhopadhyay, S.; Ray, S. Low temperature silicon oxide and nitride for surface passivation of silicon solar cells. *Sol. Energy Mater. Sol. Cells* **2002**, *71*, 197–211. [CrossRef]
26. Hendawi, R.; Ciftja, A.; Stokkan, G.; Arnberg, L.; Di Sabatino, M. The effect of preliminary heat treatment on the durability of reaction bonded silicon nitride crucibles for solar cells applications. *J. Cryst. Growth* **2020**, *542*, 125670. [CrossRef]
27. Sahouane, N.; Zerga, A. Optimization of antireflection multilayer for industrial crystalline silicon solar cells. *Energy Procedia* **2014**, *44*, 118–125. [CrossRef]
28. Torchynska, T.V.; Espinola, J.L.C.; Khomenkova, L.; Hernandez, E.V.; Adame, J.A.A.; Slaoui, A. Structural and light emitting properties of silicon-rich silicon nitride films grown by plasma enhanced-chemical vapor deposition. *Mater. Sci. Semicond. Process.* **2015**, *37*, 46–50. [CrossRef]
29. Cabañas-Tay, S.A.; Palacios-Huerta, L.; Aceves-Mijares, M.; Coyopol, A.; Morales-Morales, F.; Pérez-García, S.A.; Licea-Jiménez, L.; Domínguez-Horna, C.; Monfil-Leyva, K.; Morales-Sánchez, A. Study of narrow and intense UV electroluminescence from ITO/SRO/Si-p and ITO/SRN/SRO/Si-p based lightemitting capacitors. *J. Lumin.* **2017**, *183*, 334–340. [CrossRef]
30. Yoshinaga, S.; Ishikawa, Y.; Kawamura, Y.; Nakai, Y.; Uraoka, Y. The optical properties of silicon-rich silicon nitride prepared by plasma-enhanced chemical vapor deposition. *Mater. Sci. Semicond. Process.* **2019**, *90*, 54–58. [CrossRef]
31. Kaloyeros, A.; Pan, Y.; Goff, J.; Arkles, B. Review—Silicon Nitride and Silicon Nitride-Rich Thin Film Technologies: State-of-the-Art Processing Technologies, Properties, and Applications. *ECS J. Solid State Sci. Technol.* **2020**, *9*, 063006. [CrossRef]
32. Bommali, R.; Preet, S.; Rai, S.; Mishra, P.; Sekhar, B.; Prakash, V.G.; Srivastava, P. Excitation dependent photoluminescence study of Si-rich a-SiNx:H thin films. *J. Appl. Phys.* **2012**, *112*, 123518. [CrossRef]
33. Mamgain, S. Structural and optical study of amorphous hydrogenated silicon nitride thin film as antireflection coating on solar cell. *Adv. Mater. Proc.* **2017**, *2*, 570–574. [CrossRef]
34. Rezgui, B.; Sibai, A.; Nychyporuk, T.; Lemiti, M.; Brémond, G. *Photoluminescence and Optical Absorption Properties of Silicon Quantum Dots Embedded in Si-Rich Silicon Nitride Matrices*; Elsevier: Amsterdam, The Netherlands, 2009. [CrossRef]
35. Liao, W.; Zeng, X.; Wen, X.; Zheng, W.; Yao, W. Synthesis and Optical Properties of Si-Rich Containing Silicon Quantum Dots. *J. Electron. Mater.* **2013**, *42*, 3445–3450. [CrossRef]
36. Takeoda, S.; Fujii, M.; Hayashi, S. Size-dependent photoluminescence from surface-oxidized Si nanocrystals in a weak confinement regime. *Phys. Rev. B* **2000**, *62*, 16820–16825. [CrossRef]

37. Cheng, Y.T.; Ho, J.J.; Lee, W.J.; Tsai, S.Y.; Lu, Y.A.; Liou, J.J.; Chang, S.H.; Wang, K.L. Investigation of Low-Cost Surface Processing Techniques for Large-Size Multicrystalline Silicon Solar Cells. *Int. J. Photoenergy* **2010**, 268035. [CrossRef]
38. Liu, X.J.; Zhang, J.J.; Sun, X.W.; Pan, Y.B.; Huang, L.P.; Jin, C.Y. Growth and properties of silicon nitride films prepared by low pressure chemical vapor deposition using trichlorosilane and ammonia. *Thin Solid Films* **2004**, *460*, 72–77. [CrossRef]
39. Komarov, F.F.; Parkhomenko, I.N.; Vlasukova, L.A.; Milchain, O.V.; Togambayeva, A.K.; Kovalchuck, N.S. Annealing Effects on Photoluminescence of SiNx films grown by PECVD. In Proceedings of the 10th International Conference: "Interaction of radiation with solids", Minsk, Belarus, 24–27 September 2013; pp. 238–241, ISBN 978-985-553-141-9.
40. Sharma, V.; Tracy, C.; Schroder, D.; Heresimenka, S.; Dauksher, W.; Bowden, S. Manipulation of K Center Charge Staates in silicon nitride films to achieve excellent Surface passivation for silicon solar cells. *Appl. Phys. Lett.* **2014**, *104*, 053503. [CrossRef]
41. Kilian, S.O.; Wiggers, H. Gas-Phase Synthesis of Silicon-Rich Silicon Nitride Nanoparticles for High Performance Lithium–Ion Batteries. *Part. Part. Syst. Charact.* **2021**, *38*, 2100007. [CrossRef]
42. Zhingunov, D.M.; Popov, A.A.; Chesnokov, Y.M.; Vasiliev, A.L.; Lebedev, A.M.; Subbotin, I.A.; Yakunin, S.N.; Shalygina, O.A.; Kamenskikh, I.A. Near-IR Emitting Si Nanocrystals Fabricated by thermal Annealing of SiNx/$Si_3N_4$ Multilayers. *App. Sci.* **2019**, *9*, 4725. [CrossRef]
43. Chen, X.; Yang, W.; Yang, P.; Yuan, J.; Zhao, F.; Hao, J.; Tang, Y. Size-controlled Si Quantum dots embedded in B-doped SiNx/$Si_3N_4$ superlatice for Si quantum dot solar cells. *J. Mater. Sci. Mater. Electron.* **2017**, *28*, 1322–1327. [CrossRef]
44. So, Y.-H.; Huang, S.; Conibeer, G.; Green, M.A. Formation and Photoluminiscence of Si nanocrystals in solid films. *Thin Solid Films* **2011**, *519*, 5408–5412. [CrossRef]
45. Linnros, J.; Lalic, N.; Galeckas, A.; Grivickas, V. Analysis of the stretched exponential photoluminescence decay from nanometer-sized silicon crystals in SiO2. *J. Appl. Phys.* **1999**, *86*, 6128–6134. [CrossRef]
46. Wang, M.; Li, D.; Yuan, Z.; Yang, D.; Que, D. Photoluminescence of Si-rich silicon nitride: Defect-related states and silicon nanoclusters. *Appl. Phys. Lett.* **2007**, *90*, 131903. [CrossRef]
47. Esposito, E.M.; Mercado, L.V.; Veneri, P.D.; Lancellotti, L.; Privato, C. *Annealing Effects on PECVD-Grown Si rich aSiNx Thin Films*; Elsevier: Amsterdam, The Netherlands, 2009. [CrossRef]
48. Deshpande, S.V. Optical properties of silicon nitride films deposited by hot filament chemical vapor deposition. *J. Appl. Phys.* **1995**, *77*, 6534–6541. [CrossRef]
49. Ko, C. Annealing effects on the photoluminescence of amorphous silicon-nitride films. *J. Korean Phys. Soc.* **2006**, *48*, 1277–1280.
50. Valiei, M.; Shaibani, P.M.; Abdizadeh, H.; Kolahdouz, M.; Soleimani, E.A.; Poursaf, J. Design and optimization of single, double and multilayer anti-reflection coatings on planar and textured surface of silicon solar cells. *Mater. Today Commun.* **2022**, *32*, 104144. [CrossRef]
51. Bugaev, K.O.; Zelenina, A.A.; Volodin, V.A. Vibrational Spectroscopy of Chemical Species in Silicon and Silicon-Rich Nitride Thin Films, Russian Academy of Sciences, Novosibirsk 63090, Russia. *Int. J. Spectrosc.* **2011**, *2012*, 2011. [CrossRef]
52. Scardera, G.; Puzzer, T.; Conibeer, G.; Green, M. Fourier transform infrared spectroscopy of annealed silicon-rich silicon nitride thin films. *J. Appl. Phys.* **2008**, *104*, 104310. [CrossRef]
53. Hoyos-García, J.E. Caracterización óptica y morfológica de partículas de silicio en una matriz SiNx obtenidas por PECVD, Instituto Politécnico Nacional. 2006. Available online: http://tesis.ipn.mx (accessed on 6 September 2024).
54. Nekrashevich, S.S.; Shaposhnikov, A.V.; Gritsenko, V.A. Study of the atomic and electronic structures of amorphous silicon nitride and defects in it, institute of semiconductor physics. *JETP Lett.* **2011**, *94*, 202–205. [CrossRef]
55. Torchynska, T.; Khomenkova, L.; Slaqui, A. Modification of light Emission in Si. Rich Silicon Nitride Films Versus Stoichiometry and Excitation light Energy. *J. Electron. Mater.* **2018**, *47*, 3927–3933. [CrossRef]
56. Xie, M.; Li, D.; Wang, F.; Yang, D. Luminescence Properties of Silicon-Rich Silicon Nitride Films and Light Emitting Devices. *ECS Trans.* **2011**, *35*, 3–19. [CrossRef]
57. Ahmed, N.; Singh, C.B.; Bhattacharya, S.; Dhara, S.; Bhargav, P.B. Raman and FTIR Studies on PECVD Grown Ammonia-Free Amorphous Silicon Nitride Thin Films for Solar Cell Applications. *Conf. Pap. Sci.* **2013**, *2013*, 837676. [CrossRef]

coatings

MDPI

Article

# Impact of Substrate upon Morphology, Luminescence, and Wettability of ZnMgO Layers Deposited by Spray Pyrolysis

**Eduard V. Monaico [1,*], Vadim Morari [2], Stepan Buiucli [1], Victor V. Zalamai [1], Veaceslav V. Ursaki [1,3] and Ion M. Tiginyanu [1,3]**

1 National Center for Materials Study and Testing, Technical University of Moldova, 2004 Chisinau, Moldova; stepan.buiucli@mib.utm.md (S.B.); victor.zalamai@cnstm.utm.md (V.V.Z.); vvursaki@gmail.com (V.V.U.); tiginyanu@gmail.com (I.M.T.)

2 Institute of Electronic Engineering and Nanotechnologies "D. Ghitu", Technical University of Moldova, 2004 Chisinau, Moldova; vadim.morari@iien.utm.md

3 Academy of Sciences of Moldova, 2001 Chisinau, Moldova

* Correspondence: eduard.monaico@cnstm.utm.md

**Abstract:** In this work, we report on a comparative study of the topology, luminescence, and wettability properties of ZnMgO films prepared by a cost-effective spray pyrolysis technology on GaAs substrates with (100), (001), and (111) crystallographic orientations, as well as on Si(100) substrates. Deposition on nanostructured GaAs substrates was also considered. It was found that film growth is not epitaxial or conformal, but rather, it is granular, depending on the nucleating sites for the crystallite growth. The distribution of nucleation sites ensured the preparation of nanostructured films with good uniformity of their topology. The observed difference in columnar growth on Si substrates and pyramidal growth on GaAs ones was explained in terms of the impact of chemical bonding in substrates. The films grown on GaAs substrates with a (001) orientation were found to be made of larger crystallites compared to those deposited on substrates with a (111) orientation. These effects resulted in a difference in roughness of a factor of 1.5, which correlates with the wetting properties of films, with the most hydrophobic surface being found on films deposited on GaAs substrates with a (111) orientation. The prospects for photocatalytic and gas sensor applications of films produced on flat substrates, as well as for plasmonic and other applications of films deposited on nanostructured substrates, are discussed, taking into account the results of the analysis of their photoluminescence properties.

**Keywords:** ZnMgO thin films; substrate materials; spray pyrolysis; film roughness; luminescence; wettability

## 1. Introduction

Zinc oxide is a wide-bandgap semiconductor (~3.36 eV at room temperature) with unique physical properties, such as excellent radiation hardness [1], high exciton binding energy of ~60 meV, high electron mobility [2], and high optical transparency. Due to these properties, it has become an important multifunctional semiconductor compound used in a wide range of applications, such as photodetectors [3], gas and tactile sensors [4,5], solar cells [6,7], and transparent and flexible electronics [8–10]. In particular, nanostructured zinc oxide proved to be a versatile material and one of the best for the implementation of the random lasing mirror-less effects provided by a simultaneously amplifying and highly scattering medium with coherent and multiple scattering among random-growth-oriented nanoparticles [11,12]. In addition to ZnO nanopowders, highly disordered polycrystalline thin films, nanowires, and nanorod structures were demonstrated to be suitable for the development of random lasers [13–16]. To widen the scope of practical applications for random lasers, it is important to ensure their ability to support both optical and electrical pumping, as demonstrated on *p-n* junctions, heterojunctions, and Schottky junctions (see refs. [17,18] and references therein). Due to such abilities, ZnO-based random lasers are promising for applications in sensing, medical diagnostics and cancer treatments, biophysical characterization, and high-resolution displays [18,19]. Beyond these, networks of random lasers have been proposed for applications in signal processing in the optical domain for photonics technology [20].

Nevertheless, the bandgap fixed at a wavelength at the boundary between visible and UV light is still a limitation for ZnO's further exploration, especially for UV optoelectronic applications, e.g., in photodetectors and light-emitting devices. One widely explored possibility for controlling the material bandgap to tune the spectral range of emission or the optical sensitivity is alloying with wider-bandgap materials. Among other known materials, magnesium oxide with a bandgap of 7.8 eV at room temperature is a very suitable compound for alloying with ZnO. Alloying these two materials in ZnMgO solid solutions covers a wide ultraviolet (UV) spectral region ranging from 3.36 eV (a bandgap of ZnO) to 7.8 eV (a bandgap of MgO). Bandgap widening enabled further extension of the operation wavelengths of lasers [21,22], LEDs [23–25], and photodetectors [26] to shorter wavelengths. Particularly, electrically driven deep ultraviolet ZnMgO lasers [27] and solar blind photodetectors working at room temperature have been demonstrated [28–30], among other optoelectronic applications.

Many ZnMgO devices have been produced on thin films, which have been grown by various technological methods on a variety of substrates. ZnMgO solid-solution thin films have been prepared by Pulsed Laser Deposition (PLD) [31–35], Magnetron Sputtering [36–43], Molecular Beam Epitaxy (MBE) [44–47], Chemical Vapor Deposition (CVD) [48–51], Atomic Layer Deposition [52,53], spin-coating [54–58], spray pyrolysis [56,59–65], and other technologies. Among this variety of technological methods, spin-coating and spray pyrolysis are considered to be the simplest and most cost-effective technologies for uniform and good-quality film deposition.

As concerns the substrate materials, most ZnMgO thin film photodetectors in the metal –semiconductor–metal (MSM) configuration have been prepared on glass substrates [38,55], while silicon (Si) substrates have been used for the fabrication of devices based on *p-n* heterojunctions [28,34,39,56], since Si has advantages in terms of conductivity, with *n*- and *p*-type substrates being available. Silicon substrates are used in a wide range of applications beyond traditional electronics, including photovoltaics (solar cells), microelectromechanical systems (MEMSs), sensors, and more. Silicon has good thermal conductivity, which helps in dissipating heat generated by electronic devices. Efficient heat management is crucial for maintaining the performance and longevity of semiconductor components. On the other hand, gallium arsenide (GaAs) substrates are critically important in several high-technology fields due to their unique properties, such as higher electron mobility, which make them suitable for specific applications where silicon substrates are less effective, including microwave circuits, which are essential for wireless communication systems, satellite communications, and radar systems. In addition to this, ZnMgO thin films have been deposited on ZnO [44,46], sapphire [2,27,30,33,36,52], ITO [54,60], $ScAlMgO_4$ [45], and other substrates. The deposition of ZnMgO thin films on nanostructured substrates is also of interest from the point of view of producing nanocomposite materials for photocatalytic and gas-sensing applications [66–69].

The goal of this paper is to perform a comparative study of the influence of substrate material nature and crystallographic orientation on the surface topology, luminescence, and wettability properties of ZnMgO films prepared by spray pyrolysis, to determine their practical applications. The deposition of ZnMgO on nanostructured substrates is also considered for the preparation of nanocomposite materials.

## 2. Materials and Methods

ZnMgO films containing 14% Mg were obtained by the method of aerosol deposition on Si and GaAs substrates of different crystallographic orientations. Solutions of zinc acetate dihydrate ($Zn(CH_3CO_2)_2 \cdot 2H_2O$ with 99.999% purity and magnesium acetate tetrahydrate ($Mg(CH_3COO)_2{\cdot}4H_2O$) (0.25M) with 99% purity, purchased from Sigma-Aldrich (Sigma-Aldrich, Munich, Germany), were dissolved in 65% methyl alcohol, 25% $H_2O$ and 10% concentrated acetic acid to obtain $Zn_{1-x}Mg_xO$ films. The chemical solutions of ZnO containing 14% of Mg were mixed in an ultrasonic bath for 30 min at the temperature of 50 °C. The content of 14% of Mg was chosen to ensure the possibility of investigating photoluminescence spectra under intrinsic excitation with the 325 nm line of a He-Cd laser (Kimmon Koha Co. Ltd., Itabashi, Tokyo, Japan). The bandgap of the material with such a composition is 3.68 eV at room temperature, and 3.76 eV at 20 K, while the quantum energy of the 325 nm laser line equals 3.81 eV. In such a case, the excitation photon energy is 50 meV above the bandgap at low temperatures. The photoluminescence spectra were measured with a double SDL-1 spectrometer providing a spectral resolution better than 1 meV when using a FEU-106 photomultiplier (with an SbKNaCs photocathode working in photon-counting mode), and the samples were mounted on the cold station of an LTS-22-C-330 optical cryogenic system.

In our experiments, we used *n*-type GaAs crystals doped with silicon, possessing a carrier concentration of $2 \times 10^{18}$ cm$^{-3}$, and different crystallographic orientations: (111), (100), and (001). The as-grown crystals were purchased from Mateck GmbH (Jülich, Germany). For comparison, ZnMgO films were also deposited on Si substrates with a (100) crystallographic orientation.

Nanostructuring of the *n*-GaAs samples was carried out using electrochemical etching in a two-electrode configuration with an ELIPOR-2 potentiostat (ET&TE Etch and Technology GmbH, Kiel, Germany), where the GaAs sample served as the working electrode, and a platinum wire mesh with a surface area of 6 cm$^2$ was used as the counter electrode. Anodization was performed by applying a voltage of 3.5 V between the working and counter electrodes in 1 M $HNO_3$ electrolyte for 20 min.

The spray pyrolysis technique, schematically presented in Figure 1, involves a hot plate that heats up to 480 °C, a sprayer built from a solution reservoir and a hole to produce a fine jet, which is directed to the substrate via a quartz tube. Oxygen ($O_2$) of ultra-high purity (99.999%) was used as the carrier gas flow, to obtain high-quality ZnMgO films with minimum uncontrolled contamination, including carbon impurities, which are important for semiconductor applications. The temperature of the plate was monitored using a Cromel-Alumel thermocouple. The solution injection rate was 1.5 cc·min$^{-1}$ and the deposition time for each sample was 4 min.

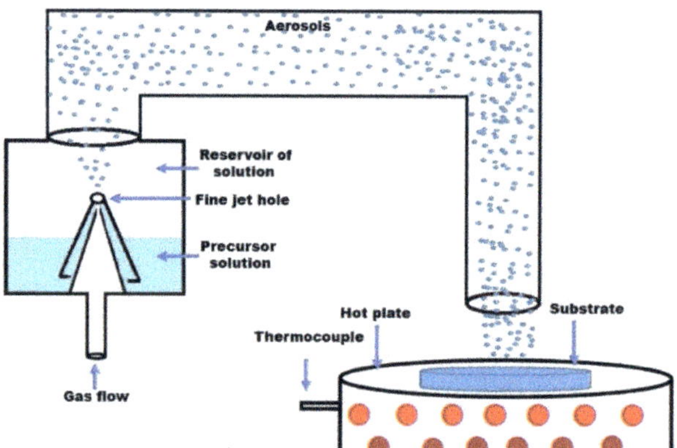


**Figure 1.** Schematic diagram of the spray pyrolysis deposition process.

The topology and roughness of the deposited films were investigated using a NanoStation II atomic force microscope (AFM) with a Si cantilever operated in non-contact mode, and the parameters were calculated using Gwyddion 2.56 software (Czech Metrology Institute, Brno, Czech Republic).

The morphology of electrochemically etched samples (top view and cross-sectional view) was investigated using TESCAN Vega TS 5130 MM (Brno, Czech Republic) scanning electron microscope (SEM) equipped with an energy dispersive X-ray (EDX) analyzer (Oxford, UK) operated at 20 kV. X-ray diffraction (XRD) measurements were performed using a Bruker AXS D8 Discover diffractometer with Cu K$\alpha$1 radiation ($\lambda$ = 0.15406 nm) operated at a beam voltage of 40 kV and a beam current of 40 mA. The wettability properties were studied by measuring the contact angle of a 3 μL droplet of distilled water on the surface of the ZnMgO films using a Drop Shape Analyzer - DSA25 (KRÜSS GmbH, Hamburg, Germany), equipped with a camera. The samples were illuminated to clearly visualize and highlight the shadow contour where the droplet contacted the film surface, thereby enabling the software to calculate the contact angle. The measurements were conducted at a temperature of 20 °C.

## 3. Results and Discussion

Figure 2 shows the AFM 3D topography (5 × 5 μm$^2$) of the $Zn_{0.86}Mg_{0.14}O$ films deposited on (100) Si and GaAs substrates with (100), (001), and (111) crystallographic orientations. The images revealed that both the chemical composition and crystallographic orientation of the substrate influenced the surface topology, in terms of the roughness and texture of the films. The Si substrate favored the growth of columnar structures, resulting in increased height of the produced crystallites, while GaAs substrates produced pyramidal structures with larger crystallites.

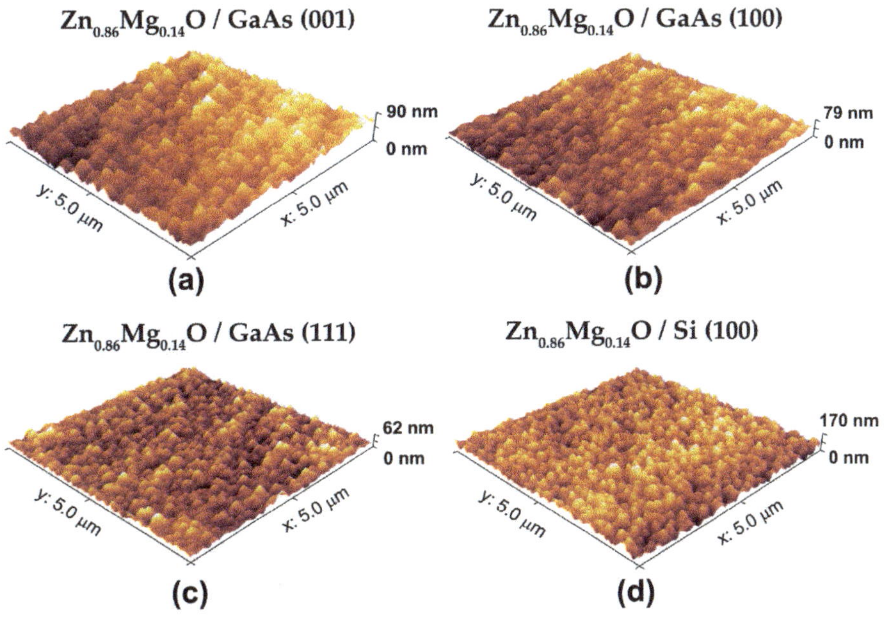


**Figure 2.** AFM 3D topography (5 × 5 μm$^2$) of $Zn_{0.86}Mg_{0.14}O$ films deposited on bulk GaAs (001) (**a**); GaAs (100) (**b**); GaAs (111) (**c**) and Si (100) (**d**) substrates.

In Figure 2, the *x*- and *y*-axes are normalized, whereas the *z*-axis is not. Images with the normalized *z*-axis are shown in Figure 3.

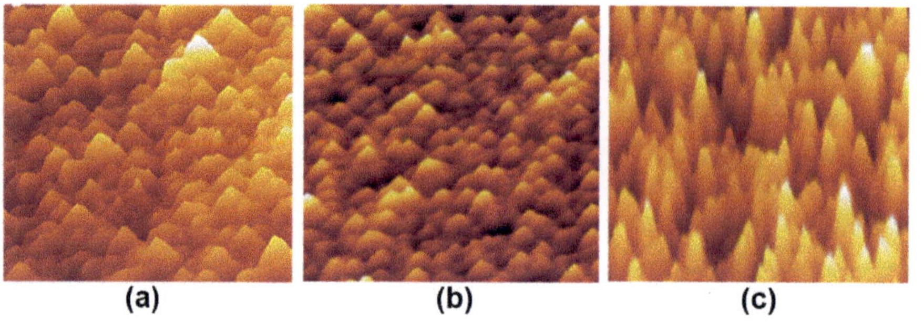


**Figure 3.** AFM 3D images with normalized z-axis of $Zn_{0.86}Mg_{0.14}O$ films deposited on bulk GaAs (001) (**a**); GaAs (111) (**b**); and Si (100) (**c**) substrates.

The roughness parameters of the films deduced from AFM measurements are summarized in Table 1. As can be seen from this analysis, the increased height of crystallites grown on the Si substrate results in a larger mean roughness compared to films deposited on GaAs substrates. On the other hand, the crystallite size and the surface roughness decrease progressively with changes in the GaAs substrate orientation, from 10.84 for GaAs (001) to 8.74 for GaAs (100), and further to 7.07 for GaAs (111). The reduced granule size and roughness of films deposited on GaAs substrates could play a critical role in enhancing the surface smoothness of the film, potentially improving its optical and electrical performance in applications where uniformity and fine structural control are essential.

**Table 1.** Surface roughness parameters of $Zn_{0.86}Mg_{0.14}O$ films deposited on Si and bulk GaAs with different crystallographic orientations.

| AFM Bulk Samples ($5 \times 5$ μm$^2$) | RMS Roughness (Sq) (nm) | Mean Roughness (Sa) (nm) | Skew (Ssk) | Excess Kurtosis | Maximum Height (Sz) (nm) |
|---|---|---|---|---|---|
| $Zn_{0.86}Mg_{0.14}O$/Si (100) | 24.02 | 19.34 | 0.2062<br>Ssk > 0 | −0.1351 | 170 |
| $Zn_{0.86}Mg_{0.14}O$/GaAs (001) | 13.32 | 10.84 | 0.0982<br>Ssk > 0 | −0.2648 | 90 |
| $Zn_{0.86}Mg_{0.14}O$/GaAs (100) | 10.95 | 8.74 | 0.2876<br>Ssk > 0 | 0.0598 | 79 |
| $Zn_{0.86}Mg_{0.14}O$/GaAs (111) | 8.93 | 7.07 | 0.5776<br>Ssk > 0 | 0.3150 | 62 |

The chemical composition of the deposited ZnMgO films was found to be nearly the same for all the substrates used within the limits of the measurement equipment's accuracy, which corresponds to a $Zn_{0.86}Mg_{0.14}O$ stoichiometry, as exemplified for a film deposited on a (111)-oriented GaAs substrate in Figure 4. The XRD analysis (Figure 4b) showed that the samples have a wurtzite crystal structure, in accordance with previous findings indicating that the wurtzite crystal structure is preserved up to a Mg concentration of 60%, and the cubic NaCl-type structure is predominant at higher Mg contents in films [56]. All the reflexes are indexed according to the PDF Card No 01-078-3032. However, the relative intensities of the XRD reflexes were different for the films deposited on different substrates. The intensity of the (002) reflex is higher for the film deposited on the Si substrate, which can be explained by the predominant columnar growth on this substrate, the columns being with the c-axis oriented perpendicular to the substrate.

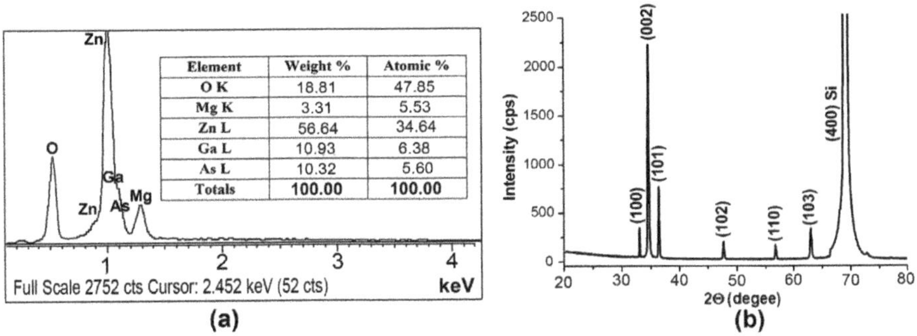


**Figure 4.** (**a**) EDX spectra of the $Zn_{0.86}Mg_{0.14}O$ films deposited on (111) GaAs substrate. (**b**) XRD spectrum of a $Zn_{0.86}Mg_{0.14}O$ film deposited on (100) Si substrate.

Because the crystallographic structures of the substrate material (diamond-type or zinc-blende-type) and that of the film (wurtzite-type) are different, the film growth is not epitaxial or conformal, and the lattice mismatch is not the main factor that influences the crystal growth. However, the chemical nature of the substrate and the crystallographic orientation determine the nucleation points for crystallite growth, resulting in columnar growth on Si substrates and pyramidal growth on GaAs substrates.

Apart from deposition on flat substrates, $Zn_{0.86}Mg_{0.14}O$ was deposited on GaAs nanowires produced by electrochemical etching of bulk GaAs substrates. The pore formation in GaAs during electrochemical anodization is strongly influenced by the anodization voltage as well as by the nature and concentration of the electrolyte used, which together control the etching process through oxidation–reduction reactions [70]. Based on the anodization curve, registered before the anodization process, the optimum voltage for GaAs nanowire formation was determined to be between 3.4 and 4 V (see inset of Figure 5a). It should be mentioned that only crystallographic pores are formed during the anodization of GaAs substrates, which exhibit a distinct triangular shape and the ability to intersect each other. As the applied voltage increased, the triangular pores widened and began to overlap, leaving unetched triangular islands, which eventually formed individual triangular nanowires. The selection of the optimized anodization voltage was critical for controlling the formation of the nanowire arrays, while the crystallographic orientation ensured the alignment of nanowire formation with respect to the substrate surface. Figure 5a shows an SEM image of an array of GaAs nanowires prepared by the electrochemical etching of a (100)-oriented GaAs substrate. This technological approach, which provides increased surface area and the possibility of engineering porous layers by design, offers substantial benefits over traditional bulk substrates [71–73]. The deposition of $Zn_{0.86}Mg_{0.14}O$ on nanowires did not produce a continuous film, but resulted in the formation of a nanocomposite material consisting of GaAs nanowires with $Zn_{0.86}Mg_{0.14}O$ inclusions, as shown in Figure 5b. The enhanced surface area of such nanocomposites and their ability to fine-tune the morphology of the nanowires and the composition of inclusions facilitate improved performance in applications, such as photodetectors, biosensors, and gas sensors. This approach effectively combined the advantages of both materials in a synergistic manner, leading to a hybrid system with enhanced performance.

Figure 5c,d presents a comparison of the SEM cross-sectional images of GaAs nanowire arrays, prepared by electrochemical etching of GaAs substrates with different crystallographic orientations and subsequently coated with $Zn_{0.86}Mg_{0.14}O$ layers. The images reveal that GaAs nanowires are aligned perpendicularly to the surface of the GaAs substrate for the (111)B orientation (Figure 5d), whereas they are tilted to the substrate surface for the (100) orientation (Figure 5c), according to previously reported studies [71]. It can be seen that, while the $Zn_{0.86}Mg_{0.14}O$ layers are deposited predominantly on the surface of nanostructured GaAs substrates, $Zn_{0.86}Mg_{0.14}O$ inclusions penetrate deeply, reaching tens of micrometers in depth. This extensive coverage indicates an effective deposition throughout the nanostructures. However, further optimization of the deposition process is recommended to enhance the uniformity of the deposition across the entire thickness of the nanowire array.

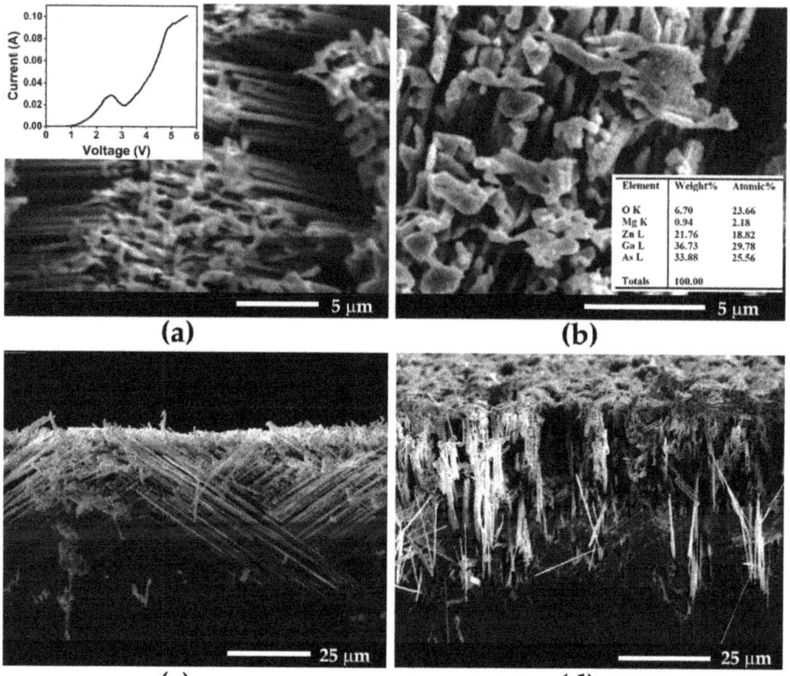


**Figure 5.** (**a**) SEM image of an array of GaAs nanowires produced by electrochemical etching of a (100) GaAs substrate. (**b**) Hybrid GaAs/ZnMgO structure after aerosol deposition. (**c**) SEM image of the cross-section of an array of GaAs nanowires produced on a (100) GaAs substrate with a deposited $Zn_{0.86}Mg_{0.14}O$ layer. (**d**) The same for a (111)B GaAs substrate. Insets: Polarization curve registered before the electrochemical etching of GaAs substrate in 1 M $HNO_3$ electrolyte at a scan rate of 50 $mV{\cdot}s^{-1}$ in (**a**) and the chemical composition analysis of the deposited ZnMgO on GaAs nanowire arrays in (**b**).

The impact of the substrate on the luminescence properties of the $Zn_{0.86}Mg_{0.14}O$ films prepared on flat substrates was similar to that on the film morphology. Figure 6 (curves 1–3) presents the photoluminescence spectra of the films deposited on flat substrates measured at room temperature. One can see that the spectra are quite similar, but the intensity of the luminescence of the films deposited on silicon is lower. The spectrum consists of an asymmetrical band in the near-bandgap region and a broader band in the visible spectral range, centered around 2.55 eV.

The luminescence of ZnO-related materials in the visible spectrum is usually attributed to either Cu impurities, or to intrinsic defects associated with oxygen vacancies ($V_O$), or to zinc vacancies ($V_{Zn}$) [74–78]. Oxygen vacancies are deep donors, while zinc vacancies are deep acceptors. Oxygen vacancies have high formation energies in *n*-type samples, while zinc vacancies have low formation energies under *n*-type conditions [79]. Therefore, it is suggested that electron transitions from a deep donor level to the valence band (D-h-type recombination) are less probable than electronic transitions from the conduction band to acceptor levels (e-A-type recombination), that is, transitions involving $V_{Zn}$ centers [76].

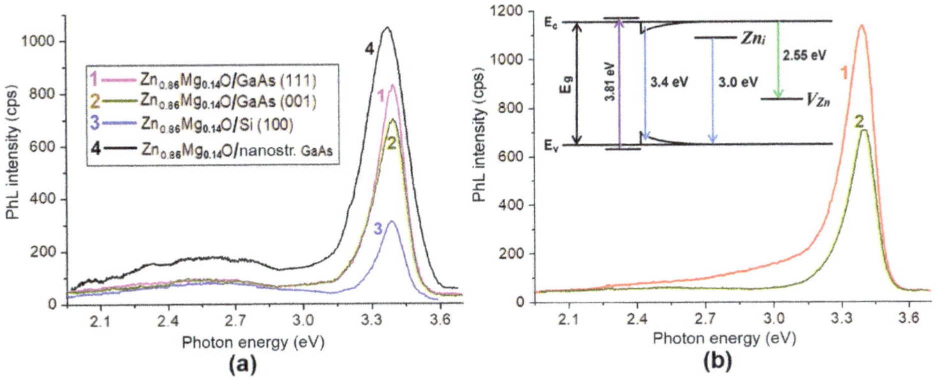


**Figure 6.** (**a**) PhL spectra of $Zn_{0.86}Mg_{0.14}O$ films deposited by spray pyrolysis on flat substrates (curves 1–3) and nanostructured GaAs substrates (curve 4) measured at room temperature. (**b**) PhL spectra of a $Zn_{0.86}Mg_{0.14}O$ film deposited on a bulk (001) GaAs substrate measured at room temperature in vacuum (curve 1) and in air (curve 2). The inset shows a scheme of the energy levels and electronic transitions responsible for the observed PhL bands.

Regarding the asymmetrical near-bandgap photoluminescence band with the characteristic feature of broadening toward the Stokes part of the emission, it has been previously shown that this is due to the formation of band tail states at the edges of the conduction and valence bands, and these large band tails in the density of states are generated by local potential fluctuations. Two mechanisms have been proposed for explaining the emergence of local potential fluctuations. According to one mechanism, the local potential fluctuations are induced by the high concentration of intrinsic defects or impurities non-homogeneously distributed at the microscopic level in heavily doped and compensated semiconductors as developed by Morgan [80], and according to another mechanism, they could be induced by local compositional fluctuations in solid solutions [61]. The scheme for the energy levels of the recombination centers and electronic transitions responsible for the observed PhL bands is presented in the inset of Figure 6b.

Assuming that the first mechanism is predominantly responsible for near-bandgap broadening, some suggestions can be made concerning the microscopic origin of the impurity responsible for the potential fluctuations. Oxygen vacancies and zinc interstitials are the most probable donor-type intrinsic defects [75,79]. As mentioned above, oxygen vacancies are deep donors, whereas interstitial zinc is shallow [79,81], that is, interstitial zinc $Zn_i$ defects can contribute to the high electron concentration in the produced $Zn_{0.86}Mg_{0.14}O$ films.

An investigation of the influence of the environmental atmosphere on the luminescence properties revealed that changing the environment from air to vacuum results in the emergence of a shoulder around 3.0 eV in the PhL spectrum of the $Zn_{0.86}Mg_{0.14}O$ film, as shown in Figure 6b. Theoretical calculations previously performed for zinc oxide pointed out that the energy gap from the interstitial zinc level to the valence band is about 2.9 eV, and the Zn interstitial level is 0.22 eV below the conductive band-edge [81]. In accordance with these theoretical statements, a violet shoulder at 415 nm (2.95 eV) in the blue emission of ZnO nanoparticles is clearly associated with interstitial Zn defects [81]. However, interstitial Zn can form complex centers with oxygen species adsorbed at the surface states of the large surface of nanostructured $Zn_{0.86}Mg_{0.14}O$ films when the vacuum is replaced by the air atmosphere. The formation of such defect complexes should result in the annihilation of the PhL band around 3.0 eV. The defect centers formed by the adsorption of oxygen species from the environment on the sample surface can act as active centers during photocatalytic oxidation processes by capturing photoinduced electrons, thus contributing to a substantial improvement in the photocatalytic activity [66,69].

The deposition of $Zn_{0.86}Mg_{0.14}O$ layers on nanostructured GaAs substrates also influences the PhL properties. As can be seen from a comparison of the PhL spectra of layers deposited on nanostructured and bulk GaAs substrates (Figure 6a), the full width at half maximum (FWHM) of the near bandgap PhL band is by a factor of 1.5 larger for the layer deposited on the nanostructured substrate than for the layer deposited on the bulk substrate (210 meV vs. 140 meV). This means that the concentration of defects or impurities is higher in the layer deposited on nanostructured GaAs substrates, assuming that the first mechanism is responsible for PhL band broadening. By using the dependence of the PhL-band FWHM on the carrier concentration deduced previously for highly conductive ZnO layers [82], one can estimate that the carrier concentration in the $Zn_{0.86}Mg_{0.14}O$ layer deposited on nanostructured GaAs substrates is approximately $5 \times 10^{20}$ cm$^{-3}$, while it is approximately $2 \times 10^{20}$ cm$^{-3}$ in the film deposited on the bulk substrate. Such high doping of the layer deposited on the nanostructured substrate could be due to the diffusion of Ga impurities from the GaAs substrate into the $Zn_{0.86}Mg_{0.14}O$ layer at the large interface of the nanostructured substrate, as discussed in a previous paper [73]. Ga impurities are shallow donors in ZnO, with a donor binding energy of 55 meV [83]. Highly doped $Zn_{0.86}Mg_{0.14}O$ inclusions in nanostructured substrates should be of interest for exploring plasmonic effects. GaAs nanowire arrays produced by anodizing (111)B GaAs substrates, which are predominantly oriented perpendicular to the substrate surface and covered by ZnO layers, have been demonstrated to be sensitive to radiation polarization [73]. The possibility of tuning the wavelength sensitivity by alloying ZnO with MgO is expected to expand the exploration of polarization-sensitive applications.

In a wider context, anisotropic light scattering and coherent backscattering of light waves with strongly polarized retroreflective light with linear polarization coinciding with that of the incident beams were found to be inherent to such structures [84]. This phenomenon was observed experimentally and argued by a theoretical model, and the calculations suggested that it can be purposefully used in optoelectronic components, particularly for enhancing the light extraction efficiency. The computational analysis of electromagnetic scattering from such anisotropic cylindrical structures numerically demonstrated their role as antennas [85,86]. When such structures are covered by magnetic materials, they exhibit magnetic anisotropy with respect to coercivity and remanence ratio [71].

In addition, the possibility of preparing various nanocomposites in a controlled manner, as in the case of those prepared on GaAs nanowires interfaced with $Zn_{0.86}Mg_{0.14}O$ inclusions illustrated in Figure 5, can provide opportunities for bandgap engineering and for the alignment of the conduction and valence bands of the nanocomposite components with respect to the HOMOs and LUMOs of organic compounds subjected to photocatalytic degradation [67,69,87]. The heterointerfaces between nanocomposite components may also play an important role in improving gas-sensing properties [68].

Figure 7 compares the temperature dependence of the photoluminescence spectra in $Zn_{0.86}Mg_{0.14}O$ films deposited on different substrates. Analysis of these spectra revealed that the crystallographic orientation of the GaAs substrate did not significantly influence the spectra. The evolution of the spectral position of the near-bandgap PhL band with increasing temperature is similar in films deposited on GaAs substrates with (001) and (111) orientations, as deduced from Figure 7a,b. The FWHM of this PhL band also increased from approximately 50 meV to 140 meV as the temperature increased from 13 K to 300 K for films deposited on both GaAs substrates.

With respect to the films deposited on the Si substrate, the spectral position of the PhL band was shifted by approximately 10 meV to lower photon energies at low temperatures, as compared to films deposited on GaAs substrates, and its FWHM decreased from 50 meV to 45 meV.

Assuming that the mechanism responsible for the near-bandgap broadening is related predominantly to local compositional fluctuations in ZnMgO solid solutions, as discussed above, it can be concluded that the columnar growth on the Si substrate leads to lower local potential fluctuations at the microscopic level in the $Zn_{0.86}Mg_{0.14}O$ films. However, assuming that the main mechanism is related to the high concentration of intrinsic defects or impurities, which, in our case, could be associated with Zn interstitial centers that are inhomogeneously distributed at the microscopic level, one can estimate the concentration of these centers in the films deposited on various substrates.

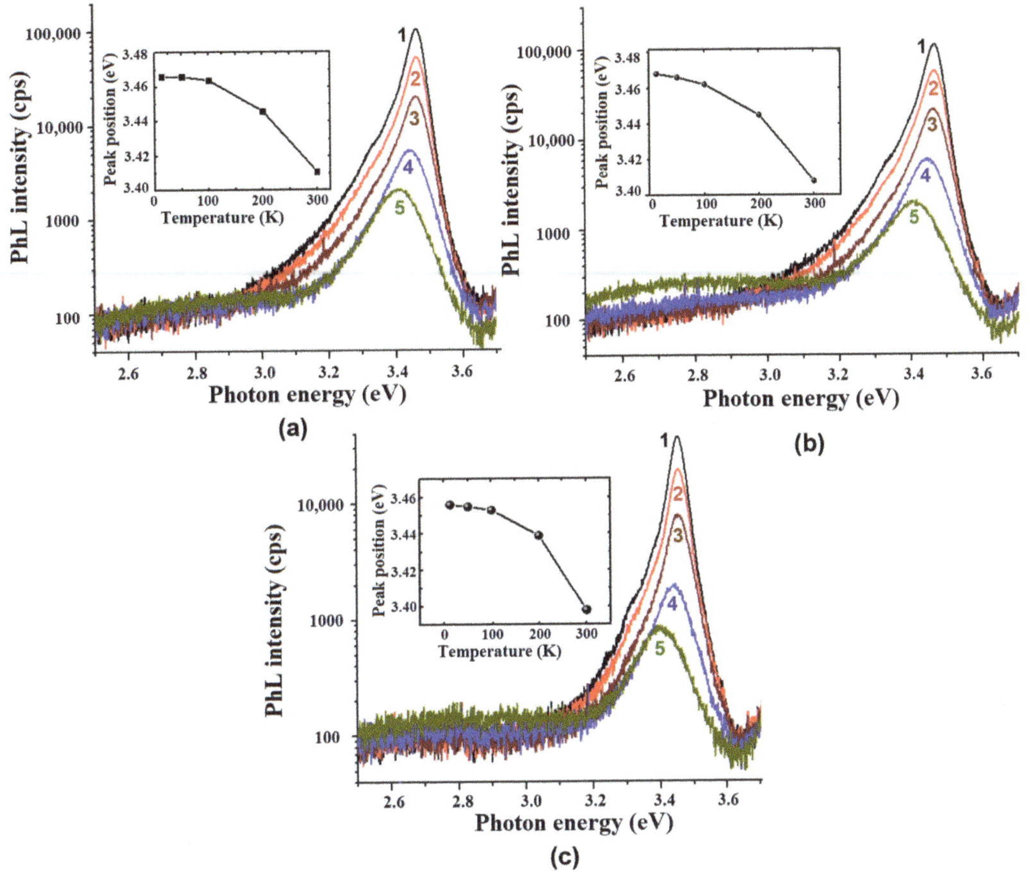


**Figure 7.** PhL spectra of $Zn_{0.86}Mg_{0.14}O$ films deposited on (001) GaAs (**a**), (111) GaAs (**b**), and (100) Si (**c**) substrates measured at temperatures of 13 K (curve 1), 50 K (curve 2), 100 K (curve 3), 200 K (curve 4), and 300 K (curve 5). The inset shows the temperature dependence of the spectral position of the near-bandgap PhL band.

By using the dependence of the PhL band FWHM on the carrier concentration as described above [82], one can estimate that the carrier concentration in $Zn_{0.86}Mg_{0.14}O$ films deposited on GaAs substrates is approximately $3 \times 10^{19}$ $cm^{-3}$ at low temperatures, and approximately $2 \times 10^{20}$ $cm^{-3}$ at room temperature, while in films deposited on Si substrates, the carrier concentration is approximately $2.5 \times 10^{19}$ $cm^{-3}$ at low temperatures.

Previous investigations demonstrated that ZnMgO films with transparence above 90% in the visible spectral range can be produced by spray pyrolysis, and the absorption edge can be moved to shorter UV wavelengths by increasing the Mg content in the films [60,62,63]. In addition, carrier concentrations as high as $1.4 \times 10^{20}$ $cm^{-3}$ have been reached [63]. The luminescence analysis described above indicates that the $Zn_{0.86}Mg_{0.14}O$ films deposited on GaAs substrates have a doping concentration of approximately $2 \times 10^{20}$ $cm^{-3}$

at room temperature and should be electrically conducting. Therefore, they are promising for application as transparent conducting coatings with the potential for scalability due to the reduced deposition costs, versatility, and environment-friendly character of the technology.

From the contact angle measurements performed on the $Zn_{0.86}Mg_{0.14}O$ films (Figure 8), it can be concluded that both the substrate nature and substrate crystallographic orientation influence the wettability properties of the deposited $Zn_{0.86}Mg_{0.14}O$ films, similar to the impact of the substrate on the film topology discussed above. The largest contact angle of 127° was measured on a film deposited on a GaAs substrate with a (111) crystallographic orientation, indicating the most hydrophobic surface. At the same time, this film was characterized by the smallest crystallite size and lowest surface roughness. It is noteworthy that the contact angle on the GaAs substrate without ZnMgO deposition is 80.4° [72], highlighting the significant increase in hydrophobicity due to the $Zn_{0.86}Mg_{0.14}O$ film.

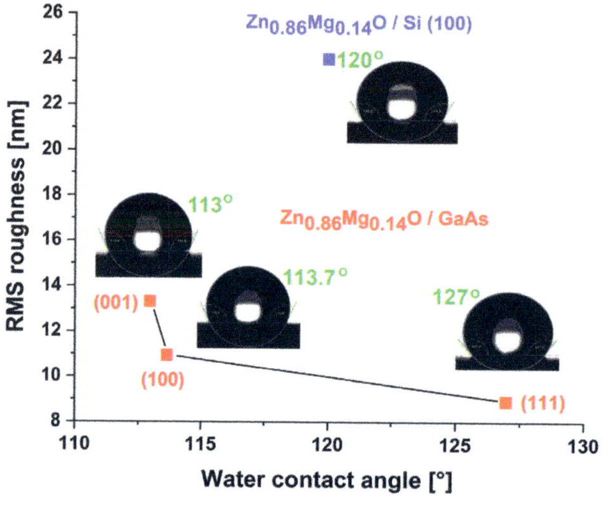


**Figure 8.** The relationship between surface roughness and water contact angle measured between a water droplet and $Zn_{0.86}Mg_{0.14}O$ films deposited on bulk GaAs substrates with (001), (100), (111) crystallographic orientations and on Si (100) substrate.

## 4. Conclusions

A systematic study of the spray pyrolysis growth of ZnMgO films on semiconductor substrates with various chemical compositions and crystallographic orientations provided insights into their impact on the produced film morphology, luminescence, and wettability properties. It was found that the film growth by spray pyrolysis was not epitaxial or conformal, but granular, depending on the nucleating sites for crystallite growth. The distribution of nucleation sites ensured the preparation of nanostructured films with good uniformity in topology. Further growth of crystallites indicated the influence of the chemical nature of the substrate on the shape of the crystallites. The columnar growth of wurtzite-type $Zn_{0.86}Mg_{0.14}O$ films with a predominant orientation of the c-axis perpendicular to the Si (100) substrate surface is supposed to be particularly determined by the covalent chemical bonds in the substrate. However, the mixed ionic and covalent character of GaAs bonding facilitates the growth of pyramidal crystallites, and the crystallite sizes are larger in films grown on substrates with (001) orientation than in films grown on (111) orientation, resulting in a difference in roughness by a factor of 1.5. The roughness of the films deposited on Si substrates was found to be by a factor of two higher than that of the films deposited on GaAs substrates, due to their columnar growth. These effects on the topology of the films correlated with their wetting properties. In particular, the largest contact angle was observed for films grown on GaAs substrates with a (111) crystallographic orientation, indicating their most hydrophobic surface. Analysis of the near-bandgap photoluminescence band in $Zn_{0.86}Mg_{014}O$ films under intrinsic excitation suggested that its spectral shape is determined by the formation of band tail states at the edges of the conduction and valence bands that are generated by local potential fluctuations. These potential fluctuations proved to be higher in films deposited on GaAs substrates than in films grown on Si substrates, indicating the better quality of the columnar structures grown on Si substrates. The results of the investigations of the influence of the environmental atmosphere on the photoluminescence properties of ZnMgO films indicated the prospects of deposition of such films onto nanostructured substrates for photocatalytic and gas sensor applications. Analysis of photoluminescence spectra of films deposited on nanostructured GaAs substrates indicates their high doping suggesting their potential for the exploration of plasmonic effects.

**Author Contributions:** Conceptualization, E.V.M.; methodology, V.M. and S.B.; validation, V.V.U. and I.M.T.; investigation, V.M., S.B., V.V.Z. and E.V.M.; resources, V.M. and E.V.M.; writing—original draft preparation, E.V.M. and V.V.U.; writing—review and editing, V.V.U., E.V.M. and I.M.T.; visualization, V.V.U., V.M. and E.V.M.; supervision, E.V.M. and I.M.T.; project administration, E.V.M. All authors have read and agreed to the published version of the manuscript.

**Funding:** This research was funded by Ministry of Education and Research from the Republic of Moldova, institutional subprogram #02.04.02 "Development of technologies and investigation of the properties of layered semiconductor compounds, hybrid nanostructures and laser sources" and #02.02.01 "Nanostructures and advanced materials for applications in spintronics, thermoelectricity and optoelectronics". Institutional financing contract no. 4/FI of 22 February 2024.

**Institutional Review Board Statement:** Not applicable.

**Informed Consent Statement:** Not applicable.

**Data Availability Statement:** The data presented in this study are available on request from the corresponding author.

**Acknowledgments:** E.V.M. acknowledges support from the Alexander von Humboldt Foundation (Bonn, Germany).

**Conflicts of Interest:** The authors declare no conflicts of interest. The funders had no role in the design of the study; in the collection, analyses, or interpretation of data; in the writing of the manuscript; or in the decision to publish the results.

## References

1. Lv, J.; Li, X. Defect Evolution in ZnO and Its Effect on Radiation Tolerance. *Phys. Chem. Chem. Phys.* **2018**, *20*, 11882–11887. [CrossRef] [PubMed]
2. Kaidashev, E.M.; Lorenz, M.; von Wenckstern, H.; Rahm, A.; Semmelhack, H.-C.; Han, K.-H.; Benndorf, G.; Bundesmann, C.; Hochmuth, H.; Grundmann, M. High Electron Mobility of Epitaxial ZnO Thin Films on C-Plane Sapphire Grown by Multistep Pulsed-Laser Deposition. *Appl. Phys. Lett.* **2003**, *82*, 3901–3903. [CrossRef]
3. Ajmal, H.M.S.; Khan, F.; Nam, K.; Kim, H.Y.; Kim, S.D. Ultraviolet Photodetection Based on High-Performance Co-Plus-Ni Doped ZnO Nanorods Grown by Hydrothermal Method on Transparent Plastic Substrate. *Nanomaterials* **2020**, *10*, 1225. [CrossRef] [PubMed]
4. Tsai, Y.-T.; Chang, S.-J.; Ji, L.-W.; Hsiao, Y.-J.; Tang, I.-T.; Lu, H.-Y.; Chu, Y.-L. High Sensitivity of NO Gas Sensors Based on Novel Ag-Doped ZnO Nanoflowers Enhanced with a UV Light-Emitting Diode. *ACS Omega* **2018**, *3*, 13798–13807. [CrossRef] [PubMed]
5. Cesini, I.; Kowalczyk, M.; Lucantonio, A.; D'Alesio, G.; Kumar, P.; Camboni, D.; Massari, L.; Pingue, P.; De Simone, A.; Fraleoni Morgera, A.; et al. Seedless Hydrothermal Growth of ZnO Nanorods as a Promising Route for Flexible Tactile Sensors. *Nanomaterials* **2020**, *10*, 977. [CrossRef]
6. Vittal, R.; Ho, K.-C. Zinc Oxide Based Dye-Sensitized Solar Cells: A Review. *Renew. Sustain. Energy Rev.* **2017**, *70*, 920–935. [CrossRef]
7. Consonni, V.; Briscoe, J.; Kärber, E.; Li, X.; Cossuet, T. ZnO Nanowires for Solar Cells: A Comprehensive Review. *Nanotechnology* **2019**, *30*, 362001. [CrossRef]

8. Jiang, L.; Li, J.; Huang, K.; Li, S.; Wang, Q.; Sun, Z.; Mei, T.; Wang, J.; Zhang, L.; Wang, N.; et al. Low-Temperature and Solution-Processable Zinc Oxide Transistors for Transparent Electronics. *ACS Omega* **2017**, *2*, 8990–8996. [CrossRef] [PubMed]
9. Natu, K.; Laad, M.; Ghule, B.; Shalu, A. Transparent and Flexible Zinc Oxide-Based Thin-Film Diodes and Thin-Film Transistors: A Review. *J. Appl. Phys.* **2023**, *134*, 190701. [CrossRef]
10. Gherendi, F.; Dobrin, D.; Nistor, M. Transparent Structures for ZnO Thin Film Paper Transistors Fabricated by Pulsed Electron Beam Deposition. *Micromachines* **2024**, *15*, 265. [CrossRef]
11. Cao, H. Review on Latest Developments in Random Lasers with Coherent Feedback. *J. Phys. A Math. Gen.* **2005**, *38*, 10497. [CrossRef]
12. Cao, H. Random Lasers: Development, Features and Applications. *Opt. Photonics News OPN* **2005**, *16*, 24–29. [CrossRef]
13. Li, H.D.; Yu, S.F.; Lau, S.P.; Leong, E.S.P. Simultaneous Formation of Visible and Ultraviolet Random Lasings in ZnO Films. *Appl. Phys. Lett.* **2006**, *89*, 021110. [CrossRef]
14. Hsu, H.-C.; Wu, C.-Y.; Hsieh, W.-F. Stimulated Emission and Lasing of Random-Growth Oriented ZnO Nanowires. *J. Appl. Phys.* **2005**, *97*, 064315. [CrossRef]
15. Ursaki, V.V.; Zalamai, V.V.; Burlacu, A.; Fallert, J.; Klingshirn, C.; Kalt, H.; Emelchenko, G.A.; Redkin, A.N.; Gruzintsev, A.N.; Rusu, E.V.; et al. A Comparative Study of Guided Modes and Random Lasing in ZnO Nanorod Structures. *J. Phys. D Appl. Phys.* **2009**, *42*, 095106. [CrossRef]
16. Ursaki, V.V.; Zalamai, V.V.; Burlacu, A.; Klingshirn, C.; Monaico, E.; Tiginyanu, I.M. Random Lasing in Nanostructured ZnO Produced from Bulk ZnSe. *Semicond. Sci. Technol.* **2009**, *24*, 085017. [CrossRef]
17. Zhang, C.; Zhang, J.; Liu, W.; Xu, H.; Hou, S.; Wang, C.; Yang, L.; Wang, Z.; Wang, X.; Liu, Y. Enhanced Ultraviolet Random Lasing from Au/MgO/ZnO Heterostructure by Introducing p-Cu2O Hole-Injection Layer. *ACS Appl. Mater. Interfaces* **2016**, *8*, 31485–31490. [CrossRef]
18. Ooi Zheng, K.; Rosli, N.; Rashid, M.; Mahadi Halim, M. Electrically Pumped Random Laser Device Based on Pd/$SiO_2$/ZnO Nanorods MIS Structure. *Results Phys.* **2024**, *64*, 107946. [CrossRef]
19. Abdulhameed, A. ZnO-Based Random Lasing and Their Sensing Applications: A Mini-Review. *Appl. Nanosci.* **2024**, *14*, 985–995. [CrossRef]
20. Consoli, A.; Caselli, N.; López, C. Networks of Random Lasers: Current Perspective and Future Challenges [Invited]. *Opt. Mater. Express OME* **2023**, *13*, 1060–1076. [CrossRef]
21. Teisseyre, H.; Jarosz, D.; Marona, L.; Bojarska, A.; Ivanov, V.; Perlin, P.; Czyszanowski, T. Homoepitaxial ZnO/ZnMgO Laser Structures and Their Properties. *Phys. Status Solidi (A)* **2021**, *218*, 2000344. [CrossRef]
22. Lin, J.-H.; Chen, G.-R.; Li, S.-J.; Song, Y.-F.; Liu, W.-R. Gain-Guiding Anisotropic Polarized Amplified Spontaneous Emissions from C-Plane ZnO/ZnMgO Multiple Quantum Wells. *Materials* **2022**, *15*, 6668. [CrossRef] [PubMed]
23. Chen, S.; Zhan, T.; Pan, X.; He, H.; Huang, J.; Lu, B.; Ye, Z. UV Electroluminescence Emissions from High-Quality ZnO/ZnMgO Multiple Quantum Well Active Layer Light-Emitting Diodes. *RSC Adv.* **2021**, *11*, 38949–38955. [CrossRef]
24. Shi, H.; Ren, Z.; Zheng, Z.; Zhou, X.; Luo, C.; Ji, H.; Zhang, Y.; Chen, H.; Wu, C.; Chen, Y. Mercaptopropionic Acid Capped ZnMgO for Efficient and Stable Quantum Dot Light-Emitting Diodes. *Laser Photonics Rev.* **2024**, 2400473. [CrossRef]
25. Shafiqul, I.M.; Yoshida, T.; Fujita, Y. P-ZnO/n-ZnMgO Nanoparticle-Based Heterojunction UV Light-Emitting Diodes. *Materials* **2022**, *15*, 8348. [CrossRef]
26. Ye, M.; Wang, D.; Jiao, S.; Chen, L. Enhanced Deep Ultraviolet Photoresponse in Ga Doped ZnMgO Thin Film. *Micromachines* **2022**, *13*, 1140. [CrossRef] [PubMed]
27. Suja, M.; Bashar, S.B.; Debnath, B.; Su, L.; Shi, W.; Lake, R.; Liu, J. Electrically Driven Deep Ultraviolet MgZnO Lasers at Room Temperature. *Sci. Rep.* **2017**, *7*, 2677. [CrossRef]
28. Hou, Y.; Mei, Z.; Liang, H.; Ye, D.; Gu, C.; Du, X. Dual-Band MgZnO Ultraviolet Photodetector Integrated with Si. *Appl. Phys. Lett.* **2013**, *102*, 153510. [CrossRef]
29. Yang, J.-L.; Liu, K.-W.; Shen, D.-Z. Recent Progress of ZnMgO Ultraviolet Photodetector*. *Chin. Phys. B* **2017**, *26*, 047308. [CrossRef]
30. Wang, L.K.; Ju, Z.G.; Zhang, J.Y.; Zheng, J.; Shen, D.Z.; Yao, B.; Zhao, D.X.; Zhang, Z.Z.; Li, B.H.; Shan, C.X. Single-Crystalline Cubic MgZnO Films and Their Application in Deep-Ultraviolet Optoelectronic Devices. *Appl. Phys. Lett.* **2009**, *95*, 131113. [CrossRef]
31. Liu, C.Y.; Xu, H.Y.; Wang, L.; Li, X.H.; Liu, Y.C. Pulsed Laser Deposition of High Mg-Content MgZnO Films: Effects of Substrate Temperature and Oxygen Pressure. *J. Appl. Phys.* **2009**, *106*, 073518. [CrossRef]
32. Polyakov, A.I.; Smirnov, N.B.; Govorkov, A.V.; Kozhukhova, E.A.; Belogorokhov, A.I.; Kim, H.S.; Norton, D.P.; Pearton, S.J. Annealing Effects on Electrical Properties of MgZnO Films Grown by Pulsed Laser Deposition. *Appl. Phys. Lett.* **2008**, *103*, 083704. [CrossRef]
33. Wang, X.; Saito, K.; Tanaka, T.; Nishio, M.; Nagaoka, T.; Arita, M.; Guo, Q. Energy Band Bowing Parameter in MgZnO Alloys. *Appl. Phys. Lett.* **2015**, *107*, 022111. [CrossRef]
34. Maity, S.; Sahu, P.P.; Bhunia, C.T. High Photo Sensing Performance With Electro-Optically Efficient Silicon Based ZnO/ZnMgO Heterojunction Structure. *IEEE Sens. J.* **2018**, *18*, 6569–6575. [CrossRef]
35. Luo, J.; Liu, H.; Deng, W.; Zhang, R.; He, C. Effect of Annealing Atmosphere on Structure and Photoluminescence of ZnMgO Thin Films. *J. Mater. Sci. Mater. Electron.* **2023**, *34*, 2172. [CrossRef]

36. Akin Sönmez, N. Cubic MgZnO Thin Films on Sapphire Substrate: Effect of Deposition Temperature. *J. Mater. Sci. Mater. Electron.* **2019**, *30*, 4104–4110. [CrossRef]
37. Xie, W.; Wang, D.; Guo, F.; Liu, T.; Jiao, S.; Wang, J.; Gao, S.; Yu, Q.; Zhang, Y.; Luan, C.; et al. The Effect of Annealing Temperature on the Optical and Electrical Properties of Cubic MgZnO Films Grown by RF Magnetron Sputtering. *J. Mater. Sci. Mater. Electron.* **2017**, *28*, 1644–1651. [CrossRef]
38. Wu, C.-Z.; Ji, L.-W.; Liu, C.-H.; Peng, S.-M.; Young, S.-J.; Lam, A.; Huang, C.-J. Ultraviolet Photodetectors Based on MgZnO Thin Films. *J. Vac. Sci. Technol. A Vac. Surf. Film.* **2011**, *29*, 03A118. [CrossRef]
39. Shiau, J.-S.; Brahma, S.; Liu, C.-P.; Huang, J.-L. Ultraviolet Photodetectors Based on MgZnO Thin Film Grown by RF Magnetron Sputtering. *Thin Solid Film.* **2016**, *620*, 170–174. [CrossRef]
40. Wang, Y.; Li, H.; Shi, X.; Guo, R.; Jin, E. Preparation of ZnMgO/ZnO Heterojunction by Magnetron Sputtering Method. In Proceedings of the International Conference on Optoelectronic Information and Functional Materials (OIFM 2023), Guangzhou, China, 14–16 April 2023; Volume 12781, pp. 831–835. [CrossRef]
41. Ferri, K.; Bachu, S.; Zhu, W.; Imperatore, M.; Hayden, J.; Alem, N.; Giebink, N.; Trolier-McKinstry, S.; Maria, J.P. Ferroelectrics Everywhere: Ferroelectricity in Magnesium Substituted Zinc Oxide Thin Films. *J. Appl. Phys.* **2021**, *130*, 044101. [CrossRef]
42. Yuan, X.; Yamada, T.; Meng, L. Strong Electro-Optic Effect in Mg Incorporated ZnO Thin Films. *Appl. Phys. Lett.* **2022**, *121*, 152903. [CrossRef]
43. Meng, L.; Chai, H.; Gao, J.; Lv, Z.; Yang, X.; Liu, W.; Zhai, T.; Yang, T. Strong Nonlinear-Polarization in ZnMgO Epitaxial Thin-Films with Li Incorporation. *J. Phys. D Appl. Phys.* **2024**, *57*, 275104. [CrossRef]
44. Liang, X.; Zhou, H.; Wang, H.-Q.; Zhang, L.; Kisslinger, K.; Kang, J. Interface and Optical Properties of $Zn_{1-x}Mg_xO$ Films with Mg Content of More than 70% Grown on the ($1\bar{2}10$)-ZnO Substrates. *AIP Adv.* **2021**, *11*, 075217. [CrossRef]
45. Kubota, M.; Onuma, T.; Tsukazaki, A.; Ohtomo, A.; Kawasaki, M.; Sota, T.; Chichibu, S. Recombination Dynamics of Excitons in $Mg_{0.11}Zn_{0.89}O$ Alloy Films Grown Using the High-Temperature-Annealed Self-Buffer Layer by Laser-Assisted Molecular-Beam Epitaxy. *Appl. Phys. Lett.* **2007**, *90*, 141903. [CrossRef]
46. Falson, J.; Kozuka, Y.; Uchida, M.; Smet, J.H.; Arima, T.; Tsukazaki, A.; Kawasaki, M. MgZnO/ZnO Heterostructures with Electron Mobility Exceeding $1 \times 10^6$ $cm^2/Vs$. *Sci. Rep.* **2016**, *6*, 26598. [CrossRef]
47. Przezdziecka, E.; Paradowska, K.M.; Jakiela, R.; Kryvyi, S.; Zielony, E.; Placzek-Popko, E.; Lisowski, W.; Sybilski, P.; Jarosz, D.; Adhikari, A.; et al. Polar and Non-Polar $Zn_{1-x}Mg_xO$:Sb Grown by MBE. *Materials* **2022**, *15*, 8409. [CrossRef] [PubMed]
48. Wei, C.; Tu, Y.; Feng, Z.; Wu, C.C.; Lin, P.; Wuu, D.-S. Resonant Raman Scattering of MgZnO Thin Films Grown on Sapphire by MOCVD. *AIP Conf. Proc.* **2010**, *1267*, 1115–1116. [CrossRef]
49. Shukla, A.; Kaushik, V.K.; Prasher, D. Growth and Characterization of MgxZn1−xO Thin Films by Aerosol-Assisted Chemical Vapor Deposition (AACVD). *Electron. Mater. Lett.* **2014**, *10*, 61–65. [CrossRef]
50. Liu, X.; Dang, G.T.; Liu, L.; Kawaharamura, T. Fabrication of $Zn_{1-x}Mg_xO$/AgyO Heterojunction Diodes by Mist CVD at Atmospheric Pressure. *Appl. Surf. Sci.* **2022**, *596*, 153465. [CrossRef]
51. Sulaiman, A.A.; Ali, G.G.; Thanon, A.I. Synthesis and Study of ZnO Thin Films Using CVD Technique For Waveguide Sensor Applications. *J. Nanostruct.* **2022**, *12*, 1–11. [CrossRef]
52. Lee, H.-Y.; Tsai, W.-H.; Lin, Y.-C.; Lee, C.-T. Performance Enhancement of MgZnO Ultraviolet Photodetectors Using Ultrathin Al2O3 Inserted Layer. *J. Vac. Sci. Technol. B Nanotechnol. Microelectron. Mater. Process. Meas. Phenom.* **2016**, *34*, 051207. [CrossRef]
53. Cui, X.; Sun, K.; Huang, J.; Sun, H.; Wang, A.; Yuan, X.; Green, M.; Hoex, B.; Hao, X. Low-Temperature Plasma-Enhanced Atomic Layer Deposition of ZnMgO for Efficient CZTS Solar Cells. *ACS Mater. Lett.* **2023**, *5*, 1456–1465. [CrossRef]
54. Zhang, B.; Luo, Y.; Mai, C.; Mu, L.; Li, M.; Wang, J.; Xu, W.; Peng, J. Effects of ZnMgO Electron Transport Layer on the Performance of InP-Based Inverted Quantum Dot Light-Emitting Diodes. *Nanomaterials* **2021**, *11*, 1246. [CrossRef] [PubMed]
55. Tsay, C.-Y.; Chen, S.-T.; Fan, M.-T. Solution-Processed Mg-Substituted ZnO Thin Films for Metal-Semiconductor-Metal Visible-Blind Photodetectors. *Coatings* **2019**, *9*, 277. [CrossRef]
56. Morari, V.; Ursaki, V.V.; Rusu, E.V.; Zalamai, V.V.; Colpo, P.; Tiginyanu, I.M. Spin-Coating and Aerosol Spray Pyrolysis Processed $Zn_{1-x}Mg_xO$ Films for UV Detector Applications. *Nanomaterials* **2022**, *12*, 3209. [CrossRef]
57. Yoshino, K.; Tominaga, H.; Komaki, Y.; Imai, M.; Higashi, T.; Minemoto, T.; Shen, Q.; Hayase, S. Low-Temperature Growth of ZnMgO Thin Films by Atmospheric Spin-Coating Using Diethylzinc Solution. *J. Electron. Mater.* **2023**, *52*, 5134–5139. [CrossRef]
58. Balta, A.K.; Ertek, Ö.; Eker, N.; Okur, İ. MgO and ZnO Composite Thin Films Using the Spin Coating Method on Microscope Glasses. *Mater. Sci. Appl.* **2015**, *6*, 40–47. [CrossRef]
59. Bouachiba, Y.; Mammeri, A.; Bouabellou, A.; Rabia, O.; Saidi, S.; Taabouche, A.; Rahal, B.; Benharrat, L.; Serrar, H.; Boudissa, M. Optoelectronic and Birefringence Properties of Weakly Mg-Doped ZnO Thin Films Prepared by Spray Pyrolysis. *J. Mater. Sci. Mater. Electron.* **2022**, *33*, 6689–6699. [CrossRef]
60. Winkler, N.; Wibowo, R.A.; Kautek, W.; Dimopoulos, T. Influence of the Aqueous Solution Composition on the Morphology of $Zn_{1-x}Mg_xO$ Films Deposited by Spray Pyrolysis. *J. Mater. Chem. C* **2019**, *7*, 3889–3900. [CrossRef]
61. Morari, V.; Pantazi, A.; Curmei, N.; Postolache, V.; Rusu, E.V.; Enachescu, M.; Tiginyanu, I.M.; Ursaki, V.V. Band Tail State Related Photoluminescence and Photoresponse of ZnMgO Solid Solution Nanostructured Films. *Beilstein J. Nanotechnol.* **2020**, *11*, 899–910. [CrossRef]
62. Bose, S.; Chevallier, C.; Ould Saad Hamady, S.; Fressengeas, N. Optical and Electrical Properties of ZnMgO with High Mg Content Elaborated by Ultrasonic Spray Pyrolysis Using Water-Based Solutions. *Phys. Status Solidi (A)* **2022**, *219*, 2200091. [CrossRef]

63. El Berjali, W.; Ould Saad Hamady, S.; Boulet, P.; Colas, V.; Gries, T.; Horwat, D.; Pierson, J.-F. ZnMgO Thin Films by Ultrasonic Spray Pyrolysis: Modulation of Optical and Electrical Properties by Post-Annealing and Magnesium Composition. *Phys. Status Solidi (A)* **2024**, *221*, 2400015. [CrossRef]
64. Prathap, P.; Revathi, N.; Reddy, A.S.; Subbaiah, Y.P.V.; Reddy, K.T.R. Synthesis of Conducting $Zn_{1-x}Mg_xO$: Al Layers by Spray Pyrolysis for Photovoltaic Application. *Thin Solid Film.* **2011**, *519*, 7592–7595. [CrossRef]
65. Lopez-Ponce, M.; Hierro, A.; Marín-Borrás, V.; Tabares, G.; Kurtz, A.; Albert, S.; Agouram, S.; Muñoz-Sanjosé, V.; Muñoz, E.; Ulloa, J.M. Optical Properties of ZnMgO Films Grown by Spray Pyrolysis and Their Application to UV Photodetection. *Semicond. Sci. Technol.* **2015**, *30*, 105026. [CrossRef]
66. Qin, X.; Cui, L.; Shao, G. Preparation of ZnO-Zn2TiO4 Sol Composite Films and Its Photocatalytic Activities. *J. Nanomater.* **2013**, *2013*, 428419. [CrossRef]
67. Mahajan, J.; Jeevanandam, P. Synthesis of $Zn_2TiO_4$@CdS Core-Shell Heteronanostructures by Novel Thermal Decomposition Approach for Photocatalytic Application. *ChemistrySelect* **2019**, *4*, 12580–12591. [CrossRef]
68. Chen, X.-Y.; Wang, X.-Z.; Liu, F.-J.; Zhang, G.-S.; Song, X.-J.; Tian, J.; Cui, H.-Z. Fabrication of Porous Zn2TiO4–ZnO Microtubes and Analysis of Their Acetone Gas Sensing Properties. *Rare Met.* **2021**, *40*, 1528–1535. [CrossRef]
69. Ciobanu, V.; Ursaki, V.V.; Lehmann, S.; Braniste, T.; Raevschi, S.; Zalamai, V.V.; Monaico, E.V.; Colpo, P.; Nielsch, K.; Tiginyanu, I.M. Aero-$TiO_2$ Prepared on the Basis of Networks of ZnO Tetrapods. *Crystals* **2022**, *12*, 1753. [CrossRef]
70. Li, X.; Guo, Z.; Xiao, Y.; Um, H.-D.; Lee, J.-H. Electrochemically Etched Pores and Wires on Smooth and Textured GaAs Surfaces. *Electrochim. Acta* **2011**, *56*, 5071–5079. [CrossRef]
71. Monaico, E.V.; Morari, V.; Ursaki, V.V.; Nielsch, K.; Tiginyanu, I.M. Core–Shell GaAs-Fe Nanowire Arrays: Fabrication Using Electrochemical Etching and Deposition and Study of Their Magnetic Properties. *Nanomaterials* **2022**, *12*, 1506. [CrossRef]
72. Monaico, E.V.; Monaico, E.I.; Ursaki, V.V.; Tiginyanu, I.M. Porous Semiconductor Compounds with Engineered Morphology as a Platform for Various Applications. *Phys. Status Solidi (RRL)—Rapid Res. Lett.* **2023**, *17*, 2300039. [CrossRef]
73. Ursaki, V.V.; Lehmann, S.; Zalamai, V.V.; Morari, V.; Nielsch, K.; Tiginyanu, I.M.; Monaico, E.V. Core–Shell Structures Prepared by Atomic Layer Deposition on GaAs Nanowires. *Crystals* **2022**, *12*, 1145. [CrossRef]
74. Özgür, Ü.; Alivov, Y.I.; Liu, C.; Teke, A.; Reshchikov, M.A.; Doğan, S.; Avrutin, V.; Cho, S.-J.; Morkoç, H. A Comprehensive Review of ZnO Materials and Devices. *J. Appl. Phys.* **2005**, *98*, 041301. [CrossRef]
75. Zhang, S.B.; Wei, S.-H.; Zunger, A. Intrinsic N-Type versus p-Type Doping Asymmetry and the Defect Physics of ZnO. *Phys. Rev. B* **2001**, *63*, 075205. [CrossRef]
76. Lupan, O.; Pauporté, T.; Tiginyanu, I.M.; Ursaki, V.V.; Heinrich, H.; Chow, L. Optical Properties of ZnO Nanowire Arrays Electrodeposited on *n*- and *p*-Type Si(1 1 1): Effects of Thermal Annealing. *Mater. Sci. Eng. B* **2011**, *176*, 1277–1284. [CrossRef]
77. Fabbri, F.; Villani, M.; Catellani, A.; Calzolari, A.; Cicero, G.; Calestani, D.; Calestani, G.; Zappettini, A.; Dierre, B.; Sekiguchi, T.; et al. Zn Vacancy Induced Green Luminescence on Non-Polar Surfaces in ZnO Nanostructures. *Sci. Rep.* **2014**, *4*, 5158. [CrossRef]
78. Gurylev, V.; Perng, T.P. Defect Engineering of ZnO: Review on Oxygen and Zinc Vacancies. *J. Eur. Ceram. Soc.* **2021**, *41*, 4977–4996. [CrossRef]
79. Janotti, A.; Van de Walle, C.G. Native Point Defects in ZnO. *Phys. Rev. B* **2007**, *76*, 165202. [CrossRef]
80. Morgan, T. Broadening of Impurity Bands in Heavily Doped Semiconductors. *Phys. Rev.—Phys. Rev. X* **1965**, *139*, 343–348. [CrossRef]
81. Zeng, H.; Duan, G.; Li, Y.; Yang, S.; Xu, X.; Cai, W. Blue Luminescence of ZnO Nanoparticles Based on Non-Equilibrium Processes: Defect Origins and Emission Controls. *Adv. Funct. Mater.* **2010**, *20*, 561–572. [CrossRef]
82. Zalamai, V.; Ursaki, V.; Rusu, E.; Arabadji, P.; Tiginyanu, I.; Sirbu, L. Photoluminescence and Resonant Raman Scattering in Highly Conductive ZnO Layers. *Appl. Phys. Lett.* **2004**, *84*, 5168–5170. [CrossRef]
83. Meyer, B.K.; Alves, H.; Hofmann, D.M.; Kriegseis, W.; Forster, D.; Bertram, F.; Christen, J.; Hoffmann, A.; Straßburg, M.; Dworzak, M.; et al. Bound Exciton and Donor–Acceptor Pair Recombinations in ZnO. *Phys. Status Solidi (B)* **2004**, *241*, 231–260. [CrossRef]
84. Gaponenko, S.V.; Monaico, E.; Sergentu, V.V.; Prislopski, S.Y.; Tiginyanu, I.M. Possible Coherent Backscattering of Lightwaves from a Strongly Absorbing Nanoporous Medium. *J. Opt.* **2018**, *20*, 075606. [CrossRef]
85. Valagiannopoulos, C.A. Study of an Electrically Anisotropic Cylinder Excited Magnetically by a Straight Strip Line. *Prog. Electromagn. Res.* **2007**, *73*, 297–325. [CrossRef]
86. Vaseghi, N.; Abrishamian, M.S. Analysis of Electromagnetic Scattering from Anisotropic Cylindrical Structures Using State Space Method. *AEU—Int. J. Electron. Commun.* **2018**, *89*, 24–33. [CrossRef]
87. Ursaki, V.; Braniste, T.; Zalamai, V.; Rusu, E.; Ciobanu, V.; Morari, V.; Podgornii, D.; Ricci, P.C.; Adelung, R.; Tiginyanu, I. Aero-ZnS Prepared by Physical Vapor Transport on Three-Dimensional Networks of Sacrificial ZnO Microtetrapods. *Beilstein J. Nanotechnol.* **2024**, *15*, 490–499. [CrossRef]

*coatings*

MDPI

*Article*

# Preparation and Properties of Thick Tungsten Coating Electrodeposited from $Na_2WO_4$-$WO_3$-KCl-NaF Molten Salt System

**Yusha Li [1,†], Xiaoxu Dong [1,†], Qing Liu [2], Yajie You [1], Zeyu Gao [1] and Yingchun Zhang [1,*]**

1 School of Materials Science and Engineering, University of Science and Technology Beijing, 30 Xueyuan Road, Haidian District, Beijing 100083, China; liyusha1011@163.com (Y.L.)

2 State Key Laboratory of Advanced Metallurgy, University of Science and Technology Beijing, 30 Xueyuan Road, Haidian District, Beijing 100083, China

* Correspondence: zycustb@163.com

† These authors contributed equally to this work.

**Abstract:** The pulsed current electrodeposition method was employed for the first time to achieve tungsten coating with a thickness of 433.72 μm on a CuCrZr alloy from $Na_2WO_4$-$WO_3$-KCl-NaF molten salt. The microstructure of the coating was observed and the coating density, porosity, hardness, bonding strength, residual stress and oxygen content were tested. The results revealed that the tungsten coating exhibited desirable characteristics such as high density, absence of impurities, excellent adhesion to the matrix (53.16 MPa), residual compressive stress as surface stress, and good stability and durability. Moreover, this thick tungsten coating possesses high density and hardness, low oxygen content and porosity. This offers a novel solution to solve the challenging issue of the connection between tungsten material and heat sink material.

**Keywords:** tungsten coating; electrodeposition; molten salt; CuCrZr alloy

## 1. Introduction

At present, the fusion energy is currently widely acknowledged as one of the most promising and ultimate sources of energy [1]. In order to optimize the utilization of fusion energy and ensure its continuous and stable release, experts have conducted extensive exploration and research, culminating in the proposal of the ITER plan, which aims to construct a controlled nuclear fusion reactor. The pursuit of controlled nuclear fusion has always been a pivotal challenge, with one of the most crucial aspects lying in the development of the plasma-oriented material [2,3]. The progress and implementation of the ITER project are anticipated to yield a breakthrough in controllable nuclear fusion technology while offering an effective mitigation strategy for the global energy crisis.

The manufacturing technology of the cladding system is also a key requirement for the construction of the ITER project. The first wall material is significantly affected by high-energy neutrons, electromagnetic radiation and high-energy particles (such as atmospheric gases, nitrogen, helium and impurities) emitted from the plasma, which consists of plasma-oriented material, intermediate heat sink material and supporting backplane material and serves as the core component of the cladding system. Currently, the CuZrCr alloy is used as the heat sink material. And tungsten's exceptional properties such as its high melting point, excellent thermal conductivity, impressive strength, low coefficient of thermal expansion, negligible vapor pressure, minimal tritium-retention capability, low sputtering yield and high sputtering threshold have made it a primary choice for plasma materials [4,5]. However, tungsten and copper exhibit significant differences in physical properties primarily attributed to their large difference in thermal expansion coefficients. This discrepancy poses challenges in combining these two materials effectively and consequently limits the application of tungsten in plasma surface materials [6,7]. Therefore, alternative approaches such as coating or other viable connection technologies are being considered. These methods involve applying a layer of tungsten coating onto a matrix material to serve as both a plasma-facing material and structural component [8,9]. Therefore, alternative approaches such as coating or other viable connection technologies are being considered. These methods involve applying a layer of tungsten coating onto a matrix material to serve as both a plasma-facing material and a structural component.

Currently, plasma spraying (PS), chemical vapor deposition (CVD), physical vapor deposition (PVD) and other coating technologies have been extensively employed for the preparation of tungsten coatings. The PS technique is known for its operational simplicity; nevertheless, it frequently yields tungsten coatings with intrinsic flaws like porosity or cracks [10]. The CVD method provides a rapid deposition rate for tungsten coatings, albeit requiring elevated temperatures [11]. Tungsten coatings via the PVD method demonstrate exceptional quality; nonetheless, depositing thick coatings utilizing this approach presents challenges and entails substantial expenses [12]. The molten salt electrodeposition (MSE) technique is emerging as a promising approach for the fabrication of metal coatings due to its cost-effectiveness and straightforward operational procedure. The coating prepared via MSE exhibits a strong adhesion to the substrate, high compactness, low porosity and tunable crystalline structure [13]. The molten salts utilized for tungsten electroplating encompass oxide and haloide-oxide compositions. The tungsten coating deposition in the oxide system does not require harsh external environmental conditions, and the preparation, purification, storage and transportation of molten salt are relatively straightforward, rendering it a flexible and practical choice for practical applications. Ningbo Sun and Fan Jiang et al. [14,15] electrodeposited tungsten coatings in $Na_2WO_4$-$WO_3$ molten salt at 1173 K. Wenxuan Qin et al. explored the reduction mechanism of tungsten ion at 1173 K by incorporating $CeO_2$, $ZrO_2$, $La_2O_3$ and $Y_2O_3$ into the $Na_2WO_4$-$WO_3$ molten salt system and subsequently achieved electrodeposition to obtain tungsten coatings with varying qualities [16]. However, due to the generally high melting point of pure oxide molten salts, it is necessary to increase the electrodeposition temperature. High temperatures will limit the choice of substrates and experimental equipment. In order to lower the electrodeposition temperature of tungsten coating, researchers have started utilizing halide molten salt systems, such as LiF-KCl-$Li_2WO_4$ [17], KF-KCl-$WO_3$ [18], NaCl-KCl-NF-$WO_3$ [19] and other molten salts. The addition of a certain amount of halide to the oxide has indeed proven effective in reducing the electrodeposition temperature.

The aim of this study is to establish a robust bonding between tungsten coating and CuCrZr alloy from the $Na_2WO_4$-$WO_3$-KCl-NaF low-temperature molten salt system using the molten salt electrodeposition method, while simultaneously increasing the thickness of the tungsten coating by extending the duration of electrodeposition. This research will offer a novel solution to solve the challenging issue of the connection between tungsten material and heat sink material, while establishing a strong foundation for their further optimization and practical application promotion.

## 2. Experiment and Parameters

### 2.1. Coating Deposition

The electrodeposition of tungsten coating from the $Na_2WO_4$-$WO_3$ molten salt system using the molten salt electrodeposition method was investigated by our group. Moreover, the incorporation of chloride ions can reduce the required temperature for electrodeposition. The $Na_2WO_4$-$WO_3$-KCl-NaF molten salt system was employed in this study to reduce the deposition temperature, and experiments were conducted based on the previously established electrodeposition parameters. $Na_2WO_4 \cdot 2H_2O$ (purity: 99.5%), $WO_3$ (purity: 99.5%), KCl (purity: 99.5%) and NaF (purity: 99.5%) in molten salt were all analytically pure. $Na_2WO_4 \cdot 2H_2O$ was dried for 24 h at 400 °C in order to remove the crystal water; $WO_3$, KCl and NaF were dried for 10 h at 200 °C. The dried $Na_2WO_4$, $WO_3$, KCl and NaF were mixed into an alumina crucible at 3:1:1:1 mol% proportion and then put into an electric furnace. The substrate was CuCrZr alloy, which served as the anode electrode. A tungsten plate (purity: 99.99%) was used as the cathode electrode. The specific elements and contents of CuCrZr alloys are given in Table 1. Prior to electrodeposition, the electrodes were mechanically polished to obtain smooth surfaces and then ultrasonically cleaned in distilled water and alcohol solutions to obtain clean surfaces. The temperature of the electric furnace increases at a rate of 5 °C/min, reaching the electrodeposition temperature. The current density was set from 30 mA/cm$^2$, which is the average current. The frequency and duty cycle were set as 1000 Hz and 30%, respectively. After electrodeposition, the samples were put in NaOH solution for ultrasonic washing to eliminate the adherent salts.

**Table 1.** Elements and contents of CuCrZr alloy (wt.%).

| Cu | Cr | Zr | Impurities |
|---|---|---|---|
| base | 0.653 | 0.132 | <0.075 |

### 2.2. *Coating Characterization*

The surface and cross-sectional morphology of the coated samples were observed by scanning electron microscopy (SEM, JSM6480LV, JEOL, Tokyo, Japan) with elemental analysis. The phase and crystal orientation of W coatings were detected by X-ray diffraction (XRD, Rigaku Industrial Co., Ltd., Tokyo, Japan, D/MAX-BB). The bonding strength was measured by tensile means using a universal testing machine. The tungsten-coating sample was attached to the mating plate with binders and the tensile rate was set to 1 mm/min until the sample fractured. The relative density was measured by the Archimedes method. The porosity of tungsten coatings was tested using a mercury injection apparatus (PM33GT-17, Quantachrome, Boynton Beach, FL, USA). The microhardness of the coating surface was measured using an MH-6 microhardness tester. The residual stress on the surface of the tungsten coating was determined through X-ray diffraction in this study. By measuring the corresponding 2θ angle at different values of ψ, and subsequently applying the least square method, we obtained the slope M from the 2θ-sin2ψ (X-Y) curve. The thermal conductivity of the tungsten coating was measured using the LFA 467HT laser thermal conductivity meter manufactured by NETZSCH in Selb, Germany.

## 3. Results and Discussions

### 3.1. *Microstructure and Crystal Structure of Thick Tungsten Coating*

The suitable temperature conditions for preparing tungsten coating in the $Na_2WO_4$-$WO_3$-KCl-NaF molten salt system were investigated in previous experiments. Through multiple experimental groups and test analyses, it was determined that an electrodeposition temperature of 780 °C yielded an ideal surface quality for electrodeposition tungsten coating. Electrodepositing thick tungsten coatings with high thermal conductivity poses a significant challenge as the coating thickness directly impacts its corrosion resistance against ionic radiation. Therefore, this paper aims to prepare thick tungsten coatings by extending the electrodeposition time. Additionally, the microstructure and properties of these thick tungsten coatings will be studied to observe their growth mechanism and performance characteristics. Table 2 provides the details on the electrodeposition parameters.

**Table 2.** Electrochemical parameters of electrodeposition thick tungsten coating.

| Deposition Time (h) | Temperature (°C) | Current Density (mA/cm$^2$) | Duty Cycle (%) | Frequency (Hz) |
|---|---|---|---|---|
| 50<br>100 | 780 | 30 | 30 | 1000 |

The surface morphology of tungsten coating, depicted in Figure 1, was achieved through electrodeposition times of 50 h and 100 h. It is evident that a layer of silver-gray coating is uniformly deposited on the CuCrZr matrix, exhibiting excellent coverage without any apparent cracks or holes. However, with the increase in electrodeposition time, the macroscopic morphology of the coating changes obviously. After 50 h electrodeposition, the overall surface of the tungsten coating is uniform and dense, and large particles begin to appear. With an extension to a deposition time of 100 h, there is a significant increase in coating thickness due to continuous growth and accumulation on the substrate during a prolonged electrodeposition process. Nevertheless, locally present visible particles contribute to a rougher surface texture primarily attributed to changes in diffusion and migration rates of tungsten ions during extended deposition processes, leading to increased non-uniformity in crystal growth.

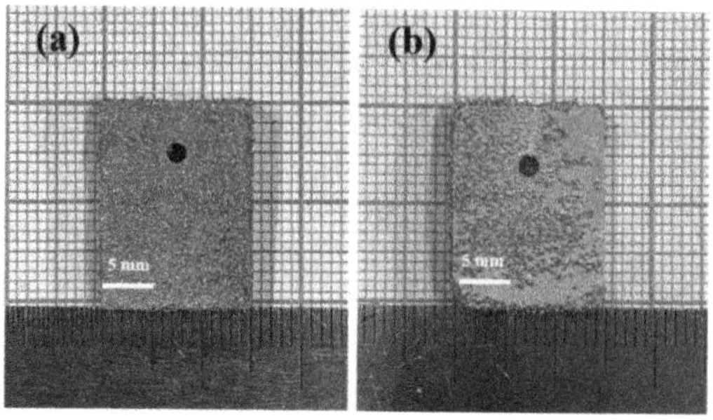

**Figure 1.** Macroscopic morphology of tungsten coatings at different deposition times: (**a**) 50 h, (**b**) 100 h.

The SEM image in Figure 2 depicts the highly compacted surface of the thick tungsten coating, exhibiting a lack of discernible gaps or cracks, which is consistent with the macroscopic morphology. When the electrodeposition time is 50 h, the grain size on the surface of the coating is large, but the arrangement is dense and the distribution is uniform. Upon reaching an electrodeposition time of 100 h, densification remains satisfactory but there is a notable increase in grain size accompanied by non-uniform distribution. This can be attributed to an extended deposition process that further retards the tungsten deposition rate while promoting grain growth and resulting in larger grain sizes and localized aggregation.

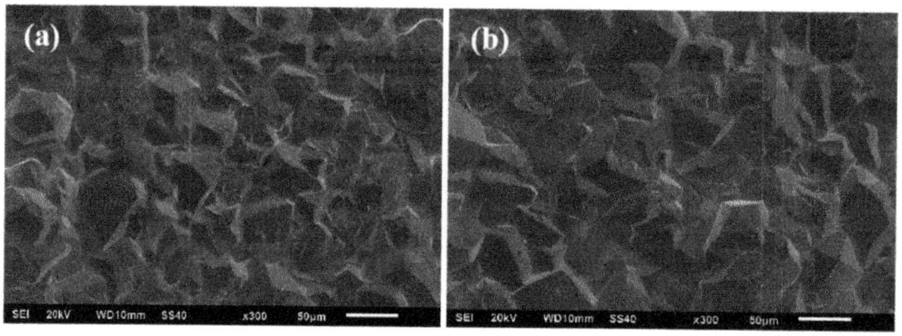

**Figure 2.** Surface SEM images of the tungsten coatings at different deposition times: (**a**) 50 h, (**b**) 100 h.

The surface roughness of the tungsten coating was analyzed using the laser confocal microscope. Image magnification of 200 times was selected to capture intricate details of the coating's surface topography, resulting in obtaining the laser confocal image shown in Figure 3. The experimental findings indicate that extending the electrodeposition time from 50 h to 100 h significantly increases the Ra value of the metal tungsten coating's surface roughness, from 7.469 μm to 10.322 μm. This indicates that with the increase in electrodeposition time, particles accumulate and grow on the surface of the coating, resulting in uneven particle distribution. Consequently, this non-uniformity results in an increase in surface roughness for the coating. However, it should be noted that while increasing electrodeposition time leads to higher surface roughness for thick coatings, such changes are not necessarily unfavorable in practical applications. In fact, under certain circumstances, deliberately increasing the roughness of coatings may enhance the bonding force with substrates or improve other specific properties. Therefore, when optimizing electrodeposition processes, striking the balance between achieving desired levels of surface roughness for coatings while considering other relevant properties based on specific application requirements becomes crucial.

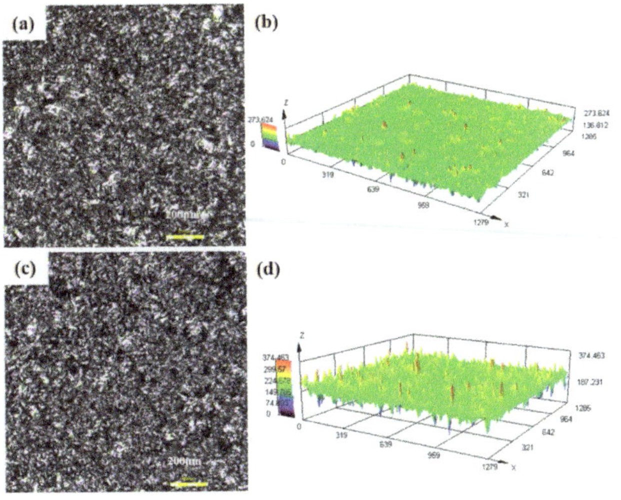


**Figure 3.** Laser confocal image of tungsten coating surface: (**a**) 50 h, 2D, (**b**) 50 h, 3D, (**c**) 100 h, 2D, (**d**) 100 h, 3D.

The X-ray diffraction (XRD) patterns of the thick tungsten coatings at different electrodeposition times are presented in Figure 4. As shown, the tungsten coating obtained after 50 h and 100 h of electrodeposition exhibits four distinct diffraction peaks located at 2θ values of 40°, 58°, 73° and 87°, respectively. These angles correspond to the crystallographic planes (110), (200), (211) and (220) of tungsten according to the standard XRD cards for tungsten. This confirms that the tungsten coating possesses exceptional crystallinity without any presence of impurity phases. Notably, both the (110) and (211) crystallographic planes exhibit relatively high diffraction peak intensities, indicating their dominance in the structure of the tungsten coatings.

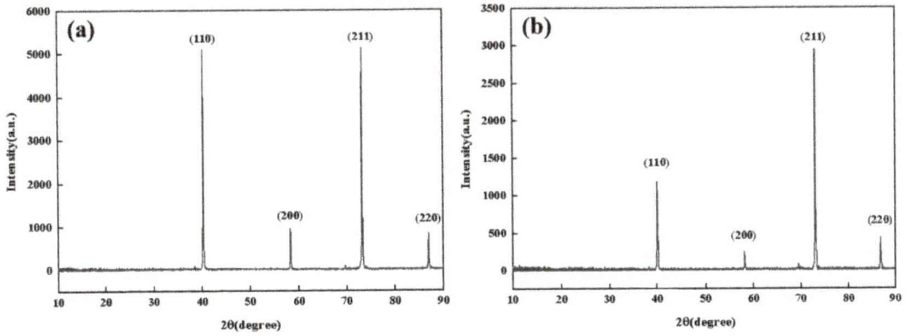


**Figure 4.** XRD patterns of tungsten coatings at different deposition times: (**a**) 50 h, (**b**) 100 h.

After comparing the XRD patterns of the tungsten coating with different electrodeposition times, it can be observed that the diffraction peak intensity of the (211) crystal plane relative to the (110) crystal plane is enhanced following a 100 h electrodeposition period. This enhancement suggests that during prolonged electrodeposition, the grains in the coating exhibit accelerated growth along the direction of the (211) crystal plane, resulting in a corresponding increase in diffraction peak intensity for this particular crystal plane. Simultaneously, these findings also indicate how electrodeposition conditions influence the crystal structure of the coating by promoting a larger proportion or more orderly arrangement of (211) crystal faces within the coatings.

The electrodeposition time exerts a significant influence on the thickness of the tungsten coating during molten salt electrodeposition preparation. The SEM diagram depicted in Figure 5 reveals the section morphology and variations in coating thickness under different deposition times. After 50 h and 100 h electrodeposition, the strong bond between the tungsten coatings and the substrate are observed without any separation phenomenon. Specifically, with a deposition time of 50 h, the coating maintains an elevated level with a thickness measuring 310.65 μm, whereas with a deposition time of 100 h, it forms a thicker tungsten coating reaching up to 433.72 μm.

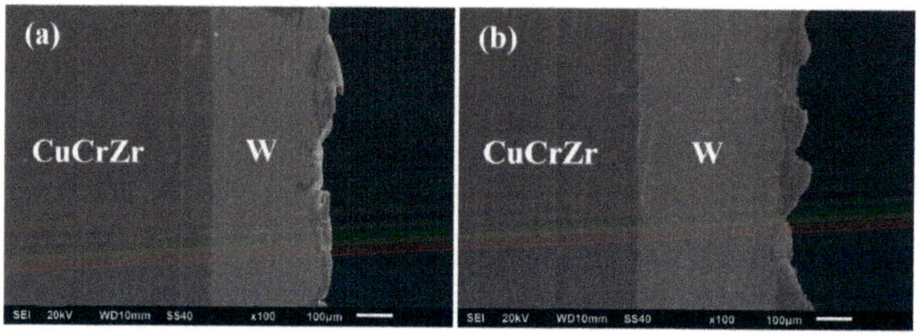


**Figure 5.** Cross-sectional SEM images of tungsten coatings at different deposition times: (**a**) 50 h, (**b**) 100 h.

The sectional line scanning of the coating electrodeposited after 100 h is illustrated in Figure 6. It can be observed from the line scanning that a negligible diffusion layer is evident, while the coating exhibits a robust adhesion to the matrix without any discernible voids or fractures.

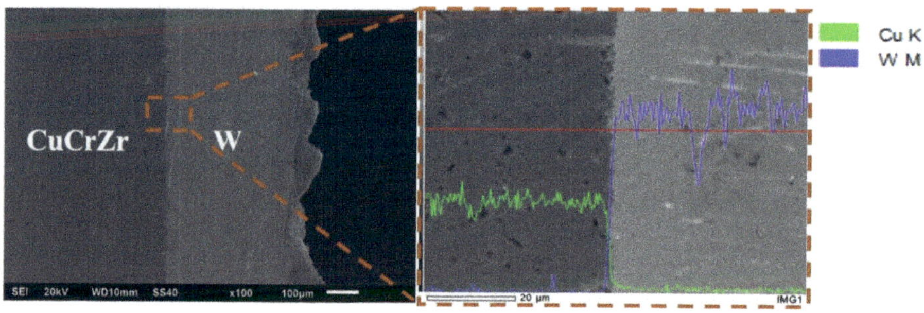


**Figure 6.** The sectional line scanning of the coating obtained after 100 h.

## 3.2. *The Performance Evaluation of Thick Tungsten Coatings*

The main physical properties of 50 h and 100 h electrodeposited tungsten coatings are presented in Table 3. Upon comparing the two datasets, it can be observed that the density of the coating increased from 18.71 g/cm$^3$ to 18.73 g/cm$^3$ as the electrodeposition time was extended from 50 h to 100 h. This indicates that with prolonged electrodeposition time, there is an enhancement in coating quality accompanied by a corresponding increase in density due to greater deposition of tungsten on the substrate, resulting in a thicker and denser coating layer. Furthermore, as the electrodeposition time increased, there was a decrease in coating porosity from 1.18% to 1.16%, along with a reduction in oxygen content from 0.032 wt.% to 0.028 wt.%. These changes can be attributed to long electrodeposition durations leading to densification of the coating and reduced porosity, thereby minimizing opportunities for oxygen interaction with tungsten metal within the coating structure. However, overall variations were relatively minor, indicating that properties obtained through long-duration electrodeposition remain stable.

**Table 3.** Physical properties of thick tungsten coatings.

| Deposition Time (h) | Density (g/cm$^3$) | Porosity (%) | Oxygen Content (%) |
|---|---|---|---|
| 50 | 18.71 | 1.18 | 0.032 |
| 100 | 18.73 | 1.16 | 0.028 |

The hardness values of the coatings obtained by electrodeposition for 50 h and 100 h are presented in Table 4. The experimental findings demonstrate that long-term electrodeposition results in relatively high coating hardness due to enhanced arrangement of tungsten particles, optimized grain structure and improved mechanical properties. At the deposition time of 50 h, the microhardness of the coating is approximately 562.37 HV, which further increases to 568.84 HV when the deposition time is extended to 100 h. The increase in deposition time enables the full growth and development of tungsten crystals within the coating, resulting in the formation of stable structures. In this experiment, the tensile method was employed to assess the adhesive strength between the thick coating and the substrate. The test results are presented in Table 4. The experimental data demonstrate that after 50 h and 100 h electrodeposition, the bonding strength between the coatings and substrate is approximately 50 MPa. Moreover, with an electrodeposition time of 100 h, the bonding strength increases to reach 53.16 MPa. Deposition time facilitates more comprehensive interfacial reactions between the coating and substrate, resulting in improved growth and arrangement of coating grains as well as reduced defects and voids. As a result, the density of the coating is enhanced, as is the interface binding force between the coating and the substrate.

**Table 4.** Mechanical properties of thick tungsten coatings.

| Deposition Time (h) | Hardness (HV) | Bonding Strength (MPa) |
|---|---|---|
| 50 | 562.37 | 49.28 |
| 100 | 568.84 | 53.16 |

The residual stress in the coating plays a crucial role in determining the quality and performance. The residual stress intensifies the phenomenon of stress concentration inside the material, which makes the coating easy to crack at the interface. This also promotes the formation and expansion of micro-cracks, ultimately impacting the service life and reliability of the coating. On the other hand, residual compressive stress helps alleviate stress concentration within the material, enhancing its fatigue performance and resistance to external stresses. Consequently, it extends the overall service life of the coating. To measure this surface residual stress on the coating, an X-ray diffraction method was employed in this experiment. Table 5 presents peak 2$\theta$ values for tungsten coatings during an electrodeposition time of 50 h along with their corresponding $\sin^2(\Psi)$ values. Figure 7 illustrates a linear fitting analysis with a slope value of 0.245. The resulting residual stress is $-309$ MPa (negative), indicating that there exists residual compressive stress within the coating. The thick tungsten coating exhibits significant surface residual compressive stress due to volume-shrinkage inconsistencies caused by rapid cooling from high-temperature molten salt conditions. Thermal stresses primarily contribute to these differences, with thicker coatings experiencing greater disparities in cooling times, leading to increased surface residual compressive stresses.

**Table 5.** The peak value and $\sin^2(\Psi)$ of tungsten coating after 50 h electrodeposition.

| $\sin^2(\Psi)$ | 0 | 0.1 | 0.2 | 0.3 | 0.4 | 0.5 | 0.6 |
|---|---|---|---|---|---|---|---|
| 2θ | 131.188 | 131.209 | 131.237 | 131.253 | 131.289 | 131.310 | 131.332 |

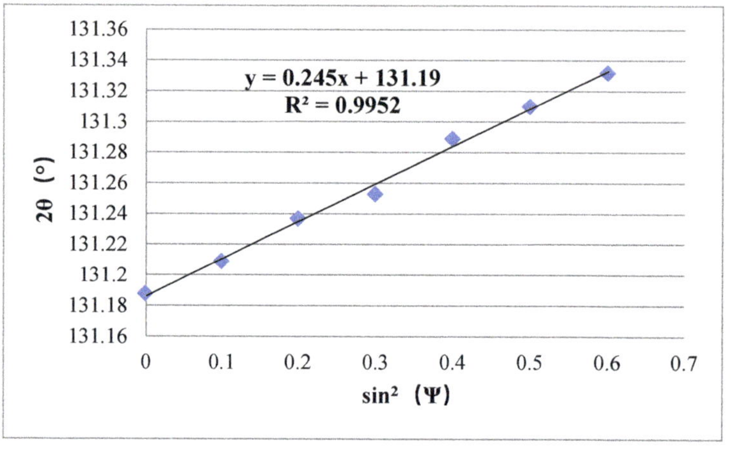


**Figure 7.** Linear fitting curve of peak 2θ and $\sin^2(\Psi)$.

## 4. Conclusions

In this paper, thick tungsten coating was obtained from $Na_2WO_4$-$WO_3$-KCl-NaF molten salt on the surface of a CuCrZr alloy at 780 °C, 30 mA/cm$^2$ current density, 30% duty cycle and 1000 Hz frequency. The microstructure and crystal structure of a thick tungsten coating were studied. The hardness, bonding strength, oxygen content, residual stress and thermal conductivity of the thick tungsten coating were measured.

A tungsten coating on a CuCrZr alloy surface can be successfully obtained from $Na_2WO_4$-$WO_3$-KCl-NaF low-temperature molten salt by the electrodeposition method. The electrodeposition time is extended to 100 h, and the thickness of the tungsten coating is increased to 433.72 μm. The adhesion strength between the thick tungsten coating and the CuCrZr alloy surface is good, measuring 53.16 MPa, while the hardness value reaches 568.84 HV. Moreover, the thick tungsten coating demonstrates high density, low oxygen content and minimal porosity. The residual stress calculations indicate that the deposited coating experiences residual compressive stress due to inconsistent volume shrinkage resulting from rapid cooling under high-temperature molten salt conditions.

**Author Contributions:** Formal analysis, Z.G.; Investigation, Q.L.; Resources, X.D.; Data curation, Y.Y.; Writing—original draft, Y.L.; Funding acquisition, Y.Z. All authors have read and agreed to the published version of the manuscript.

**Funding:** This study was supported by National MCF Energy R&D Program (No. 2019YFE03130002).

**Institutional Review Board Statement:** Not applicable.

**Informed Consent Statement:** Not applicable.

**Data Availability Statement:** Data are contained within the article.

**Conflicts of Interest:** The authors declare no conflict of interest.

## References


1. Ongena, J.; Koch, R.; Wolf, R.; Zohm, H. Magnetic-confinement fusion. *Nat. Phys.* **2016**, *12*, 398–410. [CrossRef]
2. Matthews, G. Material migration in divertor tokamaks. *J. Nuclear Mater.* **2005**, *337–339*, 1–9. [CrossRef]

3. Travere, J.; Aumeunier, M.; Joanny, M.; Loarer, T.; Firdaouss, M.; Gauthier, E.; Martin, V.; Moncada, V.; Marot, L.; Chabaud, D.; et al. Imaging Challenges for ITER Plasma-Facing Component Protection. *Fusion Sci. Technol.* **2013**, *64*, 735–740. [CrossRef]
4. Dux, R.; Bobkov, V.; Fedorczak, N.; Iraschko, K.; Kallenbach, A.; Neu, R.; Pütterich, T.; Rohde, V. Tungsten erosion at the ICRH limiters in ASDEX Upgrade. *J. Nucl. Mater.* **2007**, *363*, 112–116. [CrossRef]
5. Rohde, V.; Mertens, V.; Scarabosio, A. Gas balance in ASDEX Upgrade with tungsten first wall. *J. Nucl. Mater.* **2009**, *390*, 474–477. [CrossRef]
6. Tokunaga, K.; Yoshida, N.; Kubota, Y.; Noda, N.; Imamura, Y.; Oku, T.; Kurumada, A.; Sogabe, T.; Kato, T.; Plöchl, L. High heat flux test of actively cooled tungsten-coated carbon divertor mock-ups. *Fusion Eng. Design* **2000**, *49–50*, 371–376. [CrossRef]
7. Matějíček, J.; Koza, Y.; Weinzettl, V. Plasma sprayed tungsten-based coatings and their performance under fusion relevant conditions. *Fusion Eng. Design* **2005**, *75–79*, 395–399. [CrossRef]
8. Neu, R.; Maier, H.; Gauthier, E.; Greuner, H.; Hirai, T.; Hopf, C.; Likonen, J.; Maddaluno, G.; Matthews, G.F.; Mitteau, R.; et al. Investigation of tungsten coatings on graphite and CFC. *Phys. Scr.* **2007**, *T128*, 150–156. [CrossRef]
9. Cho, G.S.; Choe, K.H. Characterization of plasma-sprayed tungsten coating on graphite with intermediate layers. *Surf. Coat. Technol.* **2012**, *209*, 131–136. [CrossRef]
10. Moon, S.Y.; Choi, C.H.; Kim, H.S.; Oh, P.; Hong, B.G.; Kim, S.K.; Lee, D.W. Thick tungsten coating on ferritic-martensitic steel applied with a vacuum plasma spray coating method. *Surf. Coat. Technol.* **2015**, *280*, 225–231. [CrossRef]
11. Han, X.Y.; Liu, X.L.; Wang, D.Z.; Wu, Z.; Duan, B. Copper-based tungsten coating by CVD: Microstructure, thermal shock resistance and interfacial bond force. *Surf. Coat. Technol.* **2021**, *426*, 127778. [CrossRef]
12. Ganne, T.; Crépin, J.; Serror, S.; Zaoui, A. Cracking behaviour of PVD tungsten coatings deposited on steel substrates. *Acta Mater.* **2002**, *50*, 4149–4163. [CrossRef]
13. Gu, Y.K.; Liu, J.; Qu, S.X.; Deng, Y.; Han, X.; Hu, W.; Zhong, C. Electrodeposition of alloys and compounds from high-temperature molten salts. *J. Alloys Compd.* **2017**, *690*, 228–238. [CrossRef]
14. Jiang, F.; Zhang, Y.C.; Li, X.L.; Sun, N.B.; Wang, L.L. Tungsten coating prepared on V-4Cr-4Ti alloy substrate by electrodeposition from molten salt in air atmosphere. *Fusion Eng. Des.* **2014**, *89*, 83–87. [CrossRef]
15. Sun, N.B.; Zhang, Y.C.; Lang, S.T.; Jiang, F.; Wang, L.L. Tungsten coatings electro-deposited on CFC substrates from oxide molten salt. *J. Nucl. Mater.* **2014**, *455*, 450–453. [CrossRef]
16. Qin, W.X.; Xi, X.L.; Zhang, L.W.; Nie, Z.R.; Liu, C.J.; Li, R. The effect of $MO_x$ (M = Zr, Ce, La, Y) additives on the electrochemical preparation of tungsten in eutectic $Na_2WO_4$-$WO_3$ melt. *J. Electroanal. Chem.* **2023**, *935*, 117343. [CrossRef]
17. Wang, X.H.; Zhang, L.W.; Xi, X.L.; Nie, Z.R. Electrochemical Dissolution Process of Tungsten Carbide in Low Temperature Molten Salt System. *J. Electrochem. Soc.* **2022**, *169*, 051501. [CrossRef]
18. Nohira, T.; Ide, T.; Meng, X.D.; Norikawa, Y.; Yasuda, K. Electrodeposition of Tungsten from Molten KF-KCl-$WO_3$ and CsF-CsCl-$WO_3$. *J. Electrochem. Soc.* **2021**, *168*, 046505. [CrossRef]
19. Qi, Y.F.; Tang, Y.H.; Wang, B.; Zhang, M.; Ren, X.Q.; Li, Y.G.; Ma, Y.T. Characteristics of tungsten coatings deposited by molten salt electro-deposition and thermal fatigue properties of electrodeposited tungsten coatings. *Int. J. Refract. Met. Hard Mater.* **2019**, *81*, 183–188. [CrossRef]

*coatings*

MDPI

*Article*

# Structural Aspects and Adhesion of Polyurethane Composite Coatings for Surface Acoustic Wave Sensors

**Mauro dos Santos de Carvalho [1,*], Michael Rapp [2,*], Achim Voigt [2], Marian Dirschka [2] and Udo Geckle [3]**

1 Instituto de Química, Universidade Federal do Rio de Janeiro, Avenida Athos da Silveira Ramos, 149 Bloco A—4° Andar—SALA 408 Cidade Universitária, Rio de Janeiro 21941-909, RJ, Brazil
2 Institute of Microstructure Technology, Karlsruhe Institute of Technology, Hermann-von-Helmholtz-Platz 1, 76344 Eggenstein-Leopoldshafen Karlsruhe, Germany
3 Institute for Applied Materials, Karlsruhe Institute of Technology, Hermann-von-Helmholtz-Platz 1, 76344 Eggenstein-Leopoldshafen, Germany
* Correspondence: mauro@iq.ufrj.br (M.d.S.d.C.); michael.rapp@partner.kit.edu (M.R.)

**Abstract:** Surface acoustic wave-based (SAW) sensors are of great interest due to their high sensibility and fast and stable responses. They can be obtained at an overall low cost and with an intuitive and easy-to-use method. The chemical sensitization of a piezoelectric transducer plays a key role in defining the properties of SAW sensors. In this study, we investigate the structural and adhesion properties of a new class of coating material based on polyurethane polymeric composites. We used dark-field microscopy (DFM) and scanning electron microscopy (SEM) to observe the microstructure of polyurethane composite coatings on piezoelectric sensor elements and to analyze the effects of the chemical resistance and adhesion test (CAT) on the coating layers obtained with the polyurethane polymeric composites. The results of the microscopy showed that all polyurethane composite coatings exhibited excellent uniformity and stability after chemical adherence testing (CAT). All of the observations were correlated with the results of the ultrasonic analysis, which demonstrated the role of polyurethane as a binder to form the stable structure of the composites and, at the same time, as an adhesion promoter, increasing the chemical resistance and the adherence of the coating layer to the complex surface of the piezoelectric sensor element.

**Keywords:** polymeric sensing layer; spin coating; polymeric thin film

## 1. Introduction

Surface acoustic chemical (SAW) sensors have been investigated widely in recent years [1,2] due to their high sensibility, simplicity, and overall low cost in many aspects. The performance of SAW technology relies on two principal factors: First, the hardware design is responsible for the electronics on which the sensibility of the chemical transducing and the stability and precision of the sensor answer depends. And the second principal factor is the sensing interface, which will define the chemical environment that will provide the affinity towards the desired analytes [3]. In SAW technology, the sensing interface is achieved by the deposition of a suitable sensing material, which ideally should present a high affinity and selectivity to the desired analytes, together with the other properties common to all analytical methods, like precision, reproducibility, stability of analytical response, and a suitable lifetime of utilization. As hardware issues have already been conveniently addressed and the state of the art of oscillator circuit development for such high-frequency devices is currently well established, the development of SAW sensor systems that can be turned into real applications still relies on the properties of the sensing interface, and, ultimately, on the choice of the sensitization material together with the coating procedure.

Due to their diverse physical–chemical properties, polymers are materials that have been extensively explored, finding a broad range of applications in several areas. Besides their traditional uses, the vast choice in terms of chemical constitution and peculiar properties also make polymers potential candidates for applications at the molecular level, where they provide many suitable chemical environments and structural properties for several fields [4]. Among others, polymers find applications as drug transporters [5], in the support and design of catalysts [6,7], in sensor applications [8], and as sensing materials for chemical sensors [9]. Polymers enable a broad spectrum of molecular environments suitable for intermolecular interactions with analytes, provided by their diversity of chemical constitutions [10–13]. Besides their broad diversity of chemical constitutions, their variety of structural properties means that they can work alone as sensing materials or be combined with other components to form functional composites [14–16]. Polymers having three-dimensional networks enable the sorption of analytes, providing a high capacity for inclusion and detection by sensor systems [17]. These networks also exhibit suitable mechanical properties, making polymers a promising solution for the sensitization of complex surfaces [18], where they can fulfill most of the necessary requirements of coating materials [19–22].

In the sensitization of SAW sensors, the expected result of the coating process is the deposition of the desired sensitization material, forming a (usually) nanometric homogeneous layer that accounts for a minimal attenuation of wave propagation and, at the same time, possesses enough material to interact with the available quantity of the analyte, making it suitable for detection and quantification. Finally, the sensing layer should be chemically and mechanically stable for a reasonable lifetime, which is expected for the utilization of sensor systems in real analytical applications. For the coating of a SAW sensor element, some of the methods used include electrospraying, airbrushing, and spin coating [23,24].

Because of their above-mentioned advantages, polymers were used as coating materials at the beginning of the development of SAW sensors [25,26]. However, issues such as dewetting and other problems were reported [27–29], leading to a lack of utilization of polymers as sensing materials in recent years.

To address the application of polymers as sensing materials for SAW sensors, we presented a new class of polymeric composites, formed between polyurethane (PU) and a sensing polymer, as sensing materials for SAW technology [30,31]. These new coating materials are intended to take advantage of polymers' features, such as their cost and availability, along with a robust and efficient coating procedure, resulting in coating layers with increased chemical and mechanical properties compared with the pristine polymers previously used as coating materials. The new coating materials were submitted to CAT to test their chemical and mechanical stability, and the results indicated that, while the coatings of the pristine polymers were almost completely removed from the surface of the SAW sensor element, the coatings with the PU–polymer composites presented a remarkable improvement in chemical resistance and in the adhesion of their coating layers. The results showed that with the PU–polymer composites, although some loss of material was observed, most was retained over the surface, preserving the original sensor responses of their respective sensing polymers and their relative affinities against the tested organic analytes. The observations indicated that the structure of the coating layer of the composites remained over the surface, even after the application of the CAT.

To explain those results, a mechanism for the formation of the PU–polymer composites was proposed, in which the PU is responsible for both the formation of a CAT-resistant polymeric network and the increase in the adhesion of the coating layer to the complex surface of the SAW sensor element. The material loss observed after the CAT should be due to the removal of the sensing polymer units that were eventually not bonded to the composite structure. These free units of sensing polymer should then be washed out from the coating layer by the CAT. The coating layer that remains after the CAT should correspond to the structure of the PU–polymer composite, which presents improved chemical resistance and adhesion in comparison to the pristine polymers as the coating materials. The proposed mechanism of the formation of the PU-composites is schematically represented in Figures 1 and 2.

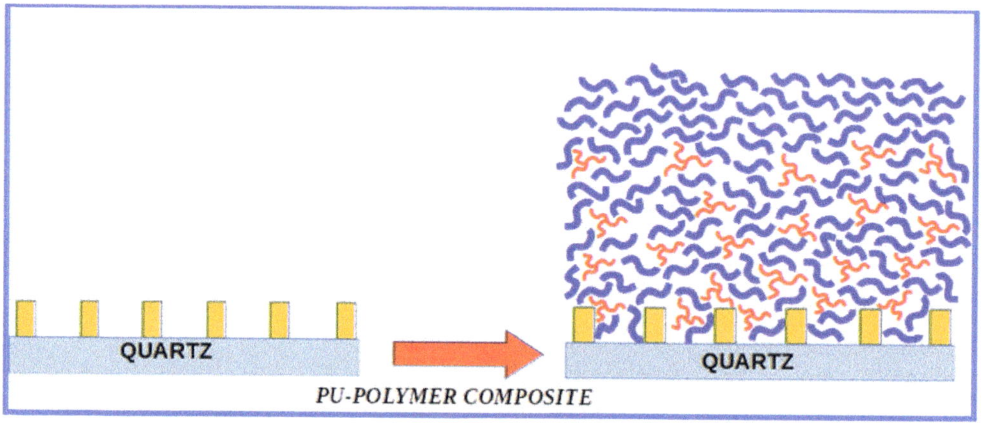


**Figure 1.** Representation of the deposition of the PU–polymer composites as the coating material over the SAW sensor element by the spin coating methodology [31]. The blue units represent the molecules of the sensing polymer, and the red ones represent the PU molecules. Some units of the sensing polymer are not involved in the structure of the PU–polymer composite, as is schematically shown at the top of the representation of the coating layer (right side of the figure).

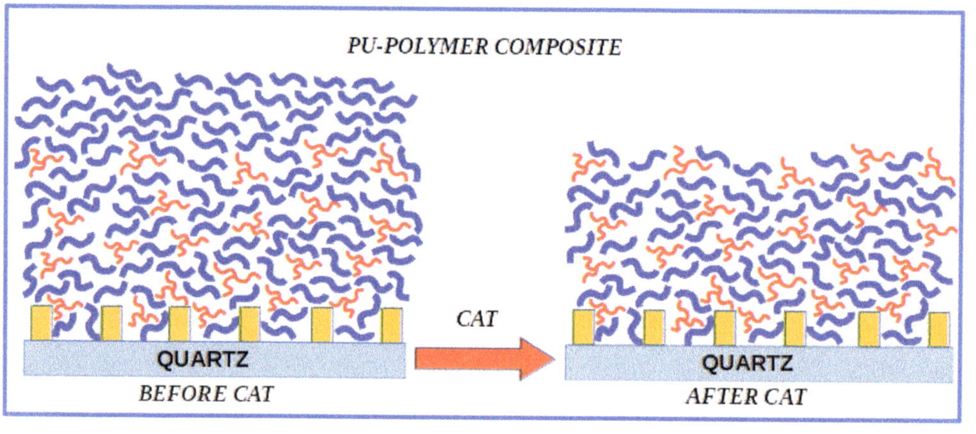


**Figure 2.** Schematic representation of the result of the CAT over a coating layer obtained by the deposition of the PU–polymer composites over the SAW sensor element [31]. The blue units represent the molecules of the sensing polymer, and the red ones represent the PU molecules. The units of the sensing polymer that are not bonded to the structure of the PU–polymer composite (seen on the top of the left side of the figure) should be removed from the coating layer by the CAT. The remaining deposited layer should correspond to the structure of the corresponding PU–polymer composite used.

In the present work, three chemically different polymers were combined with PU to form their respective PU–polymer composites: polybutylmethacrylate (PBMA), polylauryl-methacrylate (PLMA), polyisobutylene (PIB). We used DFM [32,33] and SEM to investigate the microstructure of the PU–polymer composite coating layers and the effects of the CAT to analyze the remaining structure of the PU–polymer coating materials over the SAW sensor element. The microscopy results were then correlated with the ultrasonic analysis to characterize the obtained coating layers before and after the CAT of the PU–polymer composites as the sensing materials. The results will provide more insights into the mechanism of formation of the polymeric composites and will characterize the effects of the CAT in the structure of the deposited coating materials. These structural results should be used in a future interpretation of the selectivity and long-term behavior of the sensor responses of the PU composite coating layers.

## 2. Methodology

### 2.1. Piezoelectric Sensor Elements

The piezoelectric sensor elements used in this work (Figure 3) are of high polish quartz ($SiO_2$) with a 37.5° cut orientation and dimensions of 4 mm × 8 mm and 0.5 mm thickness. Over the quartz surface, the structure of the interdigital propagators for the acoustic waves and the contact pads, made of gold, were lithographic deposited (SCD Components, Dresden, Germany).

**Figure 3.** SAW sensor element with four large contact pads at the edges (in black) and the sensor active area located in the center of the element (the continuous dark gray zone that crosses the whole element horizontally) [34].

### 2.2. Spin Coating

The spin coating solutions were prepared with concentrations of 1.6 mg/100 mL of PBMA and 0.8 mg/100 mL of PU for the PU/PBMA composite, 1.6 mg/100 mL of PLMA and 0.8 mg/100 mL of PU for the PU/PLMA composite; 0.8 mg/100 mL of PIB and 0.8 mg/100 mL of PU for the PU/PIB composite and 0.8 mg/100 mL of PU for the pristine PU deposition. For the deposition of the coating layer, 200 microliters of the spin coating solutions were dropped over the piezoelectric sensor element placed on the spin coater (Laurell MS-400B-6NPP/LITE, Lansdale, PA, USA). The rotation speed rate used was 8000 rpm for 120 s for all experiments. All parameters were precisely controlled.

### 2.3. Ultrasonic Parameters

The ultrasonic parameters of each SAW device before and after coating were recorded by comparing the HF resonance frequency transmission parameters with a network analyzer (Hewlett Packard 8712ES, Waldbronn, Germany). The ultrasonic parameters, resulting in resonance frequency shift and attenuation change ($S_{12}$ parameter), were measured.

*2.4. Scanning Electron Microscopy*

Equipment: Scanning electron microscope SR-50 A; International Scientific Instruments, Inc.; Carlsbad, CA, USA, Emission type: Wolfram-Tungsten cathode; Acceleration voltage: 0.02–30 kV; Resolution 8 nm.

*2.5. Dark Field Microscopy (DFM)*

The DFM images were taken with the microscope Axiotech 100 HD (Carl Zeiss, Jena, Germany).

*2.6. Chemicals*

Perchloroethylene (CAS 127-18-4) and toluene (CAS 108-88-3) were purchased from Sigma-Aldrich Co. (St. Louis, MO, USA), all having concentrations higher than 99%, and were used without further treatment. Polybutylmetacrylate (CAS 9003-63-8), Polylaurylmetacrylate (CAS 25719-52-2), and Polyisobutylene (CAS 9003-27-4) were purchased from Sigma-Aldrich Co. (St. Louis, MO, USA). The polyurethane (polymeric methylenediphenyldiisocianate and polyether–polyester basis polyol) was obtained from Büfa Company, Oldenburg, Germany.

*2.7. Chemical Resistance and Adhesion Test (CAT)*

The CAT is a house-made test devised to investigate the stability of the polymeric coating layers in terms of their chemical resistance to perchloroethylene and to infer the adhesion of the polymeric layer to the surface of the SAW sensor element. The method is performed by submitting the coated sensor elements to the limit condition of complete immersion in a bath of perchloroethylene for 24 h. After this, the coated sensor is left at room temperature for twelve hours to complete the evaporation of the solvent before further measurements [30].

## 3. Results and Discussion

*3.1. Ultrasonic Results*

The ultrasonic results of frequency shift and attenuation for the pristine PU and the PU–polymer composites coatings, before and after the CAT, are shown in Figures 4 and 5, respectively.

The results of the attenuation and frequency shift for the coating with the pristine PU before and after the CAT confirmed that the PU is not affected by the test, as observed in the previous works [31], reinforcing its participation in the composite formation as described in the proposed mechanism. The results of the frequency shift, which is proportional to the mass deposited over the piezoelectric sensor element, for the PU–polymer composites indicate that, although a quantity of mass of the original deposition was removed by the CAT, a significant mass of the coating layer of the composites was preserved. The attenuation results are connected to the loss of energy by the propagation of the acoustic wave over the surface of the SAW sensor, providing an integral parameter related to the material uniformity and distribution over the active area of the sensor. The uniformity of the results of the attenuation for all the PU-composites after the CAT suggests that the remaining structures are similar in respect to their homogeneity, despite the difference in their structures due to the variation in the sensing polymers in their constitutions. This observation can be explained by the fact that the composition of PU-composites leads to the formation of coating layers with more structural similarity once all of them have PU as a common component in their structures. And the observed reduction and uniformity in the attenuation results after the CAT can be attributed in part to the mass reduction (as detailed in Figure 2) but also due to the improvement of the adhesion and material distribution, as observed before for the PU–polymer composites [31,34]. To explore more deeply the individual characteristics of the coating layers formed by each PU–polymer composite, microscopic analyses using SEM and DFM were performed.

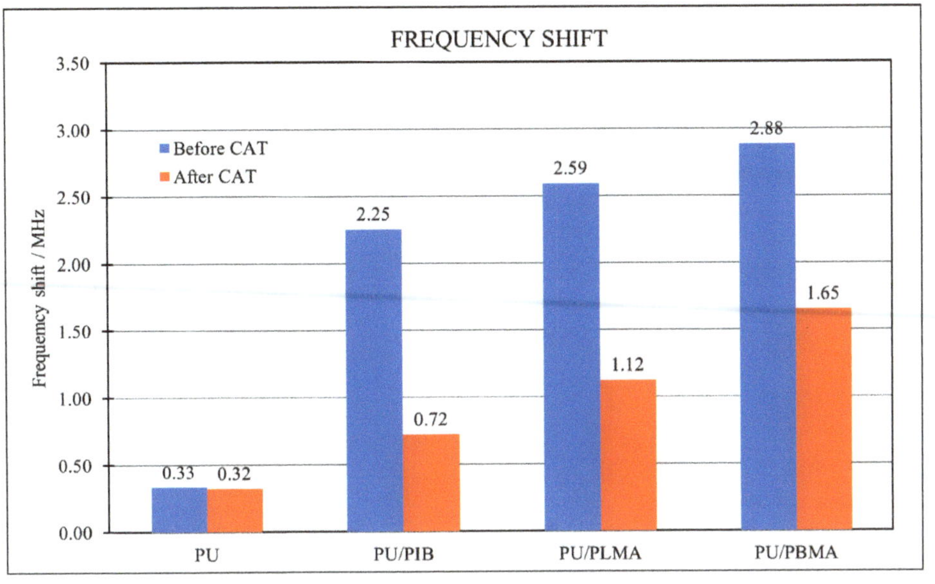


**Figure 4.** Frequency shift results of the pristine PU and the PU–polymer composites before and after the CAT.

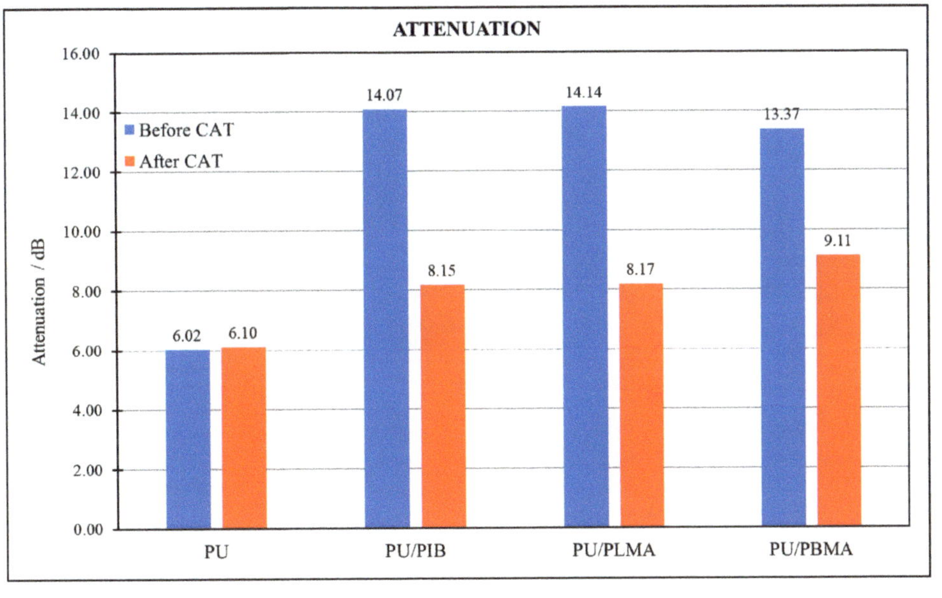


**Figure 5.** Attenuation results of the pristine PU and the PU–polymer composites before and after the CAT.

The main object of the microscopy analysis was the central part of the active area of the piezoelectric sensor element, in the location with the intersection of three different regions of the element: the area with the gold electrodes fingers, the gold continuous contact pad, and the free quartz surface, in order to investigate the interaction of the coating materials with these constitutionally distinct regions of the piezoelectric sensor surface.

## 3.2. Uncoated SAW Sensor Element

Figure 6 presents the SEM images of an uncoated SAW sensor element ready for deposition, to provide a reference about how the clean surface should look.

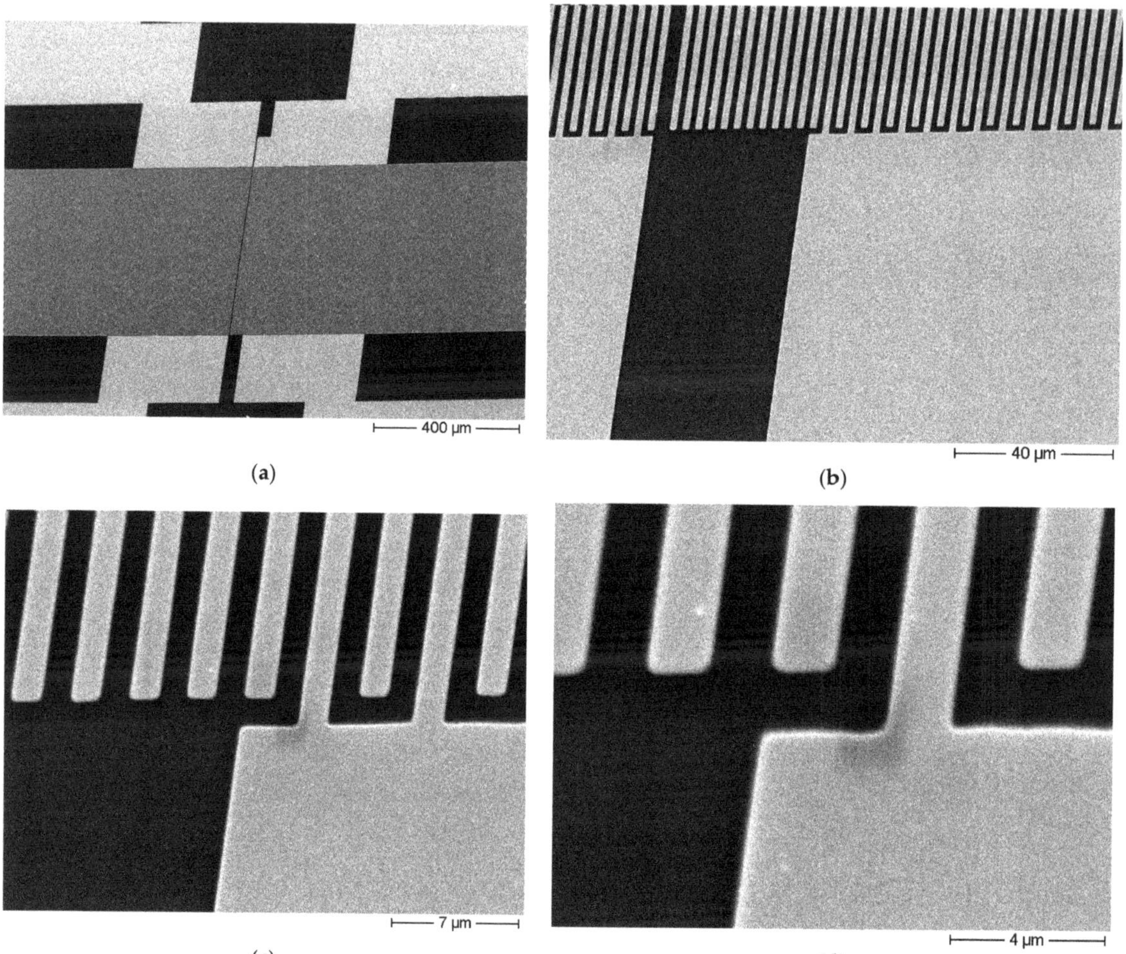


(a) (b) (c) (d)

**Figure 6.** SEM images of the surface of an uncoated piezoelectric sensor element at various magnifications. The active area of the piezoelectric sensor can be identified as the horizontal dark gray zone, which crosses the whole picture in the 100× magnification (**a**), 1000× magnification (**b**), 5000× magnification (**c**) and 10,000× magnification (**d**).

The observation of the locations presented in Figure 6c,d allows an inspection of the resulting interaction of the coating materials with the three kinds of constitutions of the surface of the sensor element to provide insights about the quality of the coverage by the coating material, its wettability to each surface, the obtained material distribution, the homogeneity of the layer, and the respective implications of these characteristics to the adhesion and mechanical stability of the resulting coating layer. From Figure 6 can be seen that all the regions of the surface of the uncoated SAW sensor element present very clean and unperturbed images by all the magnifications analyzed.

### 3.3. PU-PBMA Composite

The attenuation results indicated a homogeneous coating layer for the PU-PBMA composite (Figure 5) despite the high values of frequency shift obtained for the original deposition (before the CAT) for this coating material. The DFM images for the PU-PBMA original coating layer confirm the homogeneity of its deposition (Figure 7a). The SEM images also confirm the homogeneous aspect presented by the original coating layer of the PU-PBMA composite. By the magnifications of 5000× and 10,000× (Figure 8a,b) can be seen that the layer equally covered the quartz substrate as well as the gold fingers and electrodes in a homogeneous way.

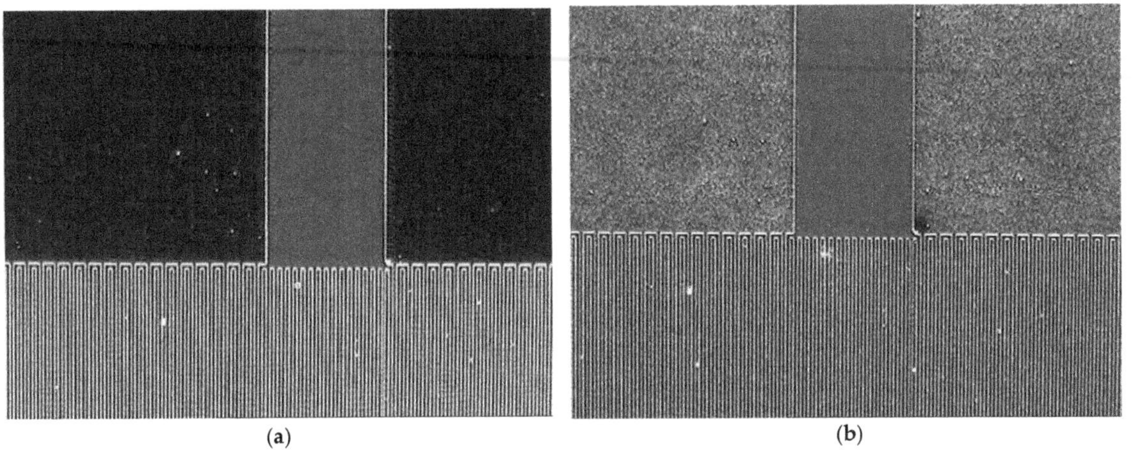

**Figure 7.** (**a**) DFM images of the coating layer with the PU-PBMA composite before and (**b**) after the CAT. Magnification of 50×.

After the application of the CAT, the DFM image reflects the modification of the structure of the coating layer (Figure 7b), indicating how the test affected the structure of the coating layer. Nevertheless, after the CAT the remaining coating layer still presents a regular aspect in terms of material distribution. This modification in the structure of the coating layer can be correlated to the frequency shift results (Figure 4), which indicates a loss of a significant amount of material by the PU-PBMA coating layer after the application of the CAT. The images of the SEM confirm the modification of the structure (Figure 8c,d), and the pattern observed in the SEM images agrees with that observed by the DFM images after the CAT.

As observed for the DFM, the SEM images show a homogeneous coverage over all the regions of the surface of the SAW element, with a very reproducible structure that should be the characteristic structure of the PU-PBMA composite that remains bound to the surface after the CAT.

### 3.4. PU-PLMA Composite

Figure 9 presents the images of DFM of the coating layer obtained with the PU-PLMA composite.

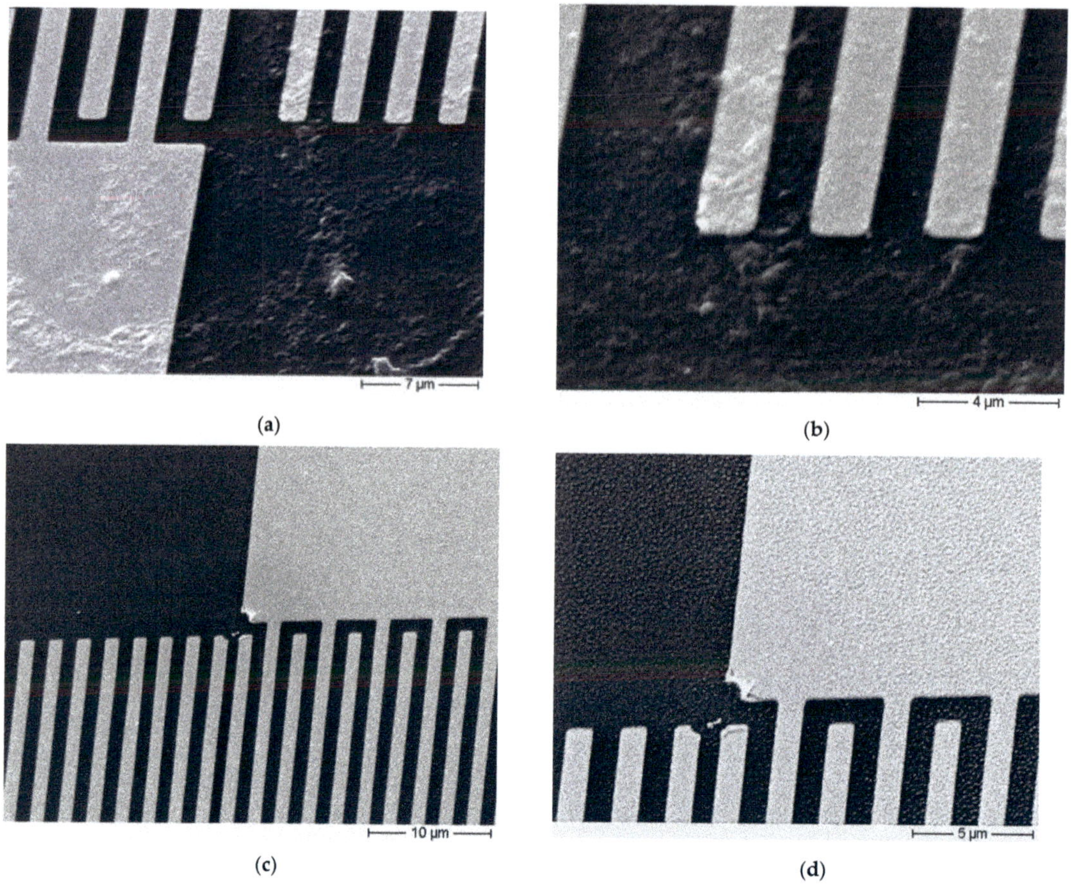


**Figure 8.** SEM images of the PU-PBMA composites before and after the CAT. (**a**) Magnification of 5000×, before CAT; (**b**) 10,000×, before CAT; (**c**) 2500×, after CAT; (**d**) 5000×, after CAT.

**Figure 9.** DFM images of the coating layer with the PU-PLMA composite, (**a**) before and (**b**) after the CAT. Magnification of 50×.

The attenuation results indicated that the original coating layer with the PU-PLMA composite should present a homogenous deposition over the surface of the saw element (Figure 5). The homogeneity of the coverage with this composite is confirmed by the DFM image (Figure 9a), where a pattern can be observed for the coating with the PU-PLMA composite. The SEM images also indicate a very homogenous deposition of this coating material (Figure 10a,b), however, showing a quite different structure in comparison with the structure obtained with the PU-PBMA composite (Figures 7a and 8a,b). The layer of the PU-PLMA composites appears to cover all regions of the surface, as also observed by the PU-PBMA composite.

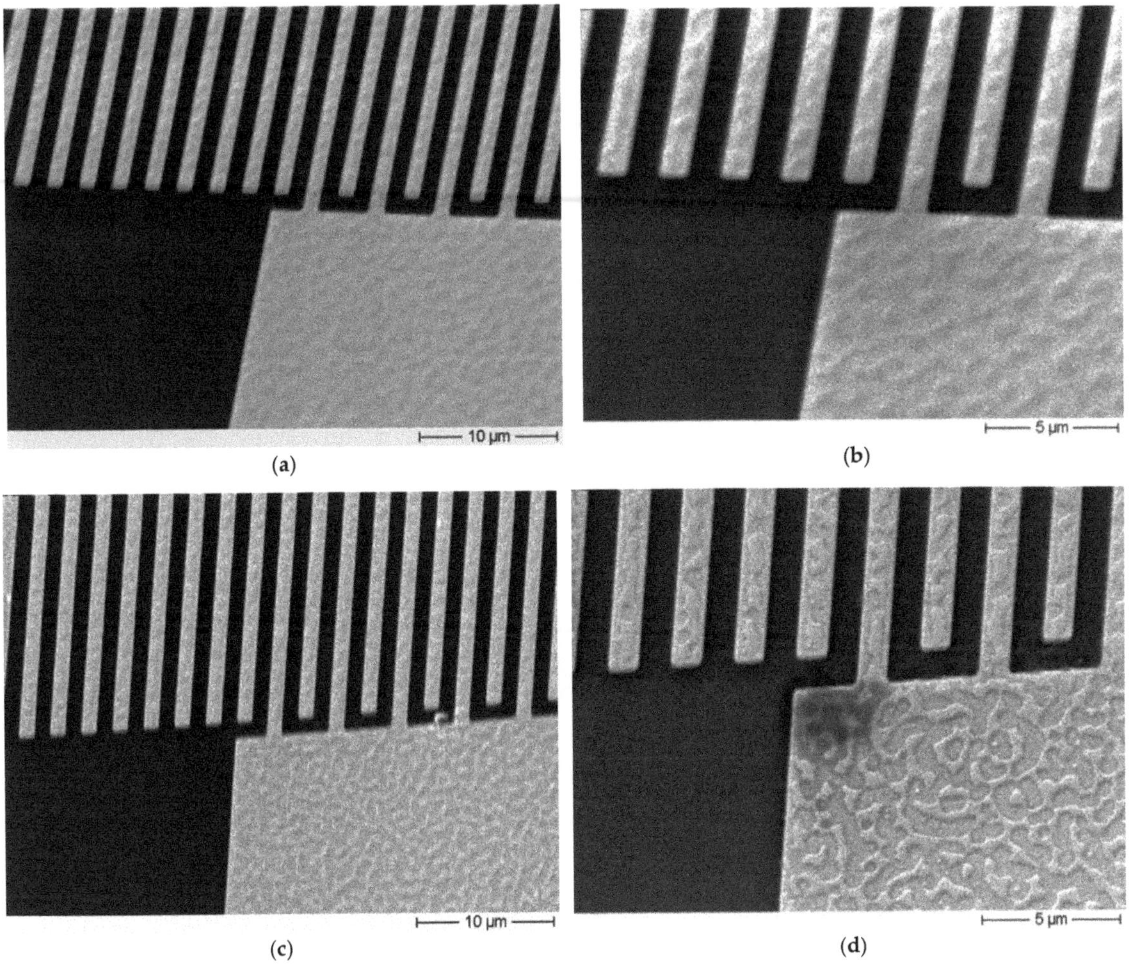


**Figure 10.** SEM images of the deposition of the PU-PLMA composite as the sensing layer. (**a**) Magnification of 2500×, before CAT; (**b**) 5000×, before CAT; (**c**) 2500×, after CAT; (**d**) 5000×, after CAT.

The frequency shift results for the PU-PBMA composite (Figure 4) also indicated a significant loss of material occurred after the application of the CAT, but again resulting in a homogeneous coating layer after the CAT, according to the attenuation results. The DFM image after the CAT (Figure 9b) appears to confirm the ultrasonic results, where the layer appears to be cleaned after the CAT, indicating the removal of part of the material from the surface. However, the remaining layer presents homogeneity. The SEM results present more clearly the effect of the CAT, evidencing the fact that some part of the material was removed, while there is a very homogeneous and reproducible structure that remains after the CAT (Figure 10c,d). These results strongly suggest that the remaining structures after the CAT should correspond to the PU-PLMA composite formed by the coating process. The remaining structures of the two composite materials, PU-PBMA and PU-PLMA, are very distinct, evidencing the constitutional differences between the two coating materials.

### 3.5. PU-PIB Composite

The coating with the PU-PIB composite presented a different behavior for its deposition. The DFM images show a regular pattern that spreads itself over all regions of the surface of the SAW sensor element, apparently covering equally all the gold and the quartz regions of the surface (Figure 11a,b).

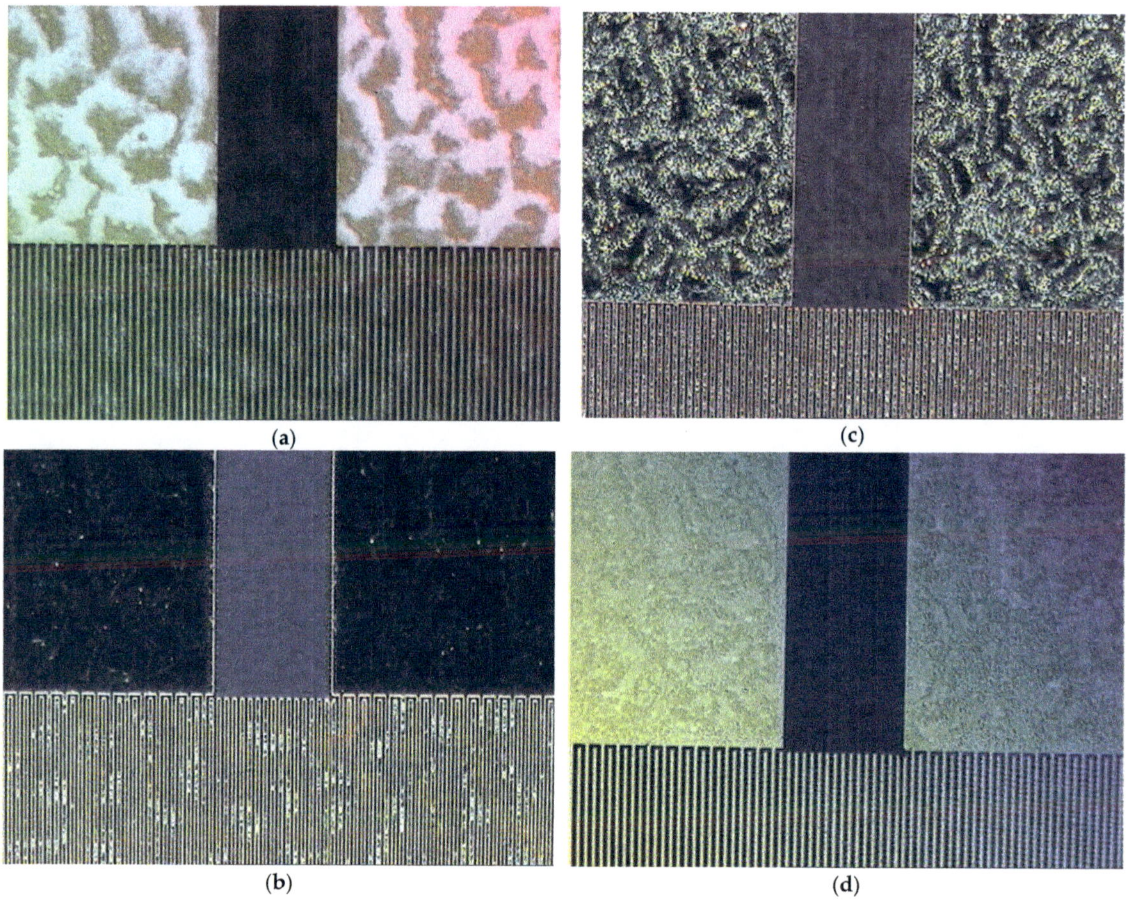

**Figure 11.** DFM images of the coating with the PU-PIB composite as the coating material, (**a**,**b**) before CAT; (**c**,**d**) after the CAT. All images were made with a magnification of 50×.

The DFM images after the CAT indicated that the deposition had been modified, suggesting some loss of the original coating (Figure 11c,d), as was observed with the other PU–polymer composites. Again, the pattern observed by the deposition before the application of the CAT seems to be maintained but apparently being cleaned after the application of the CAT.

The SEM images confirm the observed with the DFM. The images before the CAT (Figure 12a,b) show a homogeneous coverage of all regions of the surface. From the SEM images can be seen that the coating material can be found over the gold pads, over the gold fingers of the interdigital structure, in between the gold fingers on the quartz trenches, and over the quartz regions, free of the gold electrode deposition.

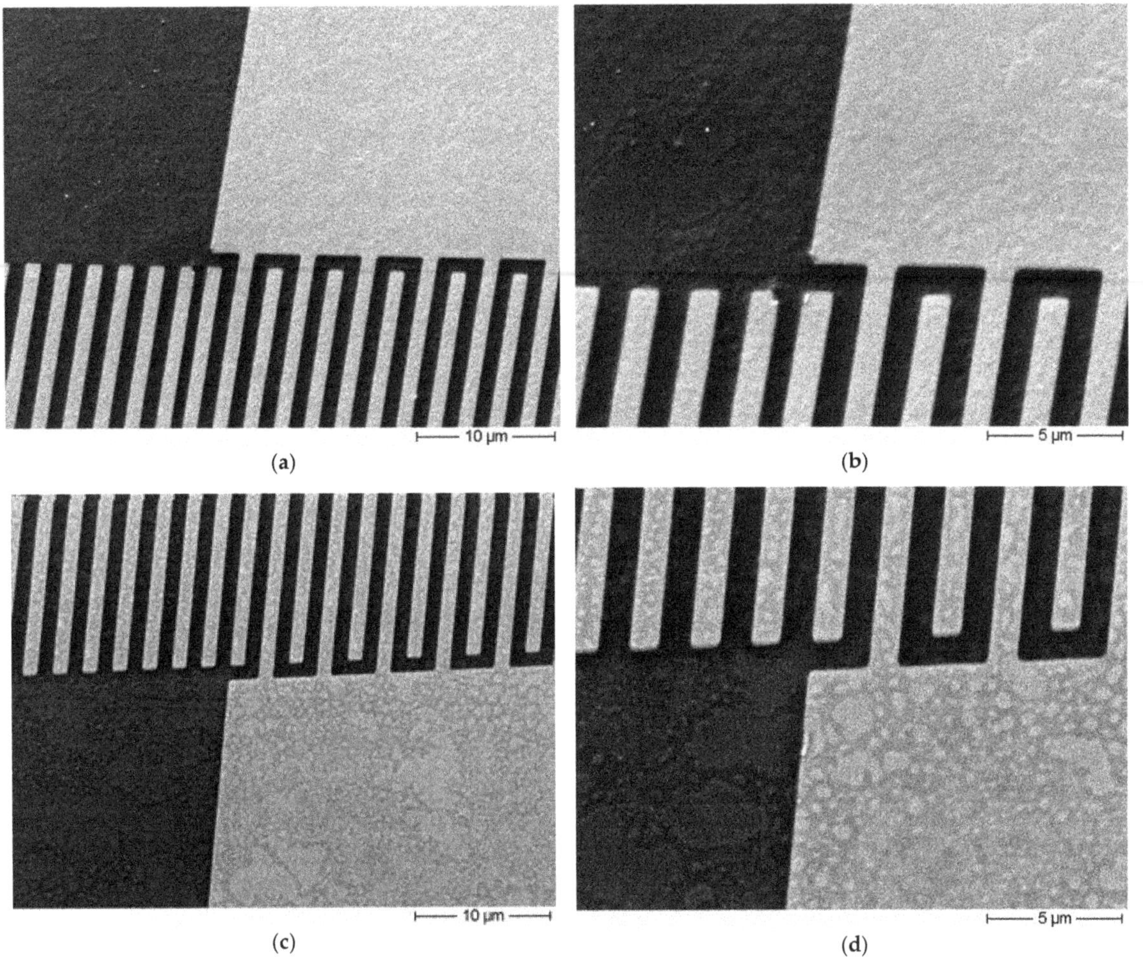


**Figure 12.** SEM images of the deposition of the PU-PIB composite as the sensing layer. (**a**) 2500×, before CAT; (**b**) 5000×, before CAT; (**c**) 2500×, after CAT; (**d**) 5000×, after CAT.

After the CAT, the SEM images (Figure 12c,d) indicate that some material was removed from the original deposition, being the removal apparently occurring equally from all the regions of the surface of the SAW element. Once again, a remaining deposition with a particular structure is observed after the CAT, suggesting that the PU-PIB composite has also resisted the CAT, as observed for the other PU–polymer composites investigated. But, again, the structure of the remaining coating layer after the CAT is distinct from those observed for the other composites, possibly due to the differences in the chemical constitution of the composites.

As seen by the results, all the PU–polymer composites exhibited the same overall behavior with respect to their formation and reaction to the CAT, and all of them seem to follow the proposal mechanism explained in Figures 1 and 2, independent of the sensing polymer combined with PU. The ultrasonic analyses are in perfect agreement with the microscopic observations for all the PU–polymer composites, revealing the role of each component in the formation of the composites. Despite the observation of the same overall behavior for all the PU–polymer composite analyzed so far, it is of crucial relevance to point out the achievements and advantages of this class of composite materials. As revealed by the microscopic analysis, each of the composites presented a characteristic microstructure and a pattern by its deposition over the complex surface of the piezoelectric sensor element, evidencing the individual influence of each polymer combined with PU. The microstructure of the coating layer will influence the selectivity against the analytes and the long-term stability of the coated sensor and, therefore, must be considered when these issues are addressed in future work. Although the composites presented different microstructures, they all resulted in highly uniform coating layers, showing great chemical resistance and adhesion improvements. And the most important feature of the PU–polymer composites is the achievement of all the improvements observed independently of the nature of the polymer used, keeping in mind that it is the polymer combined with PU that will be responsible for the quality of the sensor response in the detection process by the sensor system.

## 4. Conclusions

The ultrasonic results explained the behavior of the composites with respect to the CAT and could be correlated with the microscopy results, revealing that all the composites provided homogeneous coating layers before and after the CAT, also indicating the extension of the removal of the material from the surface by the CAT for each composite. The ultrasonic results agreed with the proposed mechanism of formation of the PU–polymer composites and are in perfect agreement with the microscopy results, reinforcing the function of each component of the composites in the properties of the obtained coating layer, where PU should act as a binder to form the polymeric composite with the sensing polymer used and, at same time, as an adhesion promoter, without interfering in the sensor response, which depends only on the characteristics of the sensing polymer forming the composite. Although seeming to follow the same mechanism of formation, the microstructures of each composite are distinct, reflecting the characteristics of the sensing polymer used in the composite. The microscopy results provided a relevant visualization of the structure and interaction of the composites with the different regions of the complex surface of the piezoelectric sensor element, allowing the evaluation of the macroscopic material distribution and the uniformity of the coating deposition. The microscopy methods could successfully detect the effects of the CAT in the coating layer of each PU–polymer composite, revealing the preserved structures that resisted the CAT, providing relevant information about the adhesion of the coating materials to the surface of the sensor that plays an important role in the long-term behavior of the SAW sensors in real applications. It was observed that the nature of the sensing polymer used strongly affects the properties of the composites in terms of their structure and their interaction with the surface of the sensor element. The results of the CAT showed that all the coating layers of the PU–polymer composites presented chemical and mechanical stability, indicating that this new class of coating materials based on the PU–polymer composites provided coating layers with improved structural properties, maintaining the chemical affinity and sensitivity of the sensing polymer used. In further work, the formation of the composite as a function of the PU–polymer stoichiometry will be investigated as well as the influence of composition of the polyurethane polymer used in the composites.

The results of this study considerably expand the choice of chemical environments for chemical sensitization since any type of polymer can virtually be used as a coating material for SAW technology. This makes this type of polymeric composite promising for other sensor technologies, like optical and biosensors, where the sensitization requires rich chemistry properties in terms of interactions, along with enhanced adhesion and other specific mechanical properties that can potentially be achieved by a tailor-made PU–polymeric composite, opening a vast and broad spectrum of possibilities in terms of chemical sensitization.

**Author Contributions:** Conceptualization, M.d.S.d.C.; Methodology, M.R. and A.V.; Software, A.V.; Validation, M.D. and U.G.; Formal analysis, M.d.S.d.C., M.D. and U.G.; Investigation, M.d.S.d.C.; Data curation, M.D. and U.G.; Writing—original draft, M.d.S.d.C.; Writing—review & editing, M.d.S.d.C. and M.R.; Supervision, M.R.; Project administration, M.R. All authors have read and agreed to the published version of the manuscript.

**Funding:** The publication was funded by the Open Access Publication Fund of the Karlsruhe Institute of Technology.

**Informed Consent Statement:** Not applicable.

**Data Availability Statement:** Data are contained within the article.

**Conflicts of Interest:** The authors declare no conflicts of interest.

## References

1. Yang, Y.; Dejous, C.; Hallil, H. Trends and Applications of Surface and Bulk Acoustic Wave Devices: A Review. *Micromachines* **2023**, *14*, 43. [CrossRef] [PubMed]
2. Mandal, D.; Banerjee, S. Surface Acoustic Wave (SAW) Sensors: Physics, Materials, and Applications. *Sensors* **2022**, *22*, 820. [CrossRef] [PubMed]
3. Länge, K. Bulk and Surface Acoustic Wave Sensor Arrays for Multi-Analyte Detection: A Review. *Sensors* **2019**, *19*, 5382. [CrossRef] [PubMed]
4. Yunas, J.; Mulyanti, B.; Hamidah, I.; Said, M.M.; Pawinanto, R.E.; Ali, W.A.F.W.; Subandi, A.; Hamzah, A.A.; Latif, R.; Majlis, B.Y. Polymer-Based MEMS Electromagnetic Actuator for Biomedical Application: A Review. *Polymers* **2020**, *12*, 1184. [CrossRef]
5. Misra, A.; Shahiwala, A. (Eds.) *Applications of Polymers in Drug Delivery*, 2nd ed.; Elsevier: Amsterdam, The Netherlands, 2020.
6. Nghiem, T.-L.; Coban, D.; Tjaberings, S.; Gröschel, A.H. Recent Advances in the Synthesis and Application of Polymer Compartments for Catalysis. *Polymers* **2020**, *12*, 2190. [CrossRef]
7. Sun, Q.; Dai, Z.; Meng, X.; Xiao, F.-S. Porous polymer catalysts with hierarchical structures. *Chem. Soc. Rev.* **2015**, *44*, 6018–6034. [CrossRef]
8. Cichosz, S.; Masek, A.; Zaborski, M. Polymer-based sensors: A review. *Polym. Test.* **2018**, *67*, 342–348. [CrossRef]
9. Di Pietrantonio, F.; Benetti, M.; Cannatà, D.; Verona, E.; Palla-Papavlu, A.; Dinca, V.; Dinescu, M.; Mattle, T.; Lippert, T. Volatile toxic compound detection by surface acoustic wave sensor array coated with chemoselective polymers deposited by laser induced forward transfer: Application to sarin. *Sens. Actuators B Chem.* **2012**, *174*, 158–167. [CrossRef]
10. Sayago, I.; Fernández, M.; Fontecha, J.; Horrillo, M.; Vera, C.; Obieta, I.; Bustero, I. New sensitive layers for surface acoustic wave gas sensors based on polymer and carbon nanotube composites. *Sens. Actuators B Chem.* **2012**, *175*, 67–72. [CrossRef]
11. Yadava, R.; Kshetrimayum, R.; Khaneja, M. Multifrequency characterization of viscoelastic polymers and vapor sensing based on SAW oscillators. *Ultrasonics* **2009**, *49*, 638–645. [CrossRef]
12. Zhang, D.; Tong, J.; Xia, B. Humidity-sensing properties of chemically reduced graphene oxide/polymer nanocomposite film sensor based on layer-by-layer nano self-assembly. *Sens. Actuators B Chem.* **2014**, *197*, 66–72. [CrossRef]
13. Tai, H.; Zhen, Y.; Liu, C.; Ye, Z.; Xie, G.; Du, X.; Jiang, Y. Facile development of high performance QCM humidity sensor based on protonated polyethylenimine-graphene oxide nanocomposite thin film. *Sens. Actuators B Chem.* **2016**, *230*, 501–509. [CrossRef]
14. Hung, T.-T.; Chung, M.-H.; Chiu, J.-J.; Yang, M.-W.; Tien, T.-N.; Shen, C.-Y. Poly(4-styrenesulfonic acid) doped polypyrrole/tungsten oxide/reduced graphene oxide nanocomposite films based surface acoustic wave sensors for NO sensing behavior. *Org. Electron.* **2020**, *88*, 106006. [CrossRef]

15. Wang, B.; Zhou, L.; Wang, X. Surface acoustic wave sensor for formaldehyde gas detection using the multi-source spray-deposited graphene/PMMA composite film. *Front. Mater.* **2023**, *9*, 1025903. [CrossRef]
16. Viespe, C.; Miu, D. Characteristics of Surface Acoustic Wave Sensors with Nanoparticles Embedded in Polymer Sensitive Layers for VOC Detection. *Sensors* **2018**, *18*, 2401. [CrossRef] [PubMed]
17. Gu, Y.; Zhao, J.; Johnson, J.A. A (Macro)Molecular-Level Understanding of Polymer Network Topology. *Trends Chem.* **2019**, *1*, 318–334. [CrossRef]
18. Tang, K.-T.; Li, C.-H.; Chiu, S.-W. An Electronic-Nose Sensor Node Based on a Polymer-Coated Surface Acoustic Wave Array for Wireless Sensor Network Applications. *Sensors* **2011**, *11*, 4609–4621. [CrossRef]
19. Su, Y.-F.; Han, G.; Kong, Z.; Nantung, T.; Lu, N. Embeddable Piezoelectric Sensors for Strength Gain Monitoring of Cementitious Materials: The Influence of Coating Materials. *Eng. Sci.* **2020**, *11*, 66–75. [CrossRef]
20. Lowdon, J.W.; Diliën, H.; Singla, P.; Peeters, M.; Cleij, T.J.; van Grinsven, B.; Eersels, K. MIPs for commercial application in low-cost sensors and assays—An overview of the current status quo. *Sens. Actuators B Chem.* **2020**, *325*, 128973. [CrossRef]
21. Navale, S.; Mane, A.; Chougule, M.; Sakhare, R.; Nalage, S.; Patil, V. Highly selective and sensitive room temperature NO2 gas sensor based on polypyrrole thin films. *Synth. Met.* **2014**, *189*, 94–99. [CrossRef]
22. Hu, J.; Qu, H.; Chang, Y.; Pang, W.; Zhang, Q.; Liu, J.; Duan, X. Miniaturized polymer coated film bulk acoustic wave resonator sensor array for quantitative gas chromatographic analysis. *Sens. Actuators B Chem.* **2018**, *274*, 419–426. [CrossRef]
23. Palla-Papavlu, A.; Voicu, S.I.; Dinescu, M. Sensitive Materials and Coating Technologies for Surface Acoustic Wave Sensors. *Chemosensors* **2021**, *9*, 105. [CrossRef]
24. Jasek, K.; Pasternak, M.; Grabka, M.; Neffe, S.; Zasada, D. Deposition of Polymer Sensor Films on SAW Surface by Electrospraying Technology. *Arch. Acoust.* **2017**, *42*, 507–513. [CrossRef]
25. Verma, P.; Yadava, R.D.S. Polymer selection for SAW sensor array based electronic noses by fuzzy c-means clustering of partition coefficients: Model studies on detection of freshness and spoilage of milk and fish. *Sens. Actuators B Chem.* **2015**, *209*, 751–769. [CrossRef]
26. Matatagui, D.; Martí, J.; Fernández, M.; Fontecha, J.; Gutiérrez, J.; Gràcia, I.; Cané, C.; Horrillo, M. Chemical warfare agents simulants detection with an optimized SAW sensor array. *Sens. Actuators B Chem.* **2011**, *154*, 199–205. [CrossRef]
27. Grate, J.W.; McGill, R.A. Dewetting Effects on Polymer-Coated Surface Acoustic Wave Vapor Sensors. *Anal. Chem.* **1995**, *67*, 4015–4019. [CrossRef]
28. Stewart, K.M.E.; Penlidis, A. Designing polymeric sensing materials: What are we doing wrong? *Polym. Adv. Technol.* **2016**, *28*, 319–344. [CrossRef]
29. Brassat, K.; Lindner, J.K.N. Nanoscale Block Copolymer Self-Assembly and Microscale Polymer Film Dewetting: Progress in Understanding the Role of Interfacial Energies in the Formation of Hierarchical Nanostructures. *Adv. Mater. Interfaces* **2020**, *7*, 1901565. [CrossRef]
30. Rapp, M.; Voigt, A.; Dirschka, M.; Carvalho, M.d.S.d. The Use of Polyurethane Composites with Sensing Polymers as New Coating Materials for Surface Acoustic Wave-Based Chemical Sensors—Part I: Analysis of the Coating Results, Sensing Responses and Adhesion of the Coating Layers of Polyurethane–Polybutylmethacrylate Composites. *Coatings* **2023**, *13*, 1911. [CrossRef]
31. Carvalho, M.d.S.d.; Rapp, M.; Voigt, A.; Dirschka, M. The Use of Polyurethane Composites with Sensing Polymers as New Coating Materials for Surface Acoustic Wave-Based Chemical Sensors—Part II: Polyurethane Composites with Polylaurylmetacrylate, Polyisobutene, and Poly(chlorotrifluoroethylene-co-vinylidene Fluoride): Coating Results, Relative Sensor Responses and Adhesion Analysis. *Coatings* **2024**, *14*, 778. [CrossRef]
32. Gao, P.F.; Lei, G.; Huang, C.Z. Dark-Field Microscopy: Recent Advances in Accurate Analysis and Emerging Applications. *Anal. Chem.* **2021**, *93*, 4707–4726. [CrossRef] [PubMed]
33. La Spina, R.; António, D.C.; Desmet, C.; Valsesia, A.; Bombera, R.; Norlén, H.; Lettieri, T.; Colpo, P. Dark Field Microscopy-Based Biosensors for the Detection of *E. coli* in Environmental Water Samples. *Sensors* **2019**, *19*, 4652. [CrossRef] [PubMed]
34. Carvalho, M.d.S.d.; Rapp, M.; Voigt, A.; Dirschka, M. The Use of Polyurethane Composites with Sensing Polymers as New Coating Materials for Surface Acoustic Wave-Based Chemical Sensors—Part III: Ultrasonic Analyses and Optical Microscopy Characterization of the Coating Results. *Coatings* **2024**, *14*, 846. [CrossRef]

*coatings*

MDPI

*Article*

# Electrodeposition of Metallic Tungsten Coating on 9Cr-ODS Steel Substrate from Binary Oxide Molten Salt

**Xiaoxu Dong †, Yusha Li †, Yajie You, Zeyu Gao and Yingchun Zhang ***

School of Materials Science and Engineering, University of Science and Technology Beijing, 30 Xueyuan Road, Haidian District, Beijing 100083, China; steven02162024@163.com (X.D.); liyusha1011@163.com (Y.L.); 15983037268@163.com (Y.Y.); 18622777816@163.com (Z.G.)

* Correspondence: zycustb@163.com

† These authors contributed equally to this work.

**Abstract:** Characteristics of electrodeposited tungsten coatings prepared at 1193 K and varying current density were investigated. The crystal structure and microstructure of tungsten coatings were characterized by scanning electron microscopy (SEM), X-ray diffraction (XRD), and photoelectron spectroscopy (XPS). The results indicated that pulsed current density significantly influence the tungsten nucleation and electro-crystallization phenomena. The average grain size of the coating becomes larger with increasing current density, which demonstrates that appropriate high cathodic current density can accelerate the growth of grains on the surface of the substrate. The micro-hardness of tungsten coatings increases with increasing thickness and then slightly decreases; the maximum micro-hardness is 589.55 HV, with the oxygen content remaining below 0.03 wt%.

**Keywords:** nuclear fusion; electrodeposition; molten salt; tungsten coating; 9Cr-ODS steel

## 1. Introduction

The technological realization of the fusion power source is critically dependent on the successful development of high-performance materials. These materials will have to be operated in extreme conditions, being subjected to complex thermal, mechanical, and chemical loads as well as strong irradiation. Tungsten is widely applied in the nuclear industries due to its high melting point (3695 K), low vapor pressure ($p = 1.3 \times 10^{-7}$ Pa), high thermal conductivity, low particle sputtering yield, and tritium retention [1–3]. However, tungsten is difficult to machine and weld; a possible solution for the utilization of tungsten at plasma facing surfaces is to coat the structural or heat sink material with a tungsten layer, which has been used as a heat sink coating for plasma oriented material (PFM). Among all the methods for realizing effective joints between tungsten armors and steel substrates, the deposition of tungsten coatings on substrates is efficient and economical [4–6]. It can be directly used for modular in situ replacement on the substrate.

Tungsten coatings can be obtained using multiple processes [7,8]. Techniques for obtaining high quality coatings, including vacuum plasma spray (VPS), physical vapor deposition (PVD), and chemical vapor deposition (CVD) have been investigated over recent decades [9,10]. But the disadvantages of the tungsten coatings prepared by normal spraying methods are obvious: weak bonding strength, high porosity, and inevitably introduced impurities. Fortunately, electrodeposition as a niche technology should have the potential to overcome these drawbacks [11]. Electrodeposition is one of the promising alternatives to fabricate functional tungsten coatings due to its simple, efficient process and because of its potential to cover complex extended surfaces at low cost.

In our prior research, we successfully achieved high-quality tungsten coatings with strong adhesion to copper alloy substrates through molten salt electrodeposition. However, a large body of literature shows that the combination of steel and tungsten is problematic [12,13], and the tungsten coating obtained by molten salt electrodeposition is expected to solve the problem of the problematic combination of steel and tungsten. Furthermore, based on our previous findings, coatings prepared by pulsed current electrodeposition are smoother compared to those obtained by direct current. Therefore, this study aims to develop tungsten coatings with enhanced adhesion to steel using pulsed current electrodeposition in molten salt.

## 2. Experiment and Parameters

### 2.1. Coating Deposition

All the chemicals were analytical reagent grade. $Na_2WO_4$–$WO_3$ binary oxide molten salt was dried in a furnace at 473 K and 773 K, respectively, for 12 h. The dried chemicals were thoroughly mixed into a eutectic composition ($Na_2WO_4$:$WO_3$ = 0.6:0.2, by mole ratio) in an alumina crucible, and then the crucible was set in an electric resistance furnace for electrodeposition. The heating rate of the mixed molten salt was 5 K/min. The working electrode was a 9Cr-ODS steel plate (C: 0.1 wt%, Cr: 9 wt%, W: 2 wt%, Mn: 0.5 wt%, Si: 0.15 wt%, V: 0.05 wt%, Ti: 0.05 wt%, Ta: 0.05 wt%, N: 0.05 wt%, O: 0.45 wt%, Y: 0.3 wt%, and Fe balance, 20 × 20 × 5 mm). The counter electrode was a tungsten plate (99.95%, 30 × 30 × 5 mm). The surfaces of electrodes were polished by mechanical and chemical methods to obtain high quality surfaces. The temperature was set as 1173 K. The average current density was set from 10 $mA/cm^2$ to 70 $mA/cm^2$. The frequency, duty cycle, and deposition duration were set as 1000 Hz, 30%, and 1 h, respectively.

### 2.2. Coating Characterization

The surface and cross-sectional morphology of the coated samples were observed by scanning electron microscopy (SEM, JSM6480LV) with elemental analysis. The phase and crystal orientation of W coatings were detected by X-ray diffraction (XRD, Rigaku Industrial Co., Ltd., Tokyo, Japan, D/MAX-BB). The oxidation state of tungsten in the coatings was characterized using X-ray photoelectron spectroscopy (XPS, Thermo Fisher Scientific Co., Ltd., Waltham, MA, USA, ESCALAB 250Xi). Micro-hardness of the coatings was measured by a EM1500L-6 micro hardness instrument with a loading force of 25 N and loading time of 10 s, and an average of five indentations was calculated. For the measurement of the oxygen content, the tungsten coating was separated from the 9Cr-ODS substrate by chemical dissolution in a dilute hydrochloric acid solution. The oxygen content was measured using a oxygen–nitrogen–hydrogen analyzer (Leco Co., Ltd., St Joseph, MI, USA, TCH600). The bonding strength was measured by tensile means, using a universal testing machine.

## 3. Results and Discussions

### 3.1. Microstructure and Crystal Structure of Thick Tungsten Coating

Figure 1 illustrates the macroscopic morphologies of tungsten coatings deposited at various pulse current densities. As depicted in the figure, silver-gray tungsten coatings can be successfully electrodeposited on the 9Cr-ODS steel substrate, exhibiting excellent coverage and a dense, smooth surface without visible cracks or pores. However, variations in macroscopic morphology are observed at different current densities. Specifically, when the current density exceeds 70 $mA \cdot cm^{-2}$, granular dendrites appear on the coating surface, which exhibit poor adhesion and tend to flake off during ultrasonic cleaning. In contrast, at current densities of 30 and 50 $mA \cdot cm^{-2}$, the coating surfaces are very smooth, with few protrusions.

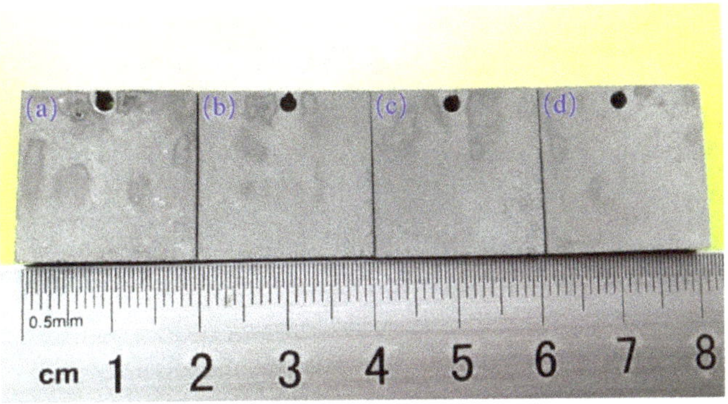


**Figure 1.** Macroscopic morphology of tungsten coatings prepared at different current densities: (**a**) 10 mA·cm$^{-2}$; (**b**) 30 mA·cm$^{-2}$; (**c**) 50 mA·cm$^{-2}$; (**d**) 70 mA·cm$^{-2}$.

The variations in the surface morphology of the tungsten coating can be further elucidated through corresponding SEM microtopography analysis. Figure 2 presents the SEM images of the tungsten coating surface under varying pulse current densities. It is evident that, across different current densities, the coating exhibits a uniform and dense structure without noticeable gaps or cracks, which aligns well with the macroscopic observations. However, despite this consistency, distinct differences in surface morphology are still apparent. The microscopic SEM images further reveal that the coating's surface smoothness is suboptimal, with grain size exhibiting significant growth at high current densities (70 mA·cm$^{-2}$).

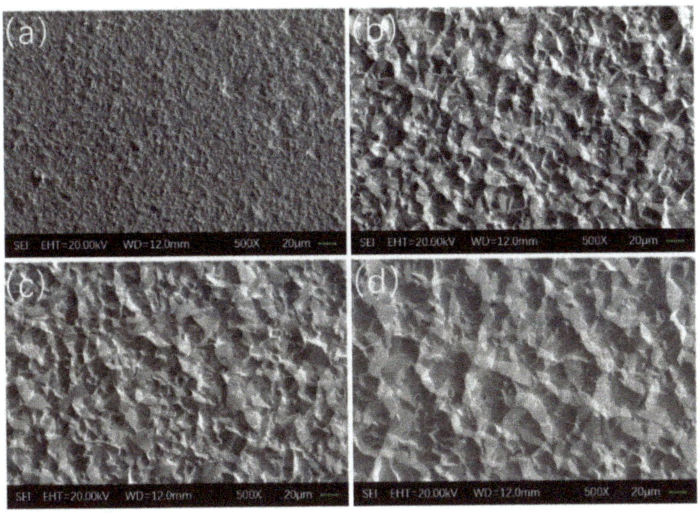


**Figure 2.** Surface SEM images of the tungsten coatings prepared at different current densities: (**a**) 10 mA·cm$^{-2}$; (**b**) 30 mA·cm$^{-2}$; (**c**) 50 mA·cm$^{-2}$; (**d**) 70 mA·cm$^{-2}$.

The Image J software was utilized to quantify the grain size in the SEM images, thereby obtaining the average tungsten grain size across various coatings. The influence of current density on grain size is illustrated in Figure 3. It can be seen that when the minimum current density is 10 mA·cm$^{-2}$, the grain size of the tungsten coating is only 1.90 μm. As the current density increases to 30 mA·cm$^{-2}$, the grain size grows to 5.36 μm, maintaining excellent distribution uniformity. Further increasing the current density to 50 mA·cm$^{-2}$ results in a grain size of 6.19 μm. At a current density of 70 mA·cm$^{-2}$, the grain size reaches its maximum of 8.03 μm. Although the grain size continues to increase, the uniformity of grain size distribution deteriorates. Grain size of the tungsten coating exhibits a linear increase as the current density rises. This finding aligns with the previously reported behavior of molten salt electrodeposition of tungsten metal coatings on CuCrZr substrates by Liu [14].

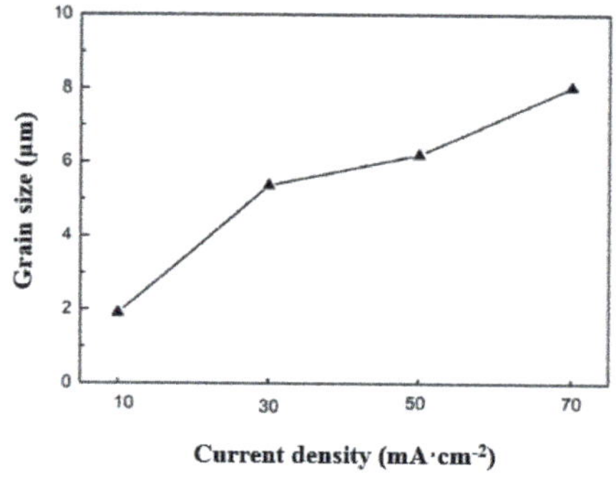


**Figure 3.** Effect of different pulse current densities on grain size of tungsten coatings.

The X-ray diffraction (XRD) patterns of the thick tungsten coatings at different electrodeposition current densities are presented in Figure 4. As shown, the tungsten coatings obtained at 10 mA·cm$^{-2}$, 30 mA·cm$^{-2}$, 50 mA·cm$^{-2}$, and 70 mA·cm$^{-2}$ exhibit four distinct diffraction peaks, located at 2θ values of 40°, 58°, 73°, and 87°, respectively. These angles correspond to the crystallographic planes (110), (200), (211), and (220) of tungsten according to the standard XRD cards for tungsten. At a 2θ angle of 58°, the most intense (200) peak was observed in all samples, indicating that the preferred orientation of the tungsten coatings was along the (200) plane. As the current density increased, the diffraction peaks corresponding to both the (110) and (211) planes progressively diminished.

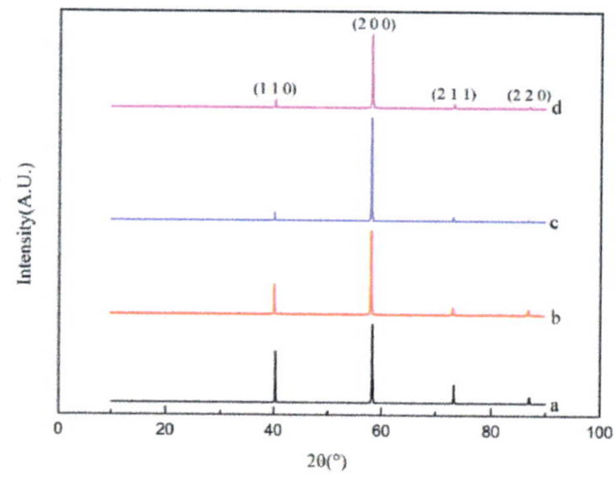


**Figure 4.** XRD patterns of tungsten coatings prepared at different current densities: (**a**) 10 mA·cm$^{-2}$; (**b**) 30 mA·cm$^{-2}$; (**c**) 50 mA·cm$^{-2}$; (**d**) 70 mA·cm$^{-2}$.

To investigate whether a zero-valence tungsten coating is deposited on the surface of 9Cr-ODS steel, Figure 5 presents the XPS spectrum of the tungsten coating prepared at a current density of 50 mA·cm$^{-2}$. The spectrum exhibits only two peaks, corresponding to W°4f7/2 at 31.5 eV and W°4f5/2 at 33.65 eV. This is similar to our previous results [11,14]. These results confirm that the tungsten in the coating is in its zero-valence state, with no other oxidation states of tungsten present. This indicates that the coating prepared by molten salt pulse electrodeposition is a single-phase metallic tungsten coating.

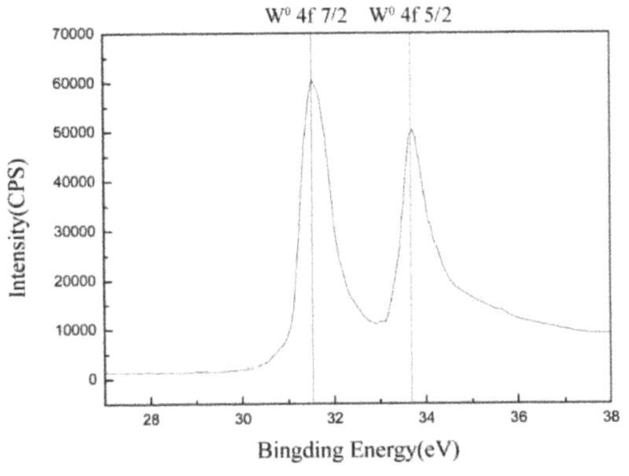


**Figure 5.** XPS spectrum of tungsten coating prepared at 50 mA·cm$^{-2}$ current density.

Figure 6 illustrates the SEM cross-sectional images of tungsten coatings deposited on a 9Cr-ODS steel substrate at varying pulse current densities. The coating thickness progressively increases with higher current densities. At a current density of 10 mA·cm$^{-2}$, the tungsten coating thickness is only 4.64 µm. This may be attributed to the low current density, which results in a reduced nucleation rate of tungsten crystal nuclei during deposition, leading to insufficient accumulation of nuclei on the substrate surface and consequently a thinner coating. As the current density increases to 30 mA·cm$^{-2}$, the coating thickness correspondingly increases to 33.54 µm. Further increasing the current density to 70 mA·cm$^{-2}$ results in a maximum coating thickness of 60.65 µm.

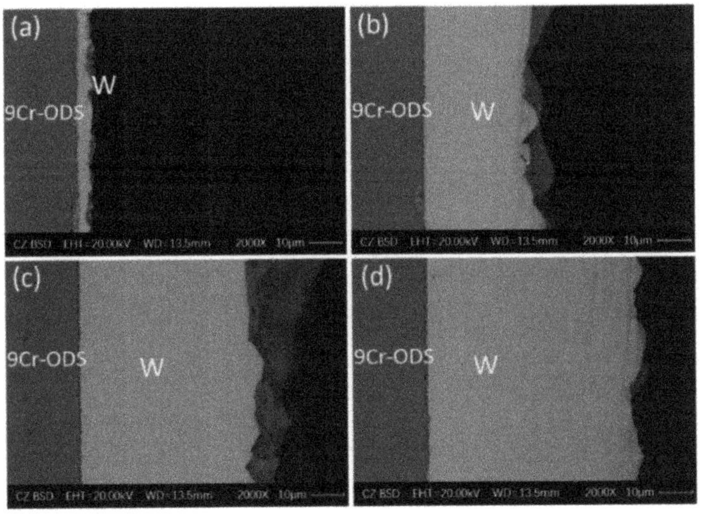


**Figure 6.** Cross-sectional SEM images of tungsten coatings prepared at different current densities: (**a**) 10 mA·cm$^{-2}$; (**b**) 30 mA·cm$^{-2}$; (**c**) 50 mA·cm$^{-2}$; (**d**) 70 mA·cm$^{-2}$.

The growth morphology of grains within the coating was characterized by fracturing the coating and analyzing its cross-sectional microstructure. Figure 7 shows the fracture SEM images of tungsten coatings prepared under different pulse current densities. It is evident that the tungsten coating fabricated via molten salt electrodeposition exhibits a columnar crystal structure similar to that produced by CVD, which contrasts with the layered structure obtained through PVD and the phased columnar crystal structure formed by VSS-W [15–17]. From Figure 7d, it can be clearly observed that a large number of equiaxed grains is present at the interface between tungsten and steel. These grains are characterized by their uniformity and fineness in this region, which is where the nucleation process of the tungsten coating predominantly occurs. As electrodeposition progresses, the diameter of the columnar crystals gradually increases, consistent with the trend of grain size evolution over electrodeposition time. Some equiaxed grains begin to grow rapidly, with their growth direction being parallel to the substrate normal but perpendicular to the substrate surface. This rapid growth leads to a preferential orientation of these grains over others, resulting in the cessation of growth for non-preferentially oriented grains, thereby establishing the dominance of the preferentially grown grains.

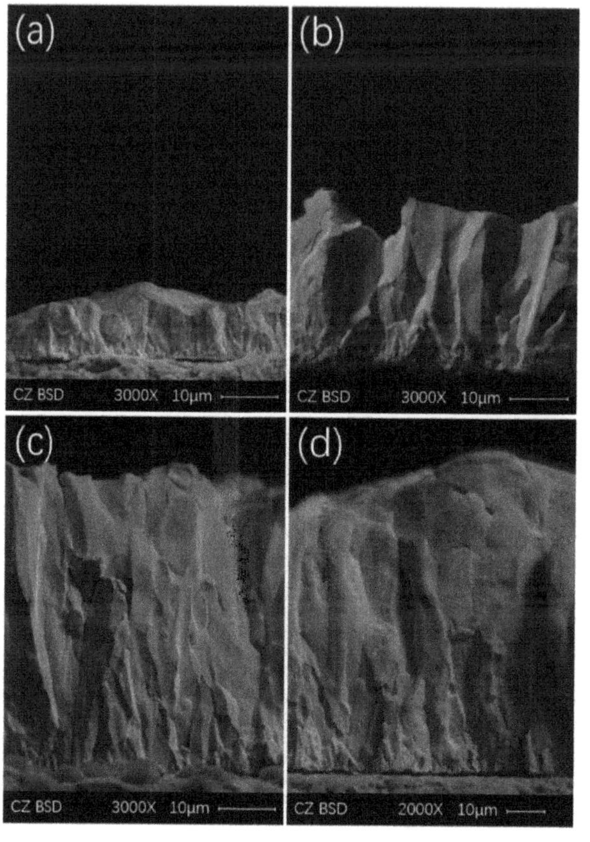


**Figure 7.** Fracture SEM images of coatings prepared at different current densities: (**a**) 10 mA·cm$^{-2}$; (**b**) 30 mA·cm$^{-2}$; (**c**) 50 mA·cm$^{-2}$; (**d**) 70 mA·cm$^{-2}$.

Figure 8 illustrates the electrodeposition waveform under pulsed power supply conditions. Here, T denotes the pulse period, defined as the time interval from the onset of one pulse to the initiation of the subsequent pulse. $t_{on}$ represents the pulse width, indicating the duration during which current is applied. $t_{off}$ signifies the pulse interval, representing the time gap during which the current is off and no electrodeposition reaction occurs. Figure 9 illustrates the schematic diagram of the electrochemical process for preparing tungsten coatings under a pulse power source. During the application of positive current ($t_{on}$), tungsten ions on the substrate surface are depleted. Conversely, when the positive current ceases ($t_{off}$), the tungsten ions released from the anode and those present in the solution diffuse to regions on the substrate surface with lower tungsten ion concentrations. This allows for timely reduction and deposition of tungsten ions onto the substrate surface at the onset of the subsequent cycle. Therefore, a flat and compact tungsten coating can be obtained on the steel surface by pulsed electrodeposition.

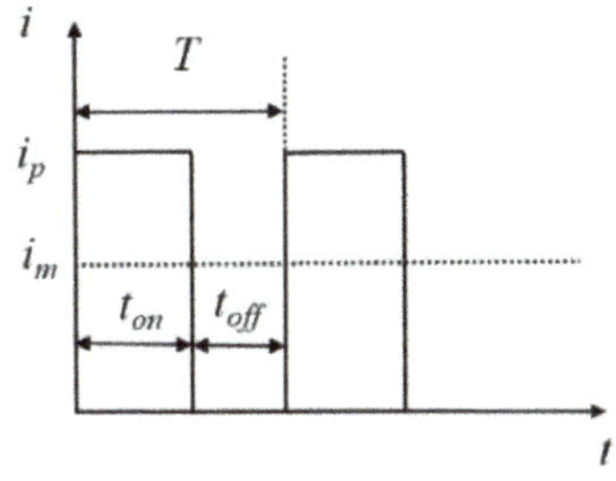


**Figure 8.** Schematic diagram of pulse electrodeposition waveform.

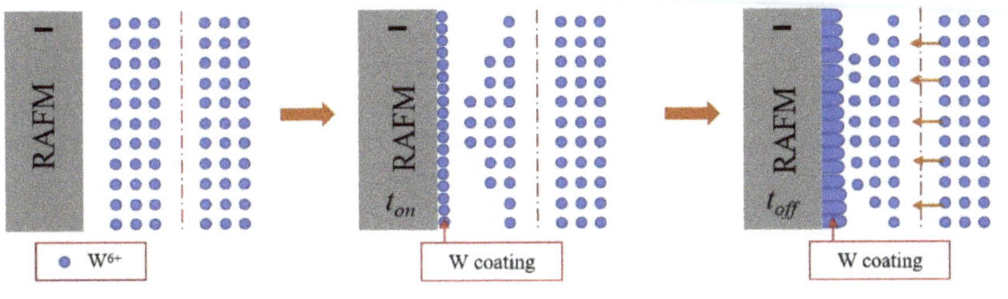


**Figure 9.** Schematic representation of the electrochemical process for the electrodeposition of a tungsten coating on the substrate surface during a single pulse period.

### 3.2. Performance Evaluation of Thick Tungsten Coatings

In this experiment, the tensile method was employed to assess the adhesive strength between the thick coating and the substrate. The bonding strengths are presented in Table 1. The bonding strength between the tungsten coating and the steel matrix, prepared via molten salt electrodeposition, exceeds 45 MPa. Notably, the bonding strength achieved at a current density of 50 mA/cm$^2$ reaches up to 51.22 MPa, which may be attributed to the denser microstructure of the coating [18].

**Table 1.** Bonding strength of tungsten coatings prepared by different current densities.

| Current Density | Bonding Strength (MPa) |
|---|---|
| 10 mA/cm$^2$ | 45.87 |
| 30 mA/cm$^2$ | 47.56 |
| 50 mA/cm$^2$ | 51.22 |
| 70 mA/cm$^2$ | 49.33 |

To characterize the mechanical properties of tungsten coatings on 9Cr-ODS steel substrates prepared by pulsed electrodeposition in molten salt, microhardness tests were conducted on polished samples. Figure 10 illustrates the trends in hardness and oxygen content of the tungsten coatings as a function of different pulse current densities. At a current density of 10 mA·cm$^{-2}$, the hardness is measured at 512.52 HV. As the current density increases to 30 mA·cm$^{-2}$, the hardness rises to 589.55 HV. Further increasing the current density to 50 mA·cm$^{-2}$ results in a decrease in hardness to 553.11 HV. When the current density reaches its maximum value of 70 mA·cm$^{-2}$, the hardness decreases to 538.75 HV.

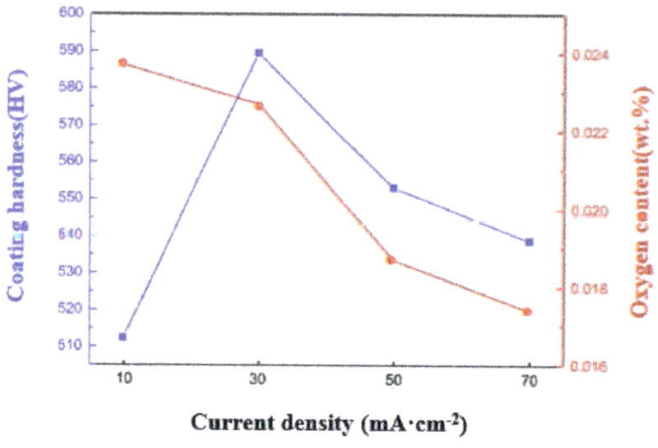


**Figure 10.** Effect of different pulse current densities on hardness and oxygen content of tungsten coatings.

The trend chart of hardness indicates that at low current densities, the substrate significantly influences the coating's hardness. As current density increases, the coating thickness also increases, thereby reducing the substrate's influence on hardness while increasing the effect of grain size. Generally, smaller metal grains result in more grain boundaries, which create greater obstacles to dislocation movement and increase resistance to deformation, leading to higher hardness and strength. Conversely, as grain size increases, the number of grain boundaries within the same region decreases, resulting in lower hardness values. Therefore, continued increases in current density lead to a decrease in hardness.

The oxygen content in the tungsten coatings on 9Cr-ODS steel substrate represents a critical performance parameter. At a current density of 10 $mA{\cdot}cm^{-2}$, the oxygen content is measured at 0.0237 wt%. When the current density is increased to 70 $mA{\cdot}cm^{-2}$, the oxygen content in the coating decreases slightly to 0.0174 wt%. The variation in current density has a minimal impact on the oxygen content, indicating that the tungsten coating prepared under these conditions exhibits consistently low oxygen levels.

## 4. Conclusions

In this paper, tungsten coatings were obtained from $Na_2WO_4$–$WO_3$ molten salt on the surface of 9Cr-ODS steel substrate at 1173 K, 30% duty cycle, and 1000 Hz frequency, with current density changing from 10 $mA/cm^2$ to 70 $mA/cm^2$. The microstructure and crystal structure of tungsten coatings were studied. The hardness and oxygen content of thick tungsten coatings were measured. The main conclusions are as follows:

(1) Dense and uniform single-phase tungsten coatings were successfully prepared on 9Cr-ODS steel substrate via pulsed electrodeposition from a $Na_2WO_4$–$WO_3$ molten salt system. The coating demonstrated superior adhesion to the steel substrate, with no observable signs of cracking or delamination, and the bonding strength exceeded 45 MPa.

(2) The microstructure of the tungsten coating is significantly influenced by variations in current density. As the current density increases, both the grain size and thickness of the tungsten coating exhibit a corresponding increase, reaching maximum values of 8.03 μm and 60.65 μm, respectively. Regardless of the specific current density, the tungsten coating consistently displays a columnar crystalline structure. As the current density increases, the hardness initially rises and then declines, while the oxygen content in the coating progressively decreases.

**Author Contributions:** Conceptualization, Y.Z.; methodology, Y.L. and X.D.; formal analysis, Y.L.; investigation, Y.L. and X.D.; resources, Y.Z.; data curation, Y.L., Z.G., Y.Y. and X.D.; writing—original draft preparation, X.D.; writing—review and editing, Y.L.; supervision, Y.L.; project administration, Y.Z.; funding acquisition, Y.Z. All authors have read and agreed to the published version of the manuscript.

**Funding:** This study was supported by the National MCF Energy R&D Program (No. 2019YFE03130002).

**Institutional Review Board Statement:** Not applicable.

**Informed Consent Statement:** Not applicable.

**Data Availability Statement:** The original contributions presented in this study are included in the article. Further inquiries can be directed to the corresponding author.

**Conflicts of Interest:** The authors declare no conflict of interest.

## References

1. Gonzalez-Arrabal, R.; Mendez-González, Y.; Perlado, J.M. Sputtering fabrication of isolated W nanocolumns: A possible al-ternative as plasma facing material for nuclear fusion reactors. *Nucl. Mater. Energy* **2024**, *40*, 101704. [CrossRef]
2. Piip, K.; Paris, P.; Hakola, A.; Bystrov, K.; De Temmerman, G.; Aints, M.; Jõgi, I.; Kozlova, J.; Laan, M.; Likonen, J.; et al. Influence of He/$d_2$ Plasma Fluxes on the Morphology and Crystallinity of Tungsten Coatings. *Phys. Scr.* **2014**, *89*, 044009. [CrossRef]
3. Ribis, J.; Bordas, E.; Trocellier, P.; Serruys, Y.; de Carlan, Y.; Legris, A. Comparison of the neutron and ion irradiation response of nano-oxides in oxide dispersion strengthened materials. *J. Mater. Res.* **2015**, *30*, 2210–2221. [CrossRef]
4. Xu, Y.; Tian, X.P.; Tan, H.Y.; Fu, H.Y.; Ni, J.J.; Luo, L.M.; Wu, Y.C. Effects of helium irradiation damage on deuterium plasma-driven permeation through tungsten coated RAFM steel. *Nucl. Fusion* **2025**, *65*, 026056. [CrossRef]
5. Lee, J.; Shin, Y.C. Crack-free tungsten coating on ods plate via a modified laser-based additive technique: Offset laser cladding. *Surf. Coat. Technol.* **2024**, *476*, 130285. [CrossRef]
6. Matthews, G.F.; Coad, P.; Greuner, H.; Hill, M.; Hirai, T.; Likonen, J.; Maier, H.; Mayer, M.; Neu, R.; Philipps, V.; et al. Development of divertor tungsten coatings for the JET ITER-like wall. *J. Nucl. Mater.* **2009**, *390–391*, 934–937. [CrossRef]
7. Wan, B.; Ding, S.; Qian, J.; Li, G.; Xiao, B.; Xu, G. Physics Design of CFETR: Determination of the Device Engineering Parameters. *IEEE Trans. Plasma Sci.* **2014**, *42*, 495–502. [CrossRef]
8. Matějíček, J.; Vilémová, M.; Mušálek, R.; Sachr, P.; Horník, J. The Influence of Interface Characteristics on the Adhesion/Cohesion of Plasma Sprayed Tungsten Coatings. *Coatings* **2013**, *3*, 108–125. [CrossRef]
9. Strekalina, D.; Dushik, V.; Shaporenkov, A.; Tatiana, L.; Bukatin, T. Prospects of tungsten CVD technology for electric vehicles. *AIP Conf. Proc.* **2024**, *3154*, 020018. [CrossRef]
10. Han, X.Y.; Liu, X.L.; Wang, D.Z.; Wu, Z.; Duan, B. Copper-based tungsten coating by CVD: Microstructure, thermal shock resistance and interfacial bond force. *Surf. Coat. Technol.* **2021**, *426*, 127778. [CrossRef]
11. Sun, N.B.; Zhang, Y.C.; Lang, S.T.; Jiang, F.; Wang, L.L. Tungsten coatings electro-deposited on CFC substrates from oxide molten salt. *J. Nucl. Mater.* **2014**, *455*, 450–453. [CrossRef]
12. Tokunaga, K.; Yoshida, N.; Kubota, Y.; Noda, N.; Imamura, Y.; Oku, T.; Kurumada, A.; Sogabe, T.; Kato, T.; Plöchl, L. High heat flux test of actively cooled tungsten-coated carbon divertor mock-ups. *Fusion Eng. Des.* **2000**, *49–50*, 371–376. [CrossRef]
13. Matějíček, J.; Koza, Y.; Weinzettl, V. Plasma sprayed tungsten-based coatings and their performance under fusion relevant conditions. *Fusion Eng. Des.* **2005**, *75–79*, 395–399. [CrossRef]
14. Liu, Y.H.; Zhang, Y.C.; Liu, Q.Z.; Li, X.L.; Jiang, F. The effects of electro-deposition current parameters on performance of tungsten coating. *Int. J. Refract. Met. Hard Mater.* **2012**, *35*, 241–245. [CrossRef]
15. Lian, Y.Y.; Liu, X.; Xu, Z.Y.; Song, J.P.; Yu, Y. Preparation and properties of CVD-W coated W/Cu FGM mock-ups. *Fusion Eng. Des.* **2013**, *88*, 1694–1698. [CrossRef]
16. Horník, J.; Tondl, D.; Sachr, P.; Anisimov, E.; Puchnin, M.; Chraska, T. The Effect of PVD Tungsten-Based Coatings on Improvement of Hardness and Wear Resistance. *Key Eng. Mater.* **2014**, *606*, 163–166. [CrossRef]
17. Deng, C.; Liu, M.; Yang, Z.; Deng, C.; Zhou, K.; Kuang, Z.; Zhang, J. Manufacture of thick VPS W coatings on relatively large CuZrCr substrate and its steady high heat load performance. *J. Nucl. Mater.* **2014**, *455*, 145–150. [CrossRef]
18. Zhang, Z.C.; Liu, Z.Q.; Cai, Y.K.; Wang, B.; Song, Q.H. Effect of Coating Thickness on the Bond Strength of Cold-Sprayed Ni-Based cBN Coating on 42CrMo Substrate. *J. Therm. Spray Technol.* **2023**, *32*, 1554–1569. [CrossRef]

*coatings*

MDPI

*Communication*

# Flexibility-Engineered $Nb_2O_5$ Carbon Nanofiber Film Anodes: Concentration-Dependent Optimization for Mechanically Robust and Stable Sodium Storage

**Xuhui Zhu †, Duiming Lin †, Siying Liu, Jufang Li, Yi Tang, Yancheng Pan, Qinglin Deng * and Lingmin Yao ***

School of Physics and Materials Science, Guangzhou University, Guangzhou 510006, China; zxh2674091377@foxmail.com (X.Z.); pyhcky@163.com (D.L.); lsy009hyhyy@163.com (S.L.); li158kqy@163.com (J.L.); tloml0210@163.com (Y.T.); pppccc6@163.com (Y.P.)

* Correspondence: qldeng@gzhu.edu.cn (Q.D.); lingminyao@gzhu.edu.cn (L.Y.)

† These authors contributed equally to this work.

**Abstract:** This study systematically investigates the correlation between the concentration of niobium oxide ($Nb_2O_5$) in carbon nanofibers (CNFs) and the microstructural characteristics and electrochemical performance of the composites. Through rational optimization of $Nb_2O_5$-to-CNFs mass ratios, we demonstrate that 50% $Nb_2O_5$ CNFs composite achieves an optimal balance between enhanced rate capability and structural stability. The composites transition from brittle to flexible with increasing $Nb_2O_5$ content, achieving unprecedented bending durability at 50% loading. This mechanical–electrochemical synergy positions $Nb_2O_5$ CNFs as viable candidates for flexible energy storage devices. Comprehensive characterization reveals that appropriate $Nb_2O_5$ incorporation significantly improves specific capacity (181 mA h $g^{-1}$ at 0.1 A $g^{-1}$) and rate performance compared to pristine CNFs. However, excessive $Nb_2O_5$ doping (>50%) induces detrimental hydrolysis reactions during synthesis, compromising both material processability and electrochemical reversibility. This work contributes to the development of flexible self-supporting frameworks that are superior to hard carbon for constructing high-performance flexible sodium-ion batteries.

**Keywords:** $Nb_2O_5$; flexible; sodium-ion batteries; anode

## 1. Introduction

Sodium-ion batteries have emerged as promising alternatives to lithium-ion systems due to sodium's natural abundance and cost-effectiveness [1–3]. Nevertheless, the development of suitable anode materials remains challenging, particularly in overcoming the sluggish Na+ diffusion kinetics and structural instability during cycling [4–9]. Niobium pentoxide ($Nb_2O_5$) has attracted growing attention owing to its intercalation pseudocapacitance behavior and suitable working potential (~1.0 V vs. $Na/Na^+$). However, its practical implementation is hindered by intrinsic low electronic conductivity and particle aggregation issues [10–16].

Recent advances propose combining $Nb_2O_5$ with conductive carbon matrices to address these limitations [17–25]. While our previous studies have systematically investigated the electrochemical properties of $Nb_2O_5$ [26–28], its role in modulating electrode flexibility remains overlooked. The inherent brittleness of carbon-based anodes severely limits their application in flexible electronics, necessitating material designs that harmonize ionic conductivity, electron transport, and mechanical resilience [29–31]. Particularly, electrospun carbon nanofibers (CNFs) offer three-dimensional conductive networks and mechanical flexibility [32–34]. Despite these merits, the critical relationship between $Nb_2O_5$ concentration and resultant electrochemical performance remains insufficiently explored.

Different from previous research work, this study bridges the gap between flexible engineering and electrochemical optimization between $Nb_2O_5$ and CNFs, providing actionable guidelines for designing flexible, binder-free electrodes. Herein, we present the systematic investigation of $Nb_2O_5$ CNFs composites with controlled mass ratios (0%–60%). Through integrated material characterization and electrochemical analysis, we establish that 50% $Nb_2O_5$ incorporation achieves an optimal sodium storage performance while maintaining structural integrity. This work elucidates the concentration-dependent behavior of metal oxide/carbon hybrids, providing essential guidelines for advanced flexible anode design.

## 2. Materials and Methods

### 2.1. Materials

Niobium pentaethoxide ($Nb(C_2H_5O)_5$, 99.95% trace metals basis), N,N-dimethylformamide (DMF, 99%), and polyacrylonitrile (PAN, average Mw = 150,000) were obtained commercially and utilized as received without additional purification.

### 2.2. Methods

The precursor solution was prepared by dissolving niobium pentaethoxide ($Nb(C_2H_5O)_5$) and 0.75 g polyacrylonitrile (PAN) in 10 mL dimethylformamide (DMF), followed by continuous magnetic stirring for 12 h at room temperature. Subsequently, the homogeneous solution was loaded into a plastic syringe equipped with a 24-gauge stainless-steel needle for electrospinning. The electrospinning process was conducted under optimized parameters: an applied voltage of 15 kV, a feeding rate of 1 mL $h^{-1}$, and a working distance of 15 cm between the needle tip and the rotating collector.

The as-spun nanofiber mats were thermally stabilized in a muffle furnace under ambient air via a two-step protocol: pre-oxidation at 250 °C for 3 h with a heating rate of 2 °C $min^{-1}$, followed by carbonization in a tubular furnace under an argon flow. The carbonization process involved ramping to 700 °C at 2 °C $min^{-1}$ and maintaining this temperature for 5 h to yield $Nb_2O_5$ embedded carbon nanofibers ($Nb_2O_5$ CNFs).

To systematically investigate the concentration-dependent effects, six samples were synthesized by varying the $Nb(C_2H_5O)_5$ mass added to the precursor solution: named as pristine CNFs, 10% $Nb_2O_5$ CNFs, 20% $Nb_2O_5$ CNFs, 30% $Nb_2O_5$ CNFs, 40% $Nb_2O_5$ CNFs, 50% $Nb_2O_5$ CNFs, and 60% $Nb_2O_5$ CNFs, respectively. The mass percentages were calculated based on the $Nb_2O_5$-to-PAN weight ratio in the precursor.

### 2.3. Characterizations

The crystal structures of all samples were examined using X-ray diffractometry (XRD, DX-2700BH, Haoyuan, Dandong, China). The microscopic morphology of the anode were investigated through field emission scanning electron microscopy (SEM, GeminiSEM 300, ZEISS, Oberkochen, Germany).

### 2.4. Electrochemical Test0073

All other samples served as working electrodes directly, without the use of current collectors, binders, or conductive additives. Self-supporting samples were directly utilized as CR-2032 half-cell anodes without employing current collectors, binders, or conductive additives. The sodium metal disk functioned as both the working and reference electrodes, with a glass fiber membrane used acting as the separator. The electrolyte comprised 1 mol $L^{-1}$ sodium perchlorate dissolved in 100% propylene carbonate (PC) and 5% fluorinated ethylene carbonate (FEC). Electrochemical charge–discharge tests were performed using the BTS-4000 battery testing system (Neware, Shenzhen, China).

## 3. Results and Discussion

The fabrication process of $Nb_2O_5$ CNFs with varying concentrations is schematically illustrated in Figure 1. Utilizing polyacrylonitrile (PAN) and niobium ethoxide as precursors, the composite nanofibers were synthesized through electrospinning followed by thermal stabilization and carbonization. Mechanical flexibility tests, as demonstrated in Figure 2, revealed a striking concentration-dependent behavior. Pristine CNFs exhibited extreme brittleness, fracturing into fragments upon manual folding. In contrast, the 10% $Nb_2O_5$ CNFs showed partial flexibility improvement, sustaining single-fold deformation despite developing permanent crease cracks. Notably, the 30% $Nb_2O_5$ CNFs achieved reversible shape recovery after three-fold creasing, while the 50% composite displayed exceptional self-recovery capability, fully restoring its original morphology even after aggressive crumpling.

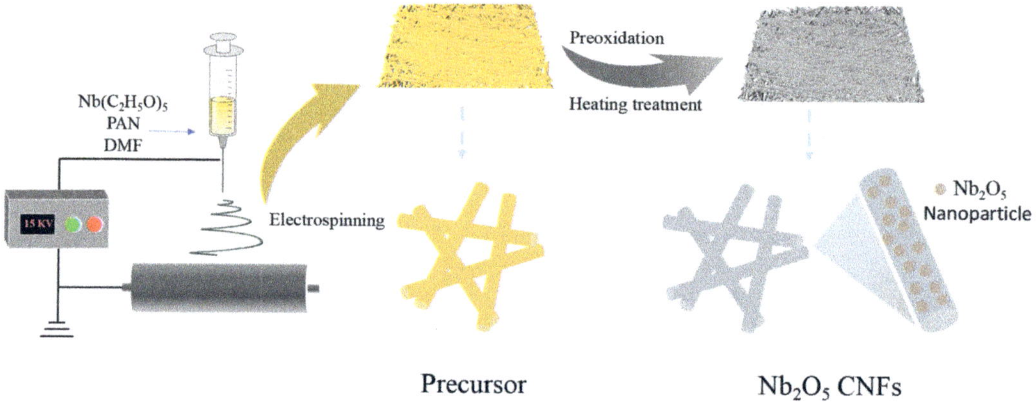


**Figure 1.** Schematic illustration of the fabrication of the $Nb_2O_5$ CNFs electrodes.

The crystalline structures of $Nb_2O_5$ CNFs composites were first analyzed by XRD (Figure 3a). All diffraction patterns corresponded to orthorhombic $Nb_2O_5$ (JCPDS 30-0873), confirming successful phase formation. Notably, the diffraction peak intensities proportionally increased with niobium ethoxide content, verifying controllable $Nb_2O_5$ crystallization within the carbon matrix. The reasons may be attributed to the uniform distribution of $Nb_2O_5$ nanoparticles acting as the buffer substances to efficiently weaken and disperse the bending stresses, and simultaneously absorb the stress energy during the folding process. In addition, Figure 3b shows the crystal structure of $Nb_2O_5$. As we can see, the Nb-ions are surrounded by six O-ions, the (001) planes of $Nb_2O_5$ have a large interplanar spacing of 0.39 nm, which could provide stable and effective tunnels for Na-ion storage and transport.

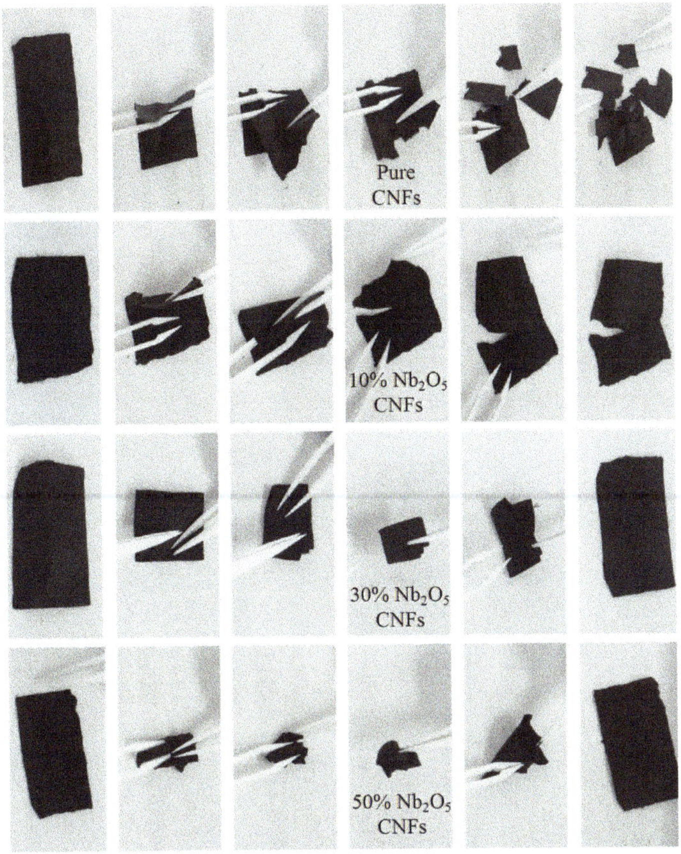


**Figure 2.** Photographs of pure CNFs and $Nb_2O_5$ CNFs (10%, 30%, 50%) which were folded many times.

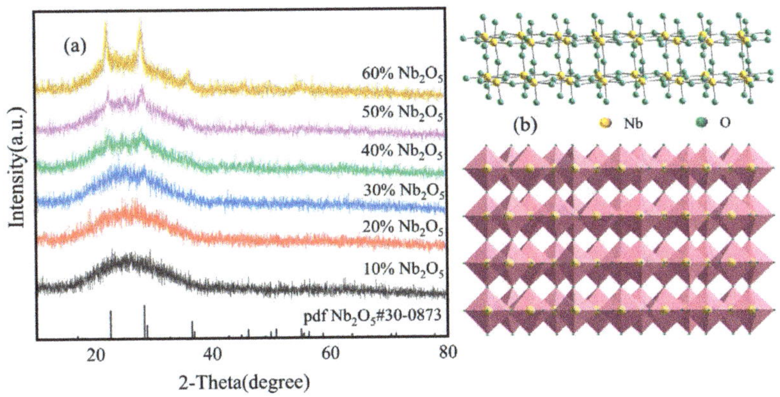


**Figure 3.** (**a**) XRD patterns of $Nb_2O_5$ CNFs with different $Nb_2O_5$ mass ratios. (**b**) The crystal structures of $Nb_2O_5$.

To probe the microstructural features, SEM observations were conducted, as shown in Figure 4. As we can see, all samples exhibited continuous network structures formed by interconnected fibers with diameters of approximately 200 nm. Remarkably, no discernible morphological differences were observed among the various compositions with the different $Nb_2O_5$ additive amount, suggesting that the electrospinning process effectively accommodated different precursor ratios.

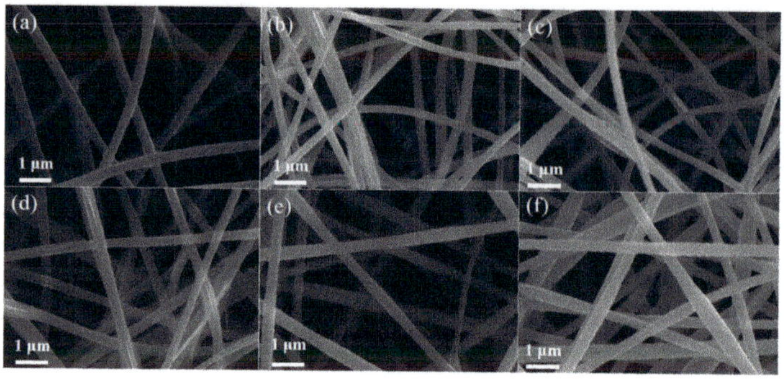


**Figure 4.** SEM images of (**a**) pure CNFs; (**b**) 10% $Nb_2O_5$ CNFs; (**c**) 30% $Nb_2O_5$ CNFs; (**d**) 40% $Nb_2O_5$ CNFs; (**e**) 50% $Nb_2O_5$ CNFs; and (**f**) 60% $Nb_2O_5$ CNFs.

Electrochemical evaluation demonstrated significant performance dependence on $Nb_2O_5$ content (Figure 5a). At 0.2 A $g^{-1}$, the 50% $Nb_2O_5$ CNFs delivered 153 mA h $g^{-1}$, outperforming pristine CNFs (124 mA h $g^{-1}$) by 23.4%. Rate capability tests demonstrated superior kinetics, with the 50% $Nb_2O_5$-CNFs retaining 50 mAh $g^{-1}$ at 10 A $g^{-1}$, surpassing the capacities of 40% (44 mA h $g^{-1}$), 60% (27 mA h $g^{-1}$), and pristine CNFs (38 mA h $g^{-1}$) counterparts. This enhancement originates from the pseudocapacitive contribution of $Nb_2O_5$ and improved ion accessibility within the hybrid structure.

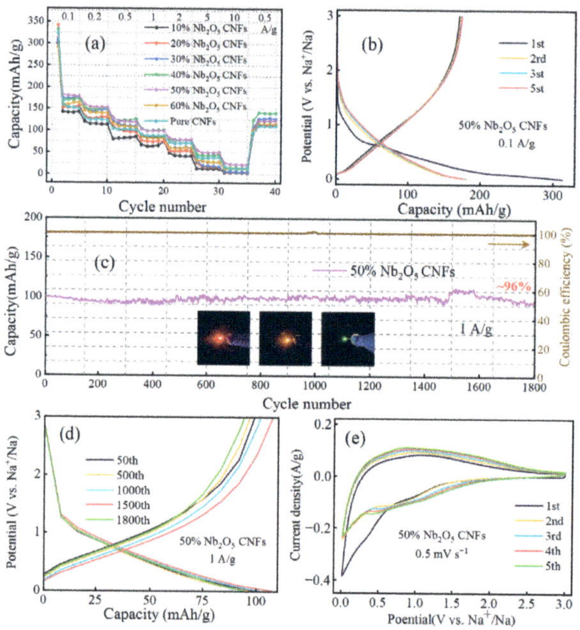


**Figure 5.** (**a**) The rate performance of $Nb_2O_5$ CNFs with different mass ratios at different current densities. (**b**) The galvanostatic charge–discharge (GCD) profile at the first four cycles of 50% $Nb_2O_5$ CNFs at 0.1 A $g^{-1}$. (**c**) Cycling stability of 50% $Nb_2O_5$ CNFs at 1 A $g^{-1}$ and the optical photos of the half-cell made up of 50% $Nb_2O_5$ CNFs with LED light. (**d**) GCD profile of 50th, 500th, 1000th, 1500th, and 1800th cycle for the 50% $Nb_2O_5$ CNFs CNFs. (**e**) The CV curves for the first five cycles of 50% $Nb_2O_5$ CNFs electrode at a scan rate of 0.5 mV $s^{-1}$.

Figure 5b presents the galvanostatic charge–discharge (GCD) profiles of the 50% $Nb_2O_5$ CNFs anode during the first five cycles at 0.1 A $g^{-1}$. For the rate performance, all anodes demonstrated a higher specific capacity and lower coulombic efficiency during the first charge–discharge cycle compared to the subsequent cycles. This can be attributed to the formation of the solid electrolyte interphase (SEI) layer and the irreversible insertion of sodium ions.

To comprehensively assess the $Na^+$ storage capabilities arising from intercomponent collaboration, extended cycling stability tests were conducted under elevated current conditions (1 A $g^{-1}$), as depicted in Figure 5c. The composite demonstrated exceptional stability across 1800 cycles, maintaining 96% capacity preservation rate with Coulombic efficiency values remaining within 99.8 ± 0.3%. Quantitative analysis of galvanostatic profiles (Figure 5d) revealed remarkable consistency in voltage plateaus and polarization evolution throughout cycling, corroborating outstanding interfacial stability and reaction reversibility.

Figure 5e displays representative cyclic voltammetry profiles for the initial five cycles of fabricated specimens recorded at a scanning rate of 0.5 mV $s^{-1}$ within an operational voltage range of 0.01–3.0 V (vs. $Na^+/Na$). Regarding the 50% $Nb_2O_5$ CNFs composite anode, apart from the irreversible processes observed during the first cycle, the subsequent cycles (second to fifth) display pronounced overlapping characteristics, demonstrating the excellent electrochemical reversibility and cycling durability of sodium ion storage mechanisms.

The optimized 50% $Nb_2O_5$ CNFs achieve a critical balance: $Nb_2O_5$ nanoparticles can achieve pseudocapacitive Na-ion storage (50 mA h $g^{-1}$ at a very large current density of 10 A $g^{-1}$), while the carbon matrix can provide sample conductivity, accompanied by excellent flexibility. The reasons may be attributed to the uniform distribution of $Nb_2O_5$ nanoparticles, which act as buffering agents to effectively mitigate and distribute the bending stresses, and simultaneously absorb the stress energy during the folding process. For the mechanism of $Nb_2O_5$ pseudocapacitive sodium storage, galvanostatic intermittent titration technique (GITT), and first-principles calculations based on density functional theory (DFT), please refer to our previous studies [26–28].

## 4. Conclusions

This work establishes a critical composition–property–performance relationship in $Nb_2O_5$ carbon nanofibers (CNFs) composites for advanced sodium-ion battery anodes. The 50% $Nb_2O_5$ CNFs composite emerges as the optimal formulation, delivering a specific capacity of 153 mA h $g^{-1}$ at 0.2 A $g^{-1}$ alongside exceptional bending durability. This performance stems from the synergistic interplay between $Nb_2O_5$'s pseudocapacitive $Na^+$ storage and the CNFs conductive 3D network, which collectively enhance ion kinetics and mechanical resilience. Furthermore, the self-supporting electrode architecture eliminates conventional binders and current collectors, achieving 96% capacity retention over 1800 cycles at 1 A $g^{-1}$, thus demonstrating the practical viability of flexible energy storage devices. We believe that 50% $Nb_2O_5$ CNFs prepared in this work can replace pure CNFs as a superior self-supporting framework, given its better flexibility and sodium storage performances.

**Author Contributions:** X.Z. and D.L. contributed equally to this work. Investigation, methodology, data curation, X.Z. and D.L.; visualization, S.L., J.L., Y.T. and Y.P.; writing—original draft preparation, X.Z. and D.L.; writing—review and editing, Q.D. and L.Y.; conceptualization, Q.D.; funding acquisition, Q.D. All authors have read and agreed to the published version of the manuscript.

**Funding:** This research was funded by the National Natural Science Foundation of China (grant no. 22109032), the Guangzhou Basic Research Program, City and University (Institute) Joint Funding Project (grant no. SL2022A03J01003), the Guangzhou University Provincial College Student Innovation Training Program (grant no. S202311078139), the Key Discipline of Materials Science and Engineering, Bureau of Education of Guangzhou (grant no. 202255464).

**Institutional Review Board Statement:** Not applicable.

**Informed Consent Statement:** Not applicable.

**Data Availability Statement:** Data are contained within the article.

**Conflicts of Interest:** The authors declare no conflicts of interest.

## References

1. Fang, H.; Gao, S.; Ren, M.; Huang, Y.; Cheng, F.; Chen, J.; Li, F. Dual-Function Presodiation with Sodium Diphenyl Ketone towards Ultra-Stable Hard Carbon Anodes for Sodium-Ion Batteries. *Angew. Chem. Int. Ed.* **2023**, *62*, e202214717. [CrossRef]
2. Sun, N.; Qiu, J.; Xu, B. Understanding of Sodium Storage Mechanism in Hard Carbons: Ongoing Development under Debate. *Adv. Energy Mater.* **2022**, *12*, 2200715. [CrossRef]
3. El Moctar, I.; Ni, Q.; Bai, Y.; Wu, F.; Wu, C. Hard Carbon Anode Materials for Sodium-Ion Batteries. *Funct. Mater. Lett.* **2018**, *11*, 1830003. [CrossRef]
4. Kim, H.; Hyun, J.C.; Kim, D.-H.; Kwak, J.H.; Lee, J.B.; Moon, J.H.; Choi, J.; Lim, H.-D.; Yang, S.J.; Jin, H.M.; et al. Revisiting Lithium- and Sodium-Ion Storage in Hard Carbon Anodes. *Adv. Mater.* **2023**, *35*, 2209128. [CrossRef]
5. Chu, Y.; Zhang, J.; Zhang, Y.; Li, Q.; Jia, Y.; Dong, X.; Xiao, J.; Tao, Y.; Yang, Q.-H. Reconfiguring Hard Carbons with Emerging Sodium-Ion Batteries: A Perspective. *Adv. Mater.* **2023**, *35*, 2212186. [CrossRef]
6. Zhao, D.; Zhang, N.; Zhao, X.; Liu, N.; Qin, B.; Qiu, M.; Wang, L. A Novel PbSe@CNTs Anode Material Based on Dual Conversion-Alloying Mechanism for Sodium-Ion Batteries. *Sci. China Mater.* **2023**, *66*, 61–68. [CrossRef]
7. Deng, Q.; Yao, L. Advanced $Nb_2O_5$ Anode towards Fast Pseudocapacitive Sodium Storage. *Coatings* **2022**, *12*, 1873. [CrossRef]
8. Zhang, X.; Bi, R.; Wang, J.; Zheng, M.; Wang, J.; Yu, R.; Wang, D. Delicate Co-Control of Shell Structure and Sulfur Vacancies in Interlayer-Expanded Tungsten Disulfide Hollow Sphere for Fast and Stable Sodium Storage. *Adv. Mater.* **2023**, *35*, 2209354. [CrossRef]
9. Suzuki, T.; Cheng, J.; Qiao, L.; Xing, Y.; Zhang, M.F.; Nishijima, H.; Yano, T.; Pan, W. Preparation of $SnO_2$ Nanotubes via a Template-Free Electrospinning Process. *RSC Adv.* **2020**, *10*, 22113–22119. [CrossRef]
10. Wu, X.; Lan, X.; Hu, R.; Yao, Y.; Yu, Y.; Zhu, M. Tin-Based Anode Materials for Stable Sodium Storage: Progress and Perspective. *Adv. Mater.* **2022**, *34*, 2106895. [CrossRef]
11. Xia, J.; Jiang, K.; Xie, J.; Guo, S.; Liu, L.; Zhang, Y.; Nie, S.; Yuan, Y.; Yan, H.; Wang, X. Tin Disulfide Embedded in N-, S-Doped Carbon Nanofibers as Anode Material for Sodium-Ion Batteries. *Chem. Eng. J.* **2019**, *359*, 1244–1251. [CrossRef]
12. Zheng, Y.; He, L.; Kong, X.; Song, Y.; Zhao, Y. Three-Dimensional Porous N-Doped Graphite Carbon with Embedded $CoS_2$ Nanoparticles as Advanced Anode for Sodium-Ion Batteries. *Appl. Surf. Sci.* **2022**, *603*, 154481. [CrossRef]
13. Lu, X.; Wang, C.; Favier, F.; Pinna, N. Electrospun Nanomaterials for Supercapacitor Electrodes: Designed Architectures and Electrochemical Performance. *Adv. Energy Mater.* **2017**, *7*, 1601301. [CrossRef]
14. Zhu, G.; Xia, G.; Pan, H.; Yu, X. Size-Controllable Nickel Sulfide Nanoparticles Embedded in Carbon Nanofibers as High-Rate Conversion Cathodes for Hybrid Mg-Based Battery. *Adv. Sci.* **2022**, *9*, 2106107. [CrossRef]
15. Tao, B.; Yang, W.; Zhou, M.; Qiu, L.; Lu, S.; Wang, X.; Zhao, Q.; Xie, Q.; Ruan, Y. Designing a Carbon Nanofiber-Encapsulated Iron Carbide Anode and Nickel-Cobalt Sulfide-Decorated Carbon Nanofiber Cathode for High-Performance Supercapacitors. *J. Colloid Interface Sci.* **2022**, *621*, 139–148. [CrossRef]
16. Ju, J.; Zhang, L.; Shi, H.; Li, Z.; Kang, W.; Cheng, B. Three-Dimensional Porous Carbon Nanofiber Loading $MoS_2$ Nanoflake-Flowerballs as a High-Performance Anode Material for Li-Ion Capacitor. *Appl. Surf. Sci.* **2019**, *484*, 392–402. [CrossRef]
17. Xiong, Y.; Qian, J.; Cao, Y.; Ai, X.; Yang, H. Electrospun $TiO_2$/C Nanofibers As a High-Capacity and Cycle-Stable Anode for Sodium-Ion Batteries. *ACS Appl. Mater. Interfaces* **2016**, *8*, 16684–16689. [CrossRef]
18. Shan, C.; Feng, X.; Yang, J.; Yang, X.; Guan, H.-Y.; Argueta, M.; Wu, X.-L.; Liu, D.-S.; Austin, D.J.; Nie, P.; et al. Hierarchical Porous Carbon Pellicles: Electrospinning Synthesis and Applications as Anodes for Sodium-Ion Batteries with an Outstanding Performance. *Carbon* **2020**, *157*, 308–315. [CrossRef]
19. Wang, N.; Gao, Y.; Wang, Y.; Liu, K.; Lai, W.; Hu, Y.; Zhao, Y.; Chou, S.; Jiang, L. Nanoengineering to Achieve High Sodium Storage: A Case Study of Carbon Coated Hierarchical Nanoporous $TiO_2$ Microfibers. *Adv. Sci.* **2016**, *3*, 1600013. [CrossRef]
20. Bai, H.; Zhu, X.; Ao, H.; He, G.; Xiao, H.; Chen, Y. Advances in Sodium-Ion Batteries at Low-Temperature: Challenges and Strategies. *J. Energy Chem.* **2024**, *90*, 518–539. [CrossRef]

21. Lígia da Silva Lima, L.; Wu, J.; Cadena, E.; Groombridge, A.S.; Dewulf, J. Towards Environmentally Sustainable Battery Anode Materials: Life Cycle Assessment of Mixed Niobium Oxide (XNO$^{TM}$) and Lithium-titanium-Oxide (LTO). *Sustain. Mater. Technol.* **2023**, *37*, e00654. [CrossRef]
22. Mu, Q.; Liu, R.; Kimura, H.; Li, J.; Jiang, H.; Zhang, X.; Yu, Z.; Sun, X.; Algadi, H.; Guo, Z.; et al. Supramolecular Self-Assembly Synthesis of Hemoglobin-like Amorphous CoP@N, P-Doped Carbon Composites Enable Ultralong Stable Cycling under High-Current Density for Lithium-Ion Battery Anodes. *Adv. Compos. Hybrid Mater.* **2023**, *6*, 23. [CrossRef]
23. Etesami, M.; Nguyen, M.T.; Yonezawa, T.; Tuantranont, A.; Somwangthanaroj, A.; Kheawhom, S. 3D Carbon Nanotubes-Graphene Hybrids for Energy Conversion and Storage Applications. *Chem. Eng. J.* **2022**, *446*, 137190. [CrossRef]
24. Xie, F.; Xu, J.; Liao, Q.; Zhang, Q.; Liu, B.; Shao, L.; Cai, J.; Shi, X.; Sun, Z.; Wong, C.-P. Progress in Niobium-Based Oxides as Anode for Fast-Charging Li-Ion Batteries. *Energy Rev.* **2023**, *2*, 100027. [CrossRef]
25. Zhao, L.; Zhang, T.; Li, W.; Li, T.; Zhang, L.; Zhang, X.; Wang, Z. Engineering of Sodium-Ion Batteries: Opportunities and Challenges. *Engineering* **2023**, *24*, 172–183. [CrossRef]
26. Zhu, X.; Lu, L.; Deng, Q.; Mai, G.; Mei, Z.; Ji, H.; Yao, L. Flexible $Nb_2O_5/MoS_2$/Carbon Nanofiber Paper Anodes for Sodium-Ion Batteries. *ACS Appl. Nano Mater.* **2025**, *8*, 200–209. [CrossRef]
27. Deng, Q.; Chen, F.; Liu, S.; Bayaguud, A.; Feng, Y.; Zhang, Z.; Fu, Y.; Yu, Y.; Zhu, C. Advantageous Functional Integration of Adsorption-Intercalation-Conversion Hybrid Mechanisms in 3D Flexible $Nb_2O_5$ @Hard Carbon@$MoS_2$ @Soft Carbon Fiber Paper Anodes for Ultrafast and Super-Stable Sodium Storage. *Adv. Funct. Mater.* **2020**, *30*, 1908665. [CrossRef]
28. Deng, Q.; Fu, Y.; Zhu, C.; Yu, Y. Niobium-Based Oxides Toward Advanced Electrochemical Energy Storage: Recent Advances and Challenges. *Small* **2019**, *15*, 1804884. [CrossRef]
29. Jiang, N.; Yu, J.; Wu, Z.; Zhao, J.; Zeng, Y.; Li, H.; Meng, M.; He, Y.; Jiao, P.; Pan, H.; et al. Surface Gradient Desodiation Chemistry in Layered Oxide Cathode Materials. *Angew. Chem. Int. Ed.* **2024**, *63*, e202410080. [CrossRef]
30. Nakayama, H.; Nose, M.; Nakanishi, S.; Iba, H. Electrochemical Reactions of Niobium Based Oxide Materials as Anodes for Sodium Ion Batteries. *Meet. Abstr.* **2014**, *MA2014-02*, 98. [CrossRef]
31. Li, Y.; Wu, F.; Li, Y.; Feng, X.; Zheng, L.; Liu, M.; Li, S.; Qian, J.; Wang, Z.; Ren, H.; et al. Multilevel Gradient-Ordered Silicon Anode with Unprecedented Sodium Storage. *Adv. Mater.* **2024**, *36*, 2310270. [CrossRef] [PubMed]
32. Li, T.; Krumeich, F.; Ono, L.K.; Guo, T.; Morimoto, R.; Ding, C.; Xu, Z.; Liu, M.; Qi, Y. Exploring Niobium Oxide-Based Materials for Fast-Charging Lithium-Ion Anodes: Insights from Structure to Property. *Mater. Sci. Eng. R Rep.* **2025**, *162*, 100887. [CrossRef]
33. Li, Z.; Wang, Y.; Wang, J.; Wu, C.; Wang, W.; Chen, Y.; Hu, C.; Mo, K.; Gao, T.; He, Y.-S.; et al. Gradient-Porous-Structured Ni-Rich Layered Oxide Cathodes with High Specific Energy and Cycle Stability for Lithium-Ion Batteries. *Nat. Commun.* **2024**, *15*, 10216. [CrossRef]
34. Kim, Y.; Yoo, J.; Lee, K. High-Performance Binder-Free Li-Ion Batteries Using Dynamically Transformed Niobium Oxide Nanochannel Electrodes. *J. Power Sources* **2025**, *630*, 236167. [CrossRef]

coatings

MDPI

*Article*

# Preparation and Characterization of Co-CuCoMnO$_x$ Solar Selective Absorption Coatings

**Xian Zeng [1], Ziqiang Long [1], Ziyong Liu [2], Qian Cao [1,*] and Xudong Cheng [1]**

[1] School of Materials Science and Engineering, Wuhan University of Technology, Wuhan 430070, China; zengxian9925@163.com (X.Z.); longzq2025@163.com (Z.L.); xdcheng54@163.com (X.C.)
[2] The 718th Research Institute of CSSC, Handan 056027, China; liuzy0711@163.com
* Correspondence: caoqian@whut.edu.cn

**Abstract:** The Co-CuCoMnO$_x$ coatings with varying proportions were prepared and investigated to develop a novel metal–ceramic solar selective absorption coating employed at high temperature. The CuCoMnO$_x$ powders were synthesized using the solid-phase reaction method. Subsequently, Co-CuCoMnO$_x$ coatings were deposited on the surface of 316L steels utilizing the atmospheric plasma spraying (APS) technique. The results showed that the synthesized CuCoMnO$_x$ powders were mainly composed of two phases, which were $Cu_{1.5}Mn_{1.5}O_4$ and $MnCo_2O_4$. The CuCoMnO$_x$ powders had a solar absorptance of 0.929 and an infrared emittance of 0.862, which was considered a good solar absorbent. The synthesized Co-CuCoMnO$_x$ coating had a typical thermal spray layered stacking structure. The chemical phases of the coatings were mainly Co, CoO, and CoCuMnO$_x$. Due to the addition of CuCoMnO$_x$ inhibiting the oxidation of Co during the thermal spraying process, the 95Co-5CuCoMnO$_x$ (wt%) coating exhibited the optimal quality factor ($\alpha/\varepsilon$) of 2.184, with a solar absorptance $\alpha$ of 0.808 and an infrared emittance $\varepsilon$ of 0.370, respectively. Moreover, this specific coating demonstrated a good thermal stability for up to 3 h when exposed to an atmospheric environment at 450 °C. The results indicate its significant potential for high-temperature solar selective absorption coating.

**Keywords:** Co-CuCoMnO$_x$; solar selective absorption coating; solid-phase reaction; atmospheric plasma spraying; high-temperature thermal stability

## 1. Introduction

In the contemporary world, numerous countries are actively engaged in research on photothermal power generation technology to effectively harness and utilize solar energy. As a critical component of solar collectors, solar selective absorption coatings play a key role [1], directly influencing the advancement of photothermal power generation technology. The principle underlying solar selective absorption coatings is their high absorptance ($\alpha$) for solar radiation energy within the ultraviolet–visible near-infrared spectrum (0.3–2.5 μm), while simultaneously exhibiting low emittance ($\varepsilon$) in the mid–far-infrared range (2.5–25 μm) [2]. Solar selective absorption coatings can be categorized into three distinct types based on their operating temperature: low-temperature solar selective absorption coatings (operating at temperatures below 100 °C), medium-temperature solar selective absorption coatings (ranging from 100 °C to 400 °C), and high-temperature solar selective absorption coatings (exceeding 400 °C). Currently, research on solar selective absorption coating for mid- and low-temperature applications has reached a relatively advanced stage. In order to reduce costs and enhance the conversion efficiency of photothermal power generation, it is essential to continuously elevate the operating temperature



https://doi.org/10.3390/coatings15050547

of these solar selective absorption coatings. Therefore, there is a pressing need for the development of high-temperature solar selective absorption coatings that can endure the highest possible temperatures [3–6].

Metal–ceramic solar selective absorption coatings demonstrate exceptional optical performance and high-temperature thermal stability, making them promising candidates for the development of advanced high-temperature solar selective absorption coatings. Generally, metal–ceramic solar selective absorption coatings are composed of transition metals and their corresponding oxides, nitrides, or carbides. Examples include Ni-$Al_2O_3$ [7], Pt-$Al_2O_3$ [8], Ti-AlN [9], Ni/TiN-$La_2O_3$ [10], Co-$Cr_3C_2$-WC [11], and Ni/Mo-TiC [12]. Among these materials, the metal–ceramic solar selective absorption coatings based on Co-WC have been the subject of extensive investigation by various researchers. Chen et al. [13] investigated three types of WC-12Co cermet materials with micron, submicron, and nanometer structures to fabricate corresponding coatings using high-velocity oxygen–fuel (HVOF) spraying. It was observed that decarburization (the decomposition of WC into $W_2C$) occurred in all three structural forms during the spraying process. Notably, the nanometer structure also resulted in the formation of new phases, such as W and $W_3O_4$. The absorptance of the coatings across these three structures ranged from 0.78 to 0.86, while their emittances varied between 0.6 and 0.7, having a quality factor ($\alpha/\varepsilon$) of 1.11~1.43. The relatively high emittance observed may be attributed to the comparatively low metal content present in these coatings. Ke et al. [14] utilized a material composition of 80% Co, 10% submicron WC, and 10% nano WC to apply a coating on stainless steel (SS) via high-velocity oxygen–fuel (HVOF) spraying. The resulting Co-WC coating exhibited an absorptance of 0.821 and an emittance of 0.434 respectively, giving a quality factor of 1.89. It is evident that the quality factor of the Co-WC type metal–ceramic solar selective absorption coating is not satisfactory.

To address the challenges associated with decarburization, oxidation, and the relatively low quality factor of Co-WC type coatings, this study contemplates the selection of a novel ceramic phase. The aim is to enhance both the optical performance and high-temperature stability of Co-containing metal–ceramic solar selective absorption coatings. Metal oxides possessing a spinel structure have demonstrated remarkable optical performance and high-temperature stability [15–21]. Consequently, they have frequently been employed as raw materials for the fabrication of solar selective absorption coatings in recent years. Among these materials, $CuCoMnO_x$ with a spinel structure has garnered significant research attention [4,22–27]. Peng et al. [28] applied $CuCoMnO_x$ coatings onto polished aluminum plates. By optimizing the coating thickness, they achieved an absorptance of 0.928 and an emittance of 0.198 for the coated surface. Rocío Bayón et al. [15] fabricated $Cu_{1.5}Mn_{1.5}O_4$ thin films on an aluminum substrate using the impregnation method, achieving an absorption rate exceeding 0.87. Subsequently, a $SiO_2$ anti-reflection layer was deposited onto the $Cu_{1.5}Mn_{1.5}O_4$ thin film via the sol–gel method, resulting in improved optical performance with an absorption rate of 0.94 and an emission rate of 0.06. George Kordas [29] incorporated CuO@$SiO_2$ microspheres into the $CuCoMnO_x$ coating, achieving an impressive absorptance/emittance of 30. Furthermore, it was concluded that the ratio of absorptance to emittance increases linearly with the concentration of CuO@$SiO_2$ microspheres. Ke et al. [14] developed a composite coating on a stainless steel (SS) substrate by successively depositing layers of $CuCoMnO_x$, $CuCoMnO_x$-$SiO_2$, and $SiO_2$ atop a WC-Co coating fabricated using HVOF. This resulted in the formation of a WC-Co/$CuCoMnO_x$/$CuCoMnO_x$-$SiO_2$/$SiO_2$ composite structure. The incorporation of $CuCoMnO_x$ was aimed at enhancing the absorption rate of the coating within the visible light spectrum while simultaneously filling in the grooves and pores present on the surface of the WC-Co layer, thereby reducing its surface roughness. The optical performance metrics for this composite coating were characterized by an absorption/emittance ratio of 0.915/0.29, demonstrating that it could maintain thermal stability for up to 50 h at 550 °C. Based on the aforementioned research, it is evident that $CuCoMnO_x$ represents a promising ceramic phase candidate for enhancing the optical performance of Co-containing solar selective absorption coatings, while also exhibiting excellent high-temperature stability.

To obtain a solar selective absorption coating with excellent optical performance and high-temperature thermal stability, $CuCoMnO_x$ spinel powder was synthesized using a combination of three powders—CuO, $Co_3O_4$, and $MnO_2$—in a muffle furnace. Then, various mass ratios of Co-$CuCoMnO_x$ coatings were designed and prepared through the atmosphere plasma spraying method. The mechanism of optical property changes in plasma-sprayed coatings with different compositions and ratios was thoroughly analyzed. Finally, the coating with the best optical performance was selected to investigate its high-temperature stability, proving that it has the potential to become a candidate for high-temperature solar selective absorption coatings.

## 2. Experiment

### 2.1. Preparation of $CuCoMnO_x$ Powders

CuO powders (with purity of 99.5%), $MnO_2$ powders (with purity of 99.5%), and $Co_3O_4$ powders (with purity of 99.9%) were sourced from Qinhuangdao Yinuo High-tech Materials Development Co., Ltd., Qinhuangdao, China. The corresponding masses of these powders were measured according to the molar ratio of Cu:Mn:Co, set at 3:3:1. Deionized water was added to the three mixed powders at a solid–liquid mass ratio of 1:0.84. Additionally, a self-prepared binder, dispersant, and several drops of defoamer (tributyl phosphate) were incorporated into the mixture. Subsequently, the mixture was introduced into a colloid mill and subjected to high-speed mixing for a duration of 2 h to ensure a homogeneous slurry. After that, the slurry was pumped into the centrifugal atomization tower to form spherical particles. The parameters for the granulation process are as follows: the air outlet temperature was set at 110 °C, while the air inlet temperature was maintained at 240 °C. The rotational speed of the atomizing disc was established at 18,000 r/min, and the feeding rate was specified as 45 mL/min. The obtained powder was sintered in a muffle furnace with a heating rate of 5 °C/min. The sinter temperature was maintained at 600 °C for 1 h to remove the binder, and at 1100 °C for 2 h to promote solid reaction. Subsequently, the material was allowed to cool within the furnace, resulting in the formation of $CuCoMnO_x$ powder.

### 2.2. Preparation of Thermal Spray Powders

The prepared $CuCoMnO_x$ powder was furtherly crushed by a ball mill to obtain tiny powders with a particle size of less than 5 μm. Co powder, with a particle size of 1.5 μm and a purity of 99.9%, was sourced from Shenzhen GeLinmei High-Tech Co., Ltd., Shenzhen, China. In this study, a total of five groups of coating materials were established, each featured different composition ratios as shown in Table 1. The raw powders were weight and mixed with deionized water, self-prepared adhesives, dispersants, and defoamer (tributyl phosphate) at a solid–liquid mass ratio of 1:1 though colloid milling to form a uniformly ground slurry. Then, the slurry was pumped into the centrifugal atomization tower to form spherical particles. The parameters for the granulation process are the same with those we have mentioned above. The powder obtained through granulation was subjected to high-temperature calcination in a tube furnace under an argon atmosphere. The heating rate was set at 5 °C/min, and the sintering temperature was maintained at 600 °C for 1 h to ensure complete burnout of the binder. Following this process, the binder-free powder was sieved to obtain particles within the size range of 45–75 μm, suitable for thermal spraying applications.

**Table 1.** Composition and proportions of the composite powder.

| Group | Material Proportion (wt%) |
|---|---|
| A | 100Co |
| B | 95Co-5CuCoMn$O_x$ |
| C | 90Co-10CuCoMn$O_x$ |
| D | 85Co-15CuCoMn$O_x$ |
| E | 80Co-20CuCoMn$O_x$ |

### 2.3. Preparation of Coatings

The substrate utilized for the spraying process was 316L stainless steel, with dimensions of 30 mm × 40 mm × 1 mm. The substrate underwent ultrasonic cleaning in acetone to eliminate oil contaminants, then was followed by ultrasonic cleaning in deionized water. Finally, it was dried in a drying oven to remove any residual moisture. The cleaned substrate surface was sandblasted using 100-mesh brown alumina sand, aiming to roughen the substrate surface and enhance the bonding strength between the substrate and the coating. Before the formal spraying process, the substrate was preheated in a drying oven at 120 °C. The absorption layer was fabricated using the APS-3000K plasma spraying equipment from the Beijing Institute of Aeronautical Manufacturing Engineering under AVIC in Beijing, China. The specific parameters for the spraying process are detailed in Table 2.

**Table 2.** Process parameters used for spraying.

| Factor | Parameter |
|---|---|
| Electrical current (A) | 350 |
| Voltage (V) | 77 |
| Ar gas flow (L/min) | 35 |
| $H_2$ gas flow (L/min) | 12 |
| Spraying distance (mm) | 140 |
| Spray gun velocity (mm/s) | 300 |
| Powder feeding quantity | 3.9 |
| Spraying times | 4 |

### 2.4. Test of High-Temperature Stability

The coated samples were placed within a tube furnace and maintained under atmospheric conditions. The heating rate was set at 5 °C/min, and the samples were held at 450 °C for a duration of 3 h before being allowed to cool alongside the furnace.

### 2.5. Characterization

The thermogravimetric curves of the powdered mixture comprising CuO, $Co_3O_4$, and $MnO_2$ after granulation were analyzed using TG-DTG (NETZSCH STA 2500, Netzsch Scientific Instruments Trading Co., Ltd., Shanghai, China). The phases of the synthesized CuCoMn$O_x$ powder, the thermal spray powders, and the coatings before and after the high-temperature heat treatment were analyzed using X-ray diffraction (XRD, Spectris Group in London, UK/EMPYREAN Smart Diffractometer/Cu target, Rigaku Corporation in Tokyo Japan/Smartlab SE/Co target). The surface and cross-sectional morphologies of the coatings were analyzed using a scanning electron microscope (SEM, JEOL in Tokyo Japan, JSM-IT300). The X-ray energy dispersive spectrometer (EDS, Oxford Instruments Technology Co., Ltd., X-MaxN20 EDS spectrometer, Shanghai, China) was utilized to perform point-scanning tests on both the powder and the coating. The surface roughness of the coatings was assessed using a surface roughness measurement instrument (Beijing Time Summit Technology Co., Ltd., Beijing, China, TIME®3221). The reflection spectra of the synthesized $CuCoMnO_x$ powder and the coating before and after the high-temperature heat treatment were measured within the wavelength ranges of 0.3–2.5 μm and 2.5–25 μm, using a UV–visible spectrophotometer (Shimadzu, UV-3600, Kyoto, Japan) and a Fourier-transform infrared spectrometer (Bruker, Tensor27, Karlsruhe, Germany), respectively. The absorptance ($\alpha$) and emittance ($\varepsilon$) are determined using Equations (1) and (2), where $I_{sol}(\lambda)$ is the irradiance of the standard solar spectrum under Air Mass 1.5 (AM1.5) conditions, $R(\lambda)$ denotes the spectral reflectance, and $I_b(\lambda)$ indicates the radiant intensity of a blackbody at room temperature [30].

$$\alpha = \frac{\int_{0.3}^{2.5} I_{sol}(\lambda)(1 - R(\lambda))d\lambda}{\int_{0.3}^{2.5} I_{sol}(\lambda)d\lambda} \quad (1)$$

$$\varepsilon = \frac{\int_{2.5}^{25} I_b(\lambda)(1 - R(\lambda))d(\lambda)}{\int_{2.5}^{25} I_b(\lambda)d(\lambda)} \quad (2)$$

## 3. Results and Discussion

### 3.1. Synthesis Mechanism and Spectral Property of $CuCoMnO_x$ Powders

In this study, $CuCoMnO_x$ powders were synthesized using a solid-phase reaction method. To elucidate the synthesis mechanism as well as the final sintering temperature of the $CuCoMnO_x$ powders, we analyzed the TG-DTG results obtained from the granulated mixture of $CuO$-$Co_3O_4$-$MnO_2$, as illustrated in Figure 1. It can be observed from Figure 1 that the granulated powder of $CuO$-$Co_3O_4$-$MnO_2$ experiences mass loss at temperatures of approximately 200 °C, 531.8 °C, 619.6 °C, 1021.2 °C, and 1051.9 °C. The maximum mass loss rates corresponding to these temperature points are as follows: 0.28%/min for 200 °C, 0.62%/min for 531.8 °C, 0.25%/min for 619.6 °C, 0.46%/min for 1021.2 °C, and 0.51%/min for 1051.9 °C. It has been reported in the literature that the decomposition of $MnO_2$ into $Mn_2O_3$ occurs at temperatures of 520 °C [31], 550 °C [32], and 596.1 °C [33]. Consequently, the weight loss observed around 531.8 °C in Figure 1 can be interpreted as the decomposition of $MnO_2$. Based on the laboratory experience, the weight loss observed near 200 °C and 619.6 °C in Figure 1 is believed to correspond to the combustion of the binder, dispersant, and defoamer incorporated into the mixed powders of $CuO$-$Co_3O_4$-$MnO_2$ during the granulation process. The primary reaction equations are presented as follows:

$$4MnO_2 \rightarrow 2Mn_2O_3 + O_2 \ (T = 531.8\ °C) \quad (3)$$

Based on the TG-DTG curve, it was concluded that the mass loss kept unchanged around 1100 °C. Therefore, the final sintering temperature of 1100 °C with a duration of 2 h were selected. The X-ray diffraction (XRD) patterns of the $CuO$-$Co_3O_4$-$MnO_2$ powder before and after sintering at 1100 °C for 2 h are shown in Figure 2. It can be seen that the phase composition of the powder before sintering consists of $CuO$, $Co_3O_4$, and $MnO_2$. After sintering, the powder mainly comprises $Cu_{1.5}Mn_{1.5}O_4$, $MnCo_2O_4$, and a small amount of $CuO$. It was coincided with the target phases of $CuCoMnO_x$ powder reported in other research [24], suggesting the successful synthetization of $CuCoMnO_x$ powder. It can also be observed from Figure 2 that the diffraction peaks of $Cu_{1.5}Mn_{1.5}O_4$ and $MnCo_2O_4$ are sharp, indicating $Cu_{1.5}Mn_{1.5}O_4$ and $MnCo_2O_4$ possess good crystallinity. The weight loss observed around 1021.2 °C and 1051.9 °C in Figure 1 is deduced to correspond to the reaction of $CuO$, $Co_3O_4$, $MnO_2$, and $Mn_2O_3$ gradually synthesizing $Cu_{1.5}Mn_{1.5}O_4$ and $MnCo_2O_4$. The equation for the synthesis reactions is as follows:

$$6CuO + 4Co_3O_4 + 4MnO_2 + 4Mn_2O_3 \rightarrow 4Cu_{1.5}Mn_{1.5}O_4 + 6MnCo_2O_4 + O_2\ (T = 1021.2\ ^\circ C, T = 1051.9\ ^\circ C) \quad (4)$$

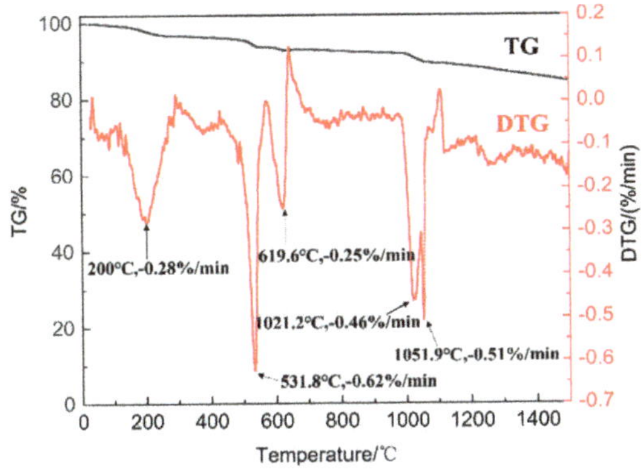


**Figure 1.** The TG-DTG diagram of the granulated $CuO\text{-}Co_3O_4\text{-}MnO_2$ powder.

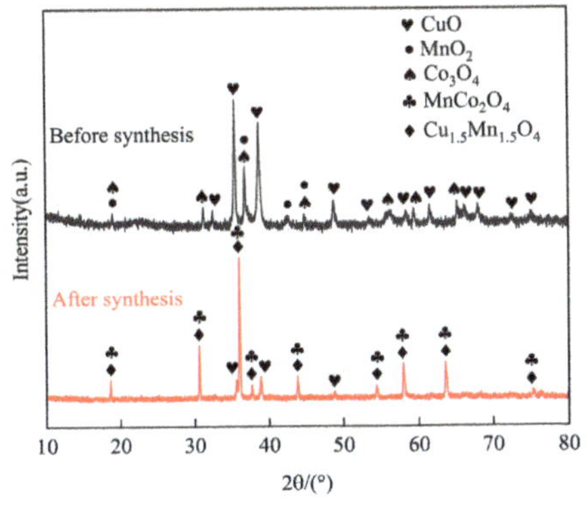


**Figure 2.** XRD patterns of $CuO\text{-}Co_3O_4\text{-}MnO_2$ powder before and after sintering at 1100 °C for 2 h.

The $CuCoMnO_x$ powder serves as a fundamental raw material for the spraying coatings, and its inherent optical performance significantly impacts the final absorptance and emittance of these coatings. To investigate the optical performance of prepared $CuCoMnO_x$ powder, the reflectance spectra were measured in the wavelength ranges of 0.3–2.5 μm and 2.5–25 μm, and is presented in Figure 3. Based on results, the absorptance of the $CuCoMnO_x$ powder was 0.929, while the emittance was 0.862, giving a quality factor of $\alpha/\varepsilon$ equaling to 1.078. Herein, it is observed that $CuCoMnO_x$ exhibits a high absorptance, while it also has a high emittance. In other words, achieving an ideal quality factor remains impossible for single $CuCoMnO_x$-based ceramic coating. Composite with low-absorption and low-emission metal Co to form a metal–ceramic coating might be an effective way to obtain good solar selective absorption performance. Thus, this paper attempted to optimize the composition of the Co-$CuCoMnO_x$ composite coatings, in order to obtain a higher quality factor of solar selective absorption performance.

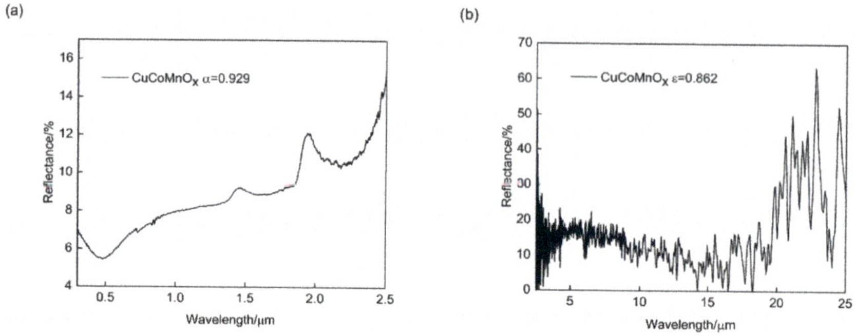


**Figure 3.** Reflectance spectra of $CuCoMnO_x$ powder in the spectral ranges of (**a**) 0.3–2.5 μm and (**b**) 2.5–25 μm.

### 3.2. Microstructure of Co-$CuCoMnO_x$ Powders and Coatings

The microstructure of prepared Co-$CuCoMnO_x$ powders for coating deposition was observed. As an example, the morphology and the element content of the Co-$CuCoMnO_x$ powder in Group B was summarized and shown in Figure 4 and Table 3, correspondingly. It can be seen that the powder for thermal spraying possessed a high degree of sphericity to ensure good flowability during the deposition. Moreover, it can be found from Table 3 that the Co-$CuCoMnO_x$ powder in Group B comprised four elements: Cu, Co, Mn, and O. The weight percentages of each element were approximately consistent with the designed content.

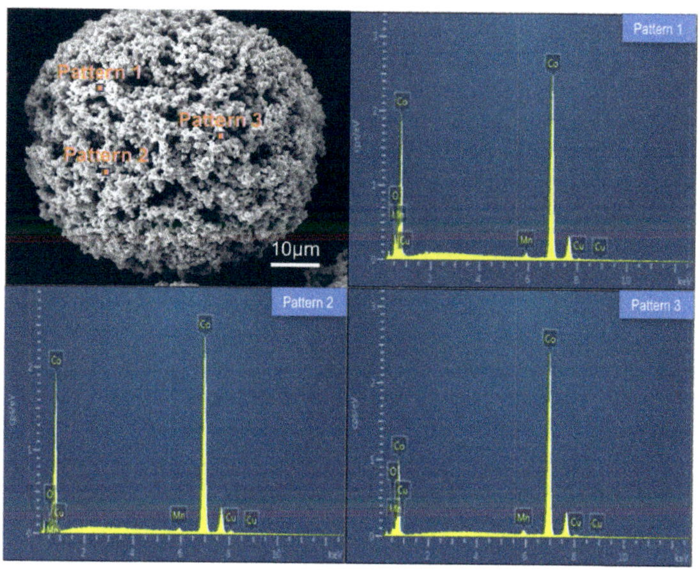


**Figure 4.** Morphology and corresponding EDS spectra of Co-$CuCoMnO_x$ powder in Group B.

To identify the phase composition of the four-groups Co-$CuCoMnO_x$ powders, Group C (10% $CuCoMnO_x$) and D (15% $CuCoMnO_x$) were subjected to X-ray diffraction (XRD) analysis. The results are presented in Figure 5. Both groups of powders mainly contain three phases—Co, $Cu_{1.5}Mn_{1.5}O_4$, and $MnCo_2O_4$—indicating that no chemical reactions occurred within the composite powders during the processes of granulation and degumming, and confirming that the phase composition aligned with the initial design specifications.

**Table 3.** Element content in different positions of Co-$CuCoMnO_x$ powder in Group B.

| Elements | wt% | | |
|---|---|---|---|
| | Pattern 1 | Pattern 2 | Pattern 3 |
| O | 6.14 | 2.71 | 1.88 |
| Mn | 1.26 | 0.82 | 1.27 |
| Co | 91.53 | 94.17 | 95.4 |
| Cu | 1.06 | 2.3 | 1.45 |

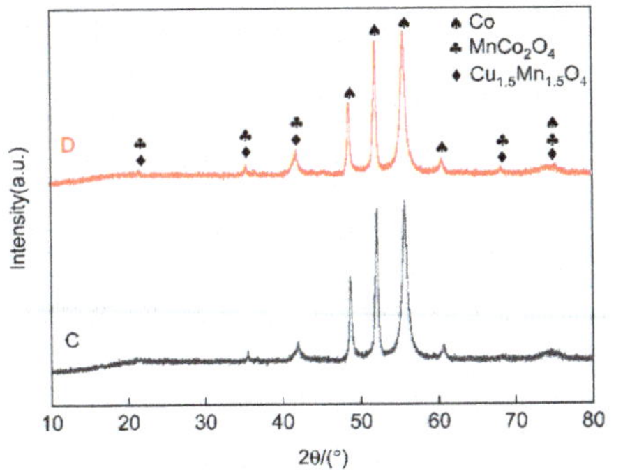


**Figure 5.** XRD patterns of Co-$CuCoMnO_x$ powders in Groups C and D (Co target). Group C: 90Co-10$CuCoMnO_x$ spray powder, Group D: 85Co-15$CuCoMnO_x$ spray powder.

To investigate the phase composition of the Co-$CuCoMnO_x$ coatings fabricated by plasma spraying, X-ray diffraction (XRD) analysis was conducted on the coating for each group, and the results are illustrated in Figure 6. According to it, both Co and CoO phases were present in all the coating samples. Furthermore, the $CoCuMnO_x$ (#47-0324) phase was also demarcated for Groups B–E. During the plasma spraying process, the temperature of plasma flame can reach as high as 10,000 °C. A portion of cobalt (Co) absorbs oxygen from the surrounding air during its transition from the spray gun to the substrate, resulting in the formation of cobalt oxide (CoO). The high-temperature flame flow inherent in plasma spraying also facilitates the mutual solubility of $Cu_{1.5}Mn_{1.5}O_4$ and $MnCo_2O_4$, leading to the generation of the $CoCuMnO_x$ phase. By comparing the relative intensities of the two diffraction peaks at 49.64° and 60.64° in Groups A–E as illustrated in Figure 6, it is evident that the intensity ratio of CoO/Co of group A is the highest. This phenomenon suggested that oxidation within the pure Co group is most pronounced. In contrast, for the other four groups, it was inferred that the addition of the $CuCoMnO_x$ powders may inhibit the oxidation of Co, since the intensity ratio of CoO/Co apparently decreased. It was thought that the added $CoCuMnO_x$ powders might consume the heat for Co oxidation, which reduced the oxidation of Co.

The surface morphologies of the five groups of coatings produced by plasma spraying are illustrated in Figure 7. All the coatings exhibited similar surface morphology, which consisted of molten zones, semi-molten zones, protrusions, and pores. The molten zone came from the fully melted droplets in the flame flow flying towards the surface of the substrate and spreading out, resulting in a relatively smooth and dense area. However, there was still a portion of the composite powder that did not undergo complete heating within the plasma flame flow prior to impacting the substrate. This incomplete heating, along with some molten materials presenting at the edge of the molten zone, collectively formed the semi-molten zone. The formation of protrusions and pores may be attributed to the rapid cooling and contraction of the droplet impacted onto the substrate. The surface roughness (Ra) of these five groups of coatings was similar, which ranged from

3.19 to 4.25 μm. When observed under high-magnification views, it was found that the local spreading conditions of the coatings of Groups A–C were relatively favorable, characterized by a smoother surface with fewer protrusions and smaller pores. In contrast, the local spreading conditions of the coatings of Groups D and E were comparatively poor, exhibiting a rougher surface with more protrusions and larger pores. The surface produced by plasma spraying was a micro-uneven structure, which appeared macroscopically flat while exhibiting microscopic irregularities. Incident light with wavelengths that are less than or equal to the dimensions of the pores or the spacing between protrusions will undergo multiple reflections within these pores or among the protrusions, leading to absorption by the coating. Conversely, incident light with wavelengths exceeding the size of the pores or the distance between protrusions will be directly reflected into the surrounding environment, akin to specular reflection. Consequently, these coatings of Groups A–C might demonstrate a lower absorption rate for ultraviolet–visible near-infrared solar radiation and a lower absorption in the mid–far-infrared range, than the coatings of Groups D–E based on the surface morphology.

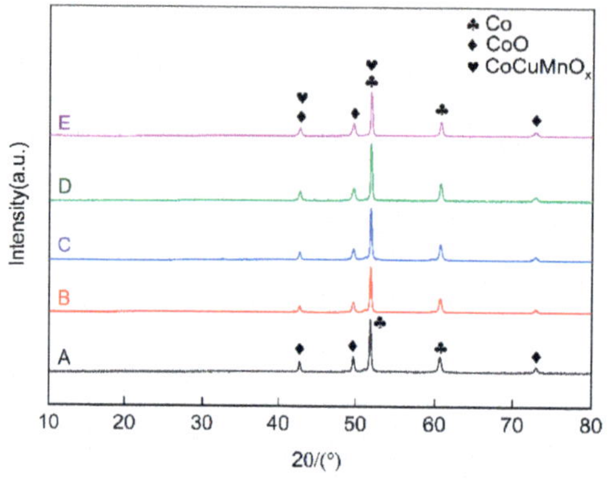


**Figure 6.** XRD diffraction patterns of the absorption coating prepared by plasma spraying (Co target). Group A: 100Co coating, Group B: 95Co-5CuCoMnO$_x$ coating, Group C: 90Co-10CuCoMnO$_x$ coating, Group D: 85Co-15CuCoMnO$_x$ coating, Group E: 80Co-20CuCoMnO$_x$ coating.

As an example, the cross-sectional morphology of the plasma-sprayed coating of Group A and E is illustrated in Figure 8. The coating thickness ranged from 10 to 20 μm. The interaction between the plasma-sprayed coating and the substrate manifested as mechanical interlocking, demonstrating a robust bond. The coating exhibited a layered structure, which resulted from the molten and semi-molten droplets interlacing and overlapping with one another. It could also be observed from Figure 8 that the plasma sprayed coating contained pores of varying sizes distributed throughout its structure.

To investigate the chemical composition differences among coatings containing CoCuMnO$_x$ and those without, the corresponding EDS point analysis of the coatings of Groups A and E was selected. The results are summarized in Table 4 and Figure 8. In the backscattering images, the brighter the phase, the larger the corresponding atomic number. Based on the results of Patterns 5 and 6, it could be inferred that the grayish-white regions within the coating of Group A mainly consisted of cobalt (Co) along with a minor presence of cobalt oxide (CoO). In addition, A small amount of matrix elements Fe and Cr are diffused in the part of Group A coating near the substrate, and there are pores inside the coating. In contrast, according to Pattern 4, the gray areas in the coating resulted from a more extensive oxidation of Co. In conjunction with the XRD results, it was inferred that the grayish-white regions marked by Pattern 7 within the coating of Group E consist of Co and a small amount of CoO. The grayish-black areas were identified as containing Co, $CoCuMnO_x$, and CoO, while the black regions not only exhibited porosity but also comprised a mixture of Co, $CoCuMnO_x$, and CoO. In addition, the presence of Cl in Group E coating might be due to a very small amount of Cl impurity in the granulation tower, which was mixed into the spherical spray powder during the spray granulation process, resulting in the inclusion of the Cl element in the coating.

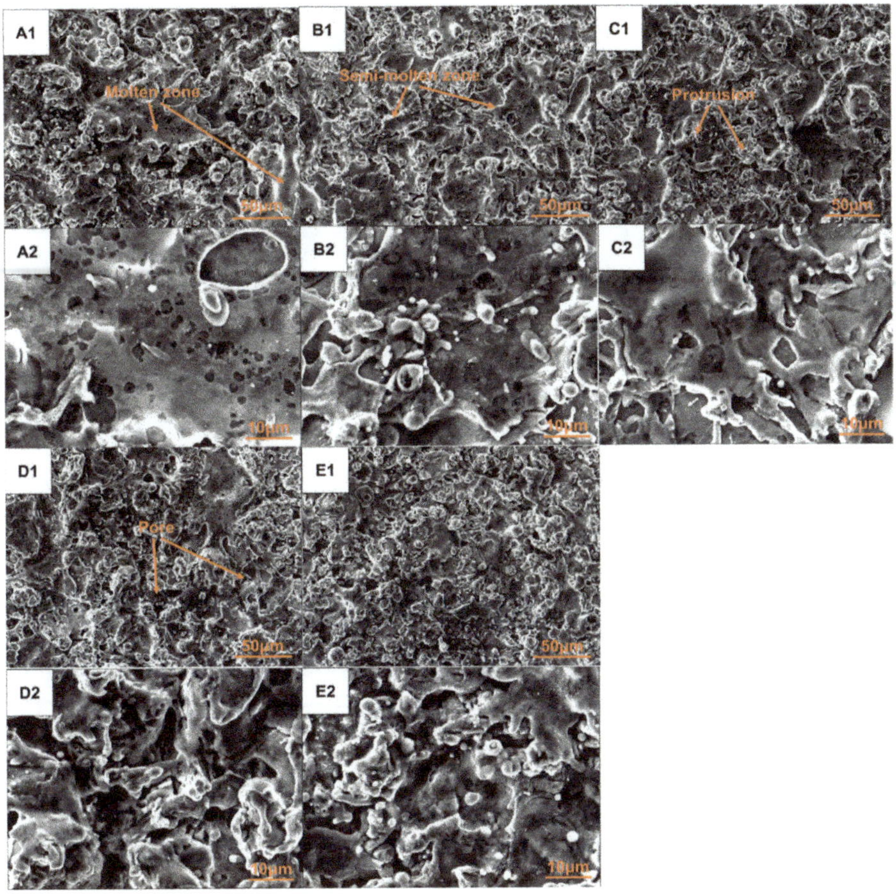


**Figure 7.** Surface morphology of plasma sprayed Co-$CuCoMnO_x$ coatings under low-magnification view (**A1–E1**) and local high-magnification view (**A2–E2**).

**Table 4.** Element content in different positions of Groups A and E coatings (wt%).

| Elements | wt% | | | | | |
|---|---|---|---|---|---|---|
| | Pattern 4 | Pattern 5 | Pattern 6 | Pattern 7 | Pattern 8 | Pattern 9 |
| Co | 73.7 | 94.75 | 92.27 | 94.93 | 89.12 | 68.81 |
| O | 26.3 | 5.25 | 3.8 | 5.07 | 7.38 | 24.8 |
| Fe | 0 | 0 | 2.87 | 0 | 0 | 0 |
| Cr | 0 | 0 | 1.06 | 0 | 0 | 0 |
| Cu | 0 | 0 | 0 | 0 | 2.38 | 1.8 |
| Mn | 0 | 0 | 0 | 0 | 1.12 | 3.2 |
| Cl | 0 | 0 | 0 | 0 | 0 | 1.4 |

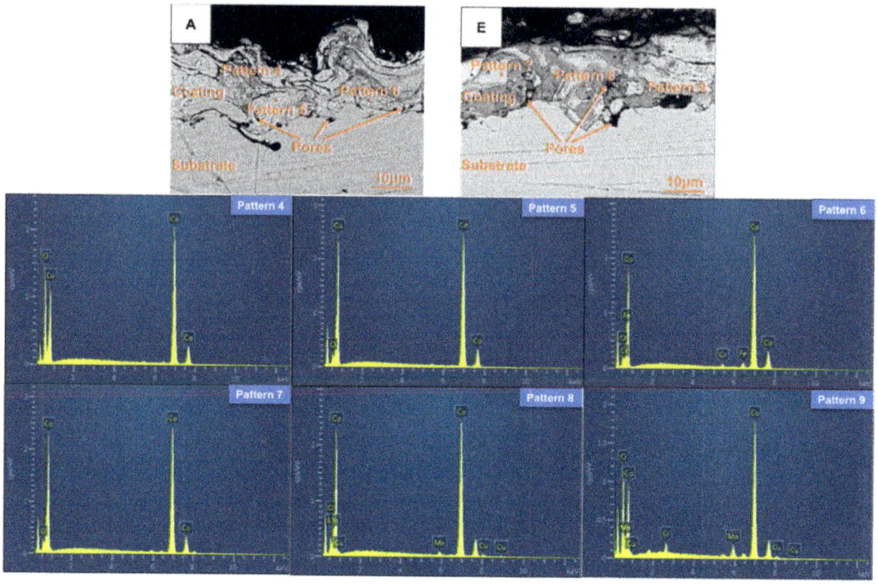


**Figure 8.** Cross-sectional morphology and corresponding EDS energy spectrum diagrams of the coatings of Groups A and E.

### 3.3. Optical Properties of Co-CuCoMnO$_x$ Coatings

The reflectance spectra for the five groups of plasma-sprayed coatings are illustrated in Figure 9, and the optical properties derived from these reflectance spectra are summarized in Table 5. According to the results, it was found that both the solar absorptance and infrared emittance of those coatings first decreased and then increased when increasing the addition amount of CuCoMnO$_x$. The highest quality factor ($\alpha/\varepsilon$) of 2.184 was obtained in the coating of Group B, whose solar absorptance and infrared emittance were 0.808 and 0.370, respectively.

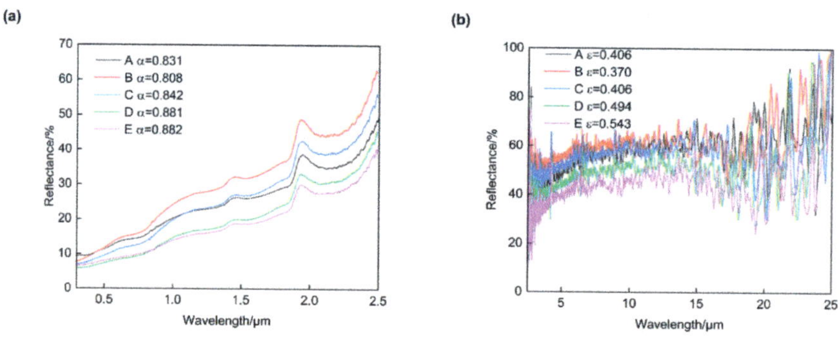


**Figure 9.** Reflectance spectra of Co-CuCoMnO$_x$ coatings in the ranges of (**a**) 0.3–2.5 μm and (**b**) 2.5–25 μm. Group A: 100Co coating, Group B: 95Co-5CuCoMnO$_x$ coating, Group C: 90Co-10CuCoMnO$_x$ coating, Group D: 85Co-15CuCoMnO$_x$ coating, Group E: 80Co-20CuCoMnO$_x$ coating.

Table 5. Optical properties of Co-CuCoMnO$_x$ coatings.

| Group | Solar Absorptance ($\alpha$) | Infrared Emittance ($\varepsilon$) | Quality Factor ($\alpha/\varepsilon$) |
|---|---|---|---|
| A | 0.831 | 0.406 | 2.047 |
| B | 0.808 | 0.370 | 2.184 |
| C | 0.842 | 0.406 | 2.074 |
| D | 0.881 | 0.494 | 1.783 |
| E | 0.882 | 0.543 | 1.624 |

Compared to the ceramic phase, pure cobalt exhibits a low solar absorptance as well as low infrared emittance [11]. However, the deposited cobalt coating (Group A) exhibited a solar selective absorption performance, which might be attributed to the oxidation and the formation of CoO during the plasma spraying. The $t_2$ state of $Co^{2+}$ ions in CoO is characterized by unstable electrons. The interactions arising from this instability lead to intrinsic absorption, which may result in an increased absorption rate and emittance of the coating [34]. Since the two Jahn–Teller ions, $Cu^{2+}$ and $Mn^{3+}$, are present in $Cu_{1.5}Mn_{1.5}O_4$, $MnCo_2O_4$, and CoCuMnO$_x$ during plasma spraying, the composite powder is rapidly heated to 10,000 °C within the plasma flame stream. The phases of $Cu_{1.5}Mn_{1.5}O_4$ and $MnCo_2O_4$ undergo mutual dissolution to form the CoCuMnO$_x$ phase. At this juncture, alterations occur in the molecular vibration and rotational modes, which subsequently induce lattice distortion and enhance the intrinsic absorption characteristics of the material while simultaneously reducing reflectivity in the infrared spectrum [35]. Therefore, it is likely that the formation of CoCuMnO$_x$ within the coating will further augment both absorption and emission rates compared to those exhibited by $Cu_{1.5}Mn_{1.5}O_4$ and $MnCo_2O_4$. Combining the XRD analysis of the coatings presented in Figure 6, it was evident that Group A contained cobalt (Co) and cobalt oxide (CoO), and also exhibits the highest concentration of cobalt oxide (CoO). The optical performance of the coating in Group A was a result of the combination of Co and CoO. However, the addition of 5 wt% CuCoMnO$_x$ to Group B resulted in a hindrance of Co oxidation; which, in other words, means that the content of Co increased. Since the added content of CuCoMnO$_x$ was relatively low, the optical performance of the coating was primarily affected by the Co content, and exhibited characteristics associated with reduced solar absorption and infrared emission. However, when a proportion of 10 wt% or more of CuCoMnO$_x$ was incorporated into the coating, the solar absorptance and infrared emittance of Groups C–E exhibited a gradual increase. This phenomenon can be attributed to the increase in CoCuMnO$_x$, which possessed high solar absorptance and high infrared emittance.

### 3.4. High Temperature Stability of Co-CuCoMnO$_x$ Coatings

To evaluate the high temperature stability of the Co-CuCoMnO$_x$ coatings, the coatings of Group B underwent a high-temperature heat treatment at 450 °C in an atmospheric environment for a duration of 3 h. The optical performance prior to and after the high-temperature heat treatment are presented in Table 6. Additionally, the X-ray diffraction (XRD) patterns of the coatings before and after the heat treatment are illustrated in Figure 10. Based on the variations in diffraction peak intensity before and after the heat treatment, it could be inferred that the content of CoO increased after the heat treatment. Consequently, the solar absorptance and infrared emittance of the coating increased after the heat treatment. The performance criterion (PC) to evaluate the high temperature stability of the coating is if the PC value described as Equation (5) [36] is below 0.05, it can be concluded that the degradation performance of the coating at elevated temperatures is not significant. Furthermore, this suggests that the coating exhibits relatively good thermal stability under such conditions [37]. After the heat treatment, it was found that no cracking or shedding happened on the coating surface. The solar absorptance of the coating reached 0.89, while its emissivity increased by 0.168 compared to pre-treatment values. The PC value of the coating after the heat treatment at 450 °C for 3 h in an atmospheric environment was calculated as 0.002, which was far below the threshold of 0.05. The attenuation of the optical performance of the coating was negligible. Therefore, it could be concluded that the coating exhibited excellent thermal stability when maintained at 450 °C in an atmospheric environment for a duration of 3 h.

$$PC = (\alpha_{\text{before heat treatment}} - \alpha_{\text{after heat treatment}}) - 0.5 \times (\varepsilon_{\text{before heat treatment}} - \varepsilon_{\text{after heat treatment}}) \quad (5)$$

**Table 6.** The absorptivity and emissivity values of Group B coatings before and after the heat treatment.

| | Before Heat Treatment | After Heat Treatment |
|---|---|---|
| Solar Absorptance ($\alpha$) | 0.808 | 0.89 |
| Infrared Emittance ($\varepsilon$) | 0.37 | 0.538 |

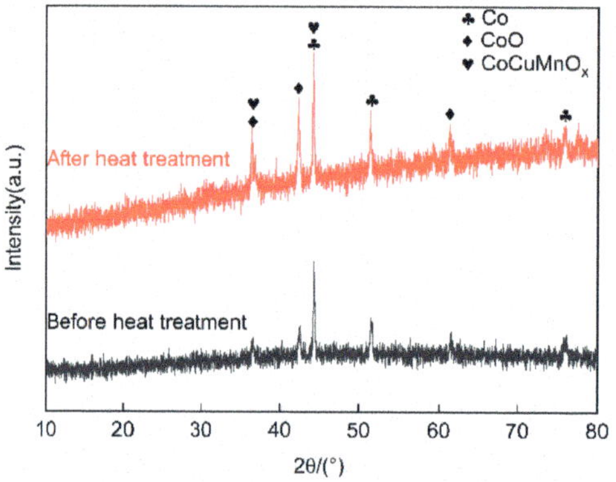


**Figure 10.** XRD patterns of the coatings in Group B before and after the heat treatment.

## 4. Conclusions

By employing the solid-phase reaction method, CuO-$Co_3O_4$-$MnO_2$ composite materials were successfully synthesized into $CuCoMnO_x$ powder at 1100 °C in a muffle furnace for 2 h. The $CuCoMnO_x$ powder predominantly consisted of $Cu_{1.5}Mn_{1.5}O_4$ and $MnCo_2O_4$. Furthermore, the $CuCoMnO_x$ powder demonstrated a solar absorptance of 0.929 and an infrared emittance of 0.862, which was considered a good absorbent. Co-$CuCoMnO_x$ spherical spray powders were prepared using spray granulation and de-gelling processes, with the phase composition of the spray powder containing $CuCoMnO_x$ being Co, $Cu_{1.5}Mn_{1.5}O_4$, and $MnCo_2O_4$, maintaining consistency with the initially designed components. Subsequently, Co-$CuCoMnO_x$ coatings were deposited on the surface of 316L steels utilizing the APS technique, aiming at developing a novel metal–ceramic solar selective absorption coating employed at high temperature. The chemical phase, microstructure, and optical performance, as well as high temperature stability of the prepared coatings, were systematically studied. The results showed that the Co-$CuCoMnO_x$ coating possessed a solar selective absorption capacity after being compound with cobalt. The synthesized Co-$CuCoMnO_x$ coating had a typical thermal spray layered stacking structure with molten zones, semi-molten zones, protrusions, and pores distributed on the coating surface, and had a thickness of 10~20 μm. The chemical phases of the coatings were mainly Co, CoO, and $CoCuMnO_x$. Here, CoO was formed by the partial oxidation of Co as it absorbed oxygen from the surrounding air during plasma spraying. The inherent high-temperature flame flow in plasma spraying promoted the mutual solubility of $Cu_{1.5}Mn_{1.5}O_4$ and $MnCo_2O_4$, leading to the formation of the $CoCuMnO_x$ phase. As the amount of $CuCoMnO_x$ added increased, both the solar absorption rate and infrared emissivity of the five groups of coatings produced by plasma spraying showed a trend of first decreasing and then increasing, while the quality factor showed a trend of first increasing and then decreasing. Due to the addition of $CuCoMnO_x$ inhibiting the oxidation of Co during the thermal spraying process, the 95Co-5$CuCoMnO_x$ (wt%) coating exhibited the optimal quality factor ($\alpha/\varepsilon$) of 2.184, with a solar absorptance $\alpha$ of 0.808 and an infrared emittance $\varepsilon$ of 0.370, respectively. Moreover, this specific coating demonstrated a good thermal stability for up to 3 h, in which the PC value was far less than 0.05 when it was exposed to an atmospheric environment at 450 °C. This finding holds significant implications for the research and development of high-temperature solar selective absorption coatings.

**Author Contributions:** X.Z. and X.C. conceived the idea. Investigation, Z.L. (Ziqiang Long) and Z.L. (Ziyong Liu). X.Z., Z.L. (Ziqiang Long) and Q.C. designed the experiment. Z.L. (Ziqiang Long) analyzed the data. X.Z. and Z.L. (Ziqiang Long) participated in the writing of the paper. All authors have read and agreed to the published version of the manuscript.

**Funding:** This work was financially supported by the National Natural Science Foundation of China (Grant No.52473280) and Guangxi Science and Technology Major Project (GUIKE AA23062025-3, China).

**Institutional Review Board Statement:** Not applicable.

**Informed Consent Statement:** Not applicable.

**Data Availability Statement:** The data presented in this study are available from the corresponding author upon request. The data are not publicly available due to privacy.

**Conflicts of Interest:** The authors declare no conflicts of interest.

## References


1. Kalogirou, S.A. Solar thermal collectors and applications. *Prog. Energy Combust. Sci.* **2004**, *30*, 231–295. [CrossRef]
2. Zäll, E.; Segervald, J.; Mahmoodi, H.; Perivoliotis, D.; Edman, L.; Wågberg, T. Achieving optically selective coatings of silica fixated carbon nanotubes for solar energy applications. *Sol. Energy Mater. Sol. Cells* **2024**, *278*, 113202. [CrossRef]
3. Zheng, L.Q.; Gao, F.Y.; Zhao, S.X.; Zhou, F.Y.; Nshimiyimana, J.P.; Diao, X.G. Optical design and co-sputtering preparation of high performance Mo–$SiO_2$ cermet solar selective absorbing coating. *Appl. Surf. Sci.* **2013**, *280*, 240–246. [CrossRef]
4. Farchado, M.; Vicente, G.S.; Barandica, N.; Sánchez-Señorán, D.; Morales, Á. High performance selective solar absorber stable in air for high temperature applications. *Sol. Energy Mater. Sol. Cells* **2024**, *271*, 112849. [CrossRef]
5. Kant, K.; Sibin, K.P.; Pitchumani, R. Novel fractal-textured solar absorber surfaces for concentrated solar power. *Sol. Energy Mater. Sol. Cells* **2022**, *248*, 112010. [CrossRef]
6. Elahi, A.N.M.T.; Crist, R.; Francoeur, M.; Park, K.; Rao, S. Characterization of high-temperature figure of merit for solar-thermal absorbers. *Sol. Energy Mater. Sol. Cells* **2023**, *257*, 112378. [CrossRef]
7. Li, Z.X.; Zhao, J.X.; Ren, L.H. Aqueous solution-chemical derived Ni–$Al_2O_3$ solar selective absorbing coatings. *Sol. Energy Mater. Sol. Cells* **2012**, *105*, 90–95. [CrossRef]
8. Tsegay, M.G.; Gebretinsae, H.G.; Sackey, J.; Arendse, C.J.; Nuru, Z.Y. Structural and optical properties of Pt-$Al_2O_3$ double cermet as selective solar absorber. *Mater. Today Proc.* **2021**, *36*, 571–575. [CrossRef]
9. Wu, Y.W.; Zheng, W.F.; Lin, L.M.; Qu, Y.; Lai, F.C. Colored solar selective absorbing coatings with metal Ti and dielectric AlN multilayer structure. *Sol. Energy Mater. Sol. Cells* **2013**, *115*, 145–150. [CrossRef]
10. Pang, X.M.; Jiang, J.X.; Li, B.; Zhao, D.Y.; Zhou, J.X. Study of wide temperature range and hard protective $La_2O_3$ doped cermet based single-layer solar selective absorbing coating by laser cladding. *Surf. Interfaces* **2021**, *27*, 101544. [CrossRef]

11. Shao, H.; Zeng, X.; Li, Q.Y.; Liu, F.S.; Liu, T.Z.; Cheng, X.D. Preparation and Properties of Co-$Cr_3C_2$-WC/$Al_2O_3$ Solar Selective Absorbing Coatings. *Surf. Technol.* **2020**, *49*, 138–145. (In Chinese)
12. Pang, X.M. High temperature solar selective absorber coating deposited by laser cladding. *Mater. Res. Express* **2017**, *4*, 095503. [CrossRef]
13. Chen, W.; Ding, Z.X.; Gao, R.Y.; Cheng, X.D. Performance Study of WC-12Co Solar Energy Selective Absorption Coatings Deposited by HVOF. *Therm. Spray Technol.* **2011**, *3*, 40–46. (In Chinese)
14. Ke, C.Z.; Zhang, X.M.; Guo, W.Y.; Li, Y.J.; Gong, D.Q.; Cheng, X.D. Solar selective coatings with multilayered structure based on thermal spraying WC-Co solar absorption layer. *Vacuum* **2018**, *152*, 114–122. [CrossRef]
15. Bayón, R.; Vicente, G.S.; Maffiotte, C.; Morales, Á. Preparation of selective absorbers based on CuMn spinels by dip-coating method. *Renew. Energy* **2008**, *33*, 348–353. [CrossRef]
16. Atchuta, S.R.; Sakthivel, S.; Barshilia, H.C. Transition metal based $Cu_xNi_yCo_{z-x-y}O_4$ spinel composite solar selective absorber coatings for concentrated solar thermal applications. *Sol. Energy Mater. Sol. Cells* **2019**, *189*, 226–232. [CrossRef]
17. Pakzad, E.; Ranjbar, Z.; Ghahari, M. Synthesized of octahedral cupper chromite spinel for spectrally selective absorber (SSA) coatings. *Prog. Org. Coat.* **2019**, *132*, 21–28. [CrossRef]
18. Atchuta, S.R.; Sakthivel, S.; Barshilia, H.C. Selective properties of high-temperature stable spinel absorber coatings for concentrated solar thermal application. *Sol. Energy* **2020**, *199*, 453–459. [CrossRef]
19. Deng, X.Q.; Xue, M.M.; Lv, Y.L.; Li, R.H.; Tong, J.M.; Shi, G.H.; Yang, Y.; Dong, Y.C. Study on spectral selective absorbing coatings with spinel structures fabricated via plasma spraying. *Vacuum* **2020**, *174*, 109214. [CrossRef]
20. Atchuta, S.R.; Sakthivel, S.; Barshilia, H.C. Nickel doped cobaltite spinel as a solar selective absorber coating for efficient photothermal conversion with a low thermal radiative loss at high operating temperatures. *Sol. Energy Mater. Sol. Cells* **2019**, *200*, 109917. [CrossRef]
21. Wei, S.C.; Xie, J.; Gao, W.; Zhan, Z.L.; Li, Z.L. Characteristics and Failure Behaviors of an $MnCo_2O_4$ Spinel Coating in High-Temperature Oxidation Processes. *Trans. Indian Inst. Met.* **2022**, *75*, 797–804. [CrossRef]
22. Aslan, E.; Emir, Ö.; Arslan, F.; Goktas, A.; Tumbul, A.; Durgun, M.; Kilic, A.; Aktacir, M.A.; Aslan, F. Improving the optical properties of $CuCoMnO_x$ spinel absorber using ZnO nanorod arrays for thermal collector and photocatalytic applications. *Ceram. Int.* **2024**, *50*, 9169–9176. [CrossRef]
23. El Mahallawy, N.; Shoeib, M.; Ali, Y. Application of $CuCoMnO_x$ coat by sol gel technique on aluminum and copper substrates for solar absorber application. *J. Coat. Technol. Res.* **2014**, *11*, 979–991. [CrossRef]
24. Vince, J.; Vuk, A.Š.; Krašovec, U.O.; Orel, B.; Köhl, M.; Heck, M. Solar absorber coatings based on $CoCuMnO_x$ spinels prepared via the sol–gel process: Structural and optical properties. *Sol. Energy Mater. Sol. Cells* **2003**, *79*, 313–330. [CrossRef]
25. Bayón, R.; Vicente, G.S.; Maffiotte, C.; Morales, Á. Characterization of copper–manganese-oxide thin films deposited by dip-coating. *Sol. Energy Mater. Sol. Cells* **2008**, *92*, 1211–1216. [CrossRef]
26. Geng, Q.F.; Zhao, X.; Gao, X.H.; Liu, G. Sol–Gel Combustion-Derived $CoCuMnO_x$ Spinels as Pigment for Spectrally Selective Paints. *J. Am. Ceram. Soc.* **2010**, *94*, 827–832. [CrossRef]
27. Farchado, M.; Vicente, G.S.; Barandica, N.; Morales, A. Novel highly performing tandem selective solar absorber for industrial heat applications. *Sol. Energy Mater. Sol. Cells* **2025**, *279*, 113249. [CrossRef]
28. Peng, Y.J.; Fang, Z.; Hu, J.P.; Wang, B.W.; Xie, P.K. Test Study on Performance of Endothermic Highly Hydrophobic Anti-Icing Materials. *J. Phys. Conf. Ser.* **2020**, *1659*, 012007. [CrossRef]
29. Kordas, G. Incorporation of spherical shaped $CuO@SiO_2$ light microtraps into $CuCoMnO_x$ spinels to enhance solar absorbance. *J. Am. Ceram. Soc.* **2019**, *103*, 1536–1541. [CrossRef]
30. Liu, H.D.; Fu, T.R.; Duan, M.H.; Wan, Q.; Chen, Y.M.; Fu, D.J.; Ren, F.; Li, Q.Y.; Cheng, X.D.; Yang, B.; et al. Structure and thermal stability of spectrally selective absorber based on AlCrON coating for solar-thermal conversion applications. *Sol. Energy Mater. Sol. Cells* **2016**, *157*, 108–116. [CrossRef]
31. He, X.M.; Pu, W.H.; Cai, Y.; Jiang, C.Y.; Wan, C.R.; Xia, D.G. Preparation of spherical $LiMn_2O_4$ for Li-ion batteries based on controlled crystallization. *J. Nonferrous Met.* **2005**, *15*, 1390–1395. (In Chinese)
32. Zhang, H.X. Desulfurization reactivity of zinc based composite oxides. *J. Fuel Chem. Technol.* **2007**, *35*, 619–623. (In Chinese)
33. Du, J.J.; Liu, Y.H.; Yao, G.C.; Zhang, X. Effect of $MnO_2$ Addition on Sintering Behavior and Magnetic Property of Nickel Ferrite Spinel. *J. Northeast Univ.* **2011**, *32*, 1452–1455. (In Chinese)
34. Zhang, X.M.; Wang, X.B.; Zhang, X.; Li, Y.J.; Cheng, X.D. Effect of multilayered CoO-$CoAl_2O_4$ films on improving solar absorptance of Co-WC solar selective absorbing coatings. *Vacuum* **2018**, *155*, 185–192. [CrossRef]
35. Geng, Q.F.; Zhao, X.; Gao, X.H.; Yu, H.C.; Yang, S.R.; Liu, G. Optimization design of $CuCr_xMn_{2-x}O_4$-based paint coatings used for solar selective applications. *Sol. Energy Mater. Sol. Cells* **2012**, *105*, 293–301. [CrossRef]

36. Köhl, M.; Heck, M.; Brunold, S.; Frei, U.; Carlsson, B.; Möller, K. Advanced procedure for the assessment of the lifetime of solar absorber coatings. *Sol. Energy Mater. Sol. Cells* **2004**, *84*, 275–289. [CrossRef]
37. Yu, H.W.; Li, J.K.; Zhang, Q.; Pang, W.; Yan, H.; Li, G.Y. Thermal Stability of Chromium-Iron Oxidation Mixture Cermet-Based Solar Selective Absorbing Coatings. *Molecules* **2020**, *25*, 1178. [CrossRef]

*coatings*

MDPI

*Article*

# Manufacturing Process and Characteristics of Silica Nanostructures for Anti-Reflection at 355 nm

**Anne Gärtner [1,2,*], Mihai-George Mureşan [3], Christian Mühlig [1], Tobias Herffurth [1], Nadja Felde [1], Hanjörg Wagner [1], Ulrike Schulz [1], Astrid Bingel [1], Sven Schröder [1], Tomáš Mocek [3] and Andreas Tünnermann [1,2]**

[1] Fraunhofer Institute for Applied Optics and Precision Engineering IOF, Albert-Einstein-Str. 7, 07745 Jena, Germany

[2] Abbe Center of Photonics, Friedrich Schiller University Jena, Albert-Einstein-Str. 6, 07745 Jena, Germany

[3] HiLASE Centre, Institute of Physics of the Czech Academy of Sciences, Za Radnici 828, 252 41 Dolní Břežany, Czech Republic; muresan@hilase.cz (M.-G.M.)

* Correspondence: anne.gaertner@iof.fraunhofer.de

**Abstract:** Recent advancements in photonics have intensified the performance requirements for optical systems and present significant challenges for optical coating technologies. Conventional interference coating systems often prove to be insufficient, especially in applications requiring large angles of light incidence or a wide wavelength range. Nanostructures, which consist of an air material mixture, offer promising alternatives. In this work, silica nanostructures are manufactured by the AR-plas2 method, in which first an organic layer is evaporated onto a substrate. This organic layer forms self-organizing nanostructures by a plasma etching step, which are subsequently coated with silica. Finally, the organic residues are removed by additional plasma etching and heat treatment steps, which results in hollow silica structures. The work examines the optical and functional properties of these structures designed for 355 nm to demonstrate their use as anti-reflective coatings for advanced optical systems.

**Keywords:** anti-reflective; nanostructures; absorption; light scattering; power spectral density function; laser-induced damage threshold

## 1. Introduction

The rapid development in photonics opens new perspectives for future technologies, particularly in areas such as quantum computing, laser material processing, or semiconductor technology. However, these innovations also require increasingly sophisticated optical systems. That means they must be suitable, e.g., for high laser powers, withstand ultra-high vacuum conditions, address multiple wavelengths simultaneously, or guarantee outstanding optical performance even at high angles of light incidence.

The challenges are particularly evident in systems that utilize ultraviolet (UV) light. Not only is the choice of materials limited, but interactions between UV light and materials can also lead to unwanted absorption, which may result in the destruction of optical components. Anti-reflective (AR) coatings in particular play a key role here, as they not only have to provide excellent optical performance but are often the weakest element in such systems [1–3]. The choice of materials and manufacturing processes is particularly important here and should be chosen carefully [4]. For example, optical systems that operate with lasers with a wavelength of 355 nm are of technical importance as they are used in many fields, e.g., semiconductor technology or material processing. That is why several research groups have developed AR coating systems for this wavelength. Hafnia,


https://doi.org/10.3390/coatings15050556

as a material with a high refractive index in combination with silica ($SiO_2$), has proven itself as a coating system with a high laser-induced damage threshold and has been investigated by different groups [1,4–8]. Juškevičius et al. [5], for example, present an AR system that consists of silica and hafnia that was applied on fused silica using e-beam coating technology. The coating system achieves a reflection of less than 0.2% at 355 nm and a laser-induced damage threshold (LIDT) of 6.8 J/cm$^2$ on non-etched fused silica (measurement wavelength: 355 nm; 1-on-1 test according to [9]; pulse duration: 3 ns; repetition rate: 15 Hz; Nd:YAG laser). Falmbigl et al. [1] investigated a double-sided coated AR coating system on fused silica, also consisting of silica and hafnia, produced by ion beam sputtering. They achieved transmission values above 99.8% at 355 nm. Absorption was measured using photothermal common-path interferometry, and they achieved a value of around 110 ppm. The LIDT value was measured with 15.7 J/cm$^2$ (measurement wavelength: 355 nm; 200-on-1 test; pulse duration: 8 ns; repetition rate: 20 Hz). However, LIDT values could only be compared with each other when the same substrate or measurement parameters were used, because they depend on various factors, including the choice of coating method, the material quality and surface condition of the substrate [10–12]. Consequently, a more meaningful evaluation of the LIDT values is achieved by comparing them with those of the uncoated substrate, an approach that is employed in this study.

The use of commercial interference coating systems to minimize unwanted reflections has its limitations, as they are often unable to meet the new requirements for optical components due to the limited choice of materials and therefore predefined refractive indices. An alternative is offered by nanostructures, in which an air material mixture creates an effective refractive index that can be optimally adapted to the design [13–19]. Several studies show that the combination of interference coatings and nanostructures or the combination of more than one nanostructure provide improved optical performance [20,21]. In addition, the LIDT can be higher than that of interference coatings. This was already demonstrated in a round robin experiment in 2010, in which wet-chemically produced sol gel coatings exhibited the highest LIDT values for a wavelength of 355 nm [4]. Another experiment was performed by Tolenis et al. They compared nanostructures that were produced using glancing angle deposition (GLAD) technology, with a coating based on silica and hafnia made by ion beam sputtering (IBS); both were deposited on fused silica [22]. They achieved a reflection of 0.2% for the GLAD structures and 0.3% for the silica/hafnia coating system at 355 nm. They showed that the LIDT of the GLAD structures was 16.8 J/cm$^2$ compared to 5.1 J/cm$^2$ for the IBS manufactured coating (measurement wavelength: 355 nm; 1-on-1 test according to [9]; pulse duration: 3.1 ns; repetition rate: 15 Hz; Nd:YAG laser). Other processes to produce nanostructures besides wet-chemical coating processes [23] are described in the literature, e.g., lithographic processes [24,25], reactive ion etching [26–28] or production using physical vapor deposition technology [7,22,29]. In this paper, nanostructures made of silica are fabricated using AR-plas2 technology, which uses a commercially available chamber for physical vapor deposition equipped with an advanced plasma source (APS). This technology is cost-effective and suitable for mass pro-duction. A major advantage of this approach is the production of layer systems within one coating process, where nanostructures can be combined with interference layer systems, but also several structures with each other. This results in improved layer systems that meet the increasing requirements, e.g., for optical broadband systems. The method was originally developed for etching polymer surfaces and later adapted to the structuring of organic layers. For this, an organic transfer layer is deposited onto an arbitrary substrate. Subsequently, self-assembling nanostructures are formed by plasma etching, which are then overcoated with silica. The organic components can be removed by additional post-treatment steps, resulting in hollow silica nanostructures.

Due to the technical importance of optical systems with lasers for 355 nm mentioned at the beginning, nanostructures with an AR effect for 355 nm were produced and investigated in this work. Since they have a higher roughness due to their morphology, which in turn leads to higher optical losses, a structural analysis was conducted in addition to the examination of optical properties. The organic material used was Xanthine, which forms nanostructures after plasma etching and has already been introduced in a previous work [30]. Most of the organic material was removed by additional treatment steps, resulting in hollow silica structures. To investigate whether residual organic material was still in the structures, absorption and LIDT measurements were conducted.

## 2. Material and Methods

### 2.1. Sample Preparation

The nanostructures were designed for the wavelength of 355 nm and double-sided coated on SK-1300 (Ohara, Japan) glass with a 1 inch diameter and 1 mm thickness. There were no specific requirements for the substrate quality. The substrates were pre-cleaned using ethanol and subsequently cleaned in an ultrasonic bath.

The commercial coating chamber SyrusPro 1100 (Bühler AG, Alzenau, Germany) was used for the AR-plas2 method. The chamber is equipped with an APS to support chemical and physical reactions, a thermal evaporator to deposit organic materials, and an electron beam gun for oxide materials. During the process, a homogeneous Xanthine (Sigma-Aldrich, Taufkirchen, Germany) layer was first deposited from a molybdenum boat onto the substrate at an initial chamber pressure of $2 \times 10^{-5}$ mbar. The deposition of the organic material was controlled by optical monitoring to achieve a thickness of 145 nm. Subsequent etching with the APS and an $O_2$/Ar gas mixture led to self-assembling nanostructure formation, which was coated with 32 nm of silica [30]. It is assumed that both chemical transformation and physical removal led to the formation of the structures. The organic part of nanostructure was removed by subsequent annealing and a standard wet-cleaning procedure, resulting in hollow silica nanostructures.

### 2.2. Topographical Measurements and Analysis

Roughness analysis was performed using an atomic force microscope (AFM) Dimension Icon (Bruker, CA, USA) and a white light interferometer (WLI) NewView 7300 (ZygoLOT Europe GmbH, Darmstadt, Germany). Using AFM measurements, roughness was analyzed for spatial frequencies from 0.02 $\mu m^{-1}$ to 256 $\mu m^{-1}$, corresponding to scan areas from $(50 \times 50)$ $\mu m^2$ to $(1 \times 1)$ $\mu m^2$. WLI measurements were performed for 5$\times$ to 50$\times$ magnification to calculate the roughness for spatial frequencies down to 0.001 $\mu m^{-1}$. By combining the different roughness measurement methods and areas, roughness information can be analyzed for a wide range of surface structures. Therefore, a quantitative evaluation of the data was performed by calculating the two-dimensional power spectral density (PSD) function for each measurement according to the following formula:

$$PSD\,(f_x, f_y) = \lim_{L\to\infty} \frac{1}{L^2} \left| \int_{-\frac{L}{2}}^{\frac{L}{2}} \int_{-\frac{L}{2}}^{\frac{L}{2}} \left( h(x,y) e^{-2\pi i (f_x x + f_y y)} dxdy \right|^2, \quad (1)$$

where h(x,y) represents the topographical surface data, L the scan size and fx and fy the spatial frequency components. The PSD function (Equation (1)) describes the relative strength of the individual roughness components in terms of spatial frequencies and thus consider not only the vertical, but also the lateral distribution of the measured topography data [31]. In addition, a scanning electron microscope (SEM) from Zeiss SIGMA SEM was used to magnify the structures.

For the ex situ characterization after laser damage tests of the samples, an OLX-5000 (Olympus, Tokyo, Japan) confocal laser scanning microscope was used, and the images were evaluated using the open-source software package Gwyddion (version 2.66) [32].

### 2.3. *Reflectance, Transmission, Absorption, and Light Scattering Measurements*

To investigate the optical performance, transmission (T) and reflection (R) have been measured with a UV/VIS spectrophotometer Lambda 950 (Perkin Elmer, Waltham, MA, USA) at an angle of light incidence of 6°. A special inset was used to measure absolute R und T values of the samples without changing its position [33]. The accuracy of the instrument is given with 0.1% in the visible and 0.2% in the UV spectral range.

Light scattering properties of the nanostructures were quantified by angle-resolved light scattering measurements (ARS; for measurement geometry, see Figure 1) using a 3D scatterometer (system MLS10, developed at Fraunhofer IOF). The ARS is defined as the scattered power ($\Delta P_s$) into the solid angle $\Delta\Omega_s$ as a function of the scattering angles $\theta_s$ and $\varphi_s$, normalized to this solid angle and the incident light power ($P_i$).

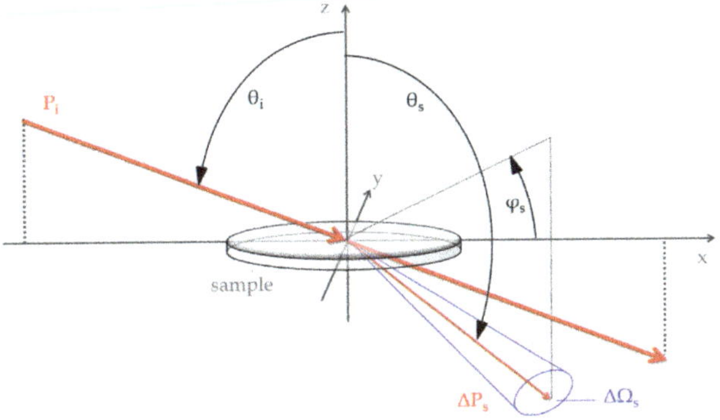


**Figure 1.** Light scattering measurement geometry in transmission direction, similar in reflection hemisphere ($\theta_i$—angle of light incidence; $\theta_s$, $\varphi_s$—polar and azimuthal scattering angles; $\Delta\Omega_s$—solid angle).

Considering the scattering over the entire hemisphere for backward and forward direction, the numerical integration of the ARS, excluding specular reflection/transmission, gives the total scattering (TS = Ps/Pi) over the whole hemisphere. TS can be calculated by the integration from the measured ARS, which is scattered into the forward or backward hemispheres. At normal incidence, scattering angles of $2° \leq \theta s \leq 85°$ must be considered for the measurement of TS (according to the ISO 13696 standard) [34]. A description of the set-up for the light scattering measurements can be found in [35]. The uncertainty is given with 10%.

Absolute absorption (A) measurements were performed using the laser-induced deflection method. This method utilizes a photothermal technique with a transverse pump sample beam configuration. For the samples, a modified measurement concept was used, the so-called thin disc vertical concept. Further details are provided in [36].

The measurements for light scattering and absorption were performed at 355 nm.

### 2.4. *Laser-Induced Damage Threshold (LIDT) Measurements*

The laser-induced damage threshold (LIDT) measurements were conducted using a previously described set-up [37]. This set-up operates in an ISO class 6 clean room under ambient conditions and uses a tripled Q-Smart 450 (Lumibird, Lannion, France) system as the laser source. A schematic of the LIDT set-up is presented in Figure 2.

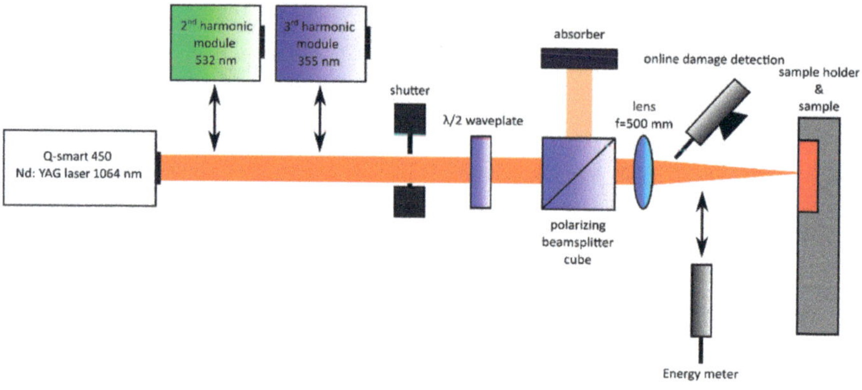


**Figure 2.** Schematics of the LIDT measurement set-up used to measure the 1-on-1, 10-on-1, and R-on-1 test types.

All tests were conducted with a linearly polarized (p-polarization) pulse with a duration of 8.5 ns in the third harmonic (355 nm) of the neodymium-doped yttrium aluminum garnet (Nd:YAG) laser and a frequency of 10 Hz. The beam diameter, which was fitted with a Gaussian profile at $1/e^2$, varied between 82 and 174 µm. The greatest contribu-tion to the overall error budget was represented by the pulse-to-pulse energy stability and the beam profiler accuracy. The combined error contribution to the tests was calculated to be approximately $\pm$ 8%. The LIDT tests were performed in compliance with ISO 21254 [9,38–40], using the 1-on-1, S-on-1, and R-on-1 methods.

## 3. Results and Discussion

### 3.1. Morphological and Surface Analysis

Figure 3 shows the SEM images of the silica nanostructures. In the cross-sectional image (Figure 3a) it can be seen, that columnar structures have been formed. The top view indicates that the structures are stochastically distributed over the surface (Figure 3b).

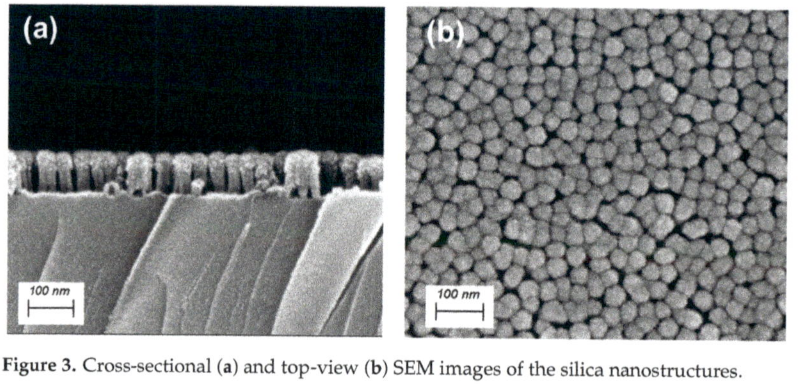


**Figure 3.** Cross-sectional (**a**) and top-view (**b**) SEM images of the silica nanostructures.

In addition to SEM images, topographical data were assessed using AFM and WLI for a quantitative evaluation of the surface by calculating the PSD functions. To capture information over a broad range of spatial frequencies, measurements were conducted in various measurement areas. Exemplarily, AFM images for a $(1 \times 1)$ $\mu m^2$ scan area are given in Figure 4.

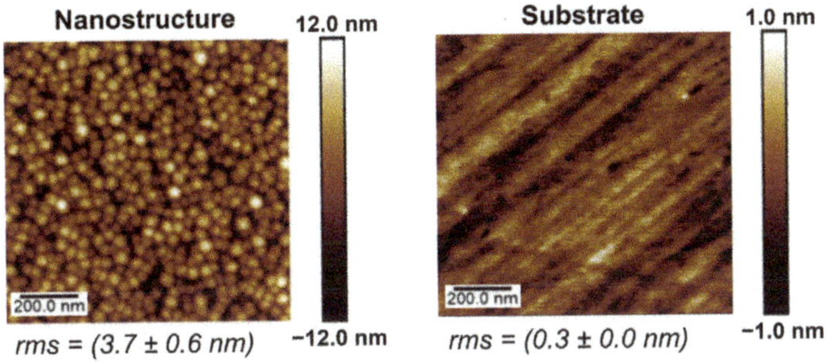


**Figure 4.** AFM images of the silica nanostructures and uncoated substrate.

The power spectral density (PSD) function is a technique for analyzing the distribution of surface roughness over different spatial frequencies. Rather than just average roughness values (e.g., average roughness (Ra) or root mean square roughness (rms)), which only provide a single value containing the vertical information of the surface, the PSD function provides a more comprehensive view by adding lateral information of the surface features in terms of their frequency components. This provides a better understanding of how the anti-reflective coating interacts with light at different wavelengths (e.g., in terms of light scattering) and can identify corresponding optical behavior. Particu-larly for light scattering the spatial frequencies are directly linked to the scattering direc-tion by the grating equations. The corresponding PSD functions were evaluated (as described in Section 2.2) and combined by geometrically averaging, resulting in a so-called Master PSD function [21]. This approach not only allowed for the analysis of the scattering-relevant spatial frequency range but also provided insights into the structures at both high and low spatial frequencies. This enabled us to evaluate the influence of the coatings on substrate roughness and, consequently, the overall morphology of the samples. The Master PSD functions for both the nanostructure and the uncoated substrate, are shown in Figure 5. Figure 5a) also shows the spatial frequency range relevant for light scattering at a wavelength of 355 nm. The Master PSD function reveals an almost perfect replication of the substrate roughness below $f = 0.1\ \mu m^{-1}$, which is typical for thin films with low intrinsic roughness. A difference to the substrate occurs in the mid- and high-spatial-frequency range, which can be attributed to the nanostructure, where the roughness becomes relevant. However, it should be noted that only particle-free areas were considered for the roughness analysis. When particles are taken into account, the roughness increases drastically, as can be seen in Figure 5b), which shows the PSD functions for particle-free and contaminated surface areas. The origin of the particles is currently unknown and is already being investigated in more detail.

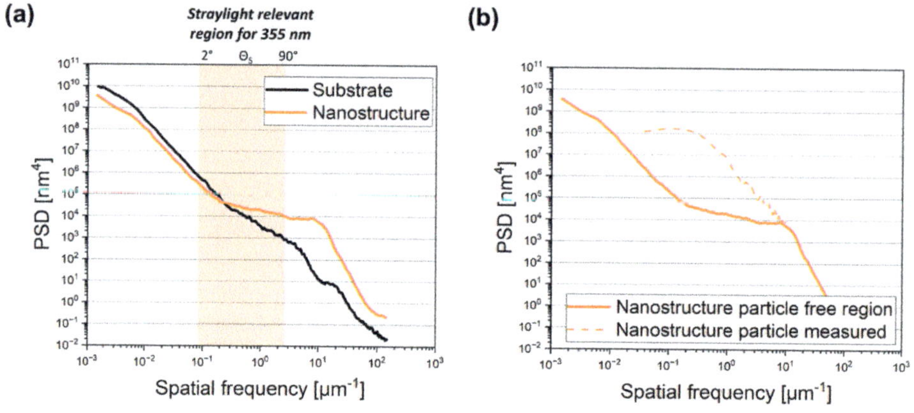


**Figure 5.** Master PSD functions of an uncoated substrate (black line) and the silica nanostructures (orange line, (**a**)). Comparison between the Master PSD functions of nanostructures on a clean position (straight line) compared to a position with a particle (dotted line, (**b**)).

### 3.2. Optical Properties and the Calculation of the Effective Refractive Index

Figure 6 shows the corresponding R and T (a) spectra of a double-sided coated substrate and an uncoated substrate, and the optical losses (b), which were calculated by subtracting R and T from 100%. The wavelength range for the measurements was between 300 nm and 400 nm.

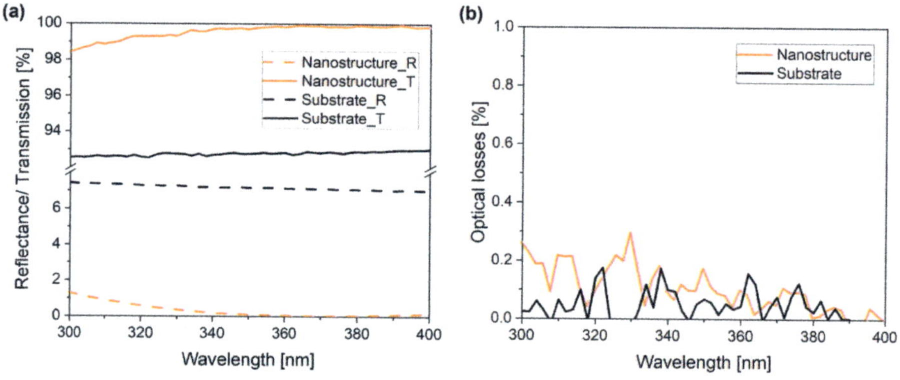


**Figure 6.** Reflectance and transmittance measurements (**a**) and optical losses (**b**) for an uncoated substrate (black line), and the double-sided coated silica nanostructures (orange line).

The nanostructures show a transmission value of 99.8% and a reflectance value of 0.1% at a design wavelength of 355 nm (see Figure 6a). This is comparable to the value of the system of interference layers given by Falmbigl et al. [1]. The minimum/maximum is shifted to 370 nm. However, the curve can be shifted to shorter wavelengths by further adjusting the coating parameters. The optical losses for the nanostructures are approximately 0.1% at 355 nm (see Figure 6b), which may contain contributions from both light scattering and absorption. However, since the noise of the measuring device prevents the exact determination of the losses, light scattering and absorption were measured separately according to the methods described in Section 2.3. The results are also given in this section.

To calculate the effective refractive index ($n_{eff}$), a three-layer model was assumed by reverse data fitting with OptiRe software (version 13.77, OptiLayer GmbH, Ismaning, Germany) using the spectra (Figure 6) and the heights extracted from microscope images taken with the SEM. For this purpose, the structures were divided into three areas and their corresponding $n_{eff}$ values were estimated (see Figure 7). A height of 40 nm with an $n_{eff}$ of 1.15 is calculated for H1, as well as a height of 20 nm and an $n_{eff}$ of 1.3 for H2 and a height of 15 nm with an $n_{eff}$ of 1.05 for H3. As the structures are not the same height and the material is not equally distributed, this calculation is subject to errors and is therefore only a first approximation.

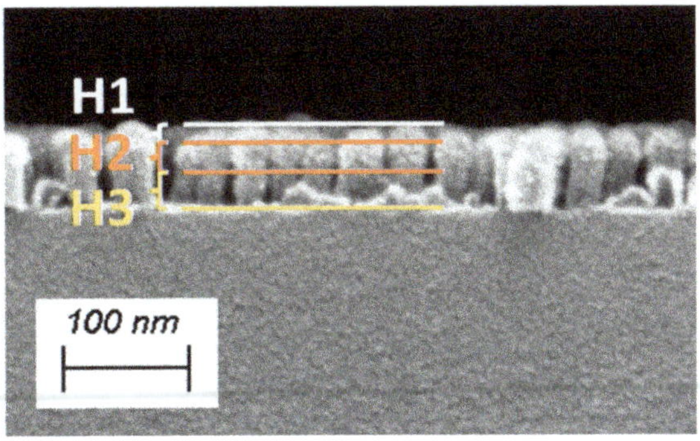


**Figure 7.** Sketch of the three-layer model to calculate the effective refractive index ($n_{eff}$) of the silica nanostructures. The calculation is a first approximation, due to measurement errors, with the follow results: H1 with 40 nm height and $n_{eff}$ of 1.15, H2 with 20 nm height and $n_{eff}$ of 1.3, and H3 with 15 nm height and $n_{eff}$ of 1.05.

Light scattering depends on different properties of the coating, e.g., the surface structure, but also volume effects or material properties (e.g., refractive index). However, the surface itself has the greatest influence. Figure 8 shows the ARS curve for the structure in-cluding the uncoated substrate. In addition, the corresponding calculated TS value for the silica nanostructures is (1652 ± 165) ppm, and it is (209 ± 21) ppm for the uncoated sub-strate.

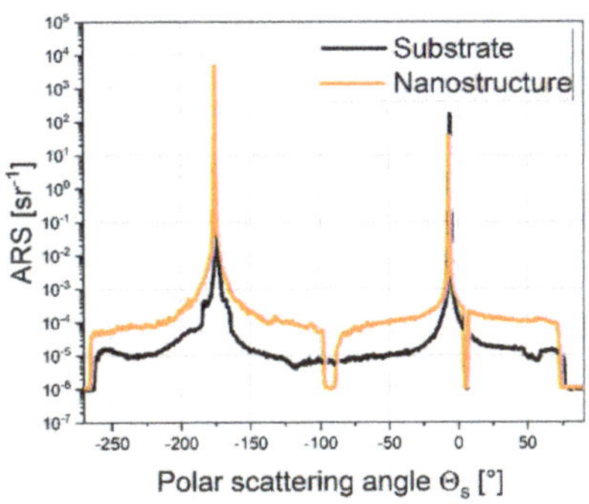


**Figure 8.** ARS measurements at 355 nm for an angle of light incidence of 6° for an uncoated substrate (black line) and the silica nanostructures (orange line).

The curves in Figure 8 display distinct peaks at −180° and 0°, corresponding to the directions of specular transmission and reflection, respectively. When evaluating the residual scatter signal in the off-specular direction, an increase in both the back and front scattering directions is observed for the nanostructures compared to the scattering from the substrate. However, based on the examination of the structural properties (see SEM images in Figure 3), a lower TS value would have been expected, as the heights of the single columns are regular and not clustered. Looking again at the ARS curves in Figure 8, oscillations can be recognized for the nanostructures both in the near-and far-angle range. This indicates light scattering induced by particles, which is particularly evident in the forward scattering [41].

The absorption value for the nanostructures was determined to be (166.8 ± 15.0) ppm. This value can be attributed to the residual organic material, which was used during the manufacturing process and can still be contained within the silica structures.

### 3.3. Laser-Induced Damage Threshold (LIDT) Results

Each sample was tested within 80% of the clear aperture using a closed hexagonal pattern with 127 positions placed 1.55 mm from each other, thus fullfilling the general recommendation for the distribution to prevent cross contamination. For the 1-on-1 and S-on-1 tests, where S is 10, 10 positions with the same fluence were tested for statistical purposes, while 24 positions were tested in the R-on-1 test. For the latter, each position was initially irradiated with a maximum of 100 pulses at a constant fluence. Subsequently, the fluence was increased by 0.5 $J/cm^2$ and the sample was irradiated once more with a maximum of 100 pulses until the surface showed visible signs of destruction. The results for all tests are given in Table 1.

**Table 1.** Summary of the LIDT values.

| Sample | 1-on-1 [$J/cm^2$] | 10-on-1 [$J/cm^2$] | R-on-1 [$J/cm^2$] |
|---|---|---|---|
| Nanostructure | 4.38 ± 0.34 | 2.87 ± 0.20 | 1.37 ± 0.42 |
| Substrate | 6.49 ± 0.50 | 4.26 ± 0.30 | 2.28 ± 0.59 |

The results of the 1-on-1 test indicate that the nanostructures show lower LIDT values to the substrate. Furthermore, the 1-on-1 methodology test provides insights into the fundamental damage mechanisms of the given material. Using high-resolution laser scanning microscopy, different damage morphologies could be distinguished between the nanostructures and the substrate, as shown in Figure 9. Only damaged positions of fluences exceeding the LIDT were considered in the study. For the silica nanostructures, a damage pattern is shown at a fluence of 11.58 $J/cm^2$, while for the uncoated substrate, damage is shown at 24.77 $J/cm^2$.

Based on the crater depth analysis in Figure 9, it can be seen that the silica nanostructure and the substrate presented different forms of melting and resolidification of the material, either from the coating or in combination with the glass substrate [42]. In addition, the material ejected from the crater is deposited in the nearby zone, visible in the slight color change, in form of a halo around the crater. For the nanostructures, the crater depth exceeded the coating thickness, reaching depths between 150 and 300 μm.

As the results of the 10-on-1 test show, the LIDT values are lower than the values observed in the 1-on-1 test, which was anticipated. Figure 10 presents the microscopic images of the damage morphologies for the samples. The damage morphology of the nanostructures shows a few small circumferential cracks besides melting. There are no cracks, which might be related to the increased absorption. The analyzed surface spot on the substrate was irradiated with an intensity of 12.15 $J/cm^2$ and exhibited indications of profound ablation that has overcome the melting phase, which is characteristic for bare substrates or crystals.

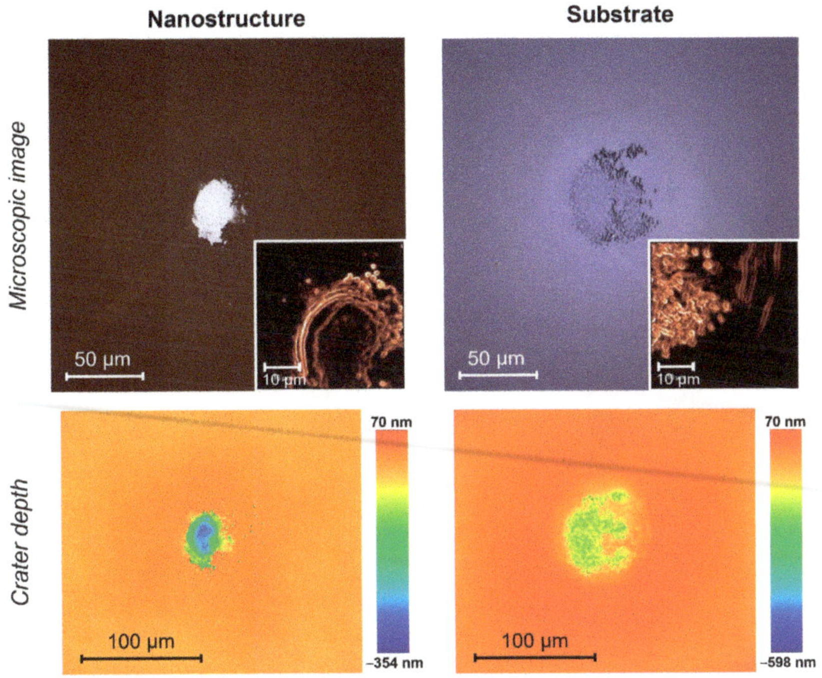


**Figure 9.** Damage morphology for the nanostructure and uncoated substrate, showing the microscope image with an inset for more details, and the crater depth for the 1-on-1 LIDT test.

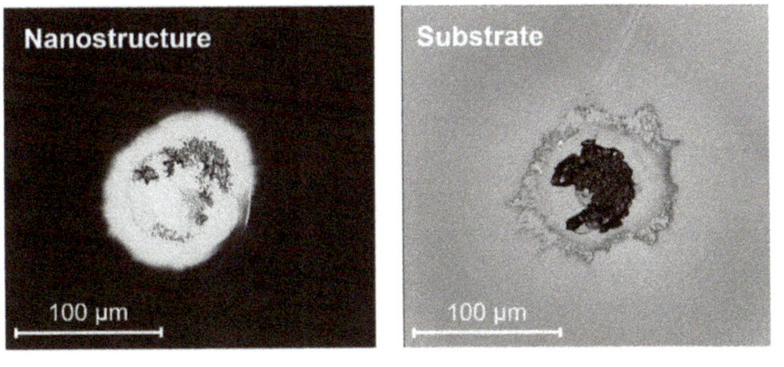


**Figure 10.** Microscopic images for the damage morphology of the silica nanostructure and the uncoated substrate after 10-on-1 LIDT test.

Although a 0.5 J/cm$^2$ step and 100 pulses for each fluence were selected for the R-on-1 test, the results shown in Table 1 represent the mean LIDT value obtained from 24 individual positions. Looking at the uncoated substrate, a standard deviation of around 25% from the mean value can be seen, whereas the standard deviation for the nanostructures is around 30%. This may be related to small inhomogeneities in the substrate, which can cause local absorption points that are known to lower the LIDT value, especially in the nanosecond regime. In addition, organic residues in the nanostructures can reduce the LIDT, which may correlate with the absorption values in Section 3.2. Figure 11 shows an overview where damage occurred after the nanostructure and the uncoated substrate were irradiated with different fluences. Additionally, damage is found on both the front and back of all samples. The front side presents marks of pulse accumulation and the respective crater formation. The craters tended to be relative large in size and presented marks of molten materials and cracks/flaking from the edges [43]. The damage on the rear side can be explained by the Fresnel diffraction by contamination particles, surface imperfections, and perhaps by the nanostructures themselves [44]. No conditional effects could be identified in the tested samples.

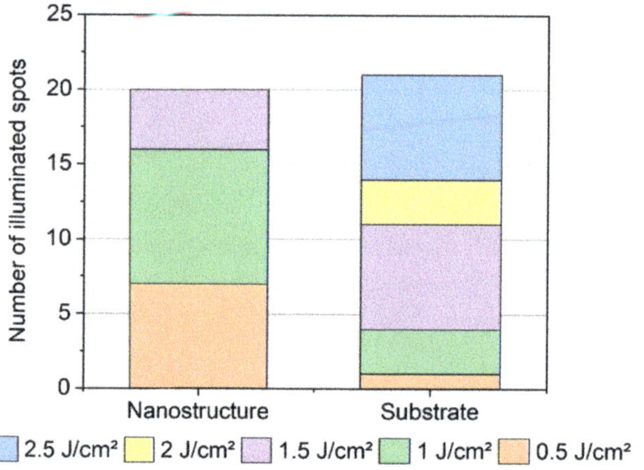


**Figure 11.** Distribution of all illumination spots where damage occurred for the R-on-1 test depending on the fluences for the silica nanostructures and uncoated substrate. The graph only consider values at which damage occurred and excludes damage below 0.5 $J/cm^2$.

Although the values are generally lower than those reported in the literature, it should be mentioned that the LIDT depends on various factors, such as the surface quality. Therefore, the values should always be compared with the LIDT of the uncoated substrate, which is the highest achievable LIDT.

## 4. Conclusions

In this study, we successfully produced hollow silica nanostructures specifically designed for a wavelength of 355 nm. The achieved transmittance of 99.8% shows very good optical performance suitable for anti-reflective coatings. The optical properties were described with a model consisting of three layers and calculated by reverse data fitting using transmittance and reflectance spectra. The surface roughness, formed by the thickened heads of the columns, exhibited an rms value of 3.7 nm. A PSD analysis revealed that the roughness of the nanostructures in the mid- and high-frequency range differs from that of the uncoated substrate, as the roughness of the nanostructures only becomes relevant in these ranges. In the low-spatial-frequency range (below 0.1 μm), the Master PSD function shows almost perfect agreement with the substrate roughness, which is typical for thin films with low intrinsic roughness. The total scattering, approximately 1652 ppm, absorption losses of around 170 ppm and the comparatively low LIDT value are presumably due to residual organics in the silica structures. Further reduction of the organics, e.g., by solvents, is currently being investigated. After optimization, the AR-plas2 process seems promising for producing UV laser-resistant AR coatings.

**Author Contributions:** Methodology, A.G., M.-G.M., C.M., N.F. and H.W.; Investigation, A.G., C.M., N.F. and H.W.; Writing—original draft, A.G., M.-G.M. and N.F.; Writing—review & editing, A.G., M.-G.M., C.M., T.H., H.W., U.S., A.B., S.S., T.M. and A.T. All authors have read and agreed to the published version of the manuscript.

**Funding:** This research was funded by the German Federal Ministry of Education and Research (BMBF) as part of the funding program Photonics Research Germany (project Qzell, contract number 13N15957), and the European Union and state budget of the Czech Republic under the project LasApp CZ.02.01.01/00/22_008/0004573.

**Institutional Review Board Statement:** Not applicable.

**Informed Consent Statement:** Not applicable.

**Data Availability Statement:** Data are contained within the article.

**Conflicts of Interest:** The authors declare no conflict of interest.

## References

1. Falmbigl, M.; Godin, K.; George, J.; Mühlig, C.; Rubin, B. Effect of annealing on properties and performance of $HfO_2/SiO_2$ optical coatings for UV-applications. *Opt. Express* **2022**, *30*, 12326–12336. [CrossRef] [PubMed]
2. Jensen, L. *Laserinduzierte Zerstörung in Oxidischen Schichtsystemen für den UV Bereich*; Gottfried Wilhelm Leibniz Universität Hannover: Hannover, Germany, 2014.
3. Arnold, B.M.; Rashvand, C.; Willis, L.; Dabney, M.S. UV fatigue of laser optics: Laser-induced contamination. *Laser-Induc. Damage Opt. Mater.* **2022**, *2022*, 12300.
4. Stolz, C.J.; Caputo, M.; Griffin, A.J.; Thomas, M.D. BDS thin film UV antireflection laser damage competition. *Laser-Induc. Damage Opt. Mater.* **2010**, *2010*, 7842.
5. Juškevičius, K.; Buzelis, R.; Abromavičius, G.; Samuilovas, R.; Abbas, S.; Belosludtsev, A.; Drazdys, R.; Kičas, S. Argon plasma etching of fused silica substrates for manufacturing high laser damage resistance optical interference coatings. *Opt. Mater. Express* **2017**, *7*, 3598–3607. [CrossRef]
6. Alam, S.; Paul, P.; Beladiya, V.; Schmitt, P.; Stenzel, O.; Trost, M.; Wilbrandt, S.; Mühlig, C.; Schröder, S.; Matthäus, G.; et al. Heterostructure films of $SiO_2$ and $HfO_2$ for high-power laser optics prepared by plasma-enhanced atomic layer deposition. *Coatings* **2023**, *13*, 278. [CrossRef]
7. Weiss, M.; Davenport, A.; Lucas, S.C.; Pamedytytė, D.; Galinis, J.; Melninkaitis, A.; Siehien, J.; Siehien, W.; Chicoine, M.; Schiettekatte, F.; et al. Laser damage of UV hafnia-based multilayer dielectric coatings at 355 nm wavelength. *Laser-Induc. Damage Opt. Mater.* **2024**, *2024*, 13190.
8. Yin, C.; Zhu, M.; Zeng, T.; Song, C.; Chai, Y.; Shao, Y.; Zhang, R.; Zhao, J.; Li, D.; Shao, J. $HfO_2/SiO_2$ anti-reflection films for UV lasers via plasma-enhanced atomic layer deposition. *J. Alloys Compd.* **2021**, *859*, 157875. [CrossRef]
9. *ISO 21254-1:2011*; Lasers and Laser-Related equipment—Test Methods for Laser-Radiation-Induced Damage Threshold—Part 1: Definitions and General Principles. International Organization for Standardization: London, UK, 2011.
10. Lyngnes, O.; Ode, A.; Ness, D.C. Anti-reflection coating damage threshold dependence on substrate material. In *Laser-Induced Damage in Optical Materials*; SPIE: Bellingham, WA, USA, 2009; Volume 7504.
11. Zhu, M.; Xing, H.; Chai, Y.; Yi, K.; Sun, J.; Wang, J.; Shao, J. Improving the laser-induced damage threshold of 532-nm antireflection coating using plasma ion cleaning. *Opt. Eng.* **2017**, *56*, 011003. [CrossRef]
12. Hobbs, D.S. Laser damage threshold measurements of anti-reflection microstructures operating in the near UV and mid-infrared. In *Laser-Induced Damage in Optical Materials*; SPIE: Bellingham, WA, USA, 2010; Volume 7842.
13. Chen, M.; Chang, H.C.; Chang, A.S.; Lin, S.Y.; Xi, J.Q.; Schubert, E.F. Design of optical path for wide-angle gradient-index antireflection coatings. *Appl. Opt.* **2007**, *46*, 6533–6538. [CrossRef]
14. Xi, J.Q.; Schubert, M.F.; Kim, J.K.; Schubert, E.F.; Chen, M.; Lin, S.Y.; Liu, W.; Smart, J.A. Optical thin-film materials with low refractive index for broadband elimination of Fresnel reflection. *Nat. Photonics* **2007**, *1*, 176–179. [CrossRef]
15. Dobrowolski, J.A.; Poitras, D.; Ma, P.; Vakil, H.; Acree, M. Toward perfect antireflection coatings: Numerical investigation. *Appl. Opt.* **2002**, *41*, 3075–3083. [CrossRef] [PubMed]
16. Minot, M.J. The angluar reflectance of single-layer gradient refractive-index films. *J. Opt. Soc. Am.* **1977**, *67*, 1046–1050. [CrossRef]
17. Southwell, W.H. Gradient-index antireflection coatings. *Opt. Lett.* **1983**, *8*, 584–586. [CrossRef] [PubMed]
18. Southwell, W.H. Pyramid-array surface-relief structures producing antireflection index matching on optical surfaces. *J. Opt. Soc. Am. A* **1991**, *8*, 549–553. [CrossRef]
19. Brunner, R.; Sandfuchs, O.; Pacholski, C.; Morhard, C.; Spatz, J. Lessons from nature: Biomimetic subwavelength structures for high-performance optics. *Laser Photonics Rev.* **2012**, *6*, 641–659. [CrossRef]
20. Bruynooghe, S.; Schulze, M.; Helgert, M.; Challier, M.; Tonova, D.; Sundermann, M.; Koch, T.; Gatto, A.; Kley, E.-B. Broadband and wide-angle hybrid antireflection coatings prepared by combining interference multilayers with subwavelength structures. *J. Nanophotonics* **2016**, *10*, 033002. [CrossRef]

21. Chhajed, S.; Schubert, M.F.; Kim, J.K.; Schubert, E.F. Nanostructured multilayer graded-index antireflection coating for Si solar cells with broadband and omnidirectional characteristics. *Appl. Phys. Lett.* **2008**, *93*, 251108. [CrossRef]
22. Tolenis, T.; Grinevičiūtė, L.; Buzelis, R.; Smalakys, L.; Pupka, E.; Melnikas, S.; Selskis, A.; Drazdys, R.; Melninkaitis, A. Sculptured anti-reflection coatings for high power lasers. *Opt. Mater. Express* **2017**, *7*, 1249–1258. [CrossRef]
23. Chi, F.; Pan, N.; Ding, C.; Wang, X.; Yi, F.; Li, X.; Lei, J. Ultraviolet laser-induced damage of freestanding silica nanoparticle films. *Appl. Surf. Sci.* **2019**, *463*, 566–572. [CrossRef]
24. Ting, C.-J.; Chen, C.-F.; Chou, C. Subwavelength structures for broadband antireflection application. *Opt. Commun.* **2009**, *282*, 434–438. [CrossRef]
25. Wang, S.; Yu, X.Z.; Fan, H.T. Simple lithographic approach for subwavelength structure antireflection. *Appl. Phys. Lett.* **2007**, *91*, 061105. [CrossRef]
26. Hobbs, D.S.; MacLeod, B.D. High laser damage threshold surface relief micro-structures for anti-reflection applications. *Laser-Induc. Damage Opt. Mater.* **2007**, *2007*, 6720.
27. Okabe, T.; Yano, T.; Yatagawa, K.; Taniguchi, J. Polyimide moth-eye nanostructures formed by oxygen ion beam etching for anti-reflection layers. *Microelectron. Eng.* **2021**, *242*, 111559. [CrossRef]
28. Barlow, A.J.; Sano, N.; Murdoch, B.J.; Portoles, J.F.; Pigram, P.J.; Cumpson, P.J. Observing the evolution of regular nanostructured indium phosphide after gas cluster ion beam etching. *Appl. Surf. Sci.* **2018**, *459*, 678–685. [CrossRef]
29. Prachachet, R.; Horprathum, M.; Eiamchai, P.; Limwichean, S.; Chananonnawathorn, C.; Samransuksamer, B.; Lertvanithphol, T.; Buranasiri, P.; Muthitamongkol, P.; Boonruang, S. A comparative study on omnidirectional anti-reflection $SiO_2$ nanostructure films coating by glancing angle deposition. *Oxide-Based Mater. Devices IX* **2018**, *2018*, 10533.
30. Gärtner, A.; Seifert, T.; Rickelt, F.; Schulz, U.; Tünnermann, A. Xanthine: A promising organic material for the development of nanostructured anti-reflective layers. *Adv. Opt. Thin Film. VII* **2021**, *2021*, 11872.
31. Duparré, A.; Ferre-Borrull, J.; Gliech, S.; Notni, G.; Steinert, J.; Bennett, J.M. Surface characterization techniques for determining the root-mean-square roughness and power spectral densities of optical components. *Appl. Opt.* **2002**, *41*, 154–171. [CrossRef]
32. Nečas, D.; Klapetek, P. Gwyddion: An open-source software for SPM data analysis. *Open Phys.* **2012**, *10*, 181–188. [CrossRef]
33. Wilbrandt, S.; Stenzel, O. In Situ and Ex Situ Spectrophotometric Characterization of Single-and Multilayer-Coatings II: Experimental Technique and Application Examples. In *Optical Characterization of Thin Solid Films*; Springer International Publishing: Cham, Switzerland, 2018; pp. 203–232.
34. Stover, J. *Optical Scattering: Measurement and Analysis*, 3rd ed.; SPIE: Bellingham, WA, USA, 2012.
35. Schröder, S.; Unglaub, D.; Trost, M.; Cheng, X.; Zhang, J.; Duparré, A. Spectral angle resolved scattering of thin film coatings. *Appl. Opt.* **2014**, *53*, A35–A41. [CrossRef]
36. Bublitz, S.; Mühlig, C. Absolute absorption measurements in optical coatings by laser induced deflection. *Coatings* **2019**, *9*, 473. [CrossRef]
37. Mydlář, M.; Vanda, J.; Mureșan, M.G.; Čech, P.; Brajer, J.; Mocek, T. Mobile LIDT. In Proceedings of the Optics and Measurement International Conference, Liberec, Czech Republic, 8–10 October 2019; p. 11385.
38. *ISO 21254-2:2011*; Lasers and Laser-Related Equipment—Test Methods for Laser-Radiation-Induced Damage Threshold—Part 2: Threshold Determination. International Organization for Standardization: London, UK, 2011.
39. *ISO 21254-3:2011*; Lasers and Laser-Related Equipment—Test Methods for Laser-Radiation-Induced Damage Threshold—Part 3: Assurance of Laser Power (Energy) Handling Capabilities. International Organization for Standardization: London, UK, 2011.
40. *ISO 21254-4:2011*; Lasers and Laser-Related Equipment—Test Methods for Laser-Radiation-Induced Damage Threshold—Part 4: Inspection, Detection and Measurement. International Organization for Standardization: London, UK, 2011.
41. Cheng, X.; Wang, Z. Defect-related properties of optical coatings. *Adv. Opt. Technol.* **2014**, *3*, 65–90. [CrossRef]
42. Xia, S.; Cai, J.; Zhang, X.; Li, J.; Jin, G.; Chang, X. Study on thermal stress of the the fused silica irradiated by millisecond–nanosecond combined pulse laser. *Pramana* **2021**, *95*, 1–8. [CrossRef]
43. Negres, R.A.; Norton, M.A.; Cross, D.A.; Carr, C.W. Growth behavior of laser-induced damage on fused silica optics under UV, ns laser irradiation. *Opt. Express* **2010**, *18*, 19966–19976. [CrossRef] [PubMed]
44. Génin, F.Y.; Feit, M.D.; Kozlowski, M.R.; Rubenchik, A.M.; Salleo, A.; Yoshiyama, J. Rear-surface laser damage on 355-nm silica optics owing to Fresnel diffraction on front-surface contamination particles. *Appl. Opt.* **2000**, *39*, 3654–3663. [CrossRef]

coatings

MDPI

*Article*

# Hygrosensitive Response and Characteristics of Copolymer Coatings with Potential for Humidity Monitoring

**Katerina Lazarova [1,*], Silvia Bozhilova [2], Martina Docheva [1], Ketrin Pavlova [1], Gergana Alexieva [1,3,*], Darinka Christova [2] and Tsvetanka Babeva [1,4]**

1 Institute of Optical Materials and Technologies "Akad. J. Malinowski", Bulgarian Academy of Sciences, Akad. G. Bonchev Str., bl. 109, 1113 Sofia, Bulgaria; mdochev@iomt.bas.bg (M.D.); kpavlova@iomt.bas.bg (K.P.); babeva@iomt.bas.bg (T.B.)

2 Institute of Polymers, Bulgarian Academy of Sciences, Akad. G. Bonchev Str., bl. 103-A, 1113 Sofia, Bulgaria; s.bozhilova@polymer.bas.bg (S.B.); dchristo@polymer.bas.bg (D.C.)

3 Faculty of Physics, University of Sofia, 5 James Bourchier Blvd, 1164 Sofia, Bulgaria

4 National Centre of Excellence Mechatronics and Clean Technologies, 8 Kliment Ohridski Blvd, 1000 Sofia, Bulgaria

* Correspondence: klazarova@iomt.bas.bg (K.L.); gerry@phys.uni-sofia.bg (G.A.); Tel.: +359-2-979-3526 (K.L.)

**Abstract**

Newly developed hygrosensitive poly(vinyl alcohol) derivatives comprising grafted poly(N,N-dimethylacrylamide) chains of varied length and graft density are presented. The optical, sensing, and hydration properties of these copolymer thin films prepared by spin-coating were systematically studied. Refractive indices (*n*), absorption coefficients (*k*), and thicknesses (*d*) were calculated via curve fitting of the reflection spectra. Reflectance measurements across a relative humidity range of 5% to 95% were used to evaluate the humidity sensing behavior. Coating swelling exceeding 100% was observed. Hydration levels under high humidity conditions were studied using a quartz crystal microbalance method. This revealed approximately 24% water content in the polymer with the higher grafting density and shorter PDMA chains compared to around 31% in the copolymer with longer PDMA brushes that were loosely grafted The potential application of these copolymers as responsive materials for advanced humidity sensing is discussed. A combined optical and gravimetric approach for characterizing the humidity sensing properties of thin nanosized coatings is demonstrated, providing opportunities for advanced characterization of new functional materials, thus broadly contributing to the state of the art of sensor technologies.

**Keywords:** copolymers; thin films; optical sensing; QCM; humidity

## 1. Introduction

Research experiments, agricultural productions, industrial processes, and many commercial and industrial applications all heavily rely on the measurement and precise regulation of environmental humidity. Numerous kinds of humidity sensors have been created based on the measurement of different effects and parameters—capacitive [1–5], resistive [5–7], optical [8–11], gravimetric [12–15], etc. However, the rapidly advancing fields of science and technology demand humidity sensors that are more sensitive, have a larger detecting range, a shorter response and recovery time, and are less expensive than those already available on the market. Moreover, it is also of importance to have a sensor with a small size and a simple operational mechanism. Therefore, a lot of work is being focused into creating extremely sensitive materials and innovative devices to deal with these problems. A sensitive thin film or coating is a key component as its interaction with water molecules determines the overall performance and which carries out the principle of humidity sensing in the majority of devices. Therefore, a lot of research is being focused on studying different sensing materials in order to achieve their optimal performance characteristics. An important aspect of the research is the technological compatibility of the obtained new material with the physical transducer. Thus, it would be a real advantage if a selected material could serve as a sensitive element both for optical and gravimetric sensors, for example.

Such research could provide valuable insights into the design and performance of polymers and other sensitive materials intended for use in optical and gravimetric sensing devices and thus broadly contribute to advanced sensor technologies and their practical implementations in various fields. In this context, we demonstrate an approach of using a homemade experimental setup for combined optical and gravimetric (QCM) testing of the humidity sensing properties of thin nanosized coatings providing opportunities for advanced characterization of new functional materials.

It is worth noting that the QCM is an extremely stable and powerful device due to its high sensitivity to mass changes on a nanogram scale, the fast response–recovery speed, miniature size, and low power consumption. Many studies [16–19] have proved that it is a promising humidity sensing device and can be an alternative approach for effective humidity detection. It allows real time monitoring of water sorption isotherms using very small amounts of samples.

As a result, findings from optical and gravimetric characterization may be complementary, offering mutually enlightening insights into the processes in sensitive nanoscale materials in conditions of changing environmental humidity levels.

Many materials have been widely explored separately for humidity detection—metal oxides [20,21], nanoparticles [2,22], zeolites [23,24], graphene [25,26], polymers [27,28], etc. Among them, polymers stand out with the noteworthy opportunities which they offer for modifying their properties. In materials research, the ability to control the material's properties is crucial. Numerous polymer materials have been extensively studied [29,30] using various techniques and methodologies, depending on the field of application. They are traditional but challenging materials because their qualities as sensitive materials can be tuned [31]. Their excellent performance, functionality, and sensing capabilities can be significantly impacted by the ability to add side chemical groups [32], offering opportunity to detect even minimum changes in the environmental conditions both optically and gravimetrically.

In our previous studies [11], we showed that modified PVA based copolymers are suitable for optical sensing of humidity. In the present study, we continued the investigation of PVA based copolymers with newly developed hygrosensitive poly(vinyl alcohol) derivatives comprising grafted poly(N,N-dimethylacrylamide) chains of varied length and graft density, to improve and expand their application possibilities. Investigation on how grafted chains affect their sensing properties and the relative humidity impact on moisture-sensitive copolymer coatings was carried out by using optical and gravimetric detection techniques. Enhancement of sensing behavior by applying this approach is demonstrated and discussed.

## 2. Materials and Methods

### 2.1. Synthesis of PVA-Graft-Poly(N,N-dimethylacrylamide)

Poly(vinyl alcohol) (PVA, approx. Mw 10000, 88% hydrolyzed, CAS 9002-89-5), N,N-dimethyacrylamide (DMA, 99%, CAS 2680-03-7), ammonium cerium(IV) nitrate (CAN, $\geq$ 98.5%, CAS 16774-21-3), nitric acid ($HNO_3$, 70%, CAS 7697-37-2) were purchased from Sigma-Aldrich (St. Louis, MO, USA). Prior to use, the monomer DMA was passed through a column of activated basic alumina for removal of the inhibitor, hydroquinone monomethyl ether, and then blown by nitrogen. Solvents and other reagents of standard laboratory reagent grade were used without additional purification.

Double hydrophilic brush copolymers studied in this work were synthesized by grafting DMA on PVA by means of redox polymerization in aqueous media using deionized water as a solvent and ammonium cerium (IV) nitrate as initiator. The procedure of redox polymerization of DMA in the presence of hydroxyl moiety was already reported in our previous studies [33, 34]. In this work PVA was applied as initiating hydroxyl moiety and a series of PVA-g-PDMA copolymers were obtained following the same general procedure after optimization. Briefly, 1 g of PVA was dissolved in 10 mL deionized water and the solution was bubbled with nitrogen for 15 min for oxygen removal. The solution was cooled down to 10 °C and then the predetermined amount of initiator CAN dissolved in 1 N $HNO_3$ was added under vigorous stirring followed by nitrogen bubbling for another 15 min. Finally, the chosen amount of the monomer DMA was added under nitrogen, and the reaction mixture was heated to 35 °C. The polymerization was carried out for 3 h at 35 °C under nitrogen atmosphere. Two brush copolymers were synthesized at monomer-to-PVA-to-initiator molar ratio 1.0:0.23:0.023 and 1.0:0.23:0.012, respectively. The obtained crude copolymer samples were purified by dialysis against deionized water and finally isolated by freeze-drying.

### 2.2. *Copolymer Characterization Methods*

#### 2.2.1. Fourier-Transform Infrared Spectroscopy

Fourier-transform infrared spectroscopy (FTIR) analysis was performed using an IRAffinity-1 spectrophotometer (Shimadzu Company, Kyoto, Japan) equipped with a MIRacleTM ATR accessory (diamond crystal; PIKE Technologies, Madison, WI, USA). Copolymer samples were scanned over a wave number range from 4000 to 700 $cm^{-1}$, performing 50 scans at a resolution of 4 $cm^{-1}$.

#### 2.2.2. Nuclear Magnetic Resonance

NMR analysis of copolymers was implemented on a Bruker Avance II+ 600 NMR spectrometer in deuterated dimethyl sulfoxide (DMSO-$d_6$) as a solvent. The copolymer compositions were estimated from the integrated proton spectra when comparing the integrals of characteristic signals assigned to PVA and PDMA repeating units.

#### 2.2.3. Dynamic Light Scattering

Dynamic light scattering measurements were performed on a NanoBrook 90Plus PALS instrument (Brookhaven Instruments Corporation, Nashua, NH, USA), Brookhaven Instruments Corporation), equipped with a 35 mW red diode laser ($\lambda$ = 640 nm). Size measurements were accomplished at a scattering angle of 90° at 25 °C. Aqueous copolymer solutions of concentration 1.25 g/L were analyzed, and each measurement was performed in triplicate.

#### 2.2.4. Thermogravimetric Analysis

Thermogravimetric analysis (TGA) was performed on a TGA-4000 PerkinElmer instrument (PerkinElmer, Waltham, MA, USA), heating the samples in an argon atmosphere at 10 °C/min within the temperature range 40–800 °C.

### 2.3. *Thin Films Deposition*

For the deposition of the copolymer coatings, the spin-coating method (WS-650-23 B, Laurel, MS, USA) was used. All samples were rotated at speed of 4000 rpm and time of

60 s using 200 µL of the copolymer solution dropped on precleaned silicon wafer substrate. Post deposition annealing was applied in air for 30 min at 60 °C. Identical coatings were deposited on quartz crystal resonators with gold electrodes and a $SiO_2$ overlayer.

### 2.4. *Characterization of Thin Films*

#### 2.4.1. Optical Characterization and Thickness Calculation

First, reflectance spectra of all samples were measured with UV-VIS-NIR (ultraviolet–visible–near infrared) spectrophotometer (Cary 5E, Varian, Australia). Then, optical constants—refractive index $n$ and absorption coefficient $k$, along with the thickness of the films $d$—were calculated via the previously developed two-stage nonlinear curve fitting method [35] from the already measured reflectance spectra.

#### 2.4.2. Sensing Properties

In order to evaluate the sensing properties of the coatings a homemade bubbler system was used [36]. Different relative humidity levels were generated by bubbling argon through distilled water maintained at 60 °C and directed to the cell containing the sample. The same bubbling system was used for both optical sensing and gravimetric measurements (Figure 1).

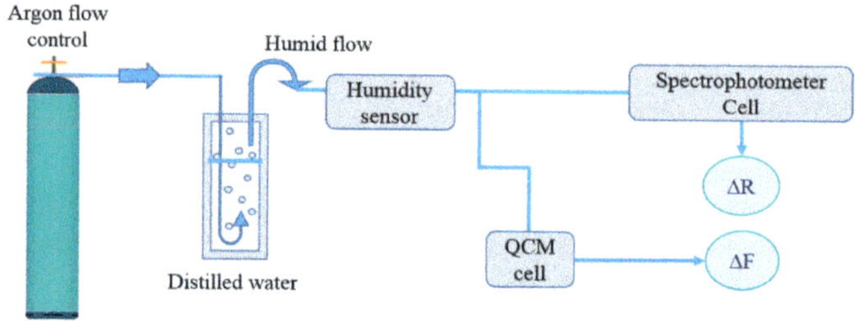


**Figure 1.** Set up for measurement of the sensing properties of the thin film/coating samples.

##### Acoustical Sensing

In order to estimate the hygrosensitive properties of the studied polymers, a technique based on the quartz crystal microbalance was used. Copolymer coatings were deposited on quartz resonators, and their hydration was continuously scanned by means of the openQCM NEXT device. We used $SiO_2$ 5 MHz resonators. Both the QCM instrument and the resonators were purchased from Novaetech, Pompei, Italy.

Assuming that the material adsorbed across the resonator's active area is rigid and uniformly distributed, and that the layer's mass is small relative to the mass of the quartz crystal, the Sauerbrey equation may be used. It describes a linear relationship between added mass and the induced frequency shift [37]:

$$\frac{\Delta f_N}{N} = -\frac{2 m_f f_0^2}{Z_q} \tag{1}$$

where $Z_q$ is acoustic impedance of quartz ($8.8 \times 10^6$ kg m$^{-2}$s$^{-1}$), $f_0$ is the fundamental resonance frequency of the quartz crystal (approximately 5 MHz for the used resonators), $N$ is the overtone number, $\frac{\Delta f_N}{N}$ is the frequency shift of the Nth overtone normalized to $N$ and $m_f$ is the areal mass (kg m$^{-2}$). This equation can be used to calculate the mass of the adsorbed material and the analyte molecules. Under the Sauerbrey approximation, the normalized frequency shifts from each overtone should be equal, and the adsorbed mass can be computed directly from (1) by taking the mean of all values derived from $\frac{\Delta f_N}{N}$.

However, in some cases, due to the rheological and/or surface properties of the applied film the coating of the quartz resonator leads to deviations from the Sauerbrey equation, and $\frac{\Delta f_N}{N}$ varies between the overtones depending on the overtone number. In this case, alternative methods for data analysis are applied [16–19,38].

According to the relationship (2), the total coating hydrated mass $m_{95\%}$ (the mass at humidity level RH = 95%) and its dry mass $m_{5\%}$ (the mass measured at RH = 5%) can be used to determine the accumulated water mass $m_{water}$ and the water content in the coating [16]:

$$water\ content\ (\%) = \frac{m95\% \ - \ m5\%}{m95\%} \times 100 = \frac{m_{water}}{m95\%} \times 100 \quad (2)$$

Prior to use, the new resonators were cleaned with water and ethanol. A standard experiment used a homemade bubbler system to initially characterize the bare resonator by using the QCM device with a temperature-controlled module in a dry Ar atmosphere at RH = 5%. Afterwards, the resonator was removed from the QCM and a thin coating was spin-coated onto the surface at a rotation speed of 4000 rpm for 60 s, applying the same procedure as for the formation of polymer coatings on Si substrates. The deposition technique was selected in order to obtain thin evenly distributed coatings over the resonator surface. After the resonator was inserted into the QCM module a homemade bubbler system connected to the QCM cell (Figure 1) was used to evaluate the sorption/desorption of water vapor by the studied polymer films. The films were exposed to continuously increasing/decreasing RH values in the range (5–95%) at constant temperature t = 20 °C.

Optical Sensing Properties

For the recording of the reflectance spectra at different values of relative humidity (*RH*) in the range from 5% to 95% RH, the samples deposited on the Si-wafer were placed one by one in a quartz cell; this was mounted onto the spectrophotometer to give the possibility to measure spectra at the desired relative humidity. Since a commercially available humidity sensor is integrated in the quartz cell, the dependence of reflectance on relative humidity (*R* versus *RH*) can also be derived. For the determination of hysteresis, reflectance spectra were measured for increasing and decreasing *RH* in the range 5–95% *RH*. To do this, a fixed wavelength ($\lambda_{max}$), preliminarily chosen as the wavelength of the highest humidity responses ($\Delta R_{max}$), was used. Determination of the wavelength for each sample, along with optical constants and thickness (and its change), was conducted by measurement of the reflectance spectra in the range 320–800 nm at 5% and 95% RH. Spectra were used to calculate the thickness at different humidity levels, from which the change of the thickness $\Delta d$ and the relative change of the thickness $\Delta d/d_{5\%}$ were found:

$$\Delta d = d_{95\%} - d_{5\%} \quad (3)$$

where $d_{95\%}$ and $d_{5\%}$ were the thicknesses of the films at dry ($RH = 5\%$) and hydrated ($RH = 95\%$) states.

Different parameters were used in order to compare optical sensing properties and quantify the samples under study—sensitivity *S*, accuracy $\Delta RH$, and hysteresis *H*. The following formula was used to determine the thin film's sensitivity *S*:

$$S = \frac{\Delta R}{RH_2 - RH_1} \quad (4)$$

where $\Delta R$ is the change of the film's reflectance in % for humidity levels changing from $RH_1$ to $RH_2$. From the already determined sensitivity *S*, the next accuracy/resolution ($\Delta RH$) of detection can be estimated. It depends on the sensitivity and measurement accuracy in the signal. The following formula was used for the calculation:

$$\Delta RH = \frac{errR\ (\%)}{S\ (\%)} \quad (5)$$

where $errR = 0.3\%$ is the experimental error (accuracy) of the reflectance $R$ and $S$ is the sensitivity, calculated from Equation (4).

Unwanted hysteresis, which is manifested as different $R$ values measured at the same humidity levels depending on the history of the humidity, can sometimes occur when $RH$ changes from low to high levels or vice versa. At best, for our sensitive material (copolymer coating), we wanted to achieve the highest detection sensitivity with the lowest possible percentage of hysteresis. So, the percentage of hysteresis ($H$) was determined by Equation (6):

$$H(\%) = \frac{\frac{1}{Q}\sum_{i=1}^{Q}\left|R_{up} - R_{down}\right|}{\Delta R}100 \quad (6)$$

where $R_{up}$ and $R_{down}$ are reflectance values measured for increasing and decreasing humidity, respectively, $\Delta R$ is the reflectance change in the entire humidity range, and $Q$ is the number of points over which the summation is performed.

## 3. Results

### 3.1. Characterization of Synthesized PVA-g-PDMA Brush Copolymers

Double hydrophilic brush copolymers, comprising relatively short PVA backbone and grafted PDMA chains, were synthesized and investigated for optical as well as acoustic sensing application. Two copolymers of similar copolymer composition but differing in grafting density and length of grafted PDMA side chains were prepared under environmentally friendly reaction conditions—water as reaction medium and moderate reaction temperature. For this, equal PVA and DMA concentration in the initial reaction mixture was applied while the initiator concentration was changed twofold. The chemical structure of the obtained copolymers is illustrated in Figure 2.

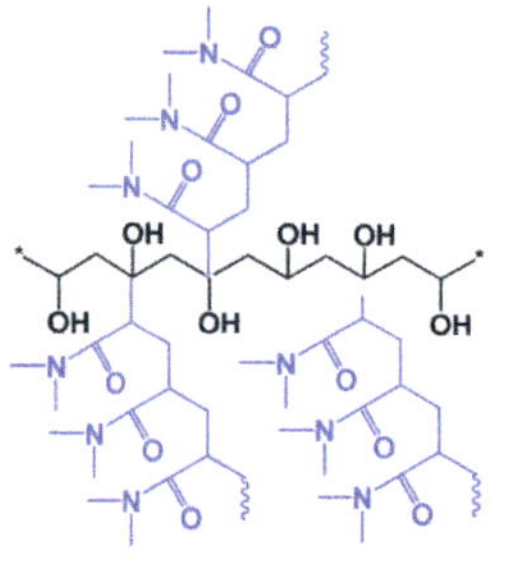

**Figure 2.** Schematic illustration of copolymer structure comprising poly(vinyl alcohol) backbone grafted with poly(N,N-dimethyl acrylamide) side chains.

Table 1 summarizes the reaction conditions and the characterization of the obtained copolymers. It can be clearly seen that copolymer yields are high and not significantly dependent on the initiator concentration. As expected, the mole fractions of grafted PDMA brushes in the copolymer composition, calculated from the integrated proton NMR spectra of G1 and G2, are close to the theoretical value set by the feed (Table 1). Consequently, the only difference between the investigated copolymers is the macromolecular architectures—G1 represents the brush copolymer of the higher grafting density and shorter PDMA graft chains, and vice versa for G2.

**Table 1.** Grafting reaction conditions.

| Sample Code | Reaction Conditions [(a)] | | | Yield [%] | PDMA Mole Fraction in Copolymer [(b)] | $D_h$ [(c)], [nm] | Đ [(c)] |
|---|---|---|---|---|---|---|---|
| | DMA Concentration, [mol/L] | DMA Mole Fraction | CAN, [mol/L] × $10^2$ | | | | |
| G1 | 3.5 | 0.80 | 9.0 | 73 | 0.78 | 595 | 0.19 |
| G2 | 3.5 | 0.80 | 4.5 | 70 | 0.84 | 341 | 0.22 |

[(a)] PVA concentration 15.0 g/L, 35 °C, 180 min; [(b)] Calculated from $^1$H NMR spectra. [(c)] Measured by DLS in aqueous solution, copolymer concentration 1.25 g/L, 25 °C.

The chemical composition of G1 and G2 was evaluated by FTIR analysis. In the FTIR spectra of G1 and G2 (Figure 3), strong carbonyl bands at 1616 cm$^{-1}$ ascribed to the C=O stretching vibrations (amide I) of PDMA segments are well visible. In addition, characteristic peaks at about 1500 cm$^{-1}$ assigned to the H–C–H bending vibrations of the PDMA methyl groups are also present. In the spectrum of pristine PVA, a broad and strong band observed between 3000–3600 cm$^{-1}$ is attributed to the stretching vibrations of the O–H bond, indicating the presence of hydroxyl groups. Other characteristic bands include those at 2915–2940 cm$^{-1}$ for stretching C–H, at 1417 cm$^{-1}$ for bending CH2, and at 1090 cm$^{-1}$ for the stretching vibration of the C–O, as well as an additional band at 1734 cm$^{-1}$ related to C=O stretching for the residual acetate groups. However, all these PVA characteristic bands are hardly detectable in the copolymer spectra due to the complex macromolecular architecture of G1 and G2 and the low PVA fraction in the copolymer composition.

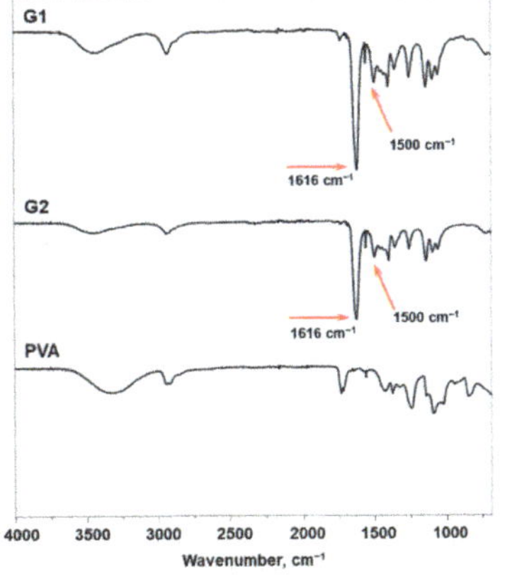


**Figure 3.** FTIR spectra of G1 and G2 compared to the spectrum of original PVA.

The data presented in Table 1 clearly indicate that the theoretical chemical compositions of copolymers G1 and G2 are identical. Since the FTIR method analyzes bond vibrations between atoms, no significant differences in the spectra of copolymers with similar compositions are expected. Indeed, the overlaid ATR-FTIR spectra of G1 and G2, shown in Figure 3, are identical and do not reveal any noticeable differences in composition. These results validate the proposed chemical composition and structure of the copolymers.

The chemical structure and composition of G1 and G2 were investigated by means of NMR spectroscopy. The proton NMR spectrum of G1 was compared to those of the original PVA in Figure 4. The proton NMR spectra of G1 and G2 are presented in Figure 4 together with those of pristine PVA. As shown in the spectrum of PVA (Figure 4a), protons of CH2 appear as a multiplet at 1.3–1.5 ppm and those of CH appear at 3.5–3.6 ppm. Protons of

OH groups can be seen as three separate peaks at 4.3, 4.5, and 4.7 ppm. In addition, the signals of methyl protons (CH3) in the residual acetate groups appear at 1.9 ppm.

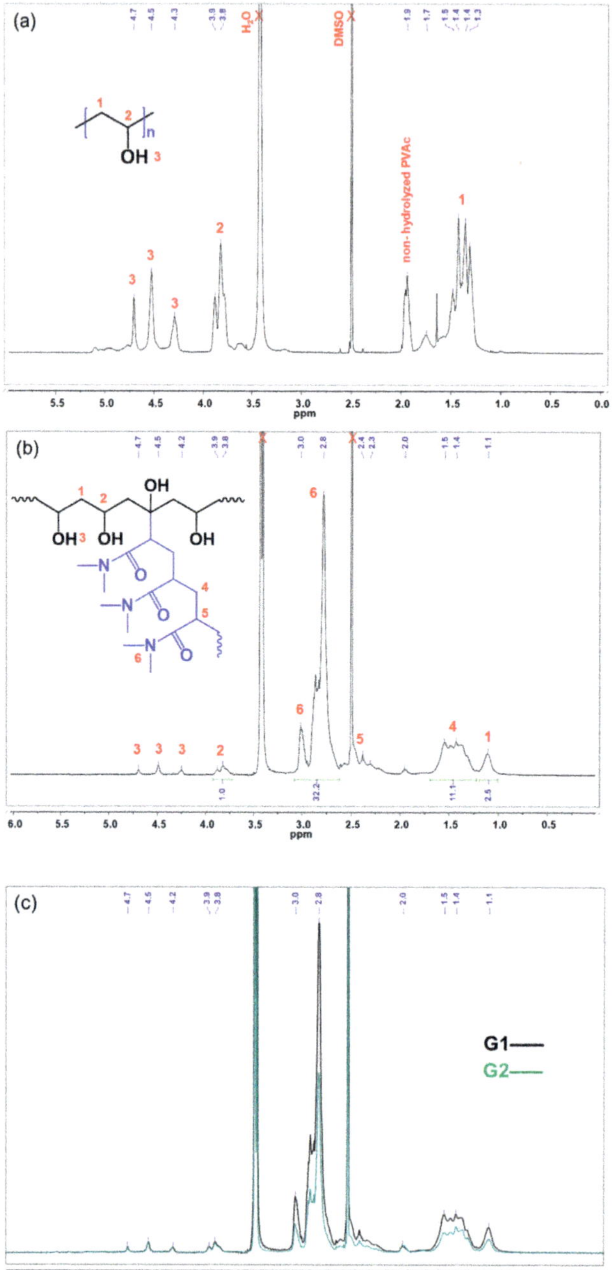

**Figure 4.** $^1$H NMR spectra (600 MHz, DMSO-$d_6$) of the following: (**a**) pristine PVA; (**b**) G1; and (**c**) overlaid spectra of G1 and G2.

In the spectrum of G1 (Figure 4b), along with the signals assigned to PVA, protons of the CH3 groups in the grafted PDMA side chains appear at 2.8–3.0 ppm. Signals of the protons of the CH2 and CH groups in PDMA can be seen as a multiplet at 1.2–1.7 ppm and shoulder peaks at 2.3–2.4 ppm. The integrated signals of methyl, methylene, and methine groups in PDMA together with the corresponding methylene and methine groups in PVA were used to estimate the copolymer composition of G1 and G2, expressed as PDMA mole fraction (Table 1).

It was emphasized above, that the copolymer composition of G1 and G2 was the same at the feed. The overlaid proton NMR spectra of G1 and G2 (shown in Figure 4c) confirm that the synthesized brush copolymers have a similar chemical composition and differ in graft density and length of the grafted PDMA side chains only.

The structural complexity of G1 and G2 and the different parameters of its macromolecular architecture affect the copolymer properties both in solution as well as in bulk. As can be seen from the DLS measurement results, listed in Table 1, the dispersity of the brush copolymer aggregates of G1 in aqueous solution is very similar to those of G2. The hydrodynamic diameter ($D_h$) of G1, however, is much larger, almost double that of G2, which is attributed to the specific macromolecular architecture of densely grafted short PDMA brushes. The much smaller hydrodynamic diameter measured for G2 is explained by the relatively longer PDMA brushes and lower grafting density, resulting in a more compact aggregate structure.

The impact of the grafting density and length of the grafted PDMA brushes is also seen in Figure 5, where the TGA profiles of G1 and G2 are compared. The mass loss of G1, densely grafted with short PDMA brushes, occurs much faster in the initial temperature range up to 150 °C as compared to G2, and is explained by the specific macromolecular assembly of the copolymers. In the next temperature intervals, the decomposition of G1 and G2 occurs simultaneously which could be expected for copolymers of similar composition.

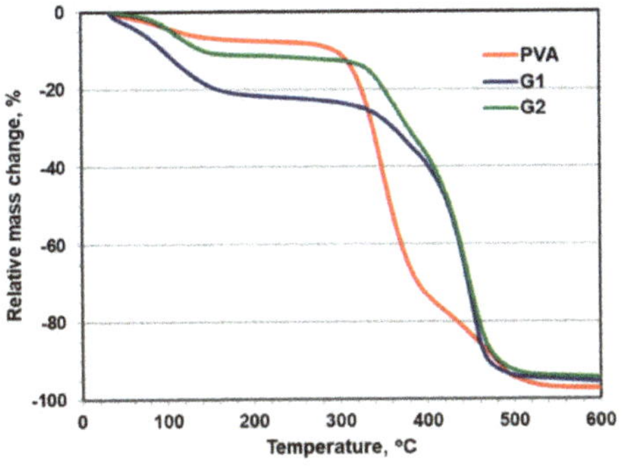


**Figure 5.** TGA curves of G1, G2, and original PVA.

## 3.2. *Optical Properties of the Copolymers Thin Film Coating*

Purposely, the 2 wt% solution of the copolymers G1 and G2 in water/methanol solvents (volume ratio of 20:80) were prepared and spin-coated on precleaned silicon substrates in order to obtain nanosized coatings/thin films. The films had a smooth surface and were transparent and free of cracks. The reflectance spectra R of the G1 and G2 copolymer films were measured at normal light incidence in the spectral range 320–800 and are presented in Figure 6a. It can be observed that, although the same concentrations of solutions were used for both films, their spectra differ. This indicates that the optical constants and thicknesses of G1 and G2 are also different. To confirm this, the reflectance

(R) spectra were used to calculate the refractive index ($n$), extinction coefficient ($k$), and film thickness ($d$) (Table 2). The dispersion curves of $n$ are presented in Figure 6b, while the thickness values, along with the $n$ and $k$ values at a wavelength of 600 nm, are summarized in Table 2.

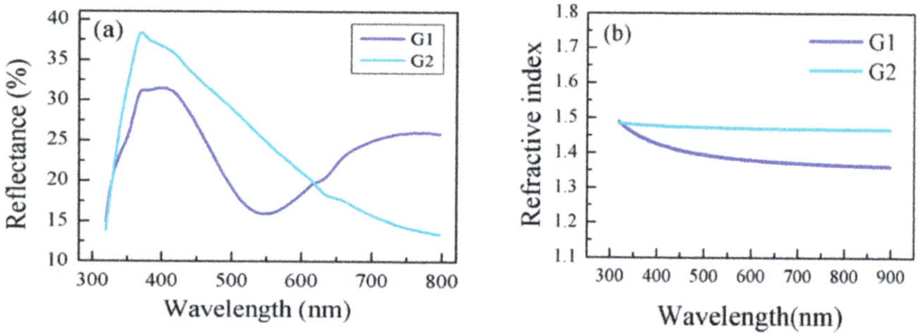


**Figure 6.** Reflectance spectra (**a**) and dispersion curves of refractive index $n$ (**b**) for G1 and G2 copolymers thin films.

**Table 2.** Refractive index ($n$), extinction coefficient ($k$), thickness ($d$) at 5 and 95% RH, absolute and relative change in $d$, and absolute change in $n$ of the studied copolymer samples G1 and G2 due to exposure from low to high relative humidity (RH).

| Sample | $n$ | $k$ | $d$ at RH 5% (nm) | $d$ at RH 95% (nm) | $\Delta d$ (nm) | $\Delta d/d_{5\%}$ (%) | $\Delta n$ |
|---|---|---|---|---|---|---|---|
| G1 | 1.40 | 0.001 | 187 | 380 | 193 | 103 | 0.19 |
| G2 | 1.47 | 0.001 | 143 | 213 | 70 | 49 | 0.08 |

As seen in Figure 6b, both thin films exhibit normal refractive index dispersion—n decreases with increasing wavelength—which indicates low light absorption in the films. The G2 film has a higher refractive index and lower thickness compared to the G1 film. Since the polymer chain density in G2 is lower, it is more likely that the chains undergo stronger compression during film deposition, resulting in a thinner coating. This increases the packing density and leads to a higher refractive index due to the higher film density. Conversely, the G1 film, with more densely distributed polymer chains, undergoes less compression, leading to a thicker coating. This lowers the packing density and results in a reduced refractive index. It is worth noting that the effective diameter of the polymer aggregates measured in solution is greater for G1 than for G2. This supports our assumption that G2 copolymers undergo stronger compression during film deposition, as smaller particles can be packed more densely.

### 3.3. Optical Humidity Sensing

The above-described experimental setup was utilized to test the coatings' sensitivity to varying humidity levels by recording the spectra of the nanosized films at various relative humidity levels. Results for the lowest (RH = 5%) and the highest studied humidity (RH = 95%) levels are shown in Figure 7.

Each spectrum was used for the determination of refractive index and the thickness of the coatings at the corresponding humidity level, followed by calculation of the absolute change in refractive index and thickness ($\Delta n$ and $\Delta d$), respectively, and the relative change of the thickness ($\Delta d/d_{5\%}$). Results are shown in Figure 8 and Table 2.

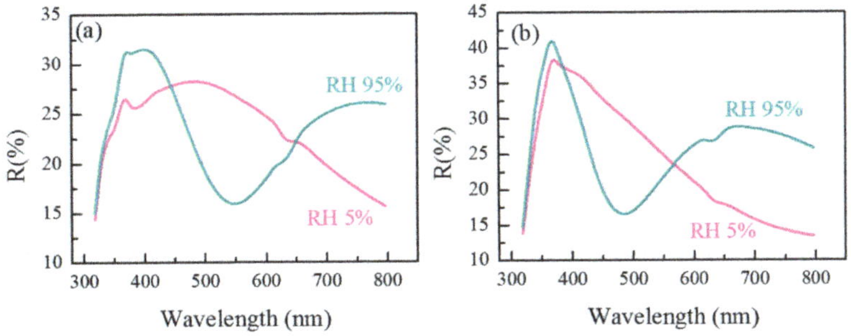


**Figure 7.** Reflectance spectra R of the copolymer coatings G1 (**a**) and G2 (**b**) measured at humidity levels of 5% and 95% RH.

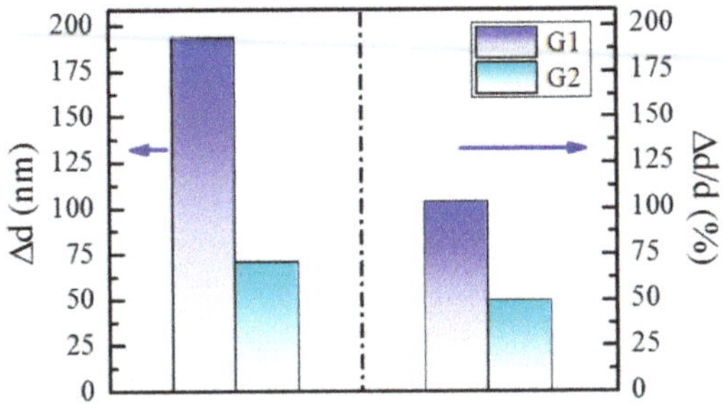


**Figure 8.** Change of the thickness Δd and relative change of the thickness $\Delta d/d_{5\%}$ of G1 and G2 copolymer coatings when exposed to humidity levels from 5% to 95%.

As can be seen from Figure 7a,b for both copolymer thin films, exposure to different humidity results in a significant change of the R spectra. Based only on the spectral shift and spectra shape change, both copolymers undergo changes in thickness and refractive index as a result of swelling and water absorption. Depending on the amount of adsorbed water from the atmosphere, the response of the two thin films is different. Thin copolymer films change their thickness due to swelling which is mostly pronounced for the copolymer films G1 with a higher grafting density. This can be explained by the fact that as grafting density decreases, brush height may not increase proportionally to polymer length. This result is not surprising as similar findings were shown, for example, by Kang, M., et al. [39]. The authors demonstrated the impact of grafting density on the studied polymer-grafted samples, which is crucial in order to determine particle assembly patterns and behavior. The authors compared variations in the brush height with the number of repeat units for each sample and showed that as the cat-CTA ratio decreases, the brush height increases as a function of polymer chain length, gradually diminishing. As the grafting density decreases, the brush height may not increase proportionally to the polymer length. This means in our case that when densely packed, the grafted chains "support" each other, remain "upright", and form a thicker layer (G1). When sparsely packed, they "lay down" and, although they may be longer, form a thinner layer (G2).

A relative change in percentage was assessed in relation to the initial film thickness at a humidity level of 5% RH in order to compare the films' sensing abilities as they differ in thickness. Figure 8 shows the change of the thickness, Δd, and the relative change of the thickness $\Delta d/d_{5\%}$ of G1 (dark blue bars) and G2 (turquoise bars) copolymer coatings when exposed to humidity levels from 5% to 95%. It can be seen that the relative change of the thickness of sample G1 is 103% compared to 49% for sample G2. That is, when copolymer G1 is applied in the form of a thin coating, the maximum swelling of the layer results in a little more than 100% increase of *d*, i.e., the layer doubles its thickness. The relative change in thickness caused by an increase in humidity from the lowest to the highest is nearly 50%, meaning that the change in thickness of sample G2 is precisely half as small. The different behavior, i.e., relative change of d of the layers is due precisely to the difference in the length and density of the brushes of the copolymers, as we demonstrated above.

Likewise, the stronger response of G1 is in line with our previous results for thin films of copolymers based on PVA with different grafting density of methyl (meth)acrylate [34], where we showed that thin films exhibit a swelling mostly pronounced in the films with a higher grafting density.

Additionally, the R-vs.-RH curves in adsorption/desorption regimes are presented in Figure 9. The measurements that show the change in reflectance, depending on the humidity level, revealed that the end of the desorption curve coincides with the beginning of the adsorption curve, which means that the thickness change in the copolymer coatings is reversible. Additional annealing for their recovery is not necessary. However, both samples exhibit hysteresis, which is smaller for G1 (13%) compared to G2 (21%).

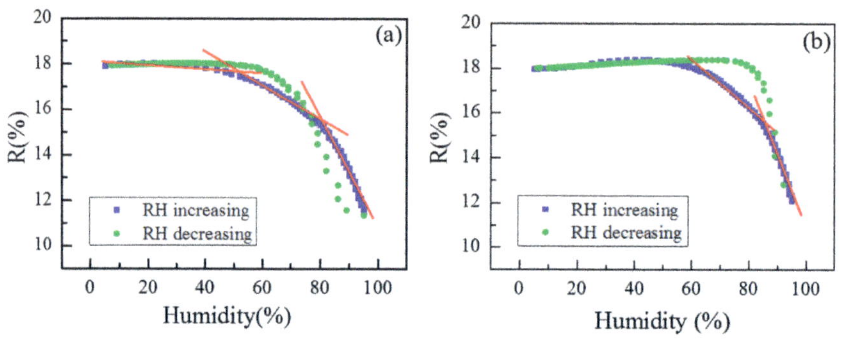


**Figure 9.** Reflectance versus relative humidity level of G1 and G2 copolymer coatings when exposed to humidity increasing from 5% to 95% (blue squares) (**a**) and vice versa—from 95% to 5% RH (green dots). Linear regions are marked with red lines (**b**).

Figure 9 demonstrates that the G1 sample is suitable for the optical sensing of the relative humidity across the entire humidity range. The R-vs.-RH curve for G1 reveals three linear regions with different slopes, indicating that RH can be measured with varying accuracy depending on the range. In contrast, the G2 film has a narrower dynamic range and is only suitable for humidity sensing at RH levels above 65%. For both polymers, the sensitivity and resolution were calculated, and the values for different RH ranges are shown in Table 3. Although the G2 sample exhibits slightly higher sensitivity than G1, the latter is preferable for optical humidity sensing due to its smaller hysteresis and full dynamic range. Nevertheless, if the application requires measuring very high humidity levels within a narrow range, G2 may be suitable, as it offers very high sensitivity—capable of resolving changes of less than 1% RH.

For each of the chosen humidity levels for both samples, color CIE coordinates were calculated from the measured reflectance spectra (Figure 10), thus giving us a visual idea of the color change of the sample as it passes through the different parts of the investigated humidity range. The color difference should ideally be noticeable to the naked eye without the need for a spectrophotometer to measure the spectrum. With the simplest possible reading by observation alone, this would enable a variety of uses for the developed copolymers in optical color sensors based solely on the colorimetry concept.

**Table 3.** Percentage of hysteresis (*H*), dynamic range, sensitivity (*S*), and accuracy for studied samples G1 and G2.

| Sample | Hysteresis H (%) | Dynamic Range (% RH) | Sensitivity S (%/% RH) | Accuracy/Resolution (% RH) |
|---|---|---|---|---|
| G1 | 13 | 5–95 | 0.01 (5–50% RH)<br>0.07 (51–80% RH)<br>0.24 (81–95% RH) | 30 (5–50% RH)<br>4 (51–80% RH)<br>1.25 (81–95% RH) |
| G2 | 21 | 65–95 | 0.11 (65–85% RH)<br>0.35 (86–95% RH) | 2.7 (65–85% RH)<br>0.85 (86–95% RH) |

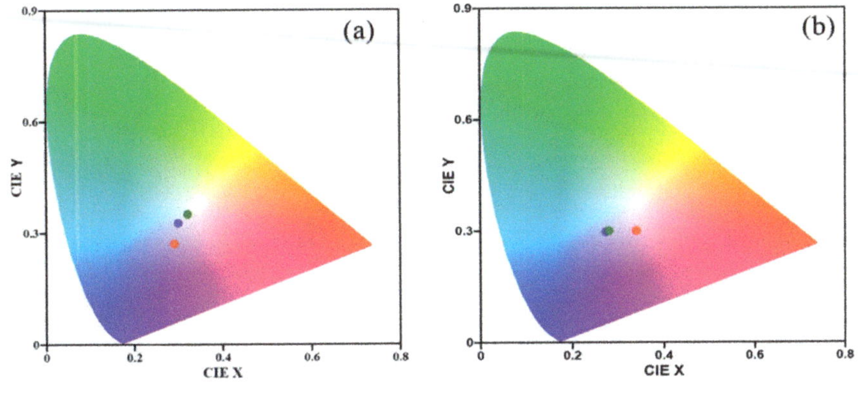


**Figure 10.** Calculated color coordinates of thin film samples of G1 (**a**) and G2 (**b**) when exposed to a different relative humidity—5% (blue dot), 50% (green dot), and 95% (red dot).

From the change in the color coordinates, it can be seen that for sample G1 the points corresponding to different humidity exhibit different colors. Although the distance between those at humidity levels 5% and 50% RH is small, i.e., the difference in color is not high, it is still distinguishable, and the points do not coincide. At high humidity of 95% RH the color is significantly changed, and the point is at a great distance from the other two. When the thin film of copolymer G2 is positioned at high humidity level 95% RH, it can be seen that the color of the sample is significantly different from the color that the coating has at humidity of 5% and 50% RH. However, in this case, the two coordinates nearly coincide at these two lower humidity levels, meaning that the samples' colors are nearly identical and thus the primary objective of differentiating them with the naked eye is impossible.

### 3.4. Acoustic Measurements and Characterization

Figure 11a,b display the frequency data for water vapor sorption/desorption processes for both being deposited on the quartz resonator surface G1 and G2 polymer films, respectively. As already mentioned, the frequency is plotted as the difference between the measured frequency and the frequency of the bare resonator divided by the order of the overtone, thus the different overtones can be compared. We see that changes in RH cause a reversible shift in the measured frequency. Additionally, we can see that the four measured harmonics are close to one another.

This behavior is characteristic of a sample close to the Sauerbrey regime and suggests that the coatings are slightly viscoelastic [19]. In order to evaluate the mass, we used the model for a single viscoelastic film in air [17,19,38]:

$$\frac{\Delta f_N}{N} = -\frac{2m_f f_0^2}{Z_q}\left(1 + \frac{Z_q^2 m_f^2 \pi^2 N^2}{3Z_f^2 m_q^2}\right) \quad (7)$$

where $m_q$ is the areal mass density of quartz (kg $m^{-2}$), and $Z_f$ is the acoustic impedance of the film (kg $m^{-2}s^{-1}$).

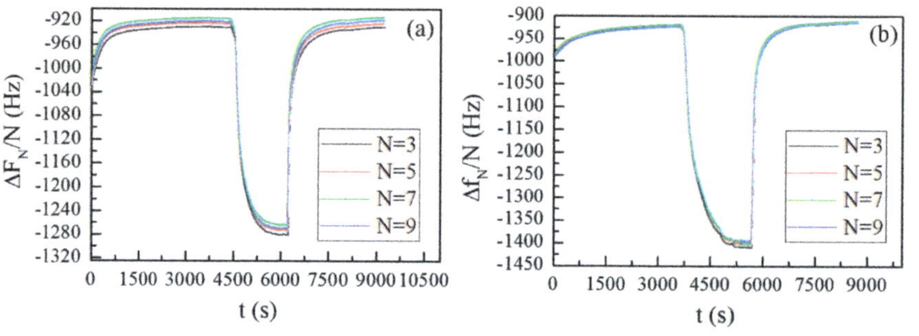


**Figure 11.** Frequency changes $\frac{\Delta f_N}{N}$ vs time for different overtones during sorption/desorption processes for G1 (**a**) and G2 (**b**) coatings.

We selected this model because in the simplest case, where there are no differences between the normalized frequency shifts of the overtones, this model simplifies to the Sauerbrey model, and if differences arise in more complex viscoelastic cases, (7) gives better accuracy for the mass calculation [16,19]. To assess the water content stored in the polymer matrix, humidity measurements were performed on a bare silica resonator surface following the same experimental procedure as with the polymer coated resonator. Consequently, the resulting water content at RH = 95% accumulated in the polymer with the more closely grafted brushes—G1 was estimated as 23.9%. Conversely, the more loosely grafted but longer PDMA brushes in the case of G2 led to a higher water content of 30.9% stored in the polymer matrix.

The higher absorption capacity of the G2 films can be attributed to their greater density, which implies a higher number of polymer chains per unit volume. This leads to an increased number of active absorption sites. However, the same high density also restricts the film's ability to swell significantly, resulting in smaller changes in thickness compared to G1 films when exposed to high relative humidity. Thus, the QCM technique was able to detect the impact of the brush architecture on the polymers' moisture sensing capabilities.

## 4. Conclusions

Double hydrophilic brush copolymers comprising a relatively short PVA backbone and grafted PDMA chains were successfully synthesized. The possibility, by varying the length and density of the brushes, to control the desired sensing properties of nanosized copolymer coatings when exposed to different humidity levels was proved optically and gravimetrically. Although the G2 sample exhibited slightly higher sensitivity and water content than G1, the latter is preferable for optical humidity sensing due to its smaller hysteresis and full dynamic range.

A possible implementation as a colorimetric humidity sensor that detects changes in the environment based only on a change in color was demonstrated and discussed, and the copolymer with higher grafting density and shorter PDMA graft chains (G1) was shown to be more suitable.

Further efforts will be dedicated to adapt the experimental setup to monitor the sorption/desorption isotherms, assess the stored water content, the humidity induced rheological changes, and the kinetic parameters at various humidity levels.

By means of the presented multimethodological approach, we studied the impact of macromolecular design and proved the efficiency of the grafting method for achieving advanced polymer coatings together with the potential of the newly synthesized poly(vinyl alcohol) derivatives for humidity sensing applications.

**Author Contributions:** Conceptualization, K.L., T.B. and D.C.; methodology, T.B., D.C., G.A. and K.L.; software, T.B. and K.L.; validation, K.L., G.A., T.B. and D.C.; formal analysis, S.B., M.D., K.P., G.A. and K.L.; investigation, M.D., K.P., S.B., G.A., K.L. and D.C.; resources, K.L., G.A., D.C. and T.B.; data curation, K.L., D.C. and T.B.; writing—original draft preparation, K.L., T.B., G.A. and D.C.; writing—review and editing, K.L., G.A. and T.B.; visualization, K.L., G.A. and D.C.; supervision, K.L. and T.B.; project administration, K.L., T.B. and D.C.; All authors have read and agreed to the published version of the manuscript.

**Funding:** The authors acknowledge the financial support from the Bulgarian National Science Fund under the Grant No. KP-06-COST/4 under COST Action CA21121 "European Network for the Mechanics of Matter at the Nano-Scale" (MecaNano).

**Institutional Review Board Statement:** Not applicable.

**Informed Consent Statement:** Not applicable.

**Data Availability Statement:** Data are contained within the article.

**Acknowledgments:** Research equipment of Distributed Research Infrastructure INFRAMAT, part of Bulgarian National Roadmap for Research Infrastructures, supported by Bulgarian Ministry of Education and Science, was used in this investigation. T.B. acknowledges the European Regional Development Fund under the Research Innovation and Digitization for Smart Transformation program 2021–2027 under the Project BG16RFPR002-1.014-0006 National Centre of Excellence Mechatronics and Clean Technologies.

**Conflicts of Interest:** The authors declare no conflicts of interest. The funders had no role in the design of the study; in the collection, analyses, or interpretation of data; in the writing of the manuscript, or in the decision to publish the results.

## References

1. Farahani, H.; Wagiran, R.; Hamidon, M.N. Humidity Sensors Principle, Mechanism, and Fabrication Technologies: A Comprehensive Review. *Sensors* **2014**, *14*, 7881–7939. [CrossRef] [PubMed]
2. Ku, C.-A.; Chung, C.-K. Advances in Humidity Nanosensors and Their Application: Review. *Sensors* **2023**, *23*, 2328. [CrossRef] [PubMed]
3. Yu, X.; Ding, X.; Yu, X.; Tang, K.; Chen, Q. A capacitive humidity sensor based on nanodiamond/silver nanoparticles (ND/Ag) nanocomposite with high stability and rapid response for respiratory monitoring. *Sens. Actuators: B. Chem.* **2024**, *416*, 136035. [CrossRef]
4. Jiang, Y.; Wu, L.; Chen, Q.; Li, N.; Tian, J. High-performance capacitive humidity sensor based on flower-like $SnS_2/Ti_3C_2$ MXene for respiration monitoring and non-contact sensing. *Sens. Actuators B Chem.* **2025**, *426*, 137012. [CrossRef]
5. Najeeb, M.A.; Ahmad, Z.; Shakoor, R.A. Organic Thin-Film Capacitive and Resistive Humidity Sensors: A Focus Review. *Adv. Mater. Interfaces* **2018**, *5*, 1800969. [CrossRef]
6. Su, P.-G.; Wang, C.-S. Novel flexible resistive-type humidity sensor. *Sens. Actuators B Chem.* **2007**, *123*, 1071–1076. [CrossRef]

7. Cho, S.; Lim, S.H.; Oh, J.; Ju, T.; Yang, S.; Kim, H.H.; Kim, Y. Synthesis-in-place of $V_2O_5$ nanobelts for wide range humidity detection. *Sens. Actuators B. Chem.* **2025**, *423*, 136711. [CrossRef]
8. Rao, X.; Zhao, L.; Xu, L.; Wang, Y.; Liu, K.; Wang, Y.; Chen, G.Y.; Liu, T.; Wang, Y. Review of Optical Humidity Sensors. *Sensors* **2021**, *21*, 8049. [CrossRef]
9. Novais, S.; Ferreira, M.S.; Pinto, J.L. Relative Humidity Fiber Sensor Based on Multimode Interferometer Coated with Agarose-Gel. *Coatings* **2018**, *8*, 453. [CrossRef]
10. Peng, J.; Zhou, J.; Sun, C.; Liu, Q. Optical Fiber Temperature and Humidity Dual Parameter Sensing Based on Fiber Bragg Gratings and Porous Film. *Sensors* **2023**, *23*, 7587. [CrossRef]
11. Lazarova, K.; Bozhilova, S.; Novakov, C.; Christova, D.; Babeva, T. Amphiphilic Poly(vinyl Alcohol) Copolymers Designed for Optical Sensor Applications—Synthesis and Properties. *Coatings* **2020**, *10*, 460. [CrossRef]
12. Wang, L.; Song, J.; Yu, C. The utilization and advancement of quartz crystal Microbalance (QCM): A mini review. *Microchem. J.* **2024**, *199*, 109967. [CrossRef]
13. Fauzi, F.; Rianjanu, A.; Santoso, I.; Triyana, K. Gas and humidity sensing with quartz crystal microbalance (QCM) coated with graphene-based materialsA mini review. *Sens. Actuators A* **2021**, *330*, 112837. [CrossRef]
14. Yao, Y.; Xue, Y. Impedance analysis of quartz crystal microbalance humidity sensors based on nanodiamond/graphene oxide nanocomposite film. *Sens. Actuators B* **2015**, *211*, 52–58. [CrossRef]
15. King, W.H. Piezoelectric Sorption Detector. *Anal. Chem.* **1964**, *36*, 1735–1739. [CrossRef]
16. Gonzalez-Martinez, J.F.; Znamenskaya Falk, Y.; Björklund, S.; Erkselius, S.; Rehnberg, N.; Sotres, J. Humidity-Induced Phase Transitions of Surfactants Embedded in Latex Coatings Can Drastically Alter Their Water Barrier and Mechanical Properties. *Polymers* **2018**, *10*, 284. [CrossRef]
17. Znamenskaya, Y.; Sotres, J.; Gavryushov, S.; Engblom, J.; Arnebrant, T.; Kocherbitov, V. Water Sorption and Glass Transition of Pig Gastric Mucin Studied by QCM-D. *J. Phys. Chem. B* **2013**, *117*, 2554–2563. [CrossRef]
18. Kelly, S.J.; Genevskiy, V.; Björklund, S.; Gonzalez-Martinez, J.F.; Poeschke, L.; Schröder, M.; Nilius, G.; Tatkov, S.; Kocherbitov, V. Water Sorption and Structural Properties of Human Airway Mucus in Health and Muco-Obstructive Diseases. *Biomacromolecules* **2024**, *25*, 1578–1591. [CrossRef]
19. Macklin, J.; Pfrang, C.; Wady, P.; Liu, W.; Davidson, R.; Milsom, A.; Squires, A. Use of Humidity Controlled Quartz Crystal Microbalance with Simultaneous Grazing Incidence Small Angle X-ray Scattering to Investigate the Self-assembly and Energetics of Lipid Thin Films. *Langmuir* **2025**, *41*, 10216–10222. [CrossRef]
20. Ma, H.; Fang, H.; Wu, W.; Zheng, C.; Wu, L.; Wang, H. A highly transparent humidity sensor with fast response speed based on $\alpha$-$MoO_3$ thin films. *RSC Adv.* **2020**, *10*, 25467–25474. [CrossRef]
21. Xie, T.; Rahman, A.F.A.; Bakar, A.A.; Arsad, A. Design and Optimization of Metal Oxide-Based Humidity Sensors: A Review on Mechanisms and Material Engineering. *J. Clust. Sci.* **2025**, *36*, 148. [CrossRef]
22. Altammar, K.A. A review on nanoparticles: Characteristics, synthesis, applications, and challenges. *Front. Microbiol.* **2023**, *14*, 1155622. [CrossRef]
23. Gao, Y.-J.; Romolini, G.; Huang, H.; Jin, H.; Saha, R.A.; Ghosh, B.; De Ras, M.; Wang, C.; Steele, J.A.; Debroye, E.; et al. Ultrasensitive turn-on luminescence humidity sensor based on a perovskite/zeolite composite. *J. Mater. Chem. C* **2022**, *10*, 12191–12196. [CrossRef]
24. Stosic, D.; Zholobenko, V. Application of Zeolite-Based Materials for Chemical Sensing of VOCs. *Sensors* **2025**, *25*, 1634. [CrossRef]
25. Nandanapalli, K.R.; Kailasa, S.K.; Lee, S. Graphene oxide-based humidity sensors. *Compr. Anal. Chem.* **2024**, *106*, 339–371. [CrossRef]
26. Laera, A.M.; Cassano, G.; Burresi, E.; Protopapa, M.L.; Penza, M.A. Graphene Oxide Flexible Sensor for Humidity Detection. *Eng. Proc.* **2023**, *48*, 44. [CrossRef]
27. Qian, J.; Tan, R.; Feng, M.; Shen, W.; Lv, D.; Song, W. Humidity Sensing Using Polymers: A Critical Review of Current Technologies and Emerging Trends. *Chemosensors* **2024**, *12*, 230. [CrossRef]
28. Pasalwad, K.A.; Baby, N.; Edjenguele, A.; Sadhasivam, S.; Palanisamy, G.; Magdum, S.; Thangarasu, S.; Oh, T.H. Progress on polymer-based materials and composites for humidity sensor applications: From materials aspects to sensor performances. *J. Mater. Chem. A*, 2025; *advance article*. [CrossRef]
29. Jakubik, W.; Wrotniak, J.; Caliendo, C.; Benetti, M.; Cannata, D.; Notargiacomo, A.; Stolarczyk, A.; Kaźmierczak-Bałata, A. SAW Humidity Sensing with rr-P3HT Polymer Films. *Sensors* **2024**, *24*, 3651. [CrossRef]
30. Guo, Y.; Xi, H.; Gu, Z.; Li, M.; Li, X.; Gao, D. A self-powered PVA-based flexible humidity sensor with humidity-related voltage output for multifunctional applications. *Colloids Surf. A Physicochem. Eng. Asp.* **2023**, *658*, 130700. [CrossRef]
31. Zhang, K.; Lu, S.; Qu, Z.; Feng, X. Tuning the Sensitivity and Dynamic Range of Optical Oxygen Sensing Films by Blending Various Polymer Matrices. *Biosensors* **2022**, *12*, 5. [CrossRef] [PubMed]

32. Oliveira, G.P.; Barboza, B.H.; Batagin-Neto, A. Polyaniline-based gas sensors: DFT study on the effect of side groups. *Comput. Theor. Chem.* **2022**, *1207*, 113526. [CrossRef]
33. Lazarova, K.; Vasileva, M.; Ivanova, S.; Novakov, C.; Christova, D.; Babeva, T. Influence of macromolecular architecture on the optical and humidity sensing properties of poly(N,N-dimethylacrylamide)-based block copolymers. *Polymers* **2018**, *10*, 769. [CrossRef] [PubMed]
34. Bozhilova, S.; Lazarova, K.; Ivanova, S.; Karashanova, D.; Babeva, T.; Christova, D. Colloidal aqueous dispersions of methyl (meth)acrylate-grafted polyvinyl alcohol designed for thin film applications. *Coatings* **2022**, *12*, 1882. [CrossRef]
35. Lazarova, K.; Vasileva, M.; Marinov, G.; Babeva, T. Optical characterization of sol-gel derived Nb2O5 thin films. *Opt. Laser Technol.* **2014**, *58*, 114–118. [CrossRef]
36. Lazarova, K.; Awala, H.; Thomas, S.; Vasileva, M.; Mintova, S.; Babeva, T. Vapor Responsive One-Dimensional Photonic Crystals from Zeolite Nanoparticles and Metal Oxide Films for Optical Sensing. *Sensors* **2014**, *14*, 12207–12218. [CrossRef]
37. Sauerbrey, G. Verwendung von Schwingquarzen zur Wägung dünner Schichten und zur Mikrowägung. *Z. Phys.* **1959**, *155*, 206–222. [CrossRef]
38. Johannsmann, D. Viscoelastic analysis of organic thin films on quartz resonators. *Macromol. Chem. Phys.* **1999**, *200*, 501–516. [CrossRef]
39. Kang, M.; Lin, P.A.; Bunch, J.A.; Lipomi, D.J.; Arya, G.; Cohen, S.M. Impact of Grafting Density on the Assembly and Mechanical Properties of Self-Assembled Metal–Organic Framework Monolayers. *J. Am. Chem. Soc.* **2025**, *147*, 6966–6973. [CrossRef]

*coatings*

MDPI

*Article*

# Electrodeposition of Fine-Grained Tungsten Coatings on CuCrZr Alloy Substrates from Relatively Low Temperature KF-KCl-$WO_3$ Molten Salt System

Xiaoxu Dong [1,†], Wenqi Liu [1,†], Yusha Li [2], Zeyu Gao [1] and Yingchun Zhang [1,*]

1 School of Materials Science and Engineering, University of Science and Technology Beijing, 30 Xueyuan Road, Haidian District, Beijing 100083, China
2 Intelligence and Information Engineering College, Tangshan University, 11, University West Road, Lubei District, Tangshan 063000, China
* Correspondence: zycustb@163.com
† These authors contributed equally to this work.

**Abstract**

To address the issue in the pure oxide molten salt system $Na_2WO_4$-$WO_3$, where the relatively high melting temperature often causes thermal corrosion of the base material and reduces electrodeposition efficiency. A new molten salt system for electrodeposition tungsten coatings on CuCrZr substrates at relatively low temperatures was investigated. The crystal structure and microstructure of the tungsten coatings were analyzed using X-ray diffraction (XRD), scanning electron microscopy (SEM), and energy-dispersive X-ray spectroscopy (EDS). The results indicate that the power supply mode, current density, and duty cycle significantly affect the microstructure, crystalline characteristics, and overall performance of the tungsten coating. Pure tungsten coatings were successfully fabricated on CuCrZr substrates at 943 K. The best electrodeposition parameters were determined to be a current density of 40 mA/cm$^2$ and a duty cycle of 40%. Moreover, after prolonged electrodeposition (60 h), the tungsten coatings retained fine grains, with sizes ranging from 2 μm to 6 μm.

**Keywords:** fusion nuclear; electrodeposition; molten salt; tungsten coating; fine grains; CuCrZr alloy

## 1. Introduction

Nuclear fusion energy is widely recognized as one of the most promising energy sources due to its advantages of abundant availability, environmental cleanliness, and high energy efficiency [1]. The material issue is the key factor determining whether controlled nuclear fusion can be ultimately achieved. Among nuclear fusion materials, plasma-facing materials and components (PFC, Plasma Facing Components) are considered to be the most important category. In the fusion reactor, PFC is directly exposed to the high-energy particles from the plasma edge. Tungsten metal has emerged as a leading candidate material for the first wall in modern tokamak devices, owing to its excellent physical properties, including a high melting point (3695 K), high thermal conductivity, minimal tritium retention, low vapor pressure ($p$ = 1.3 × $10^{-7}$ Pa), a low coefficient of thermal expansion, and superior resistance to sputtering [2–4]. Copper-based heat sink materials possess excellent thermal conductivity and machinability, and can effectively transfer the high heat load on the surface materials of PFC. Therefore, they are selected as the heat sink materials in PFC. However, significant differences exist between the thermal expansion coefficients of tungsten and CuCrZr alloy (a commonly used heat sink material), posing considerable challenges in bonding dissimilar materials. In the fabrication of tungsten coatings, techniques such as plasma spraying (PS), chemical vapor deposition (CVD), and physical vapor deposition (PVD) [5] have been widely adopted. Specifically, the PS technique [6] is known for its simplicity and high efficiency; however, the resulting coatings often exhibit porosity or cracking due to the inherent characteristics of the process. The CVD method enables rapid deposition of tungsten but requires high-temperature conditions. Although PVD can produce high-quality, dense, and uniform tungsten coatings, it faces limitations in terms of cost and technical difficulty when depositing thick coatings [7]. In contrast, direct preparation of tungsten coatings on substrate surfaces via molten salt electrodeposition is considered a promising solution to this engineering challenge.

Electrodeposition in the $Na_2WO_4$-$WO_3$ molten salt system can yield high-quality tungsten coatings with dense and uniform microstructures [8]. However, the relatively high melting temperature of this oxide-based system often leads to thermal corrosion of the substrate and reduces electrodeposition efficiency. Therefore, it is of great significance to search for a molten salt system that is compatible with more substrates to reduce the electrodeposition temperature and improve the electrodeposition efficiency for the development of tungsten coating preparation by molten salt electrodeposition. Fan Jiang et al. [9] successfully deposited tungsten coatings with low oxygen content on copper alloy substrates via pulse electroplating in the $Na_2WO_4$-$WO_3$-$NaPO_3$ molten salt system, achieving dense morphologies under various conditions. Ningbo Sun et al. [10] conducted tungsten electrodeposition in the $Na_2WO_4$-$WO_3$ molten salt at 1173 K. Yusha Li et al. [11] fabricated a 433.72 μm tungsten coating on a CuCrZr alloy substrate via pulse current electrodeposition using a $Na_2WO_4$-$WO_3$-KCl-NaF molten salt system at an electrodeposition temperature of 1053 K. This coating exhibited high density, absence of impurities, excellent adhesion to the substrate, and superior stability and durability. To reduce the electrodeposition temperature, some researchers have explored halide-based molten salt systems, such as NaCl-KCl-NF-$WO_3$ [12] and LiF-KCl-$Li_2WO_4$ [13]. In this study, a KCl-KF-$WO_3$ molten salt system was designed for tungsten electrodeposition at 943 K. By optimizing the power supply mode, current density, duty cycle, and electrodeposition time, the halide-oxide KCl-KF-$WO_3$ molten salt system successfully produced tungsten coatings with fine grains, uniform microstructure, and high density on CuCrZr substrates. A systematic investigation was also conducted to evaluate the influence of process parameters within this moderate-temperature molten salt system on the phase composition, microstructure, and properties of the resulting tungsten coatings.

## 2. Experiment and Parameters

### 2.1. Coating Deposition

In this experiment, a metallic tungsten coating was prepared using the molten salt electrodeposition technique. The molten salt system employed was composed of KF-KCl-$WO_3$. The $WO_3$ (purity: 99.5%, Tianjin FuChen Chemical Reagent Co., Ltd., Tianjin, China), KCl (purity: 99%, Macklin Biochemical Technology Co., Ltd., Shanghai, China), and KF (purity: 99%, Macklin Biochemical Technology Co., Ltd., Shanghai, China) used in the experiment were all analytical-grade reagents. KF, KCl, and $WO_3$ were individually dried at 473 K for 12 h in a vertical furnace. Subsequently, the pre-dried reagents were weighed in a predetermined molar ratio of 1.2:1:0.1 (KF:KCl:$WO_3$) and thoroughly mixed in a ball mill before being loaded into a graphite crucible. The crucible was then placed inside a reaction vessel, which was positioned in an electric furnace. The CuCrZr alloy served as the anode electrode, while a tungsten plate (purity: 99.99%, Beijing Tianlong Tungsten & Molybdenum Technology Co., Ltd., Beijing, China) was used as the cathode electrode. The crucial role of selecting tungsten as the anode is to precisely replenish tungsten ions. In the KCl-KF-$WO_3$ molten salt system, the tungsten ions required for electro-deposition mainly come from the dissolution of the anode tungsten. The reaction can be understood as follows: At the anode, the tungsten block is oxidized to tungsten ions and enters the molten salt; these ions then form active ions in the molten salt and migrate to the cathode; at the cathode, the active ions are reduced to metallic tungsten and deposited on the substrate. This is a self-consuming equilibrium process, where the dissolution rate of the anode tungsten matches the deposition rate of the cathode tungsten, which is crucial for obtaining a stable composition and uniform thickness coating. The elemental composition of the CuCrZr alloy is provided in Table 1, and the purity details of the tungsten plate are listed in Table 2. Prior to electrodeposition, both electrodes were mechanically polished using 240-, 600-, 1000-, 1500-, and 2000-mesh sandpaper sequentially to achieve a smooth surface. They were then ultrasonically cleaned in distilled water and alcohol to ensure a clean and contaminant-free surface. The furnace temperature was increased at a rate of 5 K/min, and a circulating water cooling system was activated to maintain thermal stability. Argon gas (purity $\geq$ 99.99%, Beijing Longyuan Jingcheng Gas Technology Co., Ltd., Beijing, China) was introduced as a protective atmosphere. The temperature was gradually raised according to the programmed heating schedule until the desired electrodeposition temperature was reached. The pre-treated electrodes were then carefully immersed into the molten salt, ensuring that the electrode spacing met the specified requirements. Once immersed, the power supply was turned on, and electrodeposition was carried out under the preset parameters, as detailed in Table 3. We chose the specific current density values of 20 mA/$cm^2$, 30 mA/$cm^2$, 40 mA/$cm^2$, 50 mA/$cm^2$, and 60 mA/$cm^2$ because predecessors discovered in their research on the $Na_2WO_4$-$WO_3$ molten salt system, a dense surface and high bonding strength of tungsten coatings could be obtained within the current density range of 30 mA/$cm^2$ to 40 mA/$cm^2$ [8]. Therefore, this research focused on the current density range of 30 mA/$cm^2$ to 40 mA/$cm^2$, and expanded the current density within a certain range upwards and downwards to find the suitable current density for this molten salt system. The current densities of samples 6 to 9 were selected as 40 mA/$cm^2$ because the relatively optimal current density of 40 mA/$cm^2$ was obtained in the first five sets of experiments. Therefore, based on this, we continued to explore other influencing factors. After the electrodeposition process was completed, the sample was immersed in a NaOH solution and ultrasonically cleaned to remove any residual molten salt. The above-mentioned salt ratio was determined based on the lowest melting point of this salt system.

**Table 1.** Elements and Contents of CuCrZr Alloy (wt%).

| Cu | Cr | Zr | Mg | Si |
|---|---|---|---|---|
| Balance | 0.4–1.0 | 0.02–0.1 | 0.01–0.03 | 0.02–0.04 |

**Table 2.** Elements and Contents of Metal Tungsten (wt%).

| W | O | C | Mo | Ni | Si |
|---|---|---|---|---|---|
| Balance | 0.03 | 0.02 | 0.05 | 0.01 | 0.01 |

**Table 3.** Process Parameters for Electrodeposition of Tungsten Coatings in KCl-KF-$WO_3$ Molten Salt System.

| No. | Current Density ($mA/cm^2$) | Temperature (K) | Duty Cycle (%) | Frequency (Hz) | Deposition Time (h) |
|---|---|---|---|---|---|
| 1 | 20 | 943 | 100 | 1000 | 2 |
| 2 | 30 | 943 | 100 | 1000 | 2 |
| 3 | 40 | 943 | 100 | 1000 | 2 |
| 4 | 50 | 943 | 100 | 1000 | 2 |
| 5 | 60 | 943 | 100 | 1000 | 2 |
| 6 | 40 | 943 | 30 | 1000 | 2 |
| 7 | 40 | 943 | 40 | 1000 | 2 |
| 8 | 40 | 943 | 50 | 1000 | 2 |
| 9 | 40 | 943 | 40 | 1000 | 60 |

### 2.2. Coating Characterization

High-resolution microstructural observations and elemental composition analysis of the tungsten coatings were performed using a GeminiSEM 500 scanning electron microscope (SEM) manufactured by Zeiss (Oberkochen, Germany), in conjunction with its integrated energy dispersive spectrometer (EDS). The crystallographic structure and phase composition distribution of both the substrate and the tungsten coating were systematically characterized using a Rigaku D/MAX-RBX X-ray diffractometer (XRD) produced by Rigaku Denki Co., Ltd. (Tokyo, Japan). The texture coefficient was calculated based on the XRD data, reflecting the extent to which the measured diffraction intensity of a specific crystal plane (hkl) deviates from the average intensity expected for a randomly oriented sample. The texture coefficients of the substrate and the coating were determined through XRD analysis. The corresponding calculation formula is presented below as Equation (1):

$$TC_{(hkl)} = \frac{I_{(hkl)}/I_{0(hkl)}}{\sum_{i=1}^{n} I_{(hkl)}/I_{0(hkl)}} \times 100\% \quad (1)$$

where $I_{(hkl)}$ is the cumulative intensity of a certain crystal plane (hkl) in the actual measurement results, $I_{0(hkl)}$ is the reference intensity of the corresponding crystal plane under random orientation, n is the number of independent crystal planes tested.

The microhardness of the coating was measured using an EM-1500L microhardness tester (Future-Tech, Tokyo, Japan). The test parameters were set to a load of 50 gF and a dwell time of 15 s. Five random measurement points were selected on each sample surface, and the final microhardness value was calculated as the arithmetic mean of these five readings. The density was determined using the displacement method based on Archimedes' principle. The procedure involved separating the coating from the substrate through a peeling process, followed by mass and volume measurements using a BSA224S density balance (Sartorius Group, Göttingen, Germany). The porosity of the coating was analyzed using an Autopore IV 9520 mercury porosimetry (Micromeritics Instrument Corporation, Norcross, GA) instrument manufactured in the US.

## 3. Results and Discussions

### 3.1. Surface Morphology and Crystal Structure of Tungsten Coatings Under Direct Current

Careful observation of the macroscopic surface morphology of the tungsten coatings under varying current densities is essential for understanding the influence of process parameters on coating formation. The macroscopic morphologies are illustrated in Figure 1. As the current density increases, distinct changes occur on the surface characteristics of the tungsten coatings. As shown in the figure, all six experimental groups successfully produced complete tungsten coatings on the CuCrZr substrates, each displaying a certain degree of metallic luster. At current densities of 20 mA/cm$^2$ and 30 mA/cm$^2$, the coating surfaces appear relatively smooth and flat but exhibit non-uniformity and noticeable color variations. When the current density reaches 40 mA/cm$^2$, the coating becomes uniformly distributed, with no visible pores, gaps, or color differences observed. As the current density increases further to 50 mA/cm$^2$, the metallic luster begins to diminish, and coarse grains start to appear on the surface. At a current density of 60 mA/cm$^2$, the coating uniformity deteriorates significantly, and the coarse grains continue to grow in size.

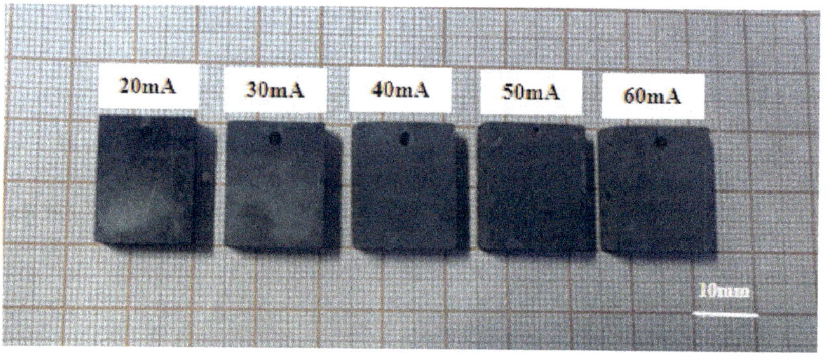


**Figure 1.** Macroscopic morphology of tungsten coatings at different DC current densities.

The surface microstructure of the tungsten coating is primarily examined using SEM images. Figure 2 presents SEM images and EDS analysis diagrams of the surface microstructure of tungsten coatings produced at various current densities under direct current. The images reveal that tungsten grains are in a polygonal, agglomerated state and are relatively evenly distributed, with some sparse and irregular pores. At a current density of 20 mA/cm$^2$, the deposition rate is slow, and tungsten nuclei have not fully developed on the substrate surface. As the current density rises to 30 mA/cm$^2$, tungsten grain nucleation occurs concurrently, and some nuclei begin to grow gradually, leading to a few areas of grain aggregation. EDS analysis indicates that the coatings produced at current densities of 20 mA/cm$^2$ and 30 mA/cm$^2$ are primarily composed of pure tungsten, though there are still some uncovered substrate areas, characterized by copper. When the current density further increases to 40 mA/cm$^2$, the higher current density hastens the nucleation process and also promotes rapid growth in some grains. Previously nucleated sites preferentially grow into larger grains, causing the coating to exhibit unevenness, yet the macroscopic surface remains relatively flat. EDS analysis confirms that the coating surface is pure tungsten, and the substrate is entirely covered. Upon reaching current densities of 50 mA/cm$^2$ and 60 mA/cm$^2$, the abnormal growth of grains and agglomeration become pronounced. It is because the current densities of 50 and 60 mA/cm$^2$ have reached or exceeded the limit diffusion current density for this system, resulting in a sharp decrease in the concentration of tungsten ions at the cathode solution interface. A significant ion concentration difference is generated between the interface and the solution body, causing a drastic change in the cathode potential. The extremely high overpotential causes the existing crystal nuclei to grow preferentially and rapidly, competing for the limited tungsten ions. Due to the lack of ions at the interface, it becomes very difficult to form new crystal nuclei, leading to abnormal grain growth. The electrodeposition tends to occur at the protrusions where ions can easily reach, forming agglomerations, and ultimately resulting in abnormal growth and agglomeration of the microstructure [14].

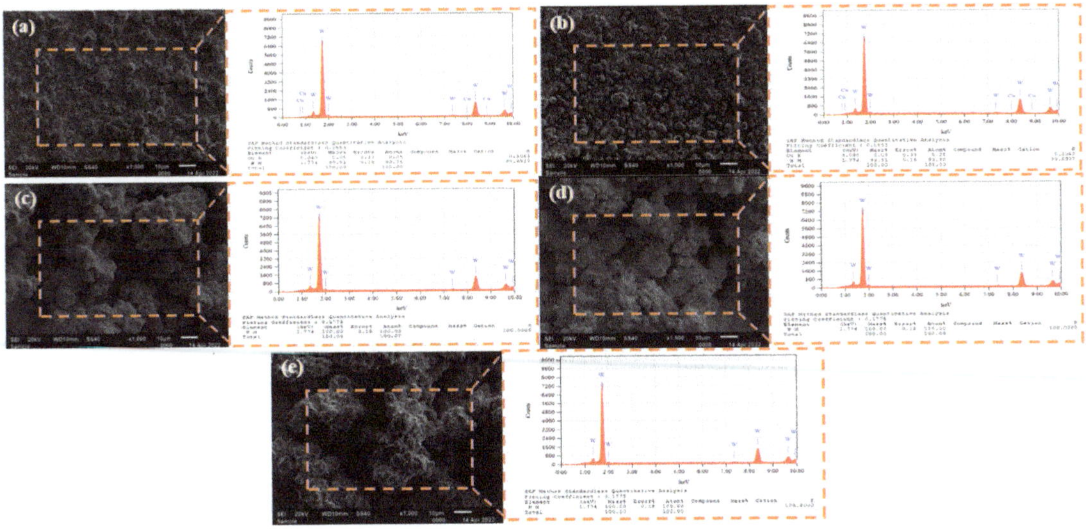


**Figure 2.** SEM and EDS images of tungsten coating surface under different DC current densities: (**a**) 20 mA/cm$^2$; (**b**) 30 mA/cm$^2$; (**c**) 40 mA/cm$^2$; (**d**) 50 mA/cm$^2$; (**e**) 60 mA/cm$^2$.

The evolution of the coating's crystal structure was analyzed using an X-ray diffraction (XRD) system, which is crucial for elucidating the mechanism underlying the relationship between current density and coating integrity. Figure 3 displays the XRD spectra of the tungsten coating surface under varying direct current densities. The experimental data indicate that the XRD spectra of tungsten coatings produced at different current densities all exhibit four distinct sets of diffraction peaks. These peaks correspond to the (111), (200), (211), and (220) crystallographic planes of tungsten, confirming that the coating possesses a tungsten phase structure and exhibits good crystallinity. At current densities of 20 mA/cm$^2$ and 30 mA/cm$^2$, diffraction peaks corresponding to copper's (111), (200), and (220) planes are also observed in the XRD spectra. The presence of these peaks can be explained as follows: at low current densities, the local electric field energy is insufficient to overcome the nucleation barrier, resulting in a smaller reduction in the critical nucleus size of tungsten grains [15,16]. Consequently, the nucleation rate decreases, and the growth of existing grains slows down, leading to the formation of larger grains. Local grain growth is limited, resulting in an insufficient coating thickness in those regions. Due to the substantial penetration depth of X-rays, some X-rays reach and reflect from the CuCrZr substrate, generating copper characteristic diffraction peaks. When the current density increases to 40–60 mA/cm$^2$, both the nucleation rate and grain growth rate of tungsten significantly increase, resulting in thicker coatings. In this case, X-rays are unable to effectively penetrate the tungsten coating, thereby shielding the CuCrZr substrate signal. As a result, there are no copper diffraction peaks on the XRD patterns.

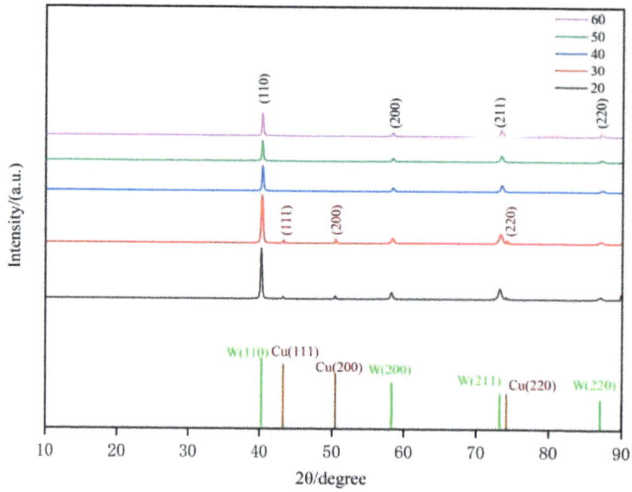


**Figure 3.** XRD patterns of tungsten coatings prepared at different current densities.

### 3.2. Surface Morphology and Crystal Structure of Tungsten Coatings Under Pulsed Current

The characterization of the tungsten coating under specified direct current conditions shows that at a current density of 40 $mA/cm^2$, the tungsten coating fully covers the substrate and achieves the highest deposition efficiency. However, the tungsten coating fabricated via direct current deposition exhibits pronounced abnormal grain growth, leading to a relatively rough surface. To address this issue, pulse electrodeposition was employed to improve coating quality. Figure 4 presents the macroscopic surface morphology of the tungsten coating at various pulse duty ratios. During pulse electrodeposition with a 30% duty cycle, the short conduction time results in low energy density, which leads to an uneven metallic luster on the coating surface. Nevertheless, the coating maintains good continuity, with no signs of substrate oxidation or localized exposure. As the duty cycle increases to 40%, the coating surface remains continuous and exhibits a uniform silver-white metallic luster. This improvement can be attributed to the extended current conduction time, which facilitates the uniform deposition of tungsten ions across the substrate surface, resulting in a denser coating. When the duty cycle is further increased to 50%, the prolonged conduction time leads to a thicker diffusion layer, insufficient ion replenishment, and a rougher coating surface, which causes the metallic luster to diminish.

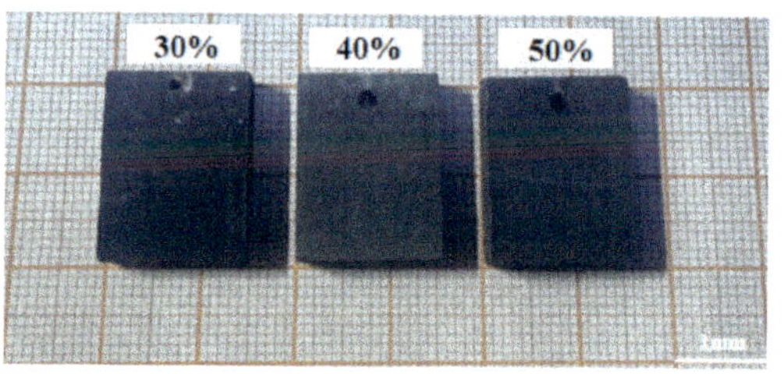


**Figure 4.** Macroscopic morphology of tungsten coatings at different pulse duty cycles.

Figure 5 illustrates the microstructural features and elemental composition of the tungsten coating produced via pulsed electrodeposition at different duty cycles. The results show that the tungsten grains are densely packed, forming a polycrystalline aggregate structure characterized by irregular surfaces, pores, and protrusions. These features may be related to the characteristics of the molten salt system. At a duty cycle of 30%, the tungsten grains are relatively fine, yet the coating surface exhibits "pits". This is attributed to the short conduction time, which results in insufficient crystal nucleus growth, leading to a mixture of dense and sparse regions and causing local non-uniformity. Increasing the duty cycle to 40% provides a moderate current conduction time, thereby extending the effective deposition period. Consequently, the ion migration and diffusion dynamics within the molten salt are significantly enhanced, promoting a higher deposition rate of tungsten ions and facilitating the relatively uniform growth of crystal grains. When the duty cycle reaches 50%, noticeable accumulation and stacking of crystal grains occur, and holes begin to appear. This phenomenon arises because a higher duty cycle corresponds to a longer current conduction time. During this stage, the continuous depletion of tungsten ions at the cathode surface leads to a sharp decrease in ion concentration near the electrode. The insufficient ion replenishment, combined with the enhanced electric field around cathode surface protrusions, promotes preferential ion deposition, resulting in dendritic growth. In contrast, the depressed areas experience slower deposition due to ion depletion, leading to the formation of holes. The resulting loose crystal structure contributes to a decline in coating performance. Energy-dispersive spectroscopy (EDS) analysis confirms that under all tested duty cycles, the coating consists of single-phase metallic tungsten with high coverage. Therefore, compared to direct current electrodeposition, pulsed electrodeposition improves deposition efficiency and effectively addresses the issue of inadequate coating coverage associated with direct current methods.

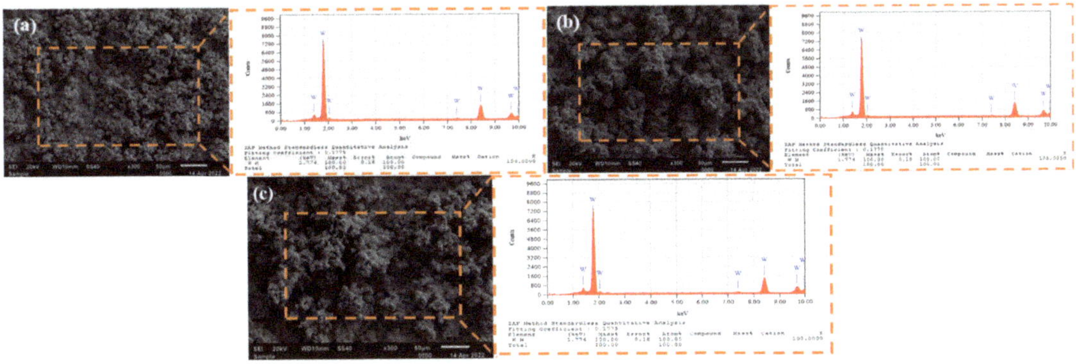


**Figure 5.** SEM and EDS images of tungsten coating surface under different pulse duty cycles: (**a**) 30%; (**b**) 40%; (**c**) 50%.

Figure 6 presents the X-ray diffraction (XRD) spectra of tungsten coatings produced under various duty cycles. The XRD results indicate that all samples exhibit the characteristic diffraction peaks of a body-centered cubic (BCC) tungsten lattice. The main diffraction peaks are located at $2\theta = 40°$, corresponding to the (110) crystal plane; at $2\theta = 58°$, corresponding to the (200) plane; at $2\theta = 73°$, corresponding to the (211) plane; and at $2\theta = 87°$, corresponding to the (220) plane. Notably, the (110) diffraction peak is significantly more intense than the others, and its full width at half maximum (FWHM) is narrower, indicating that the tungsten coating possesses a high degree of crystallinity and a low defect density along the (110) plane during the pulsed electroplating process. The favorable lattice alignment of the (110) plane with the ion migration direction in the molten salt system synergistically promotes preferential grain growth, while the relatively low surface energy of the (110) plane facilitates nucleation. When the duty cycle increases from 30% to 40%, the conduction time of the pulse current is appropriately prolonged. Near the cathode, the metal ions are rapidly consumed due to the reduction reaction, resulting in a sudden drop in local ion concentration. The appropriate prolongation of the conduction time ensures that the ion concentration is restored in time, allowing the thermodynamically most stable (110) crystal plane to grow preferentially. This is manifested in the XRD spectrum as an increase in the relative intensity of the W(110) diffraction peak. When the duty cycle increases from 40% to 50%, the high duty cycle means a longer conduction time and a higher average current density, resulting in a significant acceleration of the deposition rate. This causes the reaction ions at the cathode interface to be rapidly consumed, while the ions in the bulk solution have no time to diffuse and replenish. In this non-equilibrium state, the deposition process no longer prioritizes the crystal plane with the lowest surface energy but rather those with the fastest growth rate, leading to a weakening of the relative intensity of the W(110) peak [17].

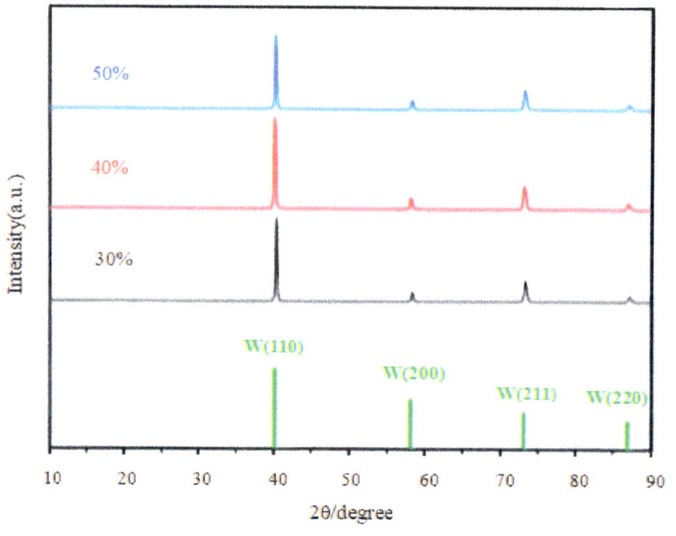


**Figure 6.** XRD patterns of tungsten coatings prepared at different pulse duty cycles.

### *3.3. The Microstructure, Crystal Structure, and Properties of the Tungsten Coating Obtained Through Long-Term (60-h) Electrodeposition*

The previous study primarily investigated the effects of two key parameters—current density and duty cycle-on the morphology and structure of tungsten coatings. It was found that under conditions of 40 mA/cm$^2$ current density and 40% duty cycle, pure metallic single-phase tungsten coating with favorable microstructural morphology and uniform coverage could be obtained. Under these optimized conditions, a tungsten coating was fabricated through long-term electrodeposition at 943 K over a period of 60 h.

Figure 7 presents the SEM and EDS images of the surface of the long-term electrodeposited tungsten coating. The surface exhibits protrusions, with an average grain size of approximately 3 μm, which is smaller than that of tungsten coatings produced within oxide-based systems. This grain refinement is attributed to the relatively limited tungsten source and the reduced energy input due to the lower deposition temperature. The figure also reveals defect characteristics of the long-term electrodeposited tungsten coating, such as grain aggregation and pore formation. In the second picture of Figure 7, the area circled by the circle represents the grain aggregation, and the area marked by the triangle represents the pores. The non-uniform grain growth is associated with uneven nucleation of tungsten grains. Over time, the initial nucleation sites consumed a significant portion of the tungsten source, leading to rapid grain growth, whereas regions without early nucleation experienced insufficient tungsten replenishment, ultimately resulting in localized grain aggregation and inhomogeneous growth. EDS analysis confirms that the coating consists of pure tungsten, and the substrate is fully covered without detectable impurities. In addition, after a long period of electro-deposition, the anode may experience passivation or uneven dissolution, which will result in uneven current distribution, local deposition interruption, and thus form defects.

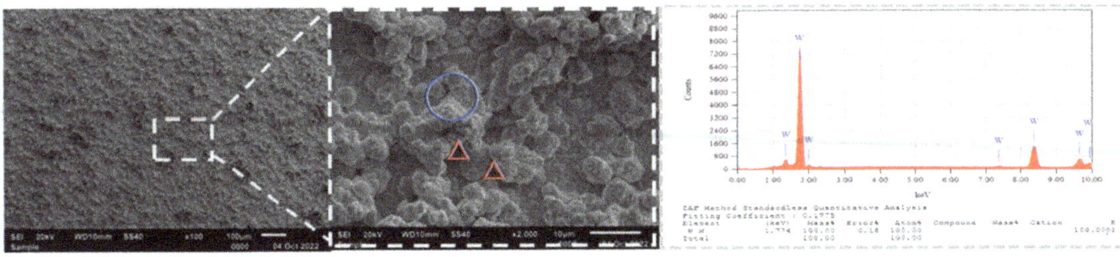

**Figure 7.** SEM and EDS images of the tungsten coating surface obtained via long-term electrodeposition.

Figure 8 shows the SEM and EDS images of the cross-section of the tungsten coating after 60 h of electro-deposition. From the SEM image of the cross-section, it can be seen that after 60 h of electro-deposition, the coating thickness is 30–50 μm. Through EDS analysis, it was found that the entire cross-section consists only of the base material and the tungsten coating, with no other impurities mixed in.

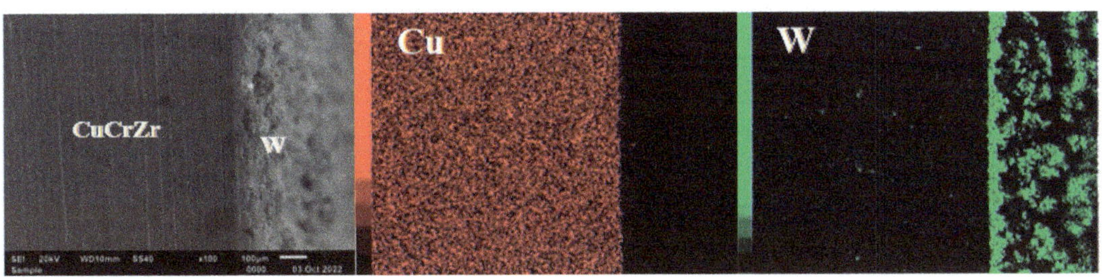


**Figure 8.** SEM and EDS images of the cross-section of the tungsten coating after 60 h of electrodeposition.

Figure 9 presents the XRD pattern of the coating surface after 60 h of electrodeposition, confirming the formation of pure metallic tungsten. Texture coefficients were calculated from the XRD data and are summarized in Table 4. The results indicate that with increasing electrodeposition time, tungsten crystals exhibit relatively uniform growth in the (110), (200), and (211) crystallographic directions, suggesting a tendency toward isotropic growth of the tungsten coatings.

Table 5 summarizes the physical properties of the tungsten coating fabricated via long-term electrodeposition over a 60-h period. The data show that the coating's density of 18.60 g/cm$^3$ is lower than the theoretical density of pure tungsten, which is 19.3 g/cm$^3$. In addition, the coating exhibits a porosity of 2.15% and a hardness of 487.50 HV, indicating the presence of localized porosity within the coating. This phenomenon is likely attributed to the limited mass transfer efficiency of the electrolyte in the low-temperature molten salt system. However, the bonding strength of this coating is 55 MPa, which indicates that there is good interface compatibility between the substrate material and the tungsten coating. In the research on coatings on the same substrate by Li et al., they believed that when the bonding strength reached 49.28 MPa, there was good interfacial compatibility between the substrate and the coating [12].

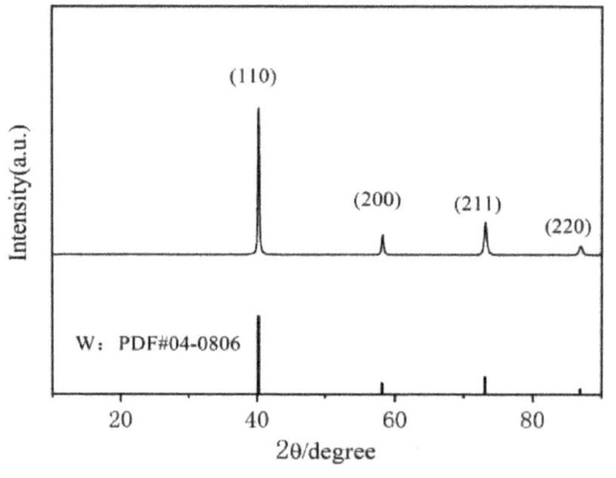


**Figure 9.** XRD patterns of the tungsten coating surface after long-term (60 h) electrodeposition.

**Table 4.** Grain texture coefficient of the surface of tungsten coating obtained through long-term electrodeposition.

| Crystal Plane | Texture Coefficient |
|---|---|
| (110) | 27.89% |
| (200) | 24.79% |
| (211) | 26.65% |
| (220) | 12.50% |

**Table 5.** Physical properties of samples with long-term electrodeposited tungsten coating.

| Sample | Density (g/cm$^3$) | Porosity (%) | Hardness (HV) | Bonding Strength (MPa) |
|---|---|---|---|---|
| Electrodeposition for 60 h | 18.60 | 2.15 | 487.50 | 55 |

Table 6 presents the porosity of all samples. It can be observed that after 2 h of electrodeposition, the porosity of the samples did not vary significantly, with the maximum being 1.24% and the minimum being 1.08%. However, after a long period of (60 h) electrodeposition, the porosity of the samples significantly increased to 2.15%. This is because as time progresses, smaller and more energetic grains gradually dissolve, while larger grains continue to grow to reduce the overall surface energy. Coarse grains mean fewer grain boundaries, which are the rapid filling channels for tungsten ion deposition. The structure after coarsening is more likely to contain larger pores and defects between the grains. From the roughness data in Table 6, it can be observed that the influence of current density on the surface roughness of the coating is greater than that of the power supply type.

Moreover, as the electrodeposition time increases, the surface roughness of the tungsten coating significantly increases.

**Table 6.** Porosity of different samples.

| Sample | Porosity (%) | Surface Roughness/μm |
|---|---|---|
| 1 (20 mA/cm$^2$, 100%, 2 h) | 1.21 | 1.443 |
| 2 (30 mA/cm$^2$, 100%, 2 h) | 1.24 | 1.736 |
| 3 (40 mA/cm$^2$, 100%, 2 h) | 1.18 | 1.474 |
| 4 (50 mA/cm$^2$, 100%, 2 h) | 1.19 | 3.358 |
| 5 (60 mA/cm$^2$, 100%, 2 h) | 1.36 | 3.342 |
| 6 (40 mA/cm$^2$, 30%, 2 h) | 1.16 | 1.375 |
| 7 (40 mA/cm$^2$, 40%, 2 h) | 1.08 | 1.267 |
| 8 (40 mA/cm$^2$, 50%, 2 h) | 1.19 | 2.642 |
| 9 (40 mA/cm$^2$, 40%, 60 h) | 2.15 | 6.544 |

## 4. Conclusions

In this study, a KF-KCl-$WO_3$ molten salt system was developed to reduce the electrodeposition temperature. Tungsten coatings were electro-deposited on CuCrZr substrates. Under varying the current density, current mode, duty cycle, and electrodeposition time, various tungsten coatings were produced. The microstructural and crystallographic characteristics of the coatings were systematically analyzed. Additionally, the density, porosity, and hardness of the tungsten coating deposited under relatively optimal conditions over an extended period were evaluated. The main findings are summarized as follows:

(1) In the KCl-KF-$WO_3$ molten salt system, a metallic tungsten coating was successfully fabricated at an electrodeposition temperature of 943 K, which is 250 K lower than the 1193 K required in the $Na_2WO_4$-$WO_3$ molten salt system.

(2) Under direct current electrodeposition, excessively low current densities resulted in slow deposition rates and insufficient grain growth, leading to incomplete coverage of the substrate. In contrast, high current densities accelerated grain nucleation but also induced abnormal growth of pre-existing grains, increasing surface roughness. Transitioning from direct current to pulse current electrodeposition improved ion migration and diffusion within the molten salt. An appropriate duty cycle further enhanced the deposition rate. Consequently, the optimal electrodeposition parameters in this system were determined to be an electrodeposition temperature of 943 K, a current density of 40 mA/cm$^2$, pulse current mode, and a duty cycle of 40%.

(3) Using the optimized parameters in the KCl-KF-$WO_3$ molten salt system, a tungsten coating was obtained after 60 h of long-term electrodeposition. The surface tungsten grains exhibited a fine and uniform morphology, with an average size of 2–6 μm. The coating had an average thickness of approximately 40 μm, a porosity of 2.15%, a hardness of 487.50 HV, and a bonding strength of 55 MPa.

**Author Contributions:** Conceptualization, X.D. and W.L.; methodology, W.L.; validation, W.L. and Y.L.; formal analysis, W.L. and Z.G.; investigation, Z.G.; resources, X.D. and W.L.; data curation, W.L.; Writing—original draft, X.D. and W.L.; writing–review & editing, W.L. and Y.L.; visualization, Z.G.; supervision, Y.Z.; project administration, Y.Z.; funding acquisition, Y.Z. All authors have read and agreed to the published version of the manuscript.

**Funding:** This study was supported by National MCF Energy R&D Program (NO. 2019YFE03130002).

**Institutional Review Board Statement:** Not applicable.

**Informed Consent Statement:** Not applicable.

**Data Availability Statement:** Data are contained within the article.

**Conflicts of Interest:** The authors declare no conflicts of interest.

## References

1. Ongena, J.; Koch, R.; Wolf, R.; Zohm, H. Magnetic-confinement fusion. *Nat. Phys.* **2016**, *12*, 398–410. [CrossRef]
2. Matthews, G.F.; Coad, P.; Greuner, H.; Hill, M.; Hitai, T.; Likonen, J. Development of divertor tungsten coatings for the JET ITER-like wall. *J. Nucl. Mater.* **2009**, *390–391*, 934–937. [CrossRef]
3. Piip, K.; Paris, P.; Hakola, A. Influence of He/$D_2$ Plasma Fluxes on the Morphology and Crystallinity of Tungsten Coatings. *Phys. Scr.* **2014**, *89*, 39–50. [CrossRef]
4. Ribis, J.; Bordas, E.; Trocellier, P. Comparison of the neutron and ion irradiation response of nano-oxides in oxide dispersion strengthened materials. *J. Mater. Res.* **2015**, *30*, 2210–2221. [CrossRef]
5. Sun, Q.L.; Mao, X.C.; Yao, Y.F.; Li, J.N.; Shen, Z.S.; Tu, K.N.; Liu, Y.X. Effect of WC thickness on the microstructure and properties of WC-C/DLC coated 304 steel. *J. Appl. Surf. Sci.* **2025**, *704*, 163418. [CrossRef]
6. Xie, J.J.; Fang, H.J.; Wang, Y.X.; Wang, W.Q.; Yu, J.H.; Li, J.C.; He, X.D.; Pu, J.B. Controlling the porosity to enhance the high-temperature tribological performance of plasma-sprayed NiCr–$Cr_3C_2$-$BaF_2$/$CaF_2$ coating by adopting axial feeding. *J. Mater. Res. Technol.* **2025**, *36*, 2577–2588. [CrossRef]
7. Ganne, T.; Crépin, J.; Serror, S.; Zaoui, A. Cracking behaviour of PVD tungsten coatings deposited on steel substrates. *Acta Mater.* **2002**, *50*, 4149–4163. [CrossRef]
8. Liu, Y.H.; Zhang, Y.C.; Li, H.L.; Liu, D.J. Analysis on Composition of Binary Molten Salt $Na_2WO_4$-$WO_3$ and Study of Electrodepositing Tungsten Coatings. *J. Mater. Her.* **2015**, *29*, 34–36+45. [CrossRef]
9. Jiang, F.; Zhang, Y.C.; Li, X.L.; Sun, N.B.; Wang, L.L. Tungsten coating prepared on V-4Cr-4Ti alloy substrate by electrodeposition from molten salt in air atmosphere. *Fusion Eng. Des.* **2014**, *89*, 83–87. [CrossRef]
10. Sun, N.B.; Zhang, Y.C.; Lang, S.T.; Jiang, F.; Wang, L.L. Tungsten coatings electro-deposited on CFC substrates from oxide molten salt. *J. Nucl. Mater.* **2014**, *455*, 450–453. [CrossRef]
11. Li, Y.S.; Dong, X.X.; Liu, Q.; You, Y.J.; Gao, Z.Y.; Zhang, Y.C. Preparation and Properties of Thick Tungsten Coating Electrodeposited from $Na_2W0_4$-$WO_3$-KCl-NaF Molten Salt System. *J. Coat.* **2024**, *14*, 1471. [CrossRef]
12. Qi, Y.F.; Tang, Y.H.; Wang, B.; Zhang, M.; Ren, X.Q.; Li, Y.G.; Ma, Y.T. Characteristics of tungsten coatings deposited by molten salt electrodeposition and thermal fatigue properties of electrodeposited tungsten coatings. *Int. J. Refract. Met. Hard Mater.* **2019**, *81*, 183–188. [CrossRef]
13. Nohira, T.; Ide, T.; Meng, X.D.; Norikawa, Y.; Yasuda, K. Electrodeposition of Tungsten from Molten KF-KCl-$WO_3$ and CsF-CsCl$WO_3$. *J. Electrochem. Soc.* **2021**, *168*, 046505. [CrossRef]
14. Zhu, D.; Zhang, W.F.; Lei, W.N. Study on Refinig Grains of Nanocrystalline Materials Based on Electrodeposition. *J. Synth. Cryst.* **2021**, *5*, 765–769. [CrossRef]
15. Pérez, H.D.; González, A.J.; Romo, R.M.; Hernández, R.S.; Mejía, E.A.; Gil, M.P.; Piñon, S.N.; Garrido, S.A.; Pardavé, P.M. Electrochemical Nucleation and Growth of Neodymium on Glassy Carbon Electrodes using Reline as a Deep Eutectic Solvent. *J. Electrochim. Acta* **2025**, *535*, 146643. [CrossRef]
16. Wang, R.Q.; Li, G.Y.; Wu, F.; Wang, S.M.; Zhang, C.Z. Cycling performance of Li metal anode in localized high concentration electrolyte and dynamic mechanism analysis. *J. Electroanal. Chem.* **2024**, *959*, 118176. [CrossRef]
17. Chen, W.T.; Wang, Q.W.; Yang, Y.F. Research on Parameter Regulation and Structural Properties of Pulse Electrodeposited Zn-Ni-PTFE Composite Coatings. *J. Surf. Eng. Remanuf.* **2025**, *25*, 1–18. [CrossRef]

*coatings*

*Review*

# Functional Coatings for Fiber Bragg Gratings: A Critical Review of Deposition Techniques for Embedded and Harsh-Environment Applications

**Cristian Vendittozzi [1,*], Emilia Di Micco [1], Michele A. Caponero [2] and Rosaria D'Amato [2]**

[1] Department of Chemical Engineering Materials Environment, Sapienza University of Rome, Via Eudossiana 18, 00184 Roma, Italy; emilia.dimicco@uniroma1.it
[2] ENEA Centro Ricerche Frascati, Via Enrico Fermi 45, 00044 Frascati, Italy; michele.caponero@enea.it (M.A.C.); rosaria.damato@enea.it (R.D.)
* Correspondence: cristian.vendittozzi@uniroma1.it

**Abstract**

Fiber Bragg Grating (FBG) sensors facilitate compact, multiplexed, and electromagnetic interference-immune monitoring in embedded and harsh environments. The removal of the polymer jacket, a measure taken to withstand elevated temperatures or facilitate integration, exposes the fragile glass. This underscores the necessity of functional coatings, which are critical for enhancing durability, calibrating sensitivity, and improving compatibility with host materials. This review methodically compares coating materials and deposition routes for FBGs, encompassing a range of techniques including top-down physical-vapor deposition (sputtering, thermal/e-beam evaporation, cathodic arc), bottom-up chemical vapor deposition (CVD)/atomic layer deposition (ALD), wet-chemical methods (sensitization/activation, electroless plating (EL), electrodeposition (ED)), fusion-based processes (casting and melt coating), and hybrid stacks (e.g., physical vapor deposition (PVD) seed → electrodeposition; gradient interlayers). The consolidation of surface-preparation best practices and quantitative trends reveals a comprehensive understanding of the interrelationships between coating material/stack, thickness/microstructure, adhesion, and sensitivity across a range of temperatures, extending from approximately 300 K to cryogenic regimes. Practical process windows and design rules are distilled to guide method selection and reliable operation across cryogenic and high-temperature regimes.

**Keywords:** fiber Bragg grating (FBG); metallic coating; sputtering; electroless plating; electrodeposition; cryogenic sensing; high-temperature sensing; embedded sensors

## 1. Introduction

Optical fiber Bragg grating (FBG) sensors are compact, multiplexable, immune to electromagnetic interference, and compatible with long-distance interrogation. These qualities have enabled the adoption of FBG sensors in biomedical monitoring, structural health monitoring, and industrial automation [1–4]. Beyond reviews specific to FBGs, comprehensive surveys of fiber-optic chemical and physical sensors, as well as distributed/networked sensing, provide a broader landscape for scalability and deployment [5–7]. In harsh or unconventional environments, such as cryogenic or high-temperature environments, radiation environments, environments with strong magnetic fields, or environments with vibration, bare gratings and standard packaging have limitations. This motivates the development of coatings and tailored processes that stabilize the optical response and extend operability [8,9]. Recent overviews focusing on cryogenic FBG temperature sensing further emphasize the importance of coating-based designs at lower temperatures [10].

Despite their versatility, bare FBGs face inherent limitations when exposed to demanding operational environments. Silica core and cladding are brittle and prone to microcracking under thermal shock or strain concentration, and the polymer jacket, typically used for protection, softens or degrades above ~150 °C [11]. At cryogenic temperatures, the low thermo-optic coefficient of silica reduces temperature sensitivity and complicates calibration, while repeated cool-down cycles induce mechanical fatigue at the grating–coating interface [12]. In aerospace applications, sensors must endure vibration, low pressure, and wide thermal excursions, while maintaining adhesion and spectral integrity under cyclic mechanical loading [13]. In energy systems, including fusion and high-temperature components, FBGs experience intense gradients and radiation that can induce refractive index drift and signal degradation [12,14]. Similarly, in structural and civil infrastructures, long-term performance requires mechanical robustness and environmental stability over millions of strain cycles [15]. Across these sectors, industries demand FBGs that maintain linearity, adhesion, and repeatability from cryogenic to ultra-high temperatures, often when embedded within metal or composite hosts. These requirements have driven the integration of functional coatings—metallic, ceramic, or hybrid—to extend FBG operability beyond the range of bare gratings and ensure compatibility with advanced structural materials [12–15]. Alternatively, FBGs can also be inscribed through polymer coatings using ultraviolet (UV) radiation; however, such gratings often suffer from poor inscription uniformity, mechanical instability, and limited thermal endurance, restricting their applicability in harsh or high-temperature environments [16–18].

Although baseline demonstrations established FBG thermometry from ~2–400 K with appropriate packaging, sensitivity and stability degrade toward the deepest cryogenic conditions without additional engineering of the grating–host interface [19,20]. Early design studies of cryogenic FBG thermometers and femtosecond (fs)-inscribed gratings highlight inscription and packaging choices that affect the coating's response at low temperatures [21,22]. Representative applications include aeronautical structural health monitoring [23], conceptual designs for cryogenic and energy systems [24], and fusion-relevant strain monitoring on tungsten divertors [25]. Each application imposes constraints on geometry, survivability, and electromagnetic compatibility that coatings can help relieve. These constraints can be addressed by adding mechanical protection, improving thermal/mechanical coupling, providing hermetic barriers, and enabling bonding to metals or ceramics, all of which can be achieved through coatings. Polymeric layers (e.g., polymethyl methacrylate, PMMA) boost apparent temperature sensitivity and ease cycling, while metallic coatings (e.g., Ni, Cu, Zn, and Sn, as well as their combinations) offer higher thermal conductivity and robust attachment. Ceramic/oxide overlayers (e.g., TiN, $TiO_2$, and ZnO) cap against oxidation and interrupt diffusion [26–30]. Multilayer and duplex designs, such as Ni/Cu/Au or Ni/Zn, are often used to balance adhesion, diffusion control, and thermo-mechanical performance across a wide temperature range [28–31]. Gradient metallic coatings, which are specifically designed for cryogenic sensitivity and cyclic stability, broaden this design space and offer an alternative to simple duplex builds [32]. Nanostructured coating approaches and comparative analyses of thermal-sensing performance further expand the design space and clarify the trade-offs between sensitivity and transient response [33,34]. A variety of deposition routes are available. Top-down physical vapor deposition (PVD), including magnetron/ion beam sputtering (IBS) and thermal/e-beam evaporation, delivers controlled thin films. Bottom-up chemical vapor deposition

(CVD)/atomic layer deposition (ALD) provides conformal oxides. Wet-chemical workflows, such as sensitization/activation and electroless (EL)/electrodeposition (ED), enable uniform metal builds and thick overcoats at low thermal budgets. Fusion-based routes, such as casting, integrate FBGs into metallic hosts. Hybrid flows, such as PVD seeding followed by ED or graded interlayers, combine the advantages of multiple routes [35].

General primers on thin films and processes complement this taxonomy by highlighting stress and roughness control on cylindrical substrates [36]. Practical metallization and re-coating procedures tailored to fibers are discussed in more detail later in the manuscript, and additional methods are cataloged in foundational surveys [37]. These methods have key engineering constraints. Strong adhesion to silica usually necessitates surface preparation and seed layers (e.g., $SnCl_2/PdCl_2$ activation for EL Ni-P). Residual stress and thickness-induced cracking impose limitations, especially for EL deposits, when the process window is exceeded [38,39]. At elevated temperatures, interdiffusion (e.g., Cu/Ni) and hydrogen uptake during plating can degrade coatings and spectra, unless barrier layers and post-bakes are employed [40]. At cryogenic temperatures, metallic overcoats increase thermal mass and can slow step response—a trade-off that must be balanced against sensitivity gains [41]. These considerations motivate the structured comparison and synthesis that follow.

### 1.1. Review Scope and Method

This review centers on coated FBGs operating from cryogenic to high-temperature regimes. Sources were identified in major databases (Scopus, Web of Science, IEEE Xplore, SPIE) using combinations of FBG with coating/metallization and process terms (PVD, ALD/CVD, EL, ED, casting) and environmental terms (cryogenic, high temperature, harsh environment). Publications from 2000 to 2025 were considered, with earlier seminal works included that were historically informative. Studies were retained when reporting coating material/stack, process details, thickness, temperature window, and sensitivity or durability metrics; purely numerical papers were used selectively to frame design rules (e.g., optimal thickness vs. temperature) [38,42]. Emphasis is placed on metallic and hybrid stacks, with polymers and oxides treated as contrasts or barriers where relevant [25,30]. Method taxonomy follows standard coating overviews to ensure consistent terminology [43,44], and general process primers offer additional context [36,37].

Figure 1 schematically maps the relationships among application domains, coating materials, and deposition routes, highlighting how environmental requirements guide both stack design and process selection. The comparative data for each regime are summarized in Table 1. Figure 2 introduces the stack archetypes referenced throughout the paper.

**Table 1.** Environmental drivers and representative coating strategies for fiber Bragg gratings. Arrow widths in Figure 1 qualitatively represent the frequency of the same combinations in the literature. The same acronyms presented in Figure 1 are used.

| Environment | Dominant Degradation Drivers | Recommended Coating Stack | Deposition Route | Key Parameters/Post-Treatments | Normalized Sensitivity $\eta_T$ (×Bare) * | Trade-Offs/Notes | Refs. |
|---|---|---|---|---|---|---|---|
| Cryogenic (2–80 K) | Low thermo-optic slope of silica, thermal shock, brittleness | Ni → Zn duplex/Ni/Sn multilayer/Ni–P + Sn | EL Ni–P → ED Zn/Sn (or alternating Ni/Sn plating) | 5–30 µm total; anneal 250 °C; $H_2$ bake | 1.5–5× | Highest $\eta^T$ gain; moderate transient penalty (~2× slower step); stress-balancing needed in multilayers | [27,29,38,41] |

**Table 1.** *Cont.*

| Environment | Dominant Degradation Drivers | Recommended Coating Stack | Deposition Route | Key Parameters/Post-Treatments | Normalized Sensitivity $\eta_T$ (×Bare) * | Trade-Offs/Notes | Refs. |
|---|---|---|---|---|---|---|---|
| High-temperature (600–1000 °C) | Oxidation, interdiffusion, drift | Ni/Cu/Au multilayer ± TiN/$Al_2O_3$ cap | PVD seed → ED Ni/Cu → PVD Au (+CVD/ALD cap) | ~1–10 µm each; $H_2$ bake before use | ≈1× (stable slope) | Excellent stability to 1000 °C; diffusion control via multilayers | [30,31,40] |
| Radiation/nuclear | Embrittlement, hydrogen accumulation, adhesion loss | Thick Ni | Evap seed → ED Ni | ≥10 µm; post-plating bake mandatory | ≈1× | Robust, but prone to $H_2$ blisters if unbaked | [45] |
| Embedded/CTE-matched | Thermal-shock delamination, CTE mismatch | Graded Mo–Cu → Ni | Sputtered gradient → ED Ni | – | 1–2× | Stress-relieved interface; process complex | [46,47] |
| Corrosive/chemical | Bath attack, oxidation in service | Ni–P or Ni–P + Zn | EL Ni–P or EL → ED | – | 1–2× | Low-cost route; limited cryo-stability | [48,49] |

* $\eta^T = S_T^{coat}/S_T^{bare}$ is the normalized temperature sensitivity.

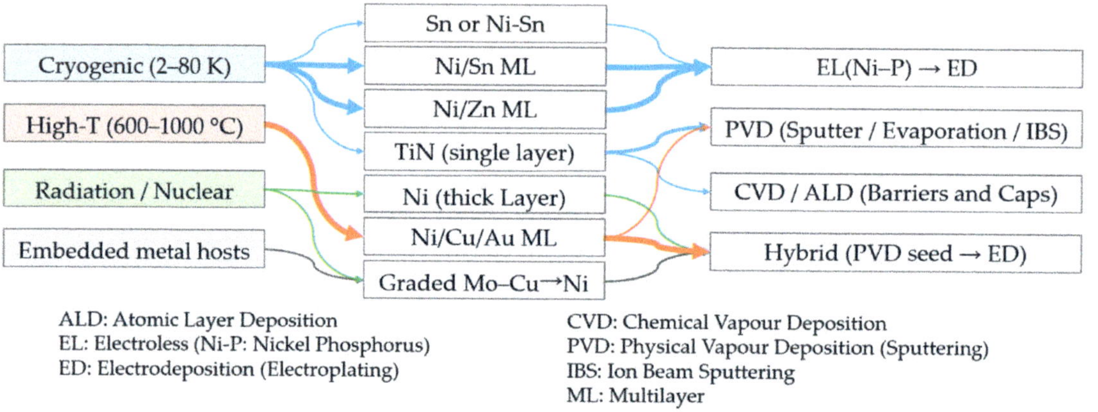


**Figure 1.** Landscape of coating-enabled fiber Bragg grating (FBG) sensing. Schematic diagram linking application domains (**left**), representative coating materials or multilayer stacks (**center**), and preferred deposition routes (**right**). Arrow thickness qualitatively represents the frequency of reported combinations in the literature. Color coding can follow temperature domain (light blue—cryogenic, orange—high-temperature, green—radiation, gray—embedded). The flow highlights how environmental requirements dictate both the choice of coating material and the deposition sequence.

### *1.2. Position of This Review Within Existing Literature*

Several recent surveys address FBG technologies from application or instrumentation standpoints [5–9,20,21]. However, no studies have consolidated coating material, process and environment relationships into unified design rules. This review focuses on coatings as a design lever, standardizes key definitions ($\eta_T$, plateau thickness, transient penalty), and integrates surface prep, deposition, and post-treatment into clear selection criteria (Tables 1–8).

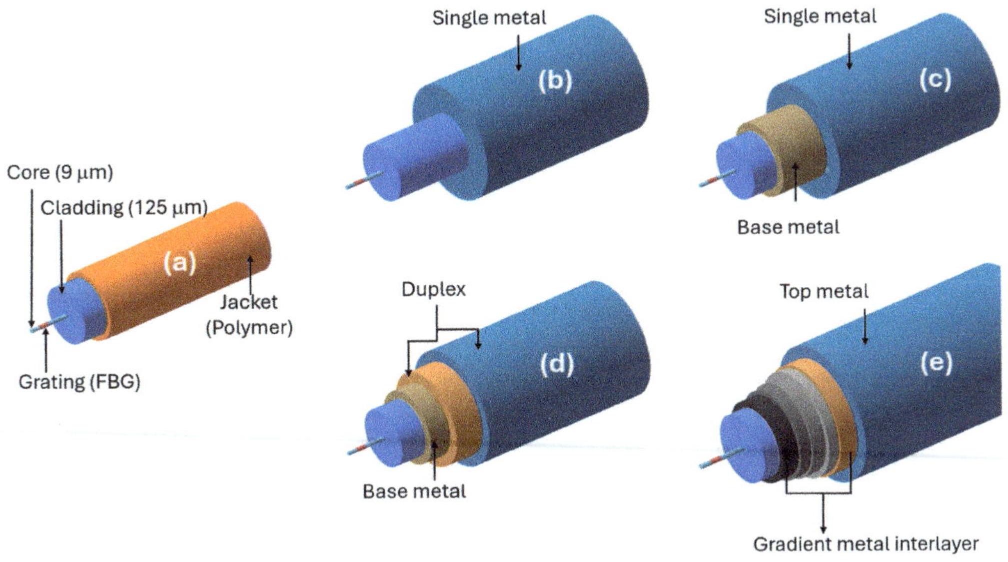


**Figure 2.** Schematic of coating-stack archetypes for FBGs: (**a**) standard 9/125 (core/cladding) fiber coated with a primary polymeric jacket; (**b**) single-layer metallic coating (e.g., Ni) providing direct thermo-mechanical coupling; (**c**) multilayer stack with a diffusion barrier (e.g., Ni/Cu/Au) designed to improve oxidation resistance and thermal durability; (**d**) duplex configuration (e.g., Ni/Cu or Ni → Zn) combining an adhesion layer with a high-CTE over-layer to enhance low-temperature sensitivity; and (**e**) graded interlayer (e.g., Mo–Cu) engineered to relieve residual stress between the fiber and the top metal. Single-layer and duplex coatings are typically used for cryogenic applications requiring strong strain transfer [27–29]; multilayers are employed for high-temperature or oxidizing environments [38,50]; and graded stacks are adopted for stress-relief and fatigue-resistant systems [51]. Common challenges across these architectures include interfacial delamination, residual stress accumulation, and spectral broadening with increasing coating thickness.

**Table 2.** Key Performance Parameters. Values are indicative ranges consolidated from [22,24,33,37,40,41,52].

| Parameter | Symbol/Expression | Physical Meaning | Typical Range or Trend |
|---|---|---|---|
| Temperature sensitivity | $S_T = \frac{d\lambda_B}{dT}$ | Apparent shift in wavelength per unit temperature | 8–12 pm $K^{-1}$ (bare), 12–50 pm $K^{-1}$ (coated) |
| Normalized sensitivity | $\eta_T = \frac{S_T^{coat}}{S_T^{bare}}$ | Sensitivity amplification due to coating | 1–5 depending on material |
| Strain transfer efficiency | $\frac{\Delta\lambda_{strain,coat}}{\Delta\lambda_{strain,bare}}$ | Degree of strain coupling between coating and fiber | 0.5–1 |
| Hysteresis | Δλ (heating vs. cooling) | Indicates elastic recovery and adhesion stability | <10 pm preferred |
| Drift | Δλ per cycle or time | Measures long-term thermal stability | <1 pm $cycle^{-1}$ typical |
| FWHM | Spectral width at ½ max | Indicates spectral broadening from strain gradients | 0.1–0.3 nm typical |
| SNR | Peak reflectivity/noise | Evaluates spectral quality | >20 dB desirable |
| Response time | Time to 95% of Δλ after ΔT | Measures coating thermal inertia | 0.1–5 s depending on thickness |

**Table 3.** Surface preparation and activation workflow (parameters and controls).

| Stage | Params | QA | Notes | Refs |
|---|---|---|---|---|
| Clean/etch | Solvent clean; acid/alkali etch | No residue; microscope | Avoid over-etching | [38,48] |
| Sensitize/activate | $SnCl_2 \rightarrow PdCl_2$ | pH/time control | Foundation for EL nucleation | [48] |
| EL Ni–P seed | ~90 °C; pH 4.5–5.0 | Continuity; spectral check | Anneal removes λ(T) hysteresis | [38,53] |
| Electroplate build | Ni/Cu/Zn/Sn plating | Thickness per step; centroid drift | Bake Ni to outgas $H_2$ | [39,40,51] |
| Barriers/caps | Thin TiN/oxides or Au | Adhesion; diffusion tests | Needed for Cu/Ni stacks | [30,54,55] |
| Post-treatments | Ni–P anneal; $H_2$ bake | Hysteresis removed; drift stabilized | Mandatory before qualification | [38,56] |

**Table 4.** Quick plating parameter reference for common metals.

| Metal | Bath | Current | Temp | Agitation | Notes | Refs |
|---|---|---|---|---|---|---|
| Ni | Watts/sulfamate | 1–5 A/dm$^2$ | 45–60 °C | Moderate | $H_2$ uptake → bake | [40,51] |
| Cu | Acid sulfate | 1–4 A/dm$^2$ | 20–30 °C | Moderate | Needs barriers/caps | [30,51] |
| Zn | Acid chloride/alkaline | 0.5–2 A/dm$^2$ | 20–35 °C | Gentle | Recipe discipline for $\lambda_0$ | [29,57,58] |
| Sn | Methanesulfonate/stannous | 0.4–0.6 A/dm$^2$ | 20–30 °C | Gentle | Whisker mitigation | [51] |

**Table 5.** Comparative synthesis of deposition routes.

| Route | Thick | Conform | Stress | Rate | Complex | Defects | Best | Refs |
|---|---|---|---|---|---|---|---|---|
| PVD | 0.01–2 μm | Good (LOS) | Low–Med | Low–Med | Med–High | Particles, stress | Seeds, TiN, thin Cu | [45,59,60] |
| CVD/ALD | 0.005–0.2 μm | Excellent | Low | Low | High | Complex precursors | Barriers/caps | [54,55,61] |
| EL (Ni-P) | 0.05–1 μm | Excellent | Med (anneal) | Med | Low | Nodules if bath aged | Uniform seeds | [40,62] |
| ED (Ni/Cu/Zn/Sn) | 1–100 μm | Excellent (rotation) | Process-dependent | High | Low–Med | $H_2$, non-uniformity | Thick builds, duplex | [41,51] |
| Hybrid (PVD → ED) | 0.1–50 μm | Good–Excellent | Balanced | Med–High | Med–High | Interface issues | Ni/Cu/Au multilayers | [30,31,63] |
| Graded interlayer | ~1–10 μm | Good | Reduced | Med | High | Process complexity | Stress relief/embedding | [46] |

To enable systematic comparison across the studies surveyed, Table 2 consolidates the key performance metrics commonly reported for coated FBG sensors. These parameters—spanning temperature sensitivity, hysteresis, drift, spectral quality, and transient response—form the quantitative framework adopted throughout this review. The same metrics are later used in Sections 3–6 to interpret coating material and process influences on performance.

**Table 6.** Recommended stacks and process windows by operating regime.

| Regime | Stack | Depo | Treat | Caps | Notes | Refs |
|---|---|---|---|---|---|---|
| Deep cryo ($\leq$80 K) | Ni → Zn duplex | Sn → Pd → EL Ni-P → Zn ED | Ni-P anneal; $H_2$ bake | — | Large slope, slower transient | [27,29,41,62] |
| 80–300 K | Ni/Cu duplex or TiN | PVD seed → Ni/Cu ED | $H_2$ bake | TiN or Au | Balanced properties | [59,63,64] |
| High-T (600–1000 °C) | Ni/Cu/Au multilayer | PVD seed → ED → PVD cap | $H_2$ bake | Ni barrier + Au cap | Diffusion/oxidation control | [30,31,40] |
| Radiation/nuclear | Thick Ni | Evap Cr/Au → Ni ED | $H_2$ bake | — | Adhesion robustness | [45] |
| Embedded in metal host | Graded Mo–Cu → Ni | Sputter gradient → Ni ED | Stress-relief soaks | As needed | Shock survival | [46,64] |

**Table 7.** Reporting template for reproducibility (per-specimen log).

| Field | Desc | Refs |
|---|---|---|
| Specimen ID | ID, grating type | Example: FBG-01 |
| Pre-coat spectral | $\lambda_0$, FWHM, SNR | 1549.876 nm, 0.18 nm, 28 dB |
| Surface prep | Clean/etch details | Acetone/IPA, HF 0.5% |
| Activation/seed | $SnCl_2$ → $PdCl_2$; EL Ni-P | 90 °C, pH 4.8, 0.15 μm |
| ED | Metal, current, thickness | Zn 0.6 A/dm$^2$, 8 min, +3 μm |
| Barriers/caps | Material, thickness | Au 80 nm PVD |
| Post-treatments | Anneal, bake | 300 °C 60 min |
| Qualification | $LN_2$ cycles; soaks | 5× 293↔77 K |
| Result summary | Slope, hysteresis | +2.3× bare |
| Failure mode | Adhesion, cracking | None after 10 cycles |

The parameters listed in Table 2 establish the quantitative basis for evaluating the impact of coating material and deposition process. Temperature sensitivity ($S_T$) and its normalized form ($\eta_T$) quantify the thermo-mechanical coupling efficiency, while hysteresis and drift capture the stability of the coating–fiber interface under cyclic loading. Full width at half maximum (FWHM) and Signal-to-noise ratio (SNR) describe the spectral quality of the reflected signal, and response time reflects the thermal inertia introduced by the coating. Together, these metrics enable direct comparison across studies that differ in coating composition, geometry, and fabrication route.

In parallel with coating strategies, polarization-maintaining fibers—particularly PANDA (Polarization-maintaining AND Absorption-reducing fiber) designs—are being explored for low-temperature and harsh-environment sensing to mitigate birefringence-related drift and enhance readout stability, including recent cryogenic demonstrations using PANDA-PMF Sagnac architectures and high-temperature PANDA-based sensors [65,66].

**Table 8.** Classification of coatings and recommended use for FBG sensors.

| Coating Type | Representative Materials | Typical Application Range | Main Advantages | Main Limitations | Typical Temperature Sensitivity/Accuracy * |
|---|---|---|---|---|---|
| Metallic (single-layer) | Ni, Cu, Al | Cryogenic (2–300 K), moderate thermal cycling | Strong strain transfer, easy deposition | Residual stress, oxidation | $\eta_T \approx 1.5$–3; ±0.5 K stability |
| Duplex/multilayer | Ni → Zn, Ni/Cu/Au, Ni/Sn | Cryogenic to 800 °C | Enhanced adhesion, tunable sensitivity | Complexity, interdiffusion | $\eta_T \approx 2$–5; ±0.3 K typical |
| Ceramic/oxide | $TiO_2$, $SiO_2$, $Al_2O_3$ | 600–1000 °C (oxidizing atmospheres) | High-T stability, oxidation resistance | Brittle, limited strain transfer | $\eta_T \approx 1$–1.5; ±1 K typical |
| Composite/hybrid | Metal + polymer, Metal + oxide | 100–500 °C, humidity/corrosion protection | Mechanical resilience, environmental sealing | Aging of polymer layers | $\eta_T \approx 1.2$–2; ±0.5 K |
| Polymeric/UV-resin | Acrylate, polyimide | <150 °C, low-cost use | Flexible, easy handling | Thermal degradation, poor uniformity | $\eta_T \approx 1$; ±1–2 K |

* Indicative ranges consolidated from [22,24,33,37,40,41,50,52].

## 2. Fundamentals and Performance Metrics for Coated FBGs

Temperature sensing with fiber Bragg gratings (FBGs) relies on the shift in the Bragg wavelength as the effective refractive index and grating period vary with temperature. For a bare grating, the Bragg condition is given by the following:

$$\lambda_B = 2n_{eff}\Lambda \quad (1)$$

where $n_{eff}$ is the effective refractive index of the fiber core and $\Lambda$ is the grating period. The wavelength–temperature response of an uncoated silica grating can be expressed as follows:

$$S_T^{bare} = \frac{d\lambda_B}{dT} = \lambda_B\left(\alpha_{SiO_2} + \xi_{SiO_2}\right) \quad (2)$$

where $\alpha_{SiO_2}$ and $\xi_{SiO_2}$ = $(1/n_{eff})(dn_{eff}/dT)$ are the thermal-expansion and thermo-optic coefficients of fused silica, respectively. This intrinsic slope typically lies around 8–12 pm $K^{-1}$ near room temperature and decreases toward cryogenic conditions [2–4].

### 2.1. Thermo-Mechanical Coupling Induced by Coatings

When a grating is metalized, the coating imposes an additional axial strain through thermo-mechanical coupling. The apparent temperature sensitivity then reflects not only the intrinsic silica response but also the coefficient of thermal expansion (CTE) mismatch, elastic modulus, thickness, and adhesion quality of the over-layer. The effective slope can be described as follows:

$$S_T^{coat} = S_T^{bare} + \lambda_B\kappa(T,\ E_i,\ \alpha_i,\ t_i,\ adesion) \quad (3)$$

where $E_i$, $\alpha_i$, and $t_i$ denote, respectively, Young's modulus, Coefficient of Thermal Expansion (CTE), and thickness of the i-th layer, and $\kappa$ is a strain-transfer function that captures mechanical coupling between the coating and the fiber core. As coating thickness increases, a greater fraction of the thermal load is transferred to the metal, and the apparent slope increases monotonically until a plateau is reached. Beyond this point, additional build yields diminishing returns while increasing residual stress, roughness, and cracking [38,42,67,68]. The height and position of the plateau depend on the metal's CTE, modulus, and interface quality. Softer and higher-CTE metals such as Zn or Sn provide large low-temperature gains at moderate thickness, whereas stiffer metals such as Ni require thicker deposits to approach the plateau but offer better adhesion and high-temperature stability [32,38,67]. Duplex and multilayer stacks exploit these contrasts, allocating thickness among layers to maximize sensitivity while avoiding localized stress.

The archetypes of coating architectures used throughout this review are illustrated in Figure 2, which distinguishes single-layer, duplex, multilayer, and graded interlayer configurations [27–29,38,50,51].

### 2.2. Normalized Temperature Sensitivity and Key Metrics

Because reported slopes vary across studies, normalized quantities are used for consistent comparison. The normalized temperature sensitivity is defined as follows:

$$\eta_T = S_T^{coat} / S_T^{bare} \tag{4}$$

This was evaluated for the same temperature range. Values of $\eta_T > 1$ denote amplification of the Bragg response due to coating; $\eta_T < 1$ indicates attenuation caused by stiff or poorly bonded films. Reported $\eta_T$ values typically fall between 1.5 and 5 for cryogenic metallic coatings such as Ni $\rightarrow$ Zn duplex or Ni/Sn multilayers [27,29,38], whereas high-temperature multilayers (Ni/Cu/Au) maintain $\eta_T \approx 1$ but provide long-term stability. To evaluate coating integrity and spectral quality, the following metrics are consistently used [38]:

- Spectral centroid shift (for mean $\lambda$ stability);
- FWHM (for spectral broadening);
- SNR (for reflectivity quality);
- Hysteresis between heating/cooling;
- Drift per thermal cycle;
- Response time to step changes in temperature.

Stable coatings preserve a narrow, high-SNR reflection with minimal centroid drift after each growth step or anneal.

The architecture illustrated in Figure 2 summarizes the evolution of coating design for FBGs. Single-layer coatings (typically Ni, Cu, or Al) are used when simple mechanical reinforcement and moderate sensitivity enhancement are sufficient. Duplex coatings introduce a seeding or adhesion layer beneath a high-CTE top layer to boost low-temperature response while maintaining fiber adhesion. Multilayer and graded coatings are developed to distribute thermal stress and prevent cracking during cyclic heating or cooling, making them suitable for high-temperature aerospace or power applications. Hybrid or composite coatings, which may include metallic, oxide, or polymeric sublayers, offer environmental sealing and compatibility with composite host materials. Each configuration represents a compromise between strain transfer efficiency, residual stress control, and environmental resistance.

### 2.3. Reporting Conventions and Comparison Framework

This review interprets the literature using consistently defined metrics and test profiles so that disparate studies can be compared on common ground [2–4]. Spectral and thermal responses are reported at standardized processing checkpoints: before coating, after seeding, after each build increment, and after post-treatments. Coating stacks are written from the glass outward, with layer thicknesses where specified; deposition routes are identified per layer (e.g., PVD seed → EL Ni–P → electroplated Ni; sputtered TiN single-layer; duplex Ni → Zn with EL seed), and post-treatments are stated explicitly (e.g., short anneal for Ni–P, vacuum bake for plated Ni). When possible, temperature slopes are normalized to the bare-fiber value from the same study; if unavailable, the closest baseline is reported. These conventions allow the comparison of $S_T^{coat}$, $\eta_T$, hysteresis, and drift across diverse process routes. Representative quantitative multipliers for different metals—approximately 1.5× for Ni, 2× for Cu, 2.5× for Zn, and 3× for Sn—are summarized later in Section 3 and Figures 2 and 3 [69].

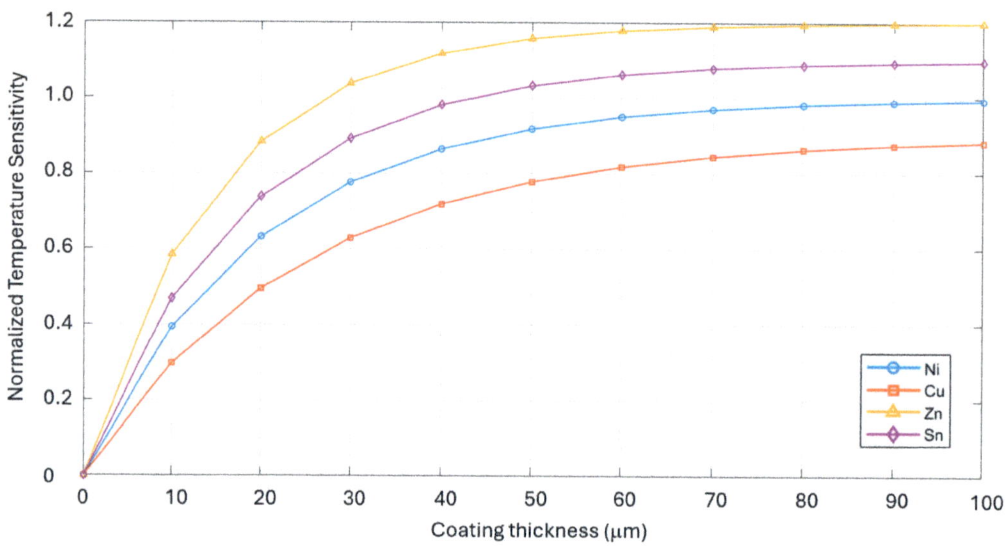


**Figure 3.** Normalized temperature sensitivity ($\eta_T = S_T^{coat}/S_T^{bare}$, see Section 2.3) as a function of coating thickness for Ni → Zn and Ni/Sn multilayers. The plateau trend denotes full thermo-mechanical coupling between the metal stack and the silica core. Representative behaviors are shown for Ni, Cu, Zn, and Sn coatings, based on trends reported in [27,38,42,63,67,68]. This figure illustrates the trade-off between coating thickness and diminishing returns in sensitivity gain. Data were consolidated from the indicated references and redrawn for clarity; curves represent representative trends rather than raw experimental data.

The described framework establishes how intrinsic FBG physics, thermo-mechanical coupling, and normalized sensitivity metrics are linked. These definitions underpin the comparative analysis of coating materials, deposition methods, and processing parameters presented in Sections 3–7, enabling the cross-study synthesis of temperature-sensitivity, stability, and durability trends [38,42,50,51,67–69].

## 3. Coating Materials and Design Constraints

The temperature response of a metalized FBG is governed by thermo-mechanical coupling between the coating and the silica fiber. As temperature changes, mismatch in

CTE and differences in elastic modulus generate axial strain in the grating; the resulting Bragg shift therefore depends on the coating's material, thickness, microstructure, and interface quality, in addition to the intrinsic thermo-optic and thermal-expansion terms of silica. Analytical and experimental studies consistently show a monotonic rise in sensitivity with thickness towards a plateau, after which additional build adds stress and defects with only marginal gain [38,42,67,68]. Design choices in this section are framed around reaching that plateau with the least residual stress and the highest spectral stability.

### 3.1. Material Families and Functional Roles

Metals are the primary workhorse because they combine high thermal conductivity, efficient strain transfer and robust attachment to glass. Nickel provides adhesion and a useful barrier/capping function; copper increases sensitivity and heat conduction but requires diffusion control at elevated temperatures; soft metals such as zinc, tin and indium excel at cryogenic temperatures thanks to larger CTE and lower modulus. In practice, duplex and multilayer stacks—e.g., Ni/Cu/Au or Ni $\rightarrow$ Zn—are adopted to balance adhesion, diffusion control and thermo-mechanical performance across wide temperature windows [27–31]. For deep-cryogenic operation, graded metallic stacks have demonstrated higher stability under cycling than single-layer counterparts while maintaining strong apparent slopes [29].

Polymeric overlayers (e.g., PMMA) can increase apparent sensitivity and ease cryogenic cycling but are limited by creep and upper-temperature stability [63,70]. Oxides and nitrides (TiN, $TiO_2$, ZnO) serve as chemically robust single layers or as thin barrier/cap layers that interrupt diffusion without imposing excessive stiffness; single-layer TiN deposited by ion-beam sputtering has shown clean, near-linear cryogenic behavior [54,55,59,61]. Functionally graded interlayers—such as Mo–Cu graded into an outer Ni coat—reduce residual stress and have demonstrated survival under thermal shock while maintaining useful slopes [46]. The broader literature on highly sensitive temperature-FBGs and nanostructured coatings reinforces these roles and clarifies the trade-offs between sensitivity and transient response [7,33].

### 3.2. Thickness Dependence and Optimal Ranges

Sensitivity increases with coating thickness until a material-dependent plateau is reached; beyond this knee, the gain per micrometer diminishes while residual stress, roughness and crack initiation risks grow [38,42,67]. EL Ni-P studies, for example, report slope growth from $\approx$12 to $\approx$19–20 pm/°C as thickness rises from $\approx$1 to $\approx$40 μm, followed by saturation; a short post-anneal removes $\lambda(T)$ hysteresis (Section 3.4) [38,71]. PVD studies show clear thickness trends for Cu up to $\approx$0.76 μm [72], and thick sputtered Cu films can be pushed to tens of micrometers when protected by barriers/caps [64]. Duplex and multilayer stacks exploit material synergy to reach the plateau at lower individual-layer thicknesses, which reduces stress concentrations.

This plateau behavior is illustrated in Figure 3, which compares the sensitivity–thickness relationship for representative metallic coatings.

The data consolidated in Figure 3 highlight the nonlinear dependence of $\eta^T$ on coating thickness, revealing two distinct regimes. At small thicknesses (<5 μm), partial strain transfer limits the apparent temperature sensitivity, and $\eta^T$ increases nearly linearly with thickness. Beyond a critical or plateau value, $\eta^T$ saturates, indicating full thermo-mechanical coupling between the coating and the silica substrate. This plateau typically occurs at 10–30 μm for Ni-based or Cu-based layers, and at smaller values for Zn and Sn coatings due to their lower elastic moduli and higher coefficients of thermal expansion. The saturation trend reflects a balance between increased thermo-elastic amplification and the onset of residual stress, which can promote microcracking and spectral broadening if coating thickness continues to increase [27,38,42,63,67,68].

### 3.3. Cryogenic Behavior and Transients

At ≤80 K, soft-metal and Ni-based stacks typically raise apparent temperature sensitivity by ~1.5–5× relative to bare gratings, depending on metal and thickness [27,29,69]. This gain comes with a transient-response penalty: metallic overcoats increase thermal mass and often slow step response by about a factor of two under $LN_2$-type profiles, an effect highlighted in comparative thermal-sensing analyses [34,42]. Representative cryogenic sensitivity multipliers for Ni, Cu, Zn, and Sn coatings are illustrated in Figure 4, highlighting the enhanced apparent slopes relative to bare gratings (of apparent temperature sensitivity relative to bare FBGs are shown for Ni (~2× [27]), Cu (~2.5× [29]), Zn (~3.5× [69]), and Sn (~4× [57]) coatings). Reported behavior during cool-down can be mildly nonlinear, with some rate dependence; qualification should therefore include dynamic as well as quasi-static profiles [69]. Early high-field cryogenic tests also indicate that metal-coated FBGs maintain function under strong magnetic fields, which is relevant to fusion and accelerator environments.

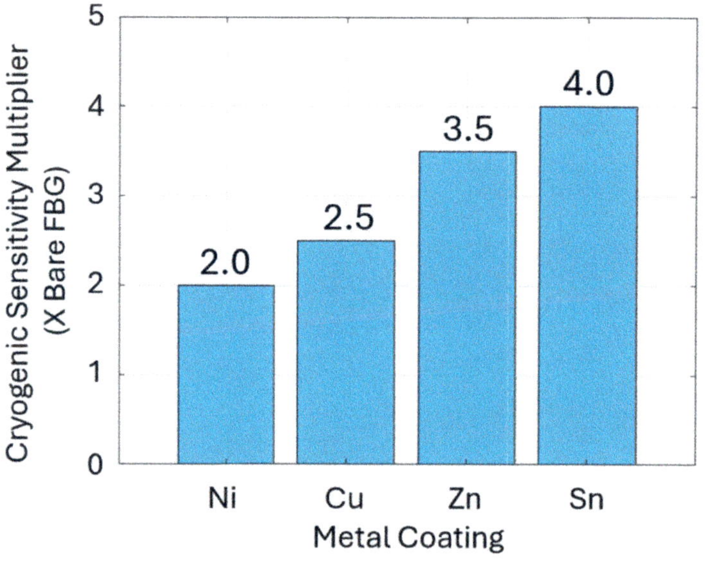


**Figure 4.** Representative enhancements of apparent temperature sensitivity relative to bare FBGs are shown for Ni (~2× [27]), Cu (~2.5× [29]), Zn (~3.5× [69]), and Sn (~4× [57]) coatings. The cryogenic sensitivity multiplier is equal to the bare FBG multiplied by a certain number. The trend reflects the larger coefficients of thermal expansion and lower modulus of soft metals at cryogenic temperatures. Literature-reported $+\eta^T$ multipliers for different coating materials and layer sequences. These are representative trends that have been redrawn for greater clarity and do not represent raw experimental data.

Figure 4 compares representative $\eta^T$ multipliers obtained from different metallic coatings and multilayer stacks. The results emphasize the strong influence of both the metal's intrinsic thermal expansion and the stacking sequence on sensitivity enhancement. Zn- and Sn-based coatings provide the highest normalized sensitivities ($\eta^T \approx 2.5$–5) at cryogenic temperatures, whereas Ni- and Cu-based films yield smaller gains ($\eta^T \approx 1.2$–2) but superior mechanical and chemical stability. Multilayer and duplex designs, such as Ni $\rightarrow$ Zn or Ni/Sn, combine these effects, producing intermediate yet tunable responses. These observations confirm that the coating's thermo-mechanical properties and architecture can be tailored to achieve a desired balance between sensitivity amplification and robustness—a theme further analyzed in Sections 4–6.

### *3.4. Deposition Route, Microstructure, and Post-Treatments*

The process route imprints microstructure and residual stress that strongly condition spectral stability. After $SnCl_2 \rightarrow PdCl_2$ activation, EL Ni–P provides a conformal conductive seed at modest thermal budgets; it commonly exhibits $\lambda(T)$ hysteresis that brief annealing ($\approx$300 °C, ~60 min) removes [38,63]. Electroplated Ni builds thickness efficiently but absorbs hydrogen during deposition; a vacuum/thermal bake is required to stabilize the spectrum and avoid embrittlement [40]. PVD routes (sputtering, evaporation, ion-beam sputtering) deliver smooth seeds (Cr/Au, Ti) and functional films; ion-beam sputtered TiN shows stable, near-linear cryogenic slopes [59]. With suitable barriers/caps, sputtered Cu can be grown to substantial thickness and then cycled to high temperature and back to cryo [64]. CVD/ALD oxides ($TiO_2$, ZnO) offer conformal barriers and caps that reduce oxidation and interdiffusion at modest temperatures [54,55,61]. Hybrid schemes—PVD seed $\rightarrow$ electroplated Ni/Cu/Zn; regenerated gratings with Ti/Ag seed and Ni plating; or graded Mo–Cu $\rightarrow$ Ni stacks—combine adhesion, sensitivity and stress management, and show improved mid- to high-temperature performance with good model–experiment agreement at the lower end [31,46]. EL nanometal strategies and low-coarsening Ni-P metallization recipes broaden the wet-chemical toolkit specifically for fiber geometries [62,73]. Head-to-head comparisons across Ni, Cu, Zn and duplex stacks underline that process history governs the spread in reported gains as much as metallurgy itself [74].

### *3.5. Adhesion, Interfaces, and Seeding*

Durable interfaces start with disciplined cleaning/etching and reliable seeding. The fiber-appropriate standard is $SnCl_2$ (sensitization) $\rightarrow$ $PdCl_2$ (activation) $\rightarrow$ EL Ni–P; Tollens' silvering and vacuum Cr/Au or Ti seeds are practical alternatives when chemistry or equipment dictates [45,48,49]. EL growth on silicate substrates is sensitive to pH, complexants and reducers; peel-test-oriented best practices on glass highlight the importance of bath housekeeping and staged deposition [50,75]. On a cylindrical dielectric, uniformity improves with rotation and gentle current-density ramps at plating start; staged builds preserve smooth interfaces and avoid stress concentrators [39,51].

### *3.6. High-Temperature Stability, Diffusion, and Barriers*

At elevated temperature, Cu/Ni interdiffusion and oxidation dominate drift and mechanical degradation. Multilayer Ni/Cu/Au stacks maintain repeatable slopes to ~1000 °C under cycling when diffusion paths are interrupted and hydrogen introduced during plating is removed beforehand [30,42]. Duplex Cu/Ni stacks are prone to Kirkendall voiding above ~700–1010 °C unless buffered; barrier and cap layers, together with appropriate thermal budgets, mitigate this risk. Regenerated gratings with Ti–Ag–Ni stacks perform to ~600 °C with model–experiment agreement in the lower range [31]. Sputtered-Cu systems subjected to high-temperature treatment can retain integrity on return to cryo when barriers/caps are engineered correctly [64]. Figure 5 summarizes these comparative limits, contrasting the maximum stable operating temperature of single Ni and Cu coatings with that of Ni/Cu/Au multilayer stacks. Broader comparative evaluations concur that multilayer designs with explicit barriers and caps outperform single-metal coats as temperature rises [74].

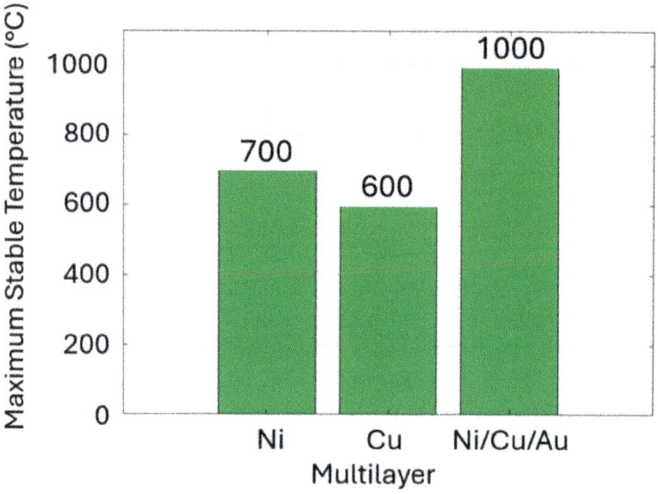


**Figure 5.** High-temperature stability of coated FBGs. Representative stability limits reported in the literature: single Ni coatings are stable up to ~700 °C before drift and embrittlement [30,40]; single Cu coatings degrade above ~600 °C unless capped [31]; Ni/Cu/Au multilayer stacks, with diffusion barriers, maintain integrity and repeatable slopes to ~1000 °C under cycling [64]. Data were consolidated from the indicated bibliographic references. These are representative trends that have been redrawn for greater clarity and do not represent raw experimental data.

### *3.7. Practical Selection Rules*

Method selection should be driven by the target environment and the thickness–sensitivity plateau. For cryogenic service, soft metals or Ni-based duplexes (e.g., Ni $\rightarrow$ Zn) are favored for high apparent slopes; transient-response penalties should be anticipated and, where necessary, mitigated by thinner, higher-modulus caps or partially metalized geometries [27,29,34,41]. For high-temperature operation ($\geq$600–1000 °C), multilayers with explicit barrier and cap layers (e.g., Ni/Cu/Au) are preferred to suppress diffusion and oxidation; hydrogen removal after plating is mandatory, and dwell at the highest temperatures should be limited or buffered [30,31,40]. Thickness should be allocated to reach the plateau without over-thickening any single constituent; modeling trends and Ni-P thickness studies provide practical guidance [38,42,67]. Interfaces should be engineered with proven seeds (Sn $\rightarrow$ Pd $\rightarrow$ EL Ni–P, or Ag/Cr–Au/Ti where appropriate), smooth transitions, staged builds and rotation for uniformity [39,45,48,49,51]. Here, diffusion control or stress relief is required; thin TiN/oxide caps, noble-metal barriers, or graded Mo–Cu interlayers are effective [46,54,59]. With these rules, stacks can be tailored to deliver the desired sensitivity while preserving spectral quality and durability across the intended temperature window.

## 4. Deposition Techniques

Coatings for FBGs are produced by a small set of process families whose physics shape adhesion, stress, microstructure, and, ultimately, spectral stability. For clarity, methods are grouped as top-down physical-vapor deposition (PVD), bottom-up CVD/ALD, wet-chemical routes (EL and ED), hybrid flows that combine these, and fusion-based approaches for direct integration into metals. This organization follows established thin-film taxonomies and usage across the optical-fiber literature and process primers [32–34,40].

### 4.1. Top-Down: PVD

PVD methods—magnetron sputtering, ion-beam sputtering and thermal/e-beam evaporation—provide dense, well-controlled films with good thickness uniformity along the fiber and around the grating region. They are widely used to create seeds that promote adhesion and conductivity before plating (e.g., Cr/Au or Ti), but they can also deliver functional single layers and multilayers in their own right [31,45]. Technique choice and parameters (pressure, power, gas, motion) govern intrinsic stress and microstructure; compressive stress from energetic bombardment improves density but can aggravate curvature-induced strain on cylindrical substrates if unmanaged. Single-layer TiN by ion-beam sputtering illustrates a PVD-only sensing coat, yielding a near-linear cryogenic slope of ≈10.7 pm/°C from 25 to −195 °C with strong chemical robustness [59]. Copper sputtering has been used as both a stand-alone sensing layer and a constituent of multilayers; thickness-dependent gains are reported up to ≈0.76 μm in simple studies [72], while thick magnetron Cu films (≈20 μm) have endured high-temperature treatments and subsequently operated down to 4.2 K when diffusion is actively managed by barrier or cap layers [43]. For high-temperature operation, multilayer Ni/Cu/Au stacks combine adhesion, conduction and diffusion control and have demonstrated repeatable slopes under cycling towards ~1000 °C when interfaces block interdiffusion pathways [30]. Cathodic-arc sources, though highly ionized, can eject macroparticles that seed roughness unless filtered or steered; magnetron and ion-beam sputtering therefore remain defaults for smooth films on fibers [60].

### 4.2. Bottom-Up: CVD and ALD

CVD and especially ALD are used to form conformal oxide caps and barriers at modest temperatures. Thin ZnO and $TiO_2$ on fiber geometries have shown stable thermal responses and illustrate the utility of conformal growth on curved substrates [54,55,61]. Growth rates are lower than in PVD and plating, but the conformality and stoichiometric control of CVD/ALD make them attractive for diffusion-barrier or oxidation-resistant over-layers above metallic stacks, or as adhesion-promoting primers when used thin enough to avoid excessive stiffness.

### 4.3. Wet-Chemical Routes: EL and ED

Wet-chemical methods dominate when thick, conformal metal builds are needed at low thermal budgets. A widely adopted workflow is $SnCl_2 \rightarrow PdCl_2$ activation on silica, deposition of an EL Ni–P seed, and ED of the target metal to the desired thickness. [48]. This route is attractive because the seed forms at a relatively low temperature and the subsequent ED step scales efficiently. Fiber-specific re-coating and metallization methods—including Tollens' silvering, tailored EL nanometal approaches, and Ni-P metallization without fiber coarsening—expand the toolbox and provide practical bath windows and housekeeping guidance for cylindrical dielectrics [52,62,64,76]. Two post-treatments are essential for spectral stability: a short anneal removes λ(T) hysteresis in Ni-P, and a vacuum/thermal bake after Ni ED outgasses hydrogen that would otherwise cause slow drift or embrittlement during cycling [38,41]. These wet-chemical routes underpin many cryo-oriented stacks. Comparative tests under dynamic cooling show metal coats raise the apparent cryogenic slope relative to bare gratings, with typical multipliers of ~1.5× (Ni), ~2× (Cu), ~2.5× (Zn) and ~3× (Sn), albeit with some rate-dependence during cool-down [69].

Ni → Zn combinations (EL Ni seed followed by Zn ED) and related zinc routes—derived from sputter-seeded or EL seeds—have been validated for low-temperature sensitivity and durability, with recipe control important to maintain $\lambda_0$ stability [29,57,58]. Where mechanical integration is desired, ED can be used to surface-mount FBGs directly onto metallic substrates, achieving linear slopes (~30.7 pm/°C) and survival to high temperatures before the joint fails; in these cases, the interface—not the grating—becomes the limiting element [77]. As with any galvanic process on small cylinders, uniformity benefits from fiber rotation and gentle current-density ramps at the start of each run; staged EL builds with spectral quality assurance (QA) checkpoints prevent roughness and voiding that accumulate during long single immersions [39].

### 4.4. Hybrid and Combination Strategies

Combining PVD with plating, or mixing metals in duplex and multilayer stacks, allows sensitivity targets to be met while controlling diffusion and stress. A graded metal approach tailored for cryogenic service offers a complementary path to sensitivity and reliability without over-thick single layers [32]. Cross-study comparisons further confirm the duplex/multilayer advantage and highlight the importance of disciplined seeding and post-treatments [74]. A representative approach is a PVD seed (Ti or noble metal) followed by electroplated Ni, which provides adhesion and rapid thickness build; regenerated gratings with Ti/Ag seeds → Ni plating exemplifies improved mid- to high-temperature performance with model–experiment agreement at the lower end of the range [32]. Duplex Ni/Cu stacks are widely reported; an etchless Cu seed plus Ni plate improves slope and repeatability over Ni-only coatings in the 30–100 °C range [63], while Ni → Zn provides a robust cryogenic path with central-wavelength stability managed by recipe control [29,57]. For very high temperatures, Ni/Cu/Au multilayers use Ni as a barrier/base, Cu for conduction and sensitivity, and Au as a cap to limit oxidation and outward diffusion; careful interface control and post-plating hydrogen removal are prerequisites for stability towards ~1000 °C [30]. Residual-stress management improves further with functionally graded interlayers. A Mo–Cu gradient beneath an outer Ni coat reduces stress concentrations and has demonstrated survival under thermal shock up to ~800 °C with slopes around 15 pm/°C [46]. Such graded stacks are particularly valuable when embedding in metallic hosts or when repeated high-temperature cycles are expected. Beyond the coating stacks themselves, broader compilations of metallic-coating effects at cryogenic temperatures confirm the thickness sensitivity and transient trends summarized here and help prioritize metals for duplexes [53].

In Section 6, Table 5 summarizes the relative merits of these process families in terms of achievable thickness, conformality, growth rate, and process complexity. It serves as a compact technical data sheet for coated FBGs.

### 4.5. Fusion-Based Approaches and Integration into Metals

A complementary path is to integrate the grating directly into a metallic component. Casting experiments embedding FBGs in aluminum alloys have shown reproducible axial response and negligible hysteresis when the grating is appropriately prepared and positioned, confirming the feasibility of fusion-based integration for structural applications [78]. Thick-metal builds on long-period gratings illustrate the stress pitfalls of very heavy deposits and motivate staged growth and stress-relief steps for any fiber-based structure [47]. Looking forward, kinetic (cold) spray—a solid-state deposition process with high rates—offers a potential route to metal over-jackets or ex situ metallization of carriers that hold the FBG, though direct application to bare fibers demands careful control of particle impact to avoid damage [56].

## 5. Surface Preparation and Activation

Reliable coatings on silica fibers start with disciplined surface preparation. Residual polymer, drawing lubricants and handling films impede adhesion and must be removed before metallization. A typical sequence begins with solvent cleaning (acetone or isopropanol with brief ultrasonics), followed by a mild chemical etch to activate the silica surface; dilute mineral acids or alkalis are selected to suit the downstream chemistry, with thorough de-ionized rinsing between stages to prevent cross-contamination. Excessive etching increases flaw size and should be avoided on sensing sections. The aim is a clean, lightly roughened surface that promotes nucleation without compromising fiber strength. General thin-film practice notes on stress and roughness control complement these steps and are particularly relevant on cylindrical substrates [48,63]. Activation of the glass for EL deposition is most commonly achieved with $SnCl_2$ sensitization followed by $PdCl_2$ activation. Tin(II) species adsorb on silica and reduce palladium in situ, leaving catalytic $Pd^0$ sites that initiate autocatalytic growth in the EL bath. Rinses between $SnCl_2$ and $PdCl_2$ are critical; chloride carry-over into the Ni–P bath degrades deposit quality. Alternative seeds are useful when chemistry or equipment dictates the following: Tollens' silvering provides a rapid aqueous conductive layer that can be over-plated immediately, and vacuum-deposited Cr/Au or Ti seeds provide very smooth, adherent underlayers that are especially valuable when later ED is planned [45,48,49,73,74,76]. For fibers that will receive thick metal builds, a conformal EL Ni–P seed is widely used because it forms uniformly at modest temperature and pH, wets glass well after Sn/Pd activation, and produces a conductive, mechanically robust surface for subsequent ED. Operating near 90 °C and pH $\approx$ 4.5–5.0 yields continuous films with low porosity; bath aging and contamination are the usual causes of nodular growth and must be monitored. Thickness should be accumulated in stages, with spectral checks between stages, because single long immersions tend to raise internal stress and roughness; practical studies on fibers show that pushing a single EL step beyond a few tenths of a micrometer increases the risk of voiding and cracking [38,39,75]. EL Ni–P coatings also exhibit $\lambda(T)$ hysteresis; short post-deposition annealing ($\approx$300 °C for ~60 min, inert/vacuum) removes this effect and stabilizes the slope for subsequent cryogenic cycling [38]. Table 3 summarizes the main stages of this workflow, including representative parameters, quality checks, and references. It provides a structured overview of surface preparation and activation for coated FBGs.

Once a conformal seed is in place, ED supplies thickness efficiently for Ni, Cu, Zn, Sn and their combinations. Uniformity on a small cylinder is improved by rotating the fiber and by using gentle current-density ramps at the start of each run to control nucleation and minimize local heating. Chemistry dictates current density, additives and agitation; for example, tin baths are commonly operated at 0.4–0.6 A $dm^{-2}$ to mitigate whisker formation without sacrificing rate, whereas Ni/Cu typically run at 1–5 A $dm^{-2}$ depending on geometry and agitation [48]. Because the current distribution is naturally non-uniform around a dielectric cylinder, auxiliary electrodes and judicious anode–cathode spacing help regularize the field. Spectral quality should be verified after each plating increment before further build-up or metal changes are attempted. Table 4 summarizes representative bath chemistry, current densities, temperatures, and agitation practices for these metals. It serves as a quick reference for commonly used ED parameters in coated FBGs.

Two post-treatments are essential for spectral stability after wet metallization. EL Ni–P requires the brief anneal mentioned above to remove hysteresis; electroplated Ni (and, to a lesser extent, Cu) benefits from a vacuum or thermal bake to outgas hydrogen absorbed during deposition. Without this step, slow spectral drift and embrittlement can appear during subsequent cycling [41,73,78]. Where Cu/Ni stacks are used and high-temperature exposure is anticipated, barrier or cap layers should be included to suppress interdiffusion and oxidation, and the thermal budget of downstream processing should be planned accordingly [30,38]. For zinc-containing stacks intended for cryogenic service, recipe discipline is important to maintain central-wavelength stability across thermal cycles [29,57]. Fiber-specific re-coating and metallization procedures provide additional, practical windows for activation, seeding and build without fiber coarsening [52,62,73,76].

Interface integrity is governed as much by process discipline as by materials choices. Cleanliness between steps, controlled rinsing and bath-renewal schedules reduce particulate defects and pinholes that would otherwise seed delamination. When PVD seeds are employed, magnetron or ion-beam sputtering is preferred for smooth films on fibers; unfiltered cathodic-arc sources eject macroparticles that can imprint roughness and stress concentrators unless they are filtered or steered appropriately [45,60]. Regardless of route, in-process spectral QA—tracking Bragg-centroid position, FWHM and SNR before coating, after activation/seed, after each thickness increment, and after any post-treatment—provides early warning of adhesion issues or stress build-up and allows recipes to be corrected before the final build is committed. When coatings are intended to operate in contact with hosts or joints, surface preparation must be coordinated with the packaging route. Electroplated surface-mounted joints can bond fibers directly onto metallic substrates and survive high temperatures before the joint fails; in such cases, the interface rather than the grating becomes the limiting element [77]. Foundational studies of apparent strain, temperature–strain discrimination and embedded configurations inform how to interpret the coated sensor's response when it is constrained by surrounding materials and how to compensate or calibrate for these effects [79–83]. Broader compilations of metallic-coating effects at cryogenic temperatures are consistent with the process-linked trends summarized here and help prioritize metals and thickness allocations for cryogenic builds [53].

## 6. Comparative Synthesis and Qualification

Comparative evaluation of coating routes shows that performance is determined as much by deposition method and process control as by material choice. Across EL, electroplated, sputtered, and hybrid stacks, reported gains vary widely, but converging patterns are clear [30,31,38,40,41,47,50,64,67,68,72]. Table 5 synthesizes these trends, comparing thickness range, conformality, growth rate, stress, and defect susceptibility for the main families. Wet-chemical approaches, including EL Ni–P and ED of Ni, Cu, Zn, or Sn, are efficient for building thickness at modest cost but require rigorous control of bath composition, additives, and current density to avoid hydrogen uptake, roughness, or coarsening [38,40,62].

Vacuum-based processes such as sputtering or ALD/CVD yield smoother, conformal films and effective barrier/cap layers. TiN, $TiO_2$, and ZnO grown by these methods show stable performance under both cryogenic and high-temperature cycling [54,55,59,61]. Hybrid approaches—PVD seed plus electroplated builds, regenerated gratings with Ti–Ag–Ni, or multilayer Ni/Cu/Au stacks—consistently extend stability and sensitivity by combining adhesion, diffusion control, and bulk builds [30,31,47,50,64]. Qualification focuses on reproducibility and stability under cycling. Metrics include slope repeatability, spectral drift, SNR, FWHM, and hysteresis, with thresholds reported across cryogenic, high-temperature, and radiation environments [27,29,34,41,44,69,80]. Table 6 consolidates tolerance ranges, while Figure 5 illustrates the comparative high-temperature stability of single Ni, single

Cu, and Ni/Cu/Au multilayers, highlighting the importance of diffusion barriers and hydrogen removal [30,31,42,64]. A recurring limitation in the literature is inconsistent reporting of processing details such as activation, seeding, thickness allocation, and annealing. Without this information, cross-study comparisons remain qualitative. To address this gap, Table 7 proposes a per-specimen reporting template, aligned with best practices from related sensor domains [39,50,60,75]. Adoption of such standards would enable reproducibility, meta-analysis, and eventual benchmarking of coating performance across laboratories. In synthesis, the comparative evidence underscores that multilayer and hybrid stacks outperform single-metal coatings, but reliable qualification depends equally on adherence to standardized metrics and reporting practices. Establishing these frameworks is essential for advancing coated FBG technologies from experimental demonstrations to deployable sensing solutions.

The comparative data in Tables 2–5 demonstrate that coating design and deposition route exert a comparable influence on performance. EL–ED sequences typically deliver higher $\eta^T$ at cryogenic conditions, whereas hybrid PVD-seeded routes offer better uniformity and stability at high temperature. These trends are consistent with the thermal and mechanical models introduced in Section 2, reinforcing the importance of tailoring both material selection and process architecture.

To consolidate the comparative discussion above, Table 8 summarizes the main coating families, their recommended application domains, and indicative accuracy ranges derived from the reviewed literature. It classifies the main coating families used for FBGs and summarizes their preferred application domains. Metallic and duplex coatings dominate cryogenic and moderate-temperature operation owing to their high strain-transfer efficiency, while ceramic and oxide films are preferred in high-temperature and oxidizing environments. Hybrid and composite coatings extend durability by combining mechanical and chemical protection. Reported normalized sensitivities ($\eta^T$) vary from $\approx$1 for polymeric layers up to $\approx$5 for optimized multilayers, corresponding to practical temperature accuracies between $\pm$0.3 K and $\pm$1 K depending on calibration and interrogation method. These comparisons provide a framework for selecting coating strategies according to environmental constraints and accuracy targets.

## 7. Design Recommendations and Outlook

The literature converges on a practical principle. The best coated-FBG design reaches the sensitivity plateau with minimal residual stress. It also suppresses diffusion and hydrogen effects that destabilize spectra. For deep-cryogenic service ($\leq$80 K), duplex designs that pair a nickel base with a soft metal provide the clearest gains. Ni $\rightarrow$ Zn and Ni + Sn routes consistently deliver large apparent-slope multipliers over bare fibers, provided recipes are disciplined and central wavelength is stabilized across cycling [27,29,57,69]. The EL Ni–P seed (after $SnCl_2 \rightarrow PdCl_2$) should be grown in staged dips, briefly annealed to remove $\lambda(T)$ hysteresis, and then over-plated; any electroplated Ni requires a vacuum/thermal bake to outgas hydrogen [38,40]. Cryogenic gains come with a transient penalty. Metal buildup increases thermal mass, which can slow steps by about two times in representative systems. Therefore, fast-transient applications may favor thinner, higher-modulus caps or partially metalized geometries to recover response time [34,41]. When maximizing cryogenic slope under aggressive cycling, graded metallic architectures are a practical alternative to very thick single-metal layers [32]. Between 80 and 300 K, the emphasis shifts to balance. Duplex Ni/Cu stacks combine nickel's adhesion and barrier role with copper's higher thermal conductivity and expansion, meeting sensitivity targets at moderate total thickness. Where fast transients dominate or chemical robustness is paramount, thin TiN (PVD) single layers offer clean, near-linear cryogenic behavior with minimal mass penalty [59]. Diffusion control should be planned even if nominal operation is below 573.15–673.15 K (300–400 °C); thin barriers or caps (e.g., TiN/oxides or Au) keep Cu/Ni interactions in check as temperature excursions increase [30,54,55,61]. For high-temperature regimes (600–1000 °C), failure modes are set by interdiffusion and oxidation. Ni/Cu/Au multilayers—Ni as base/barrier, Cu for conduction/sensitivity, Au as oxidation-resistant cap—are the most reliable pattern in the literature, achieving repeatable slopes under cycling towards ~1273.15 K (1000 °C) when diffusion paths are interrupted and when any hydrogen from wet metallization is removed beforehand [30,40]. Regenerated gratings with Ti/Ag seeds and plated Ni perform into the mid-temperature range with model–experiment agreement at the lower end [31]. Where thermal-shock or embedding is expected, functionally graded Mo–Cu → Ni interlayers reduce residual-stress concentrations while maintaining useful slopes [46]; surface-mounted electroplated joints are viable for direct bonding to metal substrates, with the joint often becoming the limiting element rather than the grating [77]. When gratings are embedded or constrained by hosts, interpretation of the apparent temperature slope should follow composite-beam logic rooted in foundational apparent-strain and discrimination studies [47,56,78,79]. Two cross-cutting choices dominate outcomes across all windows. First, thickness allocation: once the sensitivity–thickness curve approaches its knee, further metal is better spent on stress management or barriers/caps than on chasing marginal slope increases [38,42,67]. Second, process history: seeding, staged growth, and post-treatments determine microstructure and residual stress and therefore spectral stability. Benchmark comparisons emphasize that reported gains align when recipes, seeds and post-treatments are standardized across studies [74]. Fiber-specific metallization methods (re-coating procedures, EL nanometal strategies, fiber-safe Ni–P recipes) and general thin-film practice notes are what make these designs reproducible across labs [35,48,52,62,73,76]. Field overviews and cryogenic FBG surveys ensure the recommendations sit within the wider sensing landscape and application demands in aeronautics, energy and fusion [5–7,10,23–25]. Looking forward, three directions are poised to have the biggest payoff. Response-time optimization at low temperature—via thinner, higher-modulus caps, improved thermal pathways to the host, or partial metallization—can capture sensitivity gain while recovering temporal performance. Diffusion-aware design at the hot end—explicit barrier selections, noble caps, and dwell-time limits for Cu/Ni interfaces—extends stability towards the top of the range. And integration pathways that treat coatings and packaging together—graded interlayers for embedded mounts, electroplated surface joints—shrink the gap between laboratory coupons and deployable assemblies [46,77,78]. A disciplined reporting template and acceptance bands will further improve comparability and accelerate recipe transfer between groups [2,3]. A standardized reporting format is needed to improve cross-laboratory reproducibility. Table 7 provides an example of a log template for each specimen that captures essential details from surface preparation through qualification. This enables direct comparison of results across different coating routes and laboratories.

## 8. Conclusions and Outlook

A coherent picture emerges from the survey. Apparent temperature sensitivity in coated FBGs increases with metal thickness and then saturates, so the most effective designs reach a plateau with minimum residual stress and with diffusion and hydrogen effects actively managed. In practice, this means distributing thickness across duplex or multilayer stacks, engineering interfaces deliberately ($SnCl_2 \rightarrow PdCl_2$ activation; EL Ni–P seed; smooth PVD seeds where needed), and applying post-treatments—short annealing for EL Ni–P and vacuum/thermal baking after ED—that erase process memory before qualification. For deep cryogenic work, Ni combined with soft metals (Ni → Zn or Ni + Sn) yields the largest slope multipliers, at the expense of a slower step response; recipe discipline is required to stabilize central wavelengths across cycles. Between 80 and 300 K, duplex Ni/Cu provides a balanced sensitivity–transient trade-off; thin TiN offers a fast-response and PVD-only alternative with clean cryogenic behavior. At high temperature (600–1000 °C), failure modes are diffusion and oxidation; Ni/Cu/Au multilayers with explicit Ni barriers and noble caps enable repeatable slopes under cycling toward ~1000 °C, provided that the hydrogen introduced during wet metallization is removed beforehand. When joining to or embedding in metals, graded Mo–Cu → Ni interlayers control residual stress, and electroplated surface-mounted joints shift the limiting element from the grating to the interface. Evidence syntheses focused on cryogenic-coated FBGs and on thermal-sensing performance analyses corroborate the thickness–sensitivity and transient trends highlighted here. These technical conclusions sit within a broader sensing context captured by field surveys and overviews of optical-fiber sensors and networks. Application exemplars in aeronautics and energy—including aeronautical SHM, conceptual cryogenic/energy systems and fusion divertor monitoring—underscore the operational constraints that coatings help address. Foundational studies on apparent strain and temperature–strain discrimination remain central when interpreting coated FBGs in constrained or embedded configurations. General thin-film/process primers and fiber-specific metallization playbooks (re-coating procedures, EL nanometal strategies, fiber-safe Ni–P recipes) help transfer recipes between laboratories and explain why apparently similar stacks can diverge without process discipline. A recent cryogenic-FBG review further justifies the emphasis on coatings in low-temperature regimes. The case for graded cryogenic stacks and the superiority of duplex/multilayer strategies at elevated temperature is supported by both focused demonstrations and broad comparative evaluations. Overall, hybrid route planning—PVD or EN seeding, electroplated builds, thin barriers/caps, and graded interlayers where needed—sits near the Pareto front across cryogenic, ambient, and high-temperature windows. Standardizing acceptance bands and adopting per-specimen logs will accelerate comparability and deployment. The most promising near-term opportunities are (i) response-time optimization at low temperature without surrendering sensitivity; (ii) diffusion-aware designs at the hot end, with explicit dwell-time management for Cu/Ni interfaces; and (iii) integration-first stacks that treat coatings and packaging as a single design object.

*Recommendations for Coating Selection and Future Development*

Based on the comparative analysis presented, the coating selection for FBGs should primarily consider the operational environment and sensitivity required. Metallic coatings (Ni, Cu, Al) are recommended for cryogenic and moderate-temperature sensing due to their excellent strain-transfer efficiency, while duplex and multilayer systems (Ni → Zn, Ni/Sn, Ni/Cu/Au) offer tunable thermal response and improved adhesion for broader temperature spans. Ceramic and oxide coatings ($Al_2O_3$, $SiO_2$, $TiO_2$) remain the best option for long-term, high-temperature, oxidizing environments, whereas hybrid or composite stacks provide mechanical robustness and corrosion protection for embedded or structural applications.

Future development should focus on scalable, precisely controlled deposition techniques such as electrodeposition, ALD, and hybrid PVD/ED approaches to ensure uniformity and reproducibility on optical fibers. Nanoscale interlayers and diffusion barriers are essential for stress management and adhesion under thermal cycling. In parallel, integration with additive manufacturing and laser-assisted microdeposition could enable localized reinforcement and multifunctional coatings. Combining these approaches with data-driven optimization of process parameters can accelerate the transition toward next-generation industrial FBG sensors with high reliability and adaptability for aerospace, energy, and cryogenic applications.

**Author Contributions:** Conceptualization, C.V. and M.A.C.; methodology, R.D., validation, C.V. and M.A.C.; investigation, E.D.M.; formal analysis C.V. and E.D.M.; resources, E.D.M. and C.V.; data curation, E.D.M. and R.D.; writing—original draft preparation, C.V. and E.D.M.; writing—review and editing, C.V., R.D. and M.A.C.; visualization, E.D.M.; supervision, C.V.; project administration, C.V. and M.A.C.; funding acquisition, C.V. All authors have read and agreed to the published version of the manuscript.

**Funding:** This work was supported by the CREATION (CRyogenic Efficiency gAins Through optImized nanO-structured metal coatiNgs, B83C25000880005) research project.

**Acknowledgments:** The authors gratefully acknowledge the members of the CREATION consortium for constructive discussions and for sharing technical insights that helped refine the coating route classifications, stack archetypes, and qualification criteria consolidated in this review.

**Conflicts of Interest:** The authors declare no conflicts of interest. The funders had no role in the design of the study; in the collection, analyses, or interpretation of data; in the writing of the manuscript; or in the decision to publish the results.

## Abbreviations

The following abbreviations are used in this manuscript:

| | |
|---|---|
| ALD | Atomic Layer Deposition |
| CTE | Coefficient of Thermal Expansion |
| CVD | Chemical Vapor Deposition |
| ED | Electrodeposition |
| EL | Electroless |
| FBG | Fiber Bragg Grating |
| FWHM | Full-Width at Half Maximum |
| IBS | Ion Beam Sputtering |
| ML | Multilayer |
| PMMA | Poly(Methyl MethAcrylate) |
| PVD | Physical Vapor Deposition |
| QA | Quality Assurance |
| SNR | Signal-to-Noise Ratio |

## References

1. Lo Presti, D.; Massaroni, C.; D'Abbraccio, J.; Saccomandi, P.; Caponero, M.A.; Polimadei, A.; Schena, E. Fiber Bragg Gratings for Medical Applications and Future Challenges: A Review. *IEEE Access* **2020**, *8*, 156863–156888. [CrossRef]
2. Xie, J.F.; Zhang, H.; Zhu, Z.; Xu, J.N.; Hu, R.H.; Song, L.F. A Study of the Temperature Sensitivity of Fiber Bragg Gratings after Metallization. *Smart Mater. Struct.* **2007**, *16*, 1837–1842. [CrossRef]
3. Lupi, C.; Montinaro, N.; Paolozzi, A.; Giordano, M.; Russo, S.; Mastrogiacomo, F.; Campopiano, S.; Cutolo, A.; Cusano, A. Critical Issues of Double-Metal Layer Coating on FBG for Applications at High Temperatures. *Sensors* **2019**, *19*, 3824. [CrossRef]
4. Liu, Y.; Fang, J.; Jia, D.; Li, W. Temperature Characteristics of FBG Sensors with Different Coatings for High Temperature Superconductor Application. In Proceedings of the 2019 IEEE 3rd International Electrical and Energy Conference (CIEEC 2019), Beijing, China, 7–9 September 2019; Institute of Electrical and Electronics Engineers (IEEE): Piscataway, NJ, USA, 2019; pp. 1546–1550. [CrossRef]
5. Pendão, C.; Silva, I. Optical Fiber Sensors and Sensing Networks: Overview of the Main Principles and Applications. *Sensors* **2022**, *22*, 7554. [CrossRef]
6. Vendittozzi, C.; Felli, F.; Lupi, C. Modeling FBG Sensors Sensitivity from Cryogenic Temperatures to Room Temperature as a Function of Metal Coating Thickness. *Opt. Fiber Technol.* **2018**, *42*, 84–91. [CrossRef]

7. Mishra, V.; Lohar, M.; Amphawan, A. Improvement in Temperature Sensitivity of FBG by Coating of Different Materials. *Optik* **2016**, *127*, 825–828. [CrossRef]
8. Wang, X.D.; Wolfbeis, O.S. Fiber-Optic Chemical Sensors and Biosensors (2015–2019). *Anal. Chem.* **2020**, *92*, 397–428. [CrossRef] [PubMed]
9. Ramakrishnan, M.; Rajan, G.; Semenova, Y.; Farrell, G. Overview of Fiber Optic Sensor Technologies for Strain/Temperature Sensing Applications in Composite Materials. *Sensors* **2016**, *16*, 99. [CrossRef] [PubMed]
10. Wen, C.; Li, Y. Effects of Metal Coating on the Fiber Bragg Grating Temperature Sensing Characteristics. *J. Mod. Opt.* **2016**, *63*, 762–770. [CrossRef]
11. Campanella, C.E.; Cuccovillo, A.; Campanella, C.; Yurt, A.; Passaro, V.M.N. Fibre Bragg Grating Based Strain Sensors: Review of Technology and Applications. *Sensors* **2018**, *18*, 3115. [CrossRef]
12. Wang, X.; Wang, W.; Li, Q.; Xie, C.; Luo, G.N. Application of FBG sensors in tokamak tungsten divertor, component for strain and temperature monitoring. In *Optics and Photonics for Energy and the Environment*; ET5A.6. 10.1364/EE.2018.ET5A.6; Optica Publishing Group: Washington, DC, USA, 2018.
13. Hegde, G.; Asokan, S.; Hegde, G. Fiber Bragg grating sensors for aerospace applications: A review. *ISSS J. Micro Smart Syst.* **2022**, *11*, 257–275. [CrossRef]
14. Yassin, M.H.; Farhat, M.H.; Soleimanpour, R.; Nahas, M. Fiber Bragg grating (FBG)-based sensors: A review of technology and recent applications in structural health monitoring (SHM) of civil engineering structures. *Discov. Civ. Eng.* **2024**, *1*, 151. [CrossRef]
15. Lupi, C.; Vendittozzi, C.; Ciro, E.; Felli, F. FBG Spectrum Regeneration by Ni-Coating and High-Temperature Treatment. *Sensors* **2022**, *22*, 7255. [CrossRef]
16. Nogueira, R.; Oliveira, R.; Bilro, L.; Heidarialamdarloo, J. New advances in polymer fiber Bragg gratings. *Opt. Laser Technol.* **2015**, *78*, 104–109. [CrossRef]
17. Mihailov, S.J.; Grobnic, D.; Hnatovsky, C.; Walker, R.B.; Lu, P.; Coulas, D.; Ding, H. Extreme Environment Sensing Using Femtosecond Laser-Inscribed Fiber Bragg Gratings. *Sensors* **2017**, *17*, 2909. [CrossRef]
18. Leal-Junior, A.; Frizera, A.; Marques, C. Development and Characterization of UV-Resin Coated Fiber Bragg Gratings. *Sensors* **2020**, *20*, 3026. [CrossRef] [PubMed]
19. Wang, X.; Sun, X.; Hu, Y.; Zeng, L.; Liu, Q.; Duan, J. Highly-Sensitive Fiber Bragg Grating Temperature Sensors with Metallic Coatings. *Optik* **2022**, *262*, 169337. [CrossRef]
20. Deng, Y.; Jiang, J. Optical Fiber Sensors in Extreme Temperature and Radiation Environments: A Review. *IEEE Sens. J.* **2022**, *22*, 11145–11161. [CrossRef]
21. Stephen, J.; Pal, P.; Mathew, J. Advances in Cryogenic Temperature Sensing Using Fiber Bragg Grating: A Review. *IEEE Sens. J.* **2025**, *25*, 12345–12360. [CrossRef]
22. Zaynetdinov, M.; See, E.M.; Geist, B.; Ciovati, G.; Robinson, H.D.; Kochergin, V. A Fiber Bragg Grating Temperature Sensor for 2–400 K. *IEEE Sens. J.* **2015**, *15*, 1908–1912. [CrossRef]
23. Hisham, H. *Fiber Bragg Grating Sensors: Development and Applications*; Taylor & Francis Group, LLC: Boca Raton, FL, USA, 2019; ISBN 978-0-367-22485-1. [CrossRef]
24. Ciotti, M.; Nardelli, V.; Caponero, M.A.; Felli, F.; Lupi, C.; Ippoliti, L. An Optical System for Cryogenic Temperature Measurements. *Smart Mater. Struct.* **2007**, *16*, 1708–1711. [CrossRef]
25. Soman, A.C.; Chalackal, A.T.; Kanakambaran, S. Design of Cryogenic Temperature Sensors Using Copper-Coated Fiber Bragg Gratings. In Proceedings of the 2022 IEEE Region 10 Symposium (TENSYMP 2022), Kolkata, India, 29–31 May 2022; Institute of Electrical and Electronics Engineers (IEEE): Piscataway, NJ, USA, 2022. [CrossRef]
26. Chiuchiolo, A.; Bracco, C.; Rossi, F.; Bianchi, L.; Verdi, M. Cryogenic Test Facility Instrumentation with Fiber Optic and Fiber Optic Sensors for Testing Superconducting Accelerator Magnets. In *IOP Conference Series: Materials Science and Engineering*; Institute of Physics Publishing: Bristol, UK, 2017; Volume 278, p. 012082. [CrossRef]
27. Feng, Y.; Zhang, H.; Li, Y.L.; Rao, C.F. Temperature Sensing of Metal-Coated Fiber Bragg Grating. *IEEE/ASME Trans. Mechatron.* **2010**, *15*, 511–519. [CrossRef]
28. Sugino, M.; Ogata, M.; Mizuno, K.; Hasegawa, H. Development of Zinc Coating Methods on Fiber Bragg Grating Temperature Sensors. *IEEE Trans. Appl. Supercond.* **2016**, *26*, 9000606. [CrossRef]
29. Wang, X.; Sun, X.; Hu, Y.; Duan, J. Multilayer Metal-Coated Fiber Bragg Grating for High-Temperature Sensing. *Measurement* **2024**, *226*, 114132. [CrossRef]
30. Shiue, S.T.; Yang, C.H.; Chu, R.S.; Yang, T.J. Effect of the Coating Thickness and Roughness on the Mechanical Strength and Thermally Induced Stress Voids in Nickel-Coated Optical Fibers Prepared by Electroless Plating Method. *Thin Solid Films* **2005**, *485*, 169–175. [CrossRef]

31. Butt, M.A. Thin-Film Coating Methods: A Successful Marriage of High-Quality and Cost-Effectiveness—A Brief Exploration. *Coatings* **2022**, *12*, 1115. [CrossRef]
32. Muench, F. Electroless Plating of Metal Nanomaterials. *ChemElectroChem* **2021**, *8*, 3591–3600. [CrossRef]
33. Barhoum, A.; García-Betancourt, M.L.; Jeevanandam, J.; Hussien, E.A.; Mekkawy, S.A.; Mostafa, M.; Omran, M.M.; Abdalla, M.S.; Bechelany, M. Review on Natural, Incidental, Bioinspired, and Engineered Nanomaterials: History, Definitions, Classifications, Synthesis, Properties, Market, Toxicities, Risks, and Regulations. *Nanomaterials* **2022**, *12*, 177. [CrossRef] [PubMed]
34. Elsherif, M.; Salih, A.E.; Muñoz, M.G.; Alam, F.; AlQattan, B.; Savariraj, A.D.; Zaki, M.F.; Yetisen, A.K.; Park, S.; Wilkinson, T.D.; et al. Optical Fiber Sensors: Working Principle, Applications, and Limitations. *Adv. Photonics Res.* **2022**, *3*, 2100371. [CrossRef]
35. Grandal, T.; Piñeiro, E.; Asensio, A.; Rodriguez, F. Metallic Coating Techniques for Fiber Bragg Grating Sensors. In Proceedings of the 8th Iberoamerican Optics Meeting and 11th Latin American Meeting on Optics, Lasers, and Applications, Porto, Portugal, 22–26 July 2013; SPIE: Bellingham, WA, USA, 2013; p. 878538. [CrossRef]
36. Majumder, M.; Gangopadhyay, T.K.; Chakraborty, A.K.; Dasgupta, K.; Bhattacharya, D.K. Fibre Bragg Gratings in Structural Health Monitoring—Present Status and Applications. *Sens. Actuators A* **2008**, *147*, 150–164. [CrossRef]
37. Kern, W.; Schuegraf, K.K. Deposition Technologies and Applications: Introduction and Overview. In *Handbook of Thin Film Deposition Processes and Techniques*, 1st ed.; Noyes Publications: Park Ridge, NJ, USA, 1988; pp. 1–44.
38. Rao, C.; Zhang, H.; Feng, Y.; Xiao, L.; Ye, Z. Thick Metal Coating Long-Period Fiber Grating. In Proceedings of the 6th International Symposium on Advanced Optical Manufacturing and Testing Technologies: Design, Manufacturing, and Testing of Smart Structures, Micro- and Nano-Optical Devices, and Systems, Xiamen, China, 26–29 April 2012; SPIE: Bellingham, WA, USA, 2012; p. 84181Q. [CrossRef]
39. Koo, K.N.; Lee, J.H.; Kim, S.H.; Park, J.; Lee, S.H. Fabrication and Modification of Temperature FBG Sensor: Role of Optical Fiber Type and Cu Sputtered Thickness. *Phys. Scr.* **2020**, *95*, 095602. [CrossRef]
40. Tang, A.Q.; Fang, L.; Xue, S.J.; Yin, B.; Liu, L.; Zhang, P. Ni-P Coating Metallization on Fiber Bragg Grating without Fiber Coarsening and Its Temperature Sensing Property. In Proceedings of the 2011 International Conference on Optical Instruments and Technology: Optical Sensors and Applications, Beijing, China, 6–9 November 2011; SPIE: Bellingham, WA, USA, 2011; p. 819907. [CrossRef]
41. Zhao, Z.B.; Gillispie, B.A.; Smith, J.R. Coating Deposition by the Kinetic Spray Process. *Surf. Coat. Technol.* **2006**, *200*, 4746–4754. [CrossRef]
42. Scurti, F.; McGarrahan, J.; Schwartz, J. Effects of Metallic Coatings on the Thermal Sensitivity of Optical Fiber Sensors at Cryogenic Temperatures. *Opt. Mater. Express* **2017**, *7*, 1754–1764. [CrossRef]
43. Li, Y.; Yang, K.; Li, X. Temperature Sensing Characteristics of Metal Coated FBG during Dynamic Cooling Process. *Opt. Fiber Technol.* **2018**, *45*, 368–375. [CrossRef]
44. Sengupta, D.; Sai Shankar, M.; Saidi Reddy, P.; Sai Prasad, R.L.N.; Narayana, K.S.; Kishore, P. An Improved Low-Temperature Sensing Using PMMA Coated FBG. In *Optics InfoBase Conference Papers, Proceedings of the Asia Communications and Photonics Conference and Exhibition 2011, Shanghai China, 13–16 November 2011*; SPIE: Bellingham, WA, USA, 2011. [CrossRef]
45. Jin, L.; Zhang, W.; Zhang, H.; Liu, B.; Zhao, J.; Tu, Q.; Kai, G.; Dong, X. An embedded FBG sensor for simultaneous measurement of stress and temperature. *IEEE Photonics Technol. Lett.* **2006**, *18*, 154–156. [CrossRef]
46. Mattox, D.M. Arc Vapor Deposition. In *Handbook of Physical Vapor Deposition (PVD) Processing*; Elsevier: Amsterdam, The Netherlands, 2010; pp. 287–300. [CrossRef]
47. Pawar, K.; Dixit, P. A critical review of copper electroless deposition on glass substrates for microsystems packaging applications. *Surf. Eng.* **2022**, *38*, 576–617. [CrossRef]
48. Prasad, R.D.; Teli, B.; Prasad, R.S.; Prasad, R.B.; Prasad, S.R.; Sinha, P.; Sinha, A.; Sinha, P.; Saxena, M.; Prasad, R.R.; et al. A Review on Thin Film Technology and Nanomaterial Characterization Techniques. *ES Mater. Manuf.* **2024**, *25*, 1198. [CrossRef]
49. Li, C.; Liu, M.; Song, H.; Wang, J.; Wu, Y.; Chen, X. Thermal Sensing Performance Analysis, Preparation and Application of Bimetallic Layer MFBG. *Int. J. Heat Mass Transf.* **2023**, *215*, 124514. [CrossRef]
50. Listewnik, P. Temperature Fiber-Optic Sensor with ZnO ALD Coating. *Eng. Proc.* **2021**, *2*, 99. [CrossRef]
51. Zhang, P.; Wang, X.; Guan, M.; Xin, C.; Wu, W.; Lin, X.; Pei, B. Enhanced Strain and Temperature Sensing in Copper-Coated Fiber Bragg Grating Sensors across a Wide Temperature Range from Cryogenic to Elevated Levels. *Cryogenics* **2024**, *139*, 103834. [CrossRef]
52. Perry, M.; Niewczas, P.; Johnston, M.; Mackersie, J. Nickel plating of FBG strain sensors for nuclear applications. In Proceedings of the 21st International Conference on Optical Fiber Sensors, Ottawa, ON, Canada, 15–19 May 2011; Bock, W.J., Albert, J., Bao, X., Eds.; SPIE: Bellingham, WA, USA, 2011; Volume 7753. Article 77530C. [CrossRef]

53. Shen, R.-S.; Teng, R.; Li, X.-P.; Zhang, J.; Xia, D.-C.; Fan, Z.-Q.; Yu, Y.-S.; Zhang, Y.-S.; Du, G.-T. Electroless Nickel Plating and Electrodeposition on FBG Temperature Sensor. *Chem. Res. Chin. Univ.* **2008**, *24*, 635–639. [CrossRef]
54. Wang, Y.L.; Tu, Y.; Tu, S.T. Development of Highly-Sensitive and Reliable Fiber Bragg Grating Temperature Sensors with Gradient Metallic Coatings for Cryogenic Temperature Applications. *IEEE Sens. J.* **2021**, *21*, 4652–4663. [CrossRef]
55. Li, Y.; Wen, C.; Zhang, H.; Yang, J.; Yan, M.; Jiang, J. An Electrodeposition Method for Surface Mounting Optical Fiber Sensors on the Metal Substrate. *IEEE Photonics Technol. Lett.* **2016**, *28*, 1811–1814. [CrossRef]
56. Tu, Y.; Qi, Y.H.; Tu, S.T. Fabrication and thermal characteristics of multilayer metal-coated regenerated grating sensors for high-temperature sensing. *Smart Mater. Struct.* **2013**, *22*, 075026. [CrossRef]
57. Li, Y.; Hua, Z.; Yan, F.; Gang, P. Metal coating of fiber Bragg grating and the temperature sensing character after metallization. *Opt. Fiber Technol.* **2009**, *15*, 391–397. [CrossRef]
58. Mansor, N.F.; Raja Ibrahim, R.K. Temperature sensitivity of FBG coating with zinc oxide and silicon carbide. *J. Phys. Conf. Ser.* **2021**, *1892*, 012033. [CrossRef]
59. Kesavan, K.; Ravisankar, K.; Senthil, R.; Sundaram, B.A.; Parivallal, S. Studies on Apparent Strain Using FBG Strain Sensors for Different Structural Materials. *Exp. Tech.* **2014**, *38*, 31–38. [CrossRef]
60. Lindner, M.; Gaßmann, S.; Klein, T.; Roeser, F.; Möllmann, M.; Lerch, R. Fiber Bragg Sensors Embedded in Cast Aluminum Parts: Axial Strain and Temperature Response. *Sensors* **2021**, *21*, 1680. [CrossRef] [PubMed]
61. Freitas, R.; Araujo, F.; Araujo, J.; Neumann, H.; Ramalingam, R. A Study on Intermediate Buffer Layer of Coated Fiber Bragg Grating Cryogenic Temperature Sensors. In *IOP Conference Series: Materials Science and Engineering*; Institute of Physics Publishing: Bristol, UK, 2015; Volume 101, p. 012154. [CrossRef]
62. He, J.; Ding, L.; Cai, J.; Zhu, W.; Dai, J. A Novel High Temperature Resistant Mo-Cu Functional Gradient Coating for Optic Fiber Bragg Grating. *Results Phys.* **2019**, *14*, 102456. [CrossRef]
63. Weraneck, K.; Lee, S.W.; King, N.; Ward, A.D.; Haynes, M.; Sampson, P.; Allwood, D. Strain Measurement in Aluminium Alloy during the Solidification Process Using Embedded Fibre Bragg Gratings. *Sensors* **2016**, *16*, 1853. [CrossRef] [PubMed]
64. Li, Y.; Wang, Y.; Wen, C. Temperature and Strain Sensing Properties of the Zinc Coated FBG. *Optik* **2016**, *127*, 6463–6469. [CrossRef]
65. Kok, S.P.; Go, Y.I.; Wang, X.; Wong, M.D. Advances in Fiber Bragg Grating (FBG) Sensing: A Review of Conventional and New Approaches and Novel Sensing Materials in Harsh and Emerging Industrial Sensing. *IEEE Sens. J.* **2024**, *24*, 29485–29505. [CrossRef]
66. Hokkanen, A.; Salmi, A.; Vashistha, V.; Nyman, M.; Nielsen, S.K.; Jensen, T.; Jessen, M.; Wälchli, B.; Kapulainen, M.; Naulin, V.; et al. A Panda fiber temperature sensor up to 900°C. *J. Instrum.* **2022**, *17*, P07031. [CrossRef]
67. Lupi, C.; Felli, F.; Brotzu, A.; Caponero, M.A.; Paolozzi, A. Improving FBG Sensor Sensitivity at Cryogenic Temperature by Metal Coating. *IEEE Sens. J.* **2008**, *8*, 1299–1304. [CrossRef]
68. Listewnik, P.; Hirsch, M.; Struk, P.; Weber, M.; Bechelany, M.; Jędrzejewska-Szczerska, M. Preparation and Characterization of Microsphere ZnO ALD Coating Dedicated for the Fiber-Optic Refractive Index Sensor. *Nanomaterials* **2019**, *9*, 306. [CrossRef]
69. Sandlin, S.; Kinnunen, T.; Rämö, J.; Sillanpää, M. A Simple Method for Metal Re-Coating of Optical Fibre Bragg Gratings. *Surf. Coat. Technol.* **2006**, *201*, 3061–3065. [CrossRef]
70. Silva, D.; da Silva, L.F.; de Souza, R.F.; de Almeida, D.L.; de Oliveira, M.M. Sputtering Deposition of $TiO_2$ Thin Film Coatings for Fiber Optic Sensors. *Photonics* **2022**, *9*, 342. [CrossRef]
71. Lou, H.H.; Huang, Y. Electrodeposition. In *Encyclopedia of Chemical Processing*; Lee, S., Ed.; Taylor & Francis: New York, NY, USA, 2006.
72. Chan, T.H.T.; Seah, L.K.; Leung, A.Y.T.; Li, H.-N.; Lam, H.F. Fiber Bragg Grating Sensors for Structural Health Monitoring of Tsing Ma Bridge: Background and Experimental Observation. *Eng. Struct.* **2006**, *28*, 648–659. [CrossRef]
73. Hsu, C.Y.; Chiang, C.C.; Hsieh, T.S.; Hsu, H.C.; Tsai, L.; Hou, C.H. Study of Fiber Bragg Gratings with TiN-Coated for Cryogenic Temperature Measurement. *Opt. Laser Technol.* **2021**, *136*, 106768. [CrossRef]
74. Feng, Y.; Zhang, H.; Li, Y.-L.; Peng, G. Highly Sensitive Ni-Cu Duplex Metal Coated Fiber Bragg Grating Temperature Sensor. In Proceedings of the 2009 Symposium on Photonics and Optoelectronics, Wuhan, China, 14–16 August 2009; pp. 1–4. [CrossRef]
75. Li, Y.; Zhang, H.; Feng, Y.; Peng, G. A Plating Method for Metal Coating of Fiber Bragg Grating. *Chin. Opt. Lett.* **2009**, *7*, 115–117. [CrossRef]
76. Kok, S.; Go, Y.; Wang, X.; Wong, D. A Review of Nanostructure Coating Techniques to Achieve High-Precision Optical Fiber Sensing Applications. *Nanomanufacturing* **2024**, *4*, 214–240. [CrossRef]
77. Wright, R.F.; Buric, M.; Egbu, J.; Ziomek-Moroz, M.; Lu, P.; Ohodnicki, P.R. Electrolessly Coated Optical Fibers for Distributed Corrosion Monitoring. In Proceedings of the CORROSION 2019, Nashville, TN, USA, 24–28 March 2019; pp. 1–12. [CrossRef]
78. Rajini-Kumar, R.; Suesser, M.; Narayankhedkar, K.G.; Krieg, G.; Atrey, M.D. Performance Evaluation of Metal-Coated Fiber Bragg Grating Sensors for Sensing Cryogenic Temperature. *Cryogenics* **2008**, *48*, 142–147. [CrossRef]

79. Chanet, N.; Frigione, D.; Imbeaux, F.; Debieu, J.; Jacquet, P.; Mazon, D. Design and Integration of Femtosecond Fiber Bragg Gratings Temperature Probes inside Actively Cooled IT-ER-Like Plasma-Facing Components. *Fusion Eng. Des.* **2021**, *166*, 112376. [CrossRef]
80. Sala, G.; Di Landro, L.; Airoldi, A.; Bettini, P. Fibre Optics Health Monitoring for Aeronautical Applications. *Meccanica* **2015**, *50*, 2547–2567. [CrossRef]
81. Caloud, J.; Imbeaux, F.; Frigione, D.; Boulbe, C.; Jacquet, P.; Mazon, D. Conceptual Design of Fiber Bragg Grating Temperature Sensors for Heat Load Measurements in COMPASS-U Plasma-Facing Components. *Fusion Eng. Des.* **2023**, *193*, 113608. [CrossRef]
82. Wang, X.; Zhang, Y.; Li, H.; Li, S.; Huang, Y.; Liu, Y. R&D of a Strain Monitoring System for the EAST Tungsten Divertor. *Nucl. Mater. Energy* **2019**, *19*, 498–502. [CrossRef]
83. Chang, T.; Jia, L.; Sui, Q.; Cui, H.-L. High-Temperature Measurement Using Cu-Plating Fiber Bragg Grating for Metal Smart Structure Applications. In *Sensors and Smart Structures Technologies for Civil, Mechanical, and Aerospace Systems 2014*; SPIE: Bellingham, WA, USA, 2014; p. 90613P. [CrossRef]

*coatings*

MDPI

*Article*

# Langmuir and Langmuir–Blodgett Monolayers from 20 nm Sized Crystals of the Metal–Organic Framework MIL-101(Cr)

**Asen Dimov [1], George R. Ivanov [1,*], Leonard Keil [2], Andreas Terfort [2], Jinxuan Liu [3] and Velichka Strijkova [4]**

1 University Laboratory "Nanoscience and Nanotechnology", University of Architecture, Civil Engineering and Geodesy, Blvd. Hr. Smirnenski 1, 1046 Sofia, Bulgaria; adimov_fhe@uacg.bg
2 Institut für Anorganische und Analytische Chemie, Goethe-Universität Frankfurt, N160/B515 Alexander-Todd-Straße 7, 60438 Frankfurt, Germany; keil@chemie.uni-frankfurt.de (L.K.); aterfort@chemie.uni-frankfurt.de (A.T.)
3 State Key Laboratory of Fine Chemicals, Dalian University of Technology, Dalian 116024, China; jinxuan.liu@dlut.edu.cn
4 Institute of Optical Materials and Technologies, Bulgarian Academy of Sciences, 109, Acad. G. Bontchev Str., 1113 Sofia, Bulgaria; vily@iomt.bas.bg
* Correspondence: george@at-equipment.com or g_ivanov_fhe@uacg.bg; Tel.: +359-885-465-535

## Highlights

**What are the main findings?**

- A metal–organic Framework (MOF) MIL-101(Cr) with crystal sizes around 20 nm was synthesized and the results were compared to a 300 nm crystal-sized commercially available material.

**What are the implications of the main findings?**

- Smaller-sized MIL-101(Cr) crystals form a stable insoluble monolayer at the air–water interface with denser, more homogeneous water coverage and packing upon compression, and no visible micrometer-sized aggregates.
- Langmuir-Blodgett monolayers from the 20 nm MOF show six times lower surface roughness.

## Abstract

Metal–Organic Frameworks (MOFs) have diverse applications due to their tunable porosity, large surface area, and diverse chemical functionalities. Among them, one of the most researched MOFs is MIL-101(Cr), which, in addition, is very stable in water. We have used a commercially available substance with approximately 300 nm large crystals for the preparation of a sensing nano-thin layer for the emerging water contaminant PFOS, due to its high selectivity towards this compound. Here, we have synthesized 20 nm sized crystals of MIL-101(Cr), which are among the smallest reported, and compared them to the same material with 300 nm sized crystals. The material was characterized by TEM and XPS. It was possible to prepare insoluble monolayers at the air–water interface (Langmuir films), which were characterized with film compression isotherms, Brewster angle microscopy, and surface potential measurements. The Langmuir–Blodgett (LB) method was used to deposit monolayers on Si wafers and 434 MHz Surface Acoustic Wave resonator simultaneously. The LB layers were very stable over time. The smaller-sized MIL-101 (Cr) crystals exhibit denser, more homogeneous water coverage and packing upon compression, with no observable 10–100 μm aggregates. LB monolayers from the 20 nm particles have approximately six times lower surface roughness. The LB monolayer is far from being smooth, but this will allow excellent access to the MOF pores by the tested analyte in a chemical sensing application. The lack of research on depositing presynthesized MOFs using probably the best method for nanoarchitectonics—the LB method—is addressed. The 20 nm sized MOF crystals are the smallest deposited by this method so far.

**Keywords:** Metal–Organic Framework (MOF); MIL-101(Cr); Langmuir–Blodgett (LB) films; Langmuir films; Atomic Force Microscopy (AFM); monolayer isotherms; Brewster angle microscopy (BAM); surface potential; Perfluorooctane Sulfonic Acid (PFOS)

## 1. Introduction

Metal–organic frameworks (MOFs) are highly porous, crystalline materials composed of metal ions or clusters linked by organic linkers. When changing the building blocks used in MOFs, it is possible to design them to capture and store specific substances [1], drive chemical reactions [2], or for adsorption and molecular sensing [3] due to their unique properties such as large surface area, tunable pore size, and chemical versatility. This is why pioneers of the MOF synthesis were awarded the Nobel Prize in Chemistry in 2025 [4]. Beyond their established roles in catalysis and gas storage, MIL (Materials of the Institute Lavoisier)-based MOFs have emerged as highly versatile materials for chemical sensing applications. The MIL family of MOFs, encompassing systems such as MIL-101(Cr), MIL-100(Fe), MIL-53, and MIL-125, is important due to its structural diversity, large pore volumes, and chemical tunability [5–7]. The combination of robust metal–oxo clusters and functionalizable organic linkers provides a large number of active sites for interactions with analytes via physisorption, chemisorption, or coordination. These interactions underline the operation of MIL-type MOFs in various transduction modes, including gravimetric, optical, and electrochemical sensing. Gravimetric sensing, primarily implemented via quartz crystal microbalance (QCM) techniques, has been demonstrated to be effective using MIL-based thin films [8].

Despite their wide range of applications, many of which require smooth, thin-layer preparation, there are only very few papers in which the Langmuir–Blodgett (LB) method was used to deposit layers from presynthesized bulk MOFs [8–10]. In this method [11], a solution or dispersion of the tested substance is spread at the air–water interface of an LB system trough. The substance has to form an insoluble monolayer at the water surface after solvent evaporation. Afterwards the monolayer is compressed, and its surface pressure and, eventually, surface potential are recorded. The LB method is one of the very few methods for discrete layer-after-layer deposition. The surface density of defects in the monolayer is low because of its self-assembly at the air–water interface. Furthermore, there is complete control over the surface pressure of the deposition, thereby influencing the phase of the layer, molecular orientation, thickness, and layer composition. That is why this is considered the best method for supramolecular architecture [12]. There is a versatility of substrates that can be used without the need for chemical modification of the surface. While initially developed for work with biphylic molecules, such as lipids, the LB method has been successfully applied to the deposition of other structures, including nanoparticles [13], quantum dots [14], and water-soluble molecules, such as enzymes [15], among others.

An alternative approach to the LB film deposition of presynthesized MOFs is to utilize the air–water interface, which has recently emerged as an effective bottom-up route for preparing two-dimensional and ultrathin MOF assemblies with large lateral dimensions and controlled in-plane order. The liquid surface confines reactants to two dimensions and promotes oriented coordination between multi-topic organic linkers and metal centers, enabling the formation of molecularly thin, crystalline MOF nanosheets, such as the porphyrin-based systems, which are prototypical examples [5]. Investigations at the air–water interface with grazing-incidence X-ray diffraction and Brewster angle microscopy have revealed that post-injection or diffusion-controlled approaches at the subphase markedly enhance sheet domain size and uniformity. At the same time, similar methods have been adapted to assemble monolayers of preformed MOF nanoparticles or platelets [5,7,16]. Importantly, MOF assemblies formed at the air–water interface can be compressed in a Langmuir trough and transferred onto solid supports using Langmuir–Blodgett (vertical) or Langmuir–Schaefer (horizontal) techniques, yielding continuous ultrathin MOF coatings in a controlled layer-after-layer fashion [5,7]. The success of such transfers depends strongly on the MOF's hydrolytic stability, particle dispersibility, and interfacial mechanics during compression and deposition. To improve film integrity and transfer efficiency, researchers have employed the hydrophobization of MOF particles, the use of co-spreaders such as long-chain fatty acids, and employed tailored linker designs [6,7,17]. Overall, interfacial synthesis combined with Langmuir–Blodgett or Langmuir–Schaefer deposition provides a versatile platform for generating device-compatible MOF thin films. However, reproducible large-area monolayers still require the optimization of the chemistry and transfer protocol [17]. Surface-supported MOF thin films (SURMOFs) fabricated by the Langmuir–Blodgett method exhibit ultrathin layers, low roughness, and high smoothness. The thickness and crystallite size of these films can be precisely controlled, and the use of self-assembled monolayers (SAMs) as interfaces further enhances the growth of highly crystalline, oriented, and flexible multilayered architectures [18]. However, stability remains a challenge, as many MOFs are sensitive to air, water, and acidic or alkaline environments [19].

Among all MOFs, MIL-101(Cr) stands out as one of the most frequently researched and utilized in applications, with over 4500 papers published in 2021 alone [20]. This chromium-based MOF is known for its exceptionally high surface area, large pore volume, and remarkable thermal and chemical stability, making it highly versatile for applications in gas adsorption, catalysis, and separations [16,20]. Langmuir–Blodgett and drop-cast MIL-101(Cr) coatings have demonstrated high selectivity and stability in the detection of volatile organic compounds and other analytes, utilizing, for example, gravimetric signal transduction via a quartz crystal microbalance (QCM) [21–23].

For MIL-101(Cr) to be investigated at the air–water interface and subsequently deposited as an LB film, it must be stable in the aqueous environment. It is characterized by a large number of unsaturated Cr(III) sites, which contribute to its ability to bind electron-rich functional groups and confer good water stability [24–26]. The material is synthesized under harsh acidic aqueous conditions (220 °C in water), which supports its fundamental water stability [24]. The highly porous MIL-101(Cr) can withstand multiple cycles of water uptake [27]. Unfortunately, MIL-101(Cr) is too hydrophobic, and the main water uptake occurs only at high pressures [27].

Despite these advances, several challenges remain in translating MOFs into practical devices. Moisture interference, slow recovery times, and difficulties in forming uniform thin films can limit reproducibility [28,29]. Recent research has focused on integrating MOFs into polymeric or conductive matrices and employing post-synthetic modifications to enhance selectivity and stability [30,31]. The synthesis and coupling of MOF sensing systems with machine learning approaches for signal analysis and analyte classification represents a promising frontier [32,33]. Collectively, these efforts highlight the potential of MIL-based

MOFs as key materials for next-generation chemical sensors, combining high selectivity, low power consumption, and compatibility with flexible, miniaturized substrates.

Previously, we demonstrated that MIL-101(Cr) exhibits high selectivity for the emerging, highly hazardous water contaminant Perfluorooctane Sulfonic Acid (PFOS) [10]. This compound is part of a larger family of Per- and polyfluoroalkyl substances (PFAS) contaminants, also known as the so-called "forever chemicals". According to the EU directive [34] and the US EPA [35] recommendation, water intended for human consumption should be monitored and controlled to the lowest concentrations among all water contaminants (0.1 μg/L). We utilized gravimetric signal transduction via an ultra-sensitive 434 MHz two-port Surface Acoustic Wave (SAW) resonator with an LB monolayer from the 300 nm MIL-101(Cr) as the sensing layer [10]. However, our limit of detection was an order and half of magnitude above the legislative limit. In the present study, an attempt was made to improve the future sensor sensitivity to PFOS by synthesizing much smaller 20 nm MIL-101(Cr) crystals. Their performance as a monolayer at the air–water interface and as an LB film monolayer was investigated.

## 2. Materials and Methods

### 2.1. Materials

The large crystal-sized MOF MIL-101(Cr) was acquired in powder form from NovoMOF AG (Zofingen, Switzerland). It is further denoted as 300 nm MIL-101(Cr) in the text, though the crystal size varies. The nitrogen adsorption–desorption tests yielded a surface area of 3269.3 $m^2/g$ [10], which is above the average for this material and shows minimal hysteresis [10]. The X-ray diffraction pattern corresponds to previously published experimental and calculated XRD data [10,36]. The peak sharpness indicates the well-ordered crystalline structure of the MOF crystals.

The small crystal-sized MIL-101(Cr), hereafter referred to as a 20 nm MOF, was synthesized according to the following protocol, modified from a previously published report [37]. Chromium nitrate nonahydrate (($Cr(NO_3)3{\cdot}9H_2O$, 2.0 g), 1,4-benzene dicarboxylic acid ($H_2BDC$, 0.82 g), cetyltrimethylammonium bromide (CTAB, 0.36 g), and sodium acetate (NaAc, 0.10 g) were dissolved in 24 mL of deionized water by ultrasonication. The reaction mixture was transferred into a PTFE-coated steel autoclave and heated to 473 K for 24 h. After cooling to room temperature, a green powder was obtained by centrifugation. The product was washed three times with ethanol and then transferred to an Erlenmeyer flask. Under continuous shaking for one hour, 50 mL of dimethylformamide was added stepwise. After centrifugation, the product was transferred into a Teflon-coated steel autoclave, suspended in ethanol, and heated to 373 K for 20 h. After centrifugation, the product was rewashed with ethanol and dried overnight at 333 K. The MIL-101(Cr) synthesized in this manner in [37] was characterized by powder X-ray diffraction and nitrogen adsorption–desorption isotherms and was not found to differ from previously reported data for this material.

The arachidic acid (AA, $CH_3(CH_2)_{18}COOH$) used was of 98% purity (Alfa Aesar, Karlsruhe, Germany). The reagents used for the 20 nm MIL-101(Cr) synthesis were chromium nitrate nonahydrate (Aladdin, China), sodium acetate (Damao, China), CTAB (Aladdin, China), dimethylformamide (Aladdin, China), terephthalic acid (Aladdin, China), and anhydrous ethanol (Fuyu Reagent, China). All the solvents used were of high purity and were used without further purification.

The water used for both the final cleanings and for the subphase in the LB film trough was model MillyQ IQ 7003 (Merck KGaA, Darmstadt, Germany) tap water purification system. This is the highest available water purity model, featuring a resistance of 18.2

MΩ/cm and less than 2.5 ppb total organic carbon (TOC), measured in real time. The water dispenser directly pours water into the LB system trough. The dispenser and the LB system were positioned in a high-quality ISO Class 3 vertical laminar flow hood, the FlowFAST V, with complete computer control (Faster Ltd., Cornaredo (MI), Italy).

*2.2. Dispersion Preparation*

Several dispersions were tested using different solvents or solvent mixtures. Some of them are presented here. The spreading dispersion for the Langmuir films was from the 300 nm MIL-101(Cr).5 mg of the powder was dispersed in a mixture consisting of 25% methanol by volume and 75% chloroform to a solution concentration of 0.83 mg/mL. The immiscibility of chloroform with water keeps the dispersion droplets afloat at the air–water interface during Langmuir film preparation, allowing for excellent spreading and minimal loss of substance into the water volume. The very strongly polar solvent methanol improved the dispersion of polar moieties. The homogenization was carried out at room temperature (23 °C) using an ultrasonic processor (20 kHz, 1200 W; Hangzhou Dowell Ultrasonic Tech Co., Ltd., Hangzhou, China). The ultrasound mixer operated at 600 W for 60 min. Langmuir and LB films were prepared from pure 300 nm MOF or with added AA lipid to 17.7% by mass to enhance film homogeneity. For the 300 nm monolayer with AA acid added, hexane was used as a less aggressive solvent and to compare spreading behaviors using different solvents. There was no difference in the isotherms of AA when hexane or chloroform/methanol was used as a solvent. Before each spread of the MOF dispersion, it was sonicated for at least 1 h in a standard ultrasound sonicator.

For the preparation of the 20 nm MIL-101(Cr) dispersion, the same solvent mixture (25% methanol by volume and 75% chloroform) was used. The dispersion was prepared in 1.5 mL glass bottles with PTFE caps. Concentrations were from 4.9 to 6.5 mg/mL. The dispersion was heated to 70 °C and sonicated in a standard heating sonicator for several hours.

*2.3. Investigations at the Air–Water Interface and the LB Film Preparation*

For the Langmuir film measurements and LB film deposition, an LB system model Microtrough G4 (Kibron Inc. Oy, Helsinki, Finland) was used. This is the largest-sized trough model from this company, with an area between fully opened barriers of 874 $cm^2$, made from PTFE. The symmetrical compression barriers were made of a hydrophilic material to prevent film leaks. A precision Wilhelmy-type surface pressure sensor with an achievable resolution of 0.001 mN/m was used. Better results were obtained with filter paper Wilhelmy plates. The PTFE trough was cleaned with both Alconox Powder Precision Cleaner and Citranox Liquid Acid Cleaner and Detergent (Alconox Inc., New York, NY, USA) according to the company's PTFE cleaning protocol. Warm water and a brush were used. Multiple rinses with pure water followed. A new filter paper was placed on the pressure sensor and calibrated against the surface tension of pure water before each new experiment. The presence of surface-active contaminants in the subphase water was checked by compressing the barriers and monitoring the surface pressure to ensure it was zero before each layer was spread. After spreading the layer, the solvent was allowed to evaporate for at least 40 min before compression. The velocity of compression was 2 $cm^2$/min. In the LB system, a Kibron G-BAM was integrated, featuring motorized adjustment of the Brewster angle, the height above the water layer, and the focus. The field of view was 2.4 mm with 4 μm resolution. The sophisticated software enabled numerous corrections and improvements to the image in real time. The BAM received surface pressure, area, and time data in real time from the LB system, which were automatically recorded in the image file name or shown in the video recording. A low-noise MicroSpot surface potentiometer (Kibron) was also integrated into the system. An integrated motorized dip coater, LayerX (90) (Kibron), with three separate clamps, was used for LB film deposition.

All substrates for LB film deposition were cleaned for 2 min at 12 W in an air plasma using a plasma cleaner (model PDC-32G-2, Harrick Plasma, Ithaca, NY, USA) before immersion in the LB system dipping well. The air plasma simultaneously gently cleaned the surface and made it more hydrophilic, which is very important for high-quality deposition during the upstroke of the first LB layer. Before each LB film deposition, the layer was kept at a constant pressure for 5–10 min to allow it to stabilize. All isotherms were measured at 22–23 °C. Ultra-flat Si wafers (roughness of approximately 0.1 nm) were used as substrates for AFM investigation and purchased from NanoAndMore GmbH (Wetzlar, Germany). In most cases the deposition was carried out simultaneously on the Si wafer and a 434 MHz two-port Surface Acoustic Wave (SAW) resonator optimized for chemical sensing. Results from the SAW tests will be presented elsewhere.

### 2.4. Other Instruments Used

Measurements of X-ray photoelectron spectra (XPSs) studies were performed using a VG Escalab MKII electron spectrometer with achromatic AlK$\alpha$ radiation at an energy of 1486.6 eV, under a base pressure of $10^{-8}$ Pa and a total instrumental resolution of 1 eV. The binding energies (BE) were determined utilizing the C 1s line as a reference with an energy of 285.0 eV. The accuracy of the measured BE was 0.2 eV. $C_{1s}$, $O_{1s}$, $Si_{2p}$, $Cr_{2p}$, and $F_{1s}$ photoelectron lines were recorded. Atomic force microscope (AFM) imaging was performed in non-contact mode using an MFP-3D instrument (Asylum Research, Oxford Instruments, Santa Barbara, CA, USA). All measurements were conducted in air at room temperature. Silicon AFM probes (AC240TS) with a resonance frequency of 70 kHz and a nominal spring constant of 3.5 N/m were used. Morphometric characterization (surface roughness analysis) was carried out using IgorPro 6.37 software. Instruments used for the 20 nm MIL-101(Cr) synthesis were: water purification system Merck Millipore Elix high-purity water system (Merck KGaA, Darmstadt, Germany); centrifuge model H5424R (Shandong Baio Medical Technology Co., Ltd., Shandong, China); autoclave 50 mL custom-built; muffle furnace KSL-1100X (Hefei Kejing Materials Technology Co., Ltd., Hefei, China). The scanning electron microscope (SEM) is a modified 1830 model from Amray (Ada, MI, USA), equipped with a 550i controller from IXRF Systems (Austin, TX, USA). The software used was "Iridium Ultra". The working distance was 15–20 mm, and an acceleration voltage of 16–20 kV was used. Transmission electron microscopy (TEM) was performed using a JEM-200 (JEOL, Tokyo, Japan).

Generative artificial intelligence tools ChatGPT and ScienceDirect AI have been used in this paper to assist with the reference search.

## 3. Results and Discussion

### 3.1. MOF Characterization

Powder X-ray diffraction (PXRD) and $N_2$ adsorption–desorption isotherms for the commercially purchased 300 nm MIL-101(Cr) were presented in the supplemental file of [10]. Scanning electron microscopy (SEM) data for this material exhibit significant aggregation, with aggregates reaching 250 µm in size (Figure 1a). This can be due to the long-term storage of the material in powder form and its interaction with humid air. Individual MIL-101(Cr) crystals are observed at higher magnification, ranging in size from 220 nm to 450 nm. SEM and field-emission scanning electron microscopy (FESEM) analyses consistently reveal that MIL-101(Cr) exhibits a regular octahedral morphology with a smooth crystal surface, a feature repeatedly confirmed across multiple studies [38] and observed in this study. The particle size of MIL-101(Cr) varies depending on the synthesis conditions and possible modifications can reach 500 nm [39]. The MIL-101(Cr) particle sizes used for LB film deposition were those with an average size of 300 nm and 20 nm, as described below.

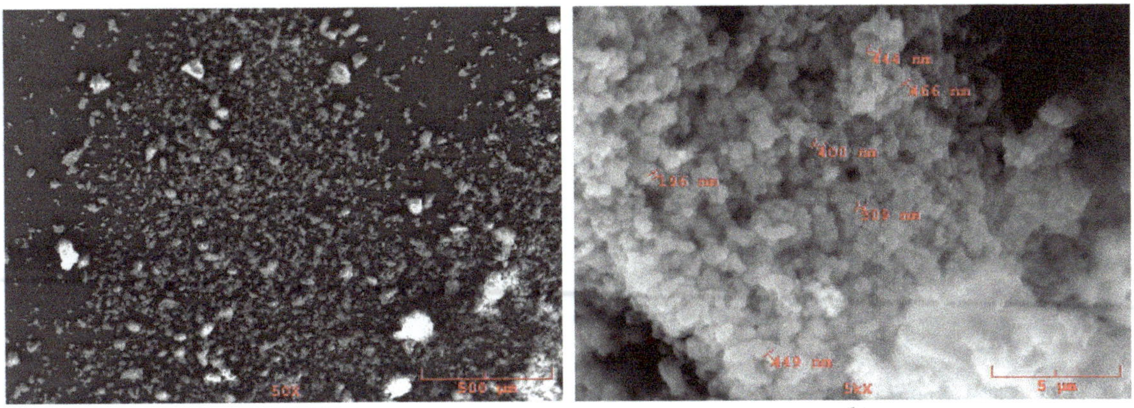


(**a**) (**b**)

**Figure 1.** SEM images at a 500 µm scale (**a**) and 5 µm scale (**b**) of the as-received commercial large crystal-sized MOF MIL-101(Cr), denoted as a 300 nm sized MIL-101(Cr). Significant substance aggregation can be seen.

TEM data for the newly synthesized 20 nm MIL-101(Cr) is shown in Figure 2. The MOF material was suspended in ethanol and homogenized by ultrasonication prior to TEM sample preparation. Though some aggregation could still be seen (Figure 2a,b), where several particles attaching to each other could be seen, it is much less compared to the 300 nm MOF material (Figure 1). Individual crystals are approximately 20–25 nm in length and have a circular or rectangular shape (Figure 2c,d). This finding is in good agreement with reported particle sizes ranging from 10 to 30 nm [40]. Nearly spherical MIL-101(Cr) crystals, 51 ± 10 nm in size, were used to deposit the presynthesized MOF via LB film deposition [8].

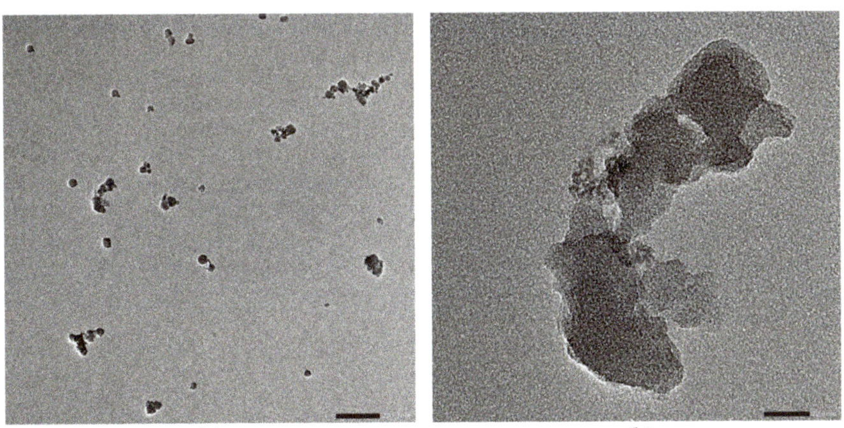

(**a**) (**b**)

**Figure 2.** *Cont.*

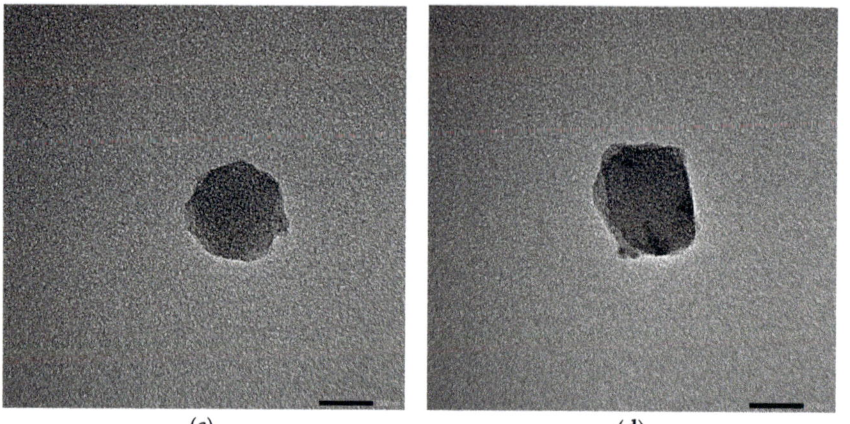

(c) (d)

**Figure 2.** TEM images of the newly synthesized MIL-101(Cr), denoted as 20 nm MOF. Scale bar in (**a**) is 200 nm. In all other cases (**b–d**) it is 20 nm.

To verify the material's chemical composition, it was deposited on a Si wafer and analyzed by XPS. The results for the two MOF materials are presented in Figure 3. The 300 nm MOF was drop-cast on the Si wafer, while the 20 nm MOF was deposited as a 1 LB monolayer at a surface pressure of 4 mN/m and is significantly thinner. This is the reason the 300 nm sample exhibits a higher-intensity signal at specific energies. Within the experimental error, the two materials are identical and correspond to the chemical composition of the MIL-101(Cr). The presence of chromium (Cr), carbon (C), and oxygen (O) peaks is clearly observed. The survey spectra typically show distinct peaks for Cr 2p, C 1s, and O 1s, confirming the elemental composition of the framework. The Cr 2p shows two prominent peaks: Cr 2p3/2: ~577.0–577.9 eV and Cr 2p1/2: ~586.7–587.3 eV. These peaks are attributed to the Cr(III) oxidation state and the presence of Cr–O bonds in the MOF nodes [41]. The peak of the C 1s spectrum is observed at 284.3, 284.8, and 285.6 eV, which are assigned to C–C, CO, and C–O, respectively [42]. In the 20 nm MOF curve, these peaks are not resolved. The O 1s spectrum contains two peaks, one at 351.6 eV, which is typical of the CrO bond, and the other at 533 eV, which stems from the hydroxyl group in absorbed $H_2O$ molecules within the framework. In the 20 nm MIL-101(Cr) XPS, the 533 peak dominates due to water trapped during the LB monolayer deposition, as evidenced by the poor vacuum [43].

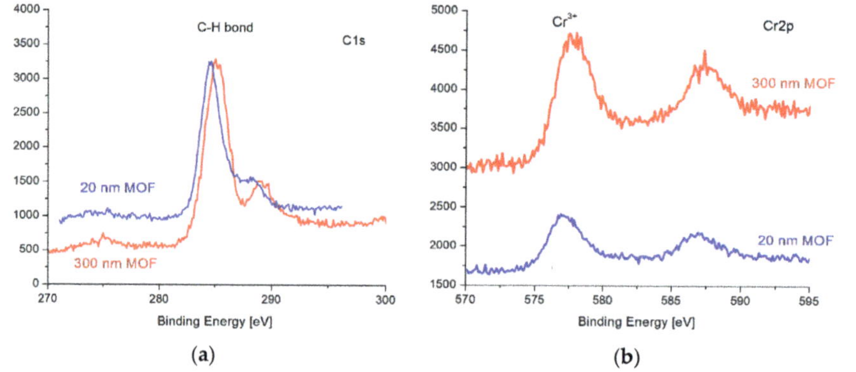


(a) (b)

**Figure 3.** *Cont.*

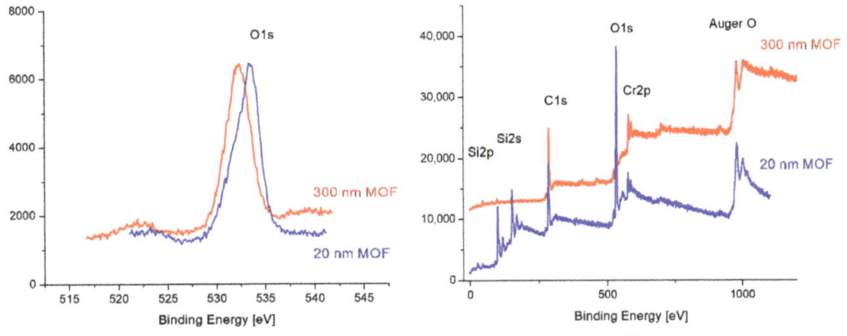


(**c**) (**d**)

**Figure 3.** XPS data and comparison of the 20 nm and 300 MIL-101(Cr) material. Within the experimental error, the two materials are identical. (**a**) C1s region of the spectra; (**b**) Cr2p region of the spectra; (**c**) O1s region of the spectra; (**d**) the entire spectra.

The presence of $Cr^{3+}$ imparts exceptional thermal and chemical stability to MIL-101(Cr), allowing it to withstand acidic and alkaline conditions (pH 0–12) and maintain its structure under various solvents. The small ionic radius, high charge, and strong polarization ability of $Cr^{3+}$ enable the formation of more covalent bonds with oxygen-containing ligands, further enhancing the framework's stability [44].

### *3.2. Investigations at the Air–Water Interface*

To measure the equilibrium spreading pressure (ESP) of the 300 nm MIL-101(Cr), a small amount of powder was placed at the air–water interface while recording the surface pressure as a function of time. There was no rise in surface pressure, so the ESP was equal to 0 mN/m at 22 °C. This means that the insoluble monolayer at the air–water interface (Langmuir film) isotherm is in a thermodynamically metastable state at all pressures at room temperature. Thus, both the isotherm and the collapse pressure depend strongly on the speed of barrier compression, so we kept it slow. Furthermore, when spreading the Langmuir film, some substance seeps under the layer, despite it taking an unusually long waiting time (40 min or more) for the solvent to evaporate. So, the area/weight axis data should be regarded with caution, which was also noted in [8] for this compound.

The compression (Figure 4) and compression–decompression (Figure 5) isotherms of the Langmuir films from MIL-101(Cr), together with surface potential data, are presented. This data will be discussed together with the simultaneously recorded BAM data (Figures 6–8). The data for 300 nm MIL-101(Cr) with 17.7% AA added in hexane solvent are presented in Figure 4 (green–blue curves) and Figure 6. The data for 300 nm MIL-101(Cr) in a chloroform/methanol solvent dispersion are presented in Figure 4 (red) and Figure 7. The data for 20 nm MIL-101(Cr) in a chloroform/methanol solvent dispersion is presented in Figure 4 (black–purple), Figures 5 and 8.

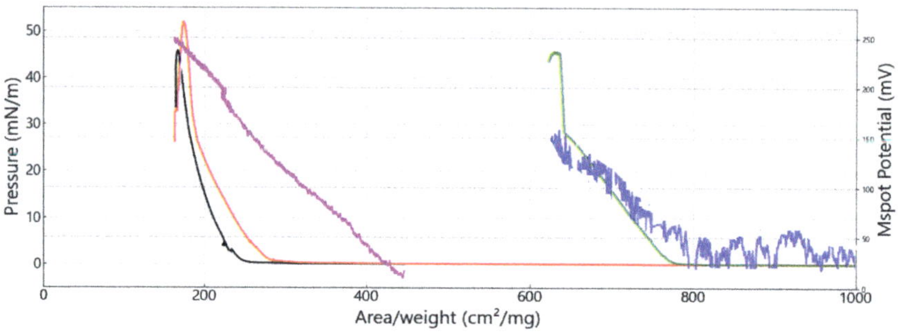


**Figure 4.** Isotherms at room temperature: black—20 nm sized MOF pressure; purple—surface potential data; LB monolayer deposition at 4 mN/m can be seen; red—300 nm sized MOF pressure data; green—300 nm sized MOF + 17% AA pressure data; and blue—surface potential data.

In all cases, stable layers were obtained with collapse pressures ranging from 45 to 50 mN/m. This is in good agreement with the previously reported data for a 51 nm crystal-sized MIL-101(Cr) isotherm [8]. The collapse pressure is significantly higher than the observed one for Fe-MIL-88B-$NH_2$, where the collapse for the 1.5 μm particles occurred at around 13 mN/m, while, for the 70 nm particles, it was around 26 mN/m [9]. No significant dependence of the collapse pressure on particle size is observed in this study (Figure 4).

In the case of the Langmuir film of the 300 nm MIL-101(Cr) with added AA of 17.7% by weight, the AA isotherm strongly dominates the isotherm. The classical isotherm of AA, characterized by the well-known liquid-expanded to liquid-condensed phase transition at 25.6 mN/m at 20 °C, was obtained. Here, the transition is slightly higher because the temperature was 23 °C. However, the AA did not increase the collapse pressure. When only 3% of AA was added to the same MOF, the surface pressure rose only slightly, to around 1.5 mN/m, while the rest of the isotherm exhibited the same 2D gas-to-solid phase transition [10]. In the compressed film BAM data, the MIL-101(Cr) aggregates, ranging in size from 10 to 100 μm, are clearly resolved as bright spots, which float in the "sea" of AA (Figure 6d). At zero surface pressure, regions with no layer appear black (Figure 6a,b). MOF aggregates stick mainly to the AA layer with which they were mixed and do not enter the no-layer zones. Two gray colors of the AA monolayer can be seen, which probably belong to different tail tilts of the fatty acid molecule. At higher pressures, this difference in gray color disappears because the molecule tails obtain uniform orientation (Figure 6d,e). A vertical stripe is visible in Figure 6e, indicating the onset of collapse in the AA monolayer at a surface pressure of 40.6 mN/m. Upon film decompression after collapse, a patchy, network-like structure with numerous no-layer zones emerges, and the MOF aggregates infiltrate these no-layer areas (Figure 6f).

In pure MIL-101(Cr) layers, the isotherm exhibits the typical behavior of a 2D gas to a 2D solid phase transition upon film compression, characterized by a sharp increase in pressure with a further slope increase (higher compressibility modulus) starting at around 27 mN/m. There was no significant difference between the isotherms of both-sized MIL-101(Cr) crystals. Upon zooming in on the surface pressure, it can be seen that it begins to increase slightly at around 360 $cm^2/mg$ (Figure 5), which corresponds to the MOF patches starting to interact (Figure 8b). This occurs well before the transition to the 2D solid phase, at around 100 $cm^2/mg$. Even at the very low pressure of around 2.5 mN/m, the monolayer appears very rigid (Figure 8c). At 4 mN/m, almost the entire surface is covered by the solid phase (Figure 8d). This was the most homogeneous part of the 20 nm MIL-101(Cr) monolayer, so we chose this pressure for the deposition of the LB monolayer.

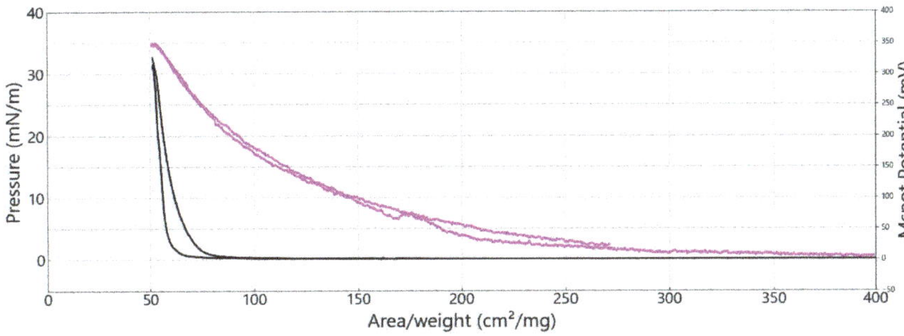


**Figure 5.** Compression and decompression isotherms of the 20 nm sized MIL-101(Cr) at room temperature: black—surface pressure; purple—surface potential.

MIL-101(Cr) exhibits pH-dependent surface charge characteristics, as determined by zeta potential measurements [45]. The material has a distinct point of zero charge (PZC) at approximately pH 9.5. Below this value, the surface carries a net positive charge. This is explained by the protonation (Cr–$OH^{2+}$) and deprotonation (CrO) equilibria of $Cr^{3+}$-hydroxyl groups, which are typically situated in the alkaline range and drive the PZC toward higher pH values [45]. At the neutral pH pure water used in this work, MIL-101(Cr) is positively charged. Consequently, the surface potential starts to increase the moment barrier compression begins, as the charge concentration below the vibrating plate electrode of the surface potentiometer increases (Figures 4 and 5, purple). This is not the case for the 300 nm MIL-101(Cr) mixed with 17.7% AA (Figure 4—blue), in which the surface potential starts to increase only after the increase in surface pressure. This increase can be partly attributed to AA's reorientation. The effect of the charged MOF should also be present, but it occupies only a small amount of the area compared to the area occupied by the AA molecules. Smaller-sized 20 nm particles cause almost a twice-higher increase in surface potential from zero to collapse pressure. This is due to much better and more homogeneous surface water coverage by the smaller particles.

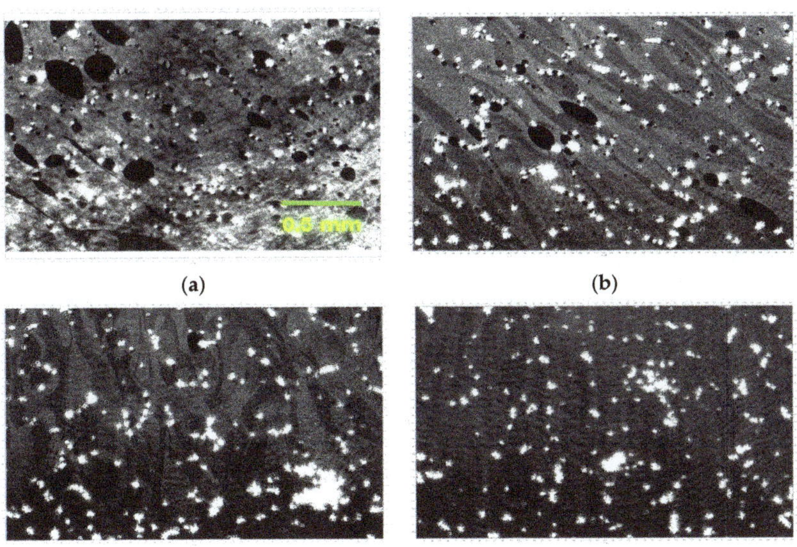


**Figure 6.** *Cont.*

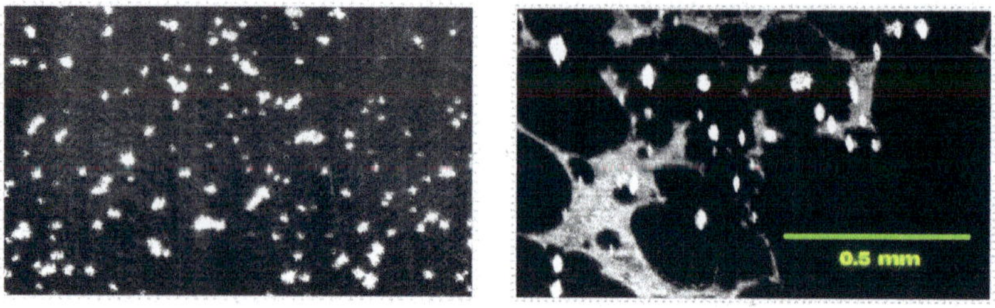


(**e**) (**f**)

**Figure 6.** BAM images of a Langmuir film from 300 nm sized MIL-101(Cr) crystal + 17.7% AA, all dispersed in hexane on film compression and decompression: (**a**) surface pressure $\pi$ = 0 mN/m, trough area $A_{tr}$ = 87,400 mm$^2$—after solvent evaporation before layer compression at fully extended trough barriers; (**b**) $\pi$ = 0.0 mN/m, $A_{tr}$ = 79,170 mm$^2$; (**c**) $\pi$ = 0.89 mN/m, $A_{tr}$ = 61,725 mm$^2$; (**d**) $\pi$ = 20.17 mN/m, $A_{tr}$ = 54,600 mm$^2$; (**e**) $\pi$ = 40.61 mN/m, $A_{tr}$ = 51,095 mm$^2$; (**f**) $\pi$ = 0.0 mN/m, $A_{tr}$ = 87,400 mm$^2$ immediately after layer decompression. Scale sizes from (**a–e**) are shown on (**a**), while the scale size of (**f**) differs.

The BAM images of the 300 nm MIL-101(Cr) appear slightly different, as the camera was positioned lower to achieve better contrast. So, the MOF aggregates appear black. It should be noted that there are many more areas uncovered by the film (Figure 7a,b), as there are no AA molecules to fill the gaps. On film compression, these areas decrease in size, but even at 11 mN/m there are still large uncovered areas. Upon further compression, collapse is reached (Figure 7f) and a multilayer is formed.

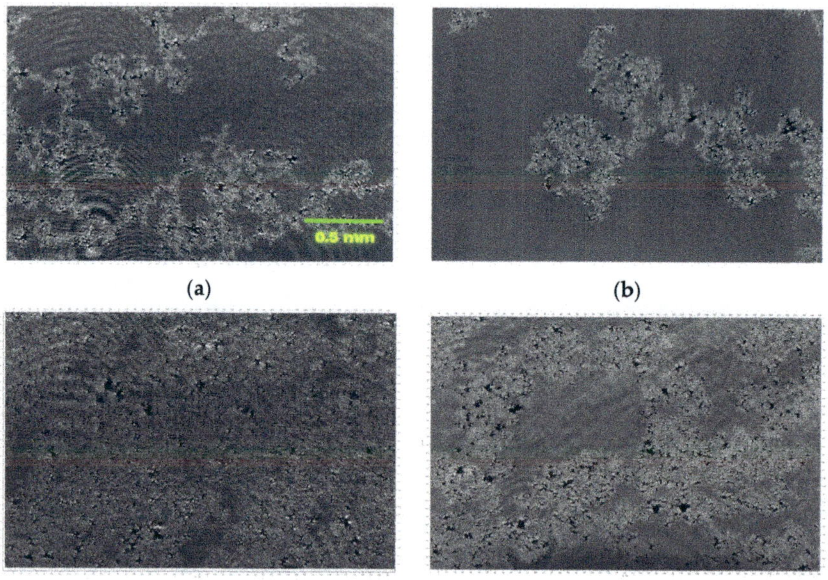


(**a**) (**b**)

(**c**) (**d**)

**Figure 7.** *Cont.*

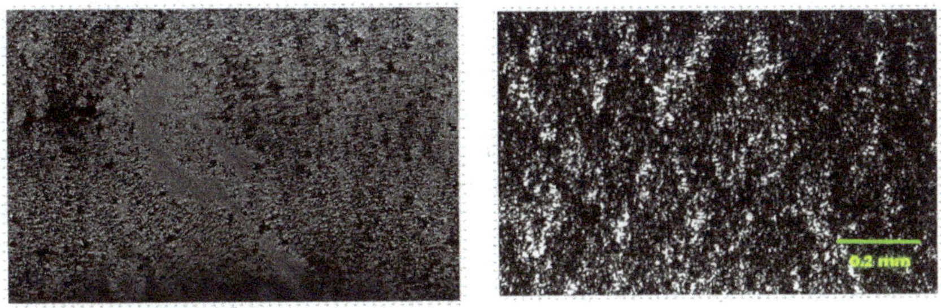


(e) (f)

**Figure 7.** BAM images of a Langmuir film from 300 nm sized MIL-101(Cr) crystal dispersed in 75% chloroform and 25% methanol solvent mixture to 0.83 mg/mL concentration on film compression: (**a**) surface pressure $\pi = 0$ mN/m, trough area $A_{tr} = 87{,}400$ mm$^2$—after solvent evaporation before layer compression at fully extended trough barriers; (**b**) $\pi = 0.66$ mN/m, $A_{tr} = 64{,}108$ mm$^2$; (**c**) $\pi = 2.93$ mN/m, $A_{tr} = 20{,}257$ mm$^2$; (**d**) $\pi = 10.95$ mN/m, $A_{tr} = 27{,}250$ mm$^2$; (**e**) $\pi = 31.09$ mN/m, $A_{tr} = 12{,}149$ mm$^2$; (**f**) $\pi = 44.28$ mN/m, $A_{tr} = 11{,}471$ mm$^2$ immediately after layer decompression. Scale sizes from (**a–e**) are shown on (**a**), while the scale size of (**f**) differs.

Compared to the larger crystal-sized MIL-101(Cr), the 20 nm MIL-101(Cr) exhibits much more dense packing, even at 0.27 mN/m surface pressure (Figure 8b). Remarkably, there are almost none of these 10–100 μm aggregates seen in the 300 nm MIL-101 (Cr) layers with or without AA. The monolayer remains patchy, even around the collapse pressure (Figure 8g). There is very little hysteresis in film compression below the collapse pressure and immediate decompression (Figure 5). This means the MOF patches can easily separate. Interestingly, after decompression to zero pressure, homogeneous-looking patches around 0.5 mm in size can be observed (Figure 8h).

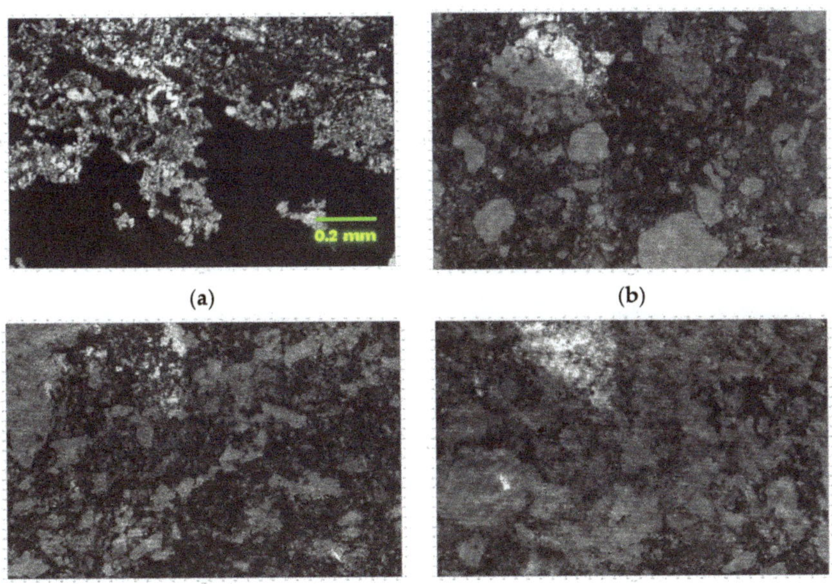


(a) (b)

(c) (d)

**Figure 8.** *Cont.*

(e) (f) (g) (h)

**Figure 8.** BAM images of a Langmuir film from 20 nm sized MIL-101(Cr) crystal dispersed in 75% chloroform and 25% methanol solvent mixture to 4.8 mg/mL concentration on film compression: (**a**) surface pressure $\pi$ = 0.08 mN/m, trough area $A_{tr}$ = 64,185 mm$^2$; (**b**) $\pi$ = 0.27 mN/m, $A_{tr}$ = 50,607 mm$^2$; (**c**) $\pi$ = 2.55 mN/m, $A_{tr}$ = 20,257 mm$^2$; (**d**) $\pi$ = 4.03 mN/m, $A_{tr}$ = 44,146 mm$^2$—LB monolayer deposition; (**e**) $\pi$ = 12.88 mN/m, $A_{tr}$ = 12,149 mm$^2$; (**f**) $\pi$ = 33.44 mN/m, $A_{tr}$ = 34,484 mm$^2$; (**g**) $\pi$ = 45.55 mN/m, $A_{tr}$ = 32,588—film collapse; (**h**) $\pi$ = 0.00 mN/m, $A_{tr}$ = 58,986 mm$^2$—immediately after layer decompression. All scale sizes are identical and are shown on (**a**).

Similarly, approximately 100 µm aggregates were observed in the BAM images of Langmuir films prepared from 1.5 µm sized Fe-MIL-88B-NH2 MOF particles [9]. At the collapse surface pressure of 14 mN/m, two distinct regions were observed. BAM images of the 70 nm sized Fe-MIL-88B-$NH_2$ reveal better surface coverage, as expected for the much larger surface-to-volume ratio of the particles. Similarly, for the 20 nm MIL-101(Cr) particles, a much better water surface coverage was observed (Figure 8a). Manipulating the MOF particle PZP potential by adjusting the water subphase pH and rendering the particles neutral can enhance particle packing [9].

### *3.3. LB Film Monolayers on Solid Support*

Simultaneously with the investigations of Langmuir films, LB monolayers were deposited on ultra-flat Si wafers for AFM investigation in non-contact mode. Here, the data for the following samples is presented: Si69—300 nm MIL-101(Cr) deposited at a surface pressure of 22 mN/m; Si71—300 nm MIL-101(Cr) deposited at a surface pressure of 3 mN/m; and Si72—20 nm MIL-101(Cr) deposited at a surface pressure of 4 mN/m. Lower surface pressure for LB film deposition was tested because, at a pressure of 22 mN/m, the mass of the deposited LB monolayer was too high, drastically decreasing the Q-factor of the 434 MHz SAW resonators. Before the LB film deposition, the monolayers were allowed to stabilize at the desired deposition pressure for 10 min. The sample notation is in line with our Nano lab standard because, simultaneously with these depositions, deposition was also carried out on other substrates. For example, G72 was deposited simultaneously with Si72, but the substrate was a two-port 434 MHz SAW resonator. Thus, data in different papers can be easily compared and data can be retrieved for further analysis or on request. The LB films were measured at least 1 day after deposition to allow them to dry. However, as shown in the XPS data above for the Si72 sample, there are some indications that trapped water remains.

Figure 9 presents the topography and phase contrast imaging data for the three LB MIL-101(Cr) monolayers tested in this study. A cross-section in the red line shows that, except for a large aggregate, the Si72 20 nm MIL-101(Cr) is much smoother. Individual aggregates are around 150 nm in size, while for the 300 nm MIL-101(Cr), aggregates are around 500 nm. Phase contrast data, which represent the mechanical properties (viscoelastic, adhesion, etc.) of the layer upon interaction with the instrument tip, also show that the 20 nm layer is significantly more homogeneous. No other AFM studies of the MIL-101(Cr) were found except for our previous work [10], where 300 nm MIL-101(Cr) was deposited as an LB monolayer at a surface pressure of 20 mN/m. Results for Si69 confirm previous findings.

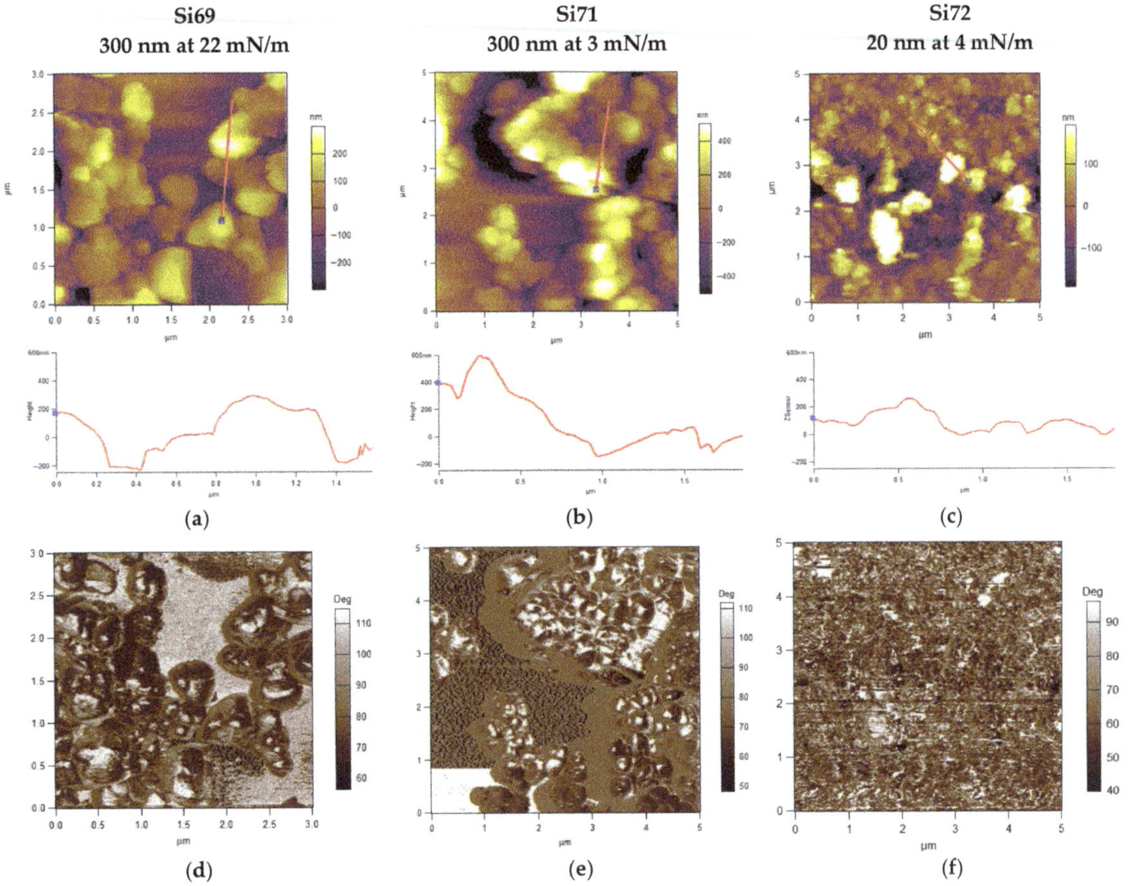


**Figure 9.** AFM images of Si69, Si71, and Si72 LB monolayers from MIL-101(Cr): (**a–c**) topography data with a cross-section across the red line; (**d–f**) phase contrast data for the corresponding topography data.

The roughness of the layers is presented in Table 1 for both the topography images in Figure 9 and larger-scale scans obtained from the same samples. It is worth noting that larger scan sizes result in greater roughness. The 20 nm MIL-101(Cr) yields a surface approximately six times smoother. As can be expected, the monolayer is far from being perfectly smooth. However, for chemical sensing applications, this is not a requirement. In fact, a well-developed surface will have a significant interaction area with the analyte being measured. We have demonstrated that a well-developed surface with 3D peaks yields four times higher sensitivity to VOCs at an equal mass of the LB film layers [46].

**Table 1.** Surface roughness of the LB film monolayers from MIL-101(Cr).

| Sample | Roughness from Scan Size 5 × 5 μm [rms] | Roughness from Scan Size 20 × 20 μm [rms] |
|---|---|---|
| **Si69** | 110.9 nm | 603.7 nm |
| **Si71** | 257.2 nm | 686.9 nm |
| **Si72** | 88.7 nm | 123.4 nm |

## 4. Conclusions

MIL-101(Cr) is one of the most frequently researched MOFs because of its interesting properties. Here, we have synthesized and characterized one of the smallest MIL-101(Cr) crystals, with an average dimension of 20 nm. This MOF also exhibits excellent stability in a water environment. This allows for investigations at the air–water interface as a Langmuir film, followed by deposition as LB layers. Recently, we have used LB monolayers deposited on a 434 MHz SAW resonator from a commercially available MIL-101(Cr) with around 300 nm crystals. A gravimetric method combined with EIS was developed to measure the emerging water contaminant PFOS [10]. Here, we compare Langmuir and LB monolayers from 300 and 20 nm sized MIL-101(Cr) particles. The smaller-sized crystal exhibits denser and more homogeneous water coverage and packing upon compression, with no observable 10–100 μm aggregates. LB monolayers from the 20 nm particles have approximately six times lower surface roughness. The monolayer is far from being smooth, but this will allow excellent access to the MOF pores by the tested analyte in a chemical sensing application. This paper addresses the lack of research on depositing presynthesized MOFs using probably the best method for nanoarchitectonics—the LB method. The 20 nm sized MOF crystals are the smallest deposited by this method so far.

**Author Contributions:** Conceptualization, G.R.I.; methodology, G.R.I., A.T., and V.S.; validation, A.D. and L.K.; formal analysis, A.D., G.R.I., and V.S.; investigation, A.D., G.R.I., and L.K.; resources, G.R.I., J.L., and V.S.; supervision, G.R.I. and J.L., writing—original draft preparation, A.D. and G.R.I.; writing—review and editing—A.D., G.R.I., L.K., A.T., and V.S.; funding acquisition, G.R.I. All authors have read and agreed to the published version of the manuscript.

**Funding:** This research was funded by the Bulgarian National Science Foundation, grant number KP-06-N88/6 (2024) and, to a lesser extent, by the Center for Research and Design of the University of Architecture, Civil Engineering and Geodesy infrastructure, grant number BN/IF–311/24.

**Institutional Review Board Statement:** Not applicable.

**Data Availability Statement:** The raw data supporting the conclusions of this article will be made available by the authors on request.

**Acknowledgments:** Thanks are due to the Rectors of the University of Architecture, Civil Engineering and Geodesy (UACEG) for providing new facilities and their complete repair to the University Lab "Nanoscience and Nanotechnology". Thanks are also due to the UACEG's administration for their assistance in acquiring the new instrumentation. The AFM used in this paper is part of the Distributed Research Infrastructure INFRAMAT, part of the Bulgarian National Roadmap for Research Infrastructures, supported by the Bulgarian Ministry of Education and Science. During the preparation of this manuscript, the authors used ChatGPT and ScienceDirect AI for reference searches. The authors have reviewed and edited the output and take full responsibility for the content of this publication.

**Conflicts of Interest:** The authors declare no conflicts of interest. The funders had no role in the design of the study, in the collection, analysis, or interpretation of data, in the writing of the manuscript, or in the decision to publish the results.

## References


1. Shams, M.; Niazi, Z.; Saeb, M.R.; Moghadam, S.M.; Mohammadi, A.K.; Fattahi, M. Tailoring the topology of ZIF-67 metal-organic frameworks (MOFs) adsorbents to capture humic acids. *Ecotox. Environ. Saf.* **2024**, *269*, 115854–115866. [CrossRef]
2. Luo, W.; Ma, Y.; Wang, M.; Wang, J.; Wu, C.; Yang, H. Three different configurations of MOFs constructed from 2,5-diketopiperazine-N, N′-diacetic acid for activating peroxymonosulfate (PMS) to degrade dyes and application in electrocatalysis. *J. Mol. Struct.* **2025**, *1336*, 142140–142151. [CrossRef]
3. Yan, Y.-T.; Fan, X.-D.; Lu, J.-L.; Yang, Z.-H.; Zhang, Y.-B.; Wu, Y.-L.; Zhang, W.-Y.; Wang, Y.-Y. A novel porous Zn-MOF based on binuclear metal clusters for fluorescence detection of Cr(VI) and adsorption of dyes. *J. Mol. Struct.* **2025**, *1322 Pt 3*, 140553–140561. [CrossRef]
4. Press Release: Nobel Prize in Chemistry 2025. Nobel Prize Outreach; The Royal Swedish Academy of Sciences. 8 October 2025. Available online: https://www.nobelprize.org/prizes/chemistry/2025/press-release/ (accessed on 31 October 2025).
5. Makiura, R.; Konovalov, O. Interfacial growth of large-area single-layer metal–organic framework nanosheets. *Sci. Rep.* **2013**, *3*, 2506. [CrossRef] [PubMed]
6. Wang, Z.; Walter, L.S.; Wang, M.; Petkov, P.S.; Liang, B.; Qi, H.; Nguyen, N.N.; Hambsch, M.; Zhong, H.; Wang, M.; et al. Interfacial synthesis of layer-oriented 2D conjugated metal–organic framework films toward directional charge transport. *J. Am. Chem. Soc.* **2021**, *143*, 13624–13632. [CrossRef] [PubMed]
7. Bétard, A.; Fischer, R.A. Metal–organic framework thin films: From fundamentals to applications. *Chem. Rev.* **2012**, *112*, 1055–1083. [CrossRef]
8. Benito, J.; Sorribas, S.; Lucas, I.; Coronas, J.; Gascon, I. Langmuir–Blodgett films of the metal–organic framework MIL-101(Cr): Preparation, characterization, and $CO_2$ adsorption study using a QCM-based setup. *ACS Appl. Mater. Interfaces* **2016**, *8*, 16486–16492. [CrossRef]
9. Benito, J.; Fenero, M.; Sorribas, S.; Zornoza, B.; Msayib, K.J.; McKeown, N.B.; Téllez, C.; Coronas, J.; Gascón, I. Fabrication of ultrathin films containing the metal organic framework Fe-MIL-88B-NH2 by the Langmuir–Blodgett technique. *Coll. Surf. A Physicochem. Eng. Aspects* **2015**, *470*, 161–170. [CrossRef]
10. Ivanov, G.R.; Venelinov, T.; Marinov, Y.G.; Hadjichristov, G.B.; Terfort, A.; David, M.; Florescu, M.; Karakuş, S. First direct gravimetric detection of perfluorooctane sulfonic acid (PFOS) water contaminants, combination with electrical measurements on the same device—Proof of concepts. *Chemosensors* **2024**, *12*, 116. [CrossRef]
11. Roberts, G.G. *Langmuir–Blodgett Films*; Springer: New York, NY, USA, 2014; ISBN 978-1-4899-3718-6. [CrossRef]
12. Petty, M.C. *Molecular Electronics: From Principles to Practice*; John Wiley & Sons, Ltd.: Chichester, UK, 2007; ISBN 978-0-470-72389-0. [CrossRef]
13. Angelov, T.; Radev, D.; Ivanov, G.; Antonov, D.; Petrov, A.G. Hydrophobic Magnetic Nanoparticles: Synthesis and LB Film Preparation. *J. Optoelectron. Adv. Mater.* **2007**, *9*, 424–426.
14. Gorbachev, I.A.; Smirnov, A.V.; Glukhovskoy, E.G.; Kolesov, V.V.; Ivanov, G.R.; Kuznetsova, I.E. Morphology of Mixed Langmuir and Langmuir–Schaefer Monolayers with Covered CdSe/CdS/ZnS Quantum Dots and Arachidic Acid. *Langmuir* **2021**, *37*, 14105–14113. [CrossRef] [PubMed]
15. Gorbachev, I.A.; Smirnov, A.V.; Ivanov, G.R.; Venelinov, T.I.; Amova, A.A.; Datsuk, E.V.; Anisimkin, V.I.; Kuznetsova, I.E.; Kolesov, V.V. Langmuir–Blodgett Films with Immobilized Glucose Oxidase Enzyme Molecules for Acoustic Glucose Sensor Application. *Sensors* **2023**, *23*, 5290. [CrossRef] [PubMed]
16. Chen, J.; Tang, M.; Nie, S.; Xiao, P.; Zhao, T.; Chen, Y. Synthesis Methods, Performance Optimization, and Application Progress of Metal–Organic Framework Material MIL-101(Cr). *Chemistry* **2025**, *7*, 78. [CrossRef]
17. Sahabudeen, H.; Dong, R.; Feng, X. Interfacial Synthesis of Structurally Defined Organic Two-Dimensional Materials: Progress and Perspectives. *Chimia* **2019**, *73*, 480. [CrossRef]
18. Chuhadiya, S.; Himanshu; Suthar, D.; Patel, S.L.; Dhaka, M.S. Metal organic frameworks as hybrid porous materials for energy storage and conversion devices: A review. *Coord. Chem. Rev.* **2021**, *446*, 214115–214140. [CrossRef]
19. Zhu, Y.; Li, J.; Gu, J.; Kan, L.; Liu, Y. A three-dimensional Cu-MOF with strong $\pi$-$\pi$ interactions exhibiting high water and chemical stability. *Inorg. Chem. Commun.* **2019**, *99*, 108–112. [CrossRef]

20. Zou, M.; Dong, M.; Zhao, T. Advances in Metal–Organic Frameworks MIL-101(Cr). *Int. J. Mol. Sci.* **2022**, *23*, 9396. [CrossRef]
21. Haghighi, E.; Zeinali, S. Nanoporous MIL-101(Cr) as a Sensing Layer Coated on a Quartz Crystal Microbalance (QCM) Nanosensor to Detect Volatile Organic Compounds (VOCs). *RSC Adv.* **2019**, *9*, 24103–24112. [CrossRef]
22. Haghighi, E.; Zeinali, S. Formaldehyde Detection Using Quartz Crystal Microbalance (QCM) Nanosensor Coated by Nanoporous MIL-101(Cr) Film. *Microporous Mesoporous Mater.* **2020**, *300*, 110065. [CrossRef]
23. Wang, L. Metal-organic frameworks for QCM-based gas sensors: A review. *Sens. Actuators A Phys.* **2020**, *307*, 111984. [CrossRef]
24. Castell, A.; Arroyo-Manzanares, N.; Viñas, P.; López-García, I.; Campillo, N. Advanced materials for magnetic solid-phase extraction of mycotoxins: A review. *TrAC Trends Anal. Chem.* **2024**, *178*, 117826. [CrossRef]
25. Song, Y.; Zheng, S.; Ji, Y.; He, B.; Wang, M. Fabrication of novel mixed matrix composite membranes with polyethyleneimine-modified MIL-101 for $CO_2/CH_4$ separation. *Chem. Eng. J.* **2025**, *518*, 164526. [CrossRef]
26. Evans, J.D.; Garai, B.; Reinsch, H.; Li, W.; Dissegna, S.; Bon, V.; Senkovska, I.; Kaskel, S.; Janiak, C.; Stock, N.; et al. Metal–organic frameworks in Germany: From synthesis to function. *Coord. Chem. Rev.* **2019**, *380*, 378–418. [CrossRef]
27. Ehrenmann, J.; Henninger, S.K.; Janiak, C. Water adsorption characteristics of MIL-101 for heat-transformation applications of MOFs. *Eur. J. Inorg. Chem.* **2011**, *4*, 471–474. [CrossRef]
28. Zhang, W.; Li, X.; Ding, X.; Hua, K.; Sun, A.; Hu, X.; Nie, Z.; Zhang, Y.; Wang, J.; Li, R.; et al. Progress and Opportunities for Metal–Organic Framework Composites in Electrochemical Sensors. *RSC Adv.* **2023**, *13*, 15442–15461. [CrossRef]
29. Meng, Y.; Huang, Y.; Huang, G.; Song, Y. TPN-COF@Fe-MIL-100 Composite Used as an Electrochemical Aptasensor for Detection of Trace Tetracycline Residues. *RSC Adv.* **2023**, *13*, 28148–28157. [CrossRef]
30. Benseghir, Y.; Tsang, M.Y.; Schöfbeck, F.; Hetey, D.; Kitao, T.; Uemura, T.; Shiozawa, H.; Reithofer, M.R.; Chin, J.M. Electric-Field Assisted Spatioselective Deposition of MIL-101(Cr)/PEDOT to Enhance Electrical Conductivity and Humidity Sensing Performance. *J. Colloid Interface Sci.* **2025**, *678*, 979–986. [CrossRef]
31. He, W.; Lv, D.; Guan, Y.; Yu, S. Post-Synthesis Modification of Metal–Organic Frameworks: Synthesis, Characteristics, and Applications. *J. Mater. Chem. A* **2023**, *11*, 45. [CrossRef]
32. Du, R.; Xin, R.; Wang, H.; Zhu, W.; Li, R.; Liu, W. Machine learning: An accelerator for the exploration and application of advanced metal-organic frameworks. *Chem. Eng. J.* **2024**, *490*, 151828. [CrossRef]
33. Wei, D.; Xu, C.; Wang, Y.; Feng, W.; Deng, C.; Wu, X.; Deng, Y.; Yang, Z.; Zhang, Z. Machine Learning-Driven Sensor Array Based on Luminescent Metal–Organic Frameworks for Simultaneous Discrimination of Multiple Anions. *Chem. Eng. J.* **2025**, *512*, 162796. [CrossRef]
34. Directive (EU) 2020/2184 of the European Parliament and of the Council of 16 December 2020 on the Quality of Water Intended for Human Consumption. Official Journal of the European Union, L 435, 23 December 2020. Available online: https://eur-lex.europa.eu/eli/dir/2020/2184/oj (accessed on 29 October 2025).
35. Drinking Water Health Advisories for PFOA and PFOS. United States Environmental Protection Agency (US EPA). 15 June 2022. Available online: https://www.epa.gov/sdwa/drinking-water-health-advisories-pfoa-and-pfos (accessed on 30 October 2025).
36. Yuan, M.; Ge, C.; Yang, W.; Wang, J.; Yin, C. Enhanced $CO_2$ adsorption performance of amino-modified MIL-101(Cr–Al) materials: Synthesis, characterization, and mechanistic insights. *Therm. Sci. Eng. Prog.* **2025**, *64*, 103794. [CrossRef]
37. Shen, T.; Luo, J.; Zhang, S.; Luo, X. Hierarchically mesostructured MIL-101 metal–organic frameworks with different mineralizing agents for adsorptive removal of methyl orange and methylene blue from aqueous solution. *J. Environ. Chem. Eng.* **2015**, *3*, 1372–1383. [CrossRef]
38. Hosseini, S.A.; Kazemeini, M.; Mohammadi, A. A dual-functional Ni–MCr LDH nanostructure utilized for degradation of meropenem antibiotic pollutant: Synthesis and physicochemical characterizations toward development of a novel chemical mechanism. *Adv. Powder Technol.* **2025**, *36*, 105084. [CrossRef]
39. Férey, G.; Mellot-Draznieks, C.; Serre, C.; Millange, F.; Dutour, J.; Surblé, S.; Margiolaki, I. A chromium terephthalate-based solid with unusually large pore volumes and surface area. *Science* **2005**, *309*, 2040–2042. [CrossRef]
40. Parak, P.; Nikseresht, A.; Mohammadi, M.; Emaminia, M.S. Application of MIL-101(Cr) for biofuel dehydration and process optimization using the central composite design method. *Nanoscale Adv.* **2024**, *6*, 4625–4634. [CrossRef]
41. You, M.; Tang, Y.; Liang, X.; Deng, T.; Peng, H.; Tian, J.; Du, J. Efficient photocatalytic degradation of tetracycline hydrochloride by Fe–$TiO_2$/MIL-101(Cr) nanocomposites under visible light. *J. Water Process Eng.* **2025**, *69*, 106590. [CrossRef]
42. Li, Z.; Hu, J.; Xiao, Y.; Zha, Q.; Zeng, L.; Zhu, M. Surfactant-assisted Cr–metal–organic framework for the detection of bisphenol A in dust from e-waste recycling area. *Anal. Chim. Acta* **2021**, *1146*, 174–183. [CrossRef]
43. Vo, T.K.; Kim, J.-H.; Kwon, H.T.; Kim, J. Cost-effective and eco-friendly synthesis of MIL-101(Cr) from waste hexavalent chromium and its application for carbon monoxide separation. *J. Ind. Eng. Chem.* **2019**, *80*, 345–351. [CrossRef]

44. Zhi, G.; Qi, X.; Li, Y.; Wang, J.; Wang, J. Efficient treatment of smelting wastewater: 3D nickel foam @MOF shatters the previous limitation, enabling high-throughput selective capture of arsenic to form non-homogeneous nuclei. *Separ. Purif. Technol.* **2024**, *328*, 124927–124945. [CrossRef]
45. Han, J.; Albadarin, A.B.; Guo, C.; Annath, H.; Mangwandi, C. Solubility enhancement of BCS class II drugs via in-situ loading onto MIL-101(Cr) in a green solvent system. *Separ. Purif. Technol.* **2026**, *380*, 135225. [CrossRef]
46. Avramov, I.D.; Ivanov, G.R. Layer by Layer Optimization of Langmuir–Blodgett Films for Surface Acoustic Wave (SAW) Based Sensors for Volatile Organic Compounds (VOC) Detection. *Coatings* **2022**, *12*, 669. [CrossRef]

MDPI AG
Grosspeteranlage 5
4052 Basel
Switzerland
Tel.: +41 61 683 77 34

*Coatings* Editorial Office
E-mail: coatings@mdpi.com
www.mdpi.com/journal/coatings

www.ingramcontent.com/pod-product-compliance
Lightning Source LLC
Chambersburg PA
CBHW061811210926
51850CB00006B/669